NETWORKED CONTROL SYSTEMS

NETWORKED CONTROL SYSTEMS

Cloud Control and Secure Control

MAGDI S. MAHMOUD
Systems Engineering Department
King Fahd University of Petroleum and Minerals
Dhahran, Saudi Arabia

YUANQING XIA
School of Automation
Beijing Institute of Technology
Beijing, China

ELSEVIER
Butterworth-Heinemann
An imprint of Elsevier

Butterworth-Heinemann is an imprint of Elsevier
The Boulevard, Langford Lane, Kidlington, Oxford OX5 1GB, United Kingdom
50 Hampshire Street, 5th Floor, Cambridge, MA 02139, United States

Library of Congress Cataloging-in-Publication Data
A catalog record for this book is available from the Library of Congress

British Library Cataloguing-in-Publication Data
A catalogue record for this book is available from the British Library

ISBN: 978-0-12-816119-7

For information on all Butterworth-Heinemann publications
visit our website at https://www.elsevier.com/books-and-journals

Publisher: Mara Conner
Acquisition Editor: Sonnini R. Yura
Editorial Project Manager: Peter Adamson
Production Project Manager: Bharatwaj Varatharajan
Designer: Victoria Pearson

Typeset by VTeX

DEDICATION

This book is dedicated to our families. With tolerance, patience and wonderful frame of mind, they have encouraged and supported us for many years.

Magdi S. Mahmoud, Yuanqing Xia

CONTENTS

ABOUT THE AUTHORS

Magdi S. Mahmoud is a Distinguished Professor at KFUPM, Dhahran, Saudi Arabia. He obtained his BSc (Honors) degree in communication engineering, MSc in electronic engineering and PhD in systems engineering, all from Cairo University in 1968, 1972 and 1974, respectively. He has been a professor of engineering since 1984. He was a faculty member at different universities worldwide including Egypt (CU, AUC), Kuwait (KU), UAE (UAEU), UK (UMIST), USA (Pitt, Case Western), Singapore (NTU) and Australia (Adelaide). He lectured in Venezuela (Caracas), Germany (Hanover), UK (Kent), USA (UoSA), Canada (Montreal) and China (BIT, Yanshan, USTB). He is the principal author of 23 books, 18 book-chapters and the author/co-author of more than 575 peer-reviewed papers.

Professor Mahmoud is the recipient of 1978 and 1986 Science State Incentive Prizes for outstanding research in engineering (Egypt); the Abdul-Hameed Showman Prize for Young Arab Scientists in engineering sciences, 1986 (Jordan); the Prestigious Award for Best Researcher at Kuwait University, 1992 (Kuwait); the State Medal of Science and Arts-first class, 1979 (Egypt); and the State Distinguished Award-first class, 1995 (Egypt).

He was listed in the 1979 edition of Who's Who in Technology Today (USA). Professor Mahmoud was also the vice-chairman of the IFAC-SECOM working group on large-scale systems methodology and applications (1981–1986), an associate editor of LSS Journal (1985–1988) and editor-at-large of the EEE series, Marcel-Dekker, USA. He is an associate editor of the International Journal of Parallel and Distributed Systems of Networks, IASTED, since 1997, as well as a member of the New York Academy of Sciences.

Professor Mahmoud is currently actively engaged in teaching and research developing modern methodologies to distributed control and filtering, networked-control systems, triggering mechanisms in dynamical systems, renewable-energy systems and microgrid control systems.

He is a fellow of the IEE, a senior member of the IEEE, a member of Sigma Xi, the CEI (UK), the Egyptian Engineers society, the Kuwait Engineers society and a registered consultant engineer of information engineering and systems (Egypt).

Yuanqing Xia was born in Anhui Province, China in 1971, and graduated from the Department of Mathematics, Chuzhou University, China in 1991. He received his MSc degree in Fundamental Mathematics from Anhui University, China in 1998, and PhD degree in control theory and control engineering from Beijing University of Aeronautics and Astronautics, China in 2001. From 1991 to 1995, he was a teacher at Tongcheng Middle-School, China. From January 2002 to November 2003, he was a postdoctoral research associate at the Institute of Systems Science, Academy of Mathematics and System Sciences, Chinese Academy of Sciences, China, where he worked on navigation, guidance and control. From November 2003 to February 2004, he was with the National University of Singapore as a Research Fellow, where he worked on variable structure control. From February 2004 to February 2006, he was with the University of Glamorgan, UK, as a research fellow, where he worked on networked control systems. From February 2007 to June 2008, he was a guest professor with Innsbruck Medical University, Austria, where he worked on biomedical signal processing. Since July 2004, he has been with the School of Automation, Beijing Institute of Technology, Beijing, first as an associate professor, then, since 2008, as a professor. And in 2012, he was appointed as Xu Teli distinguished professor at the Beijing Institute of Technology, then in 2016, as chair professor. In 2012, he was awarded by the National Science Foundation for Distinguished Young Scholars of China, and in 2016, he was honored as the Yangtze River Scholar Distinguished Professor and was supported by National High Level Talents Special Support Plan ("Million People Plan") by the Organization Department of the CPC Central Committee. He is now the dean of School of Automation, Beijing Institute of Technology. He has published 8 monographs in Springer, John Wiley and CRC, and more than 100 papers in international scientific journals. He is a deputy editor of Journal of Beijing Institute of Technology, an associate editor of Acta Automatica Sinica, Control Theory and Applications, International Journal of Innovative Computing, Information and Control, International Journal of Automation and Computing. He obtained the Second Award of the Beijing Municipal Science and Technology (No. 1) in 2010 and 2015, the Second National Award for Science and Technology (No. 2) in 2011, and the Second Natural Science Award of the Ministry of Education (No. 1) in 2012. His research interests include networked control systems, robust control and signal processing, active disturbance rejection control and flight control.

ACKNOWLEDGMENTS

Special thanks are due to the Elsevier team; particularly Acquisitions Editor Sonnini R. Yura, Past Editorial Project Manager Katie Chan and Present Editorial Project Manager Peter Adamson for guidance and dedication throughout the publishing process. The production team headed by Manager Bharatwaj Varatharajan have done a wonderful job in producing the book.

We are grateful to all the anonymous referees for carefully reviewing and selecting the appropriate topics for the final version during this process. Portions of this volume were developed and upgraded while offering the graduate courses **SCE-612-171**, **SCE-701-171**, **SCE-612-172**, **SCE-701-172** at KFUPM, Saudi Arabia. The first author would like to thank the research support opportunities afforded by the School of Automation at Beijing Institute of Technology during regular technical visits over the past several years.

This work is fully supported by the Deanship of Scientific Research (DSR) at KFUPM through distinguished professorship research project no. **BW 171004**.

Magdi S. Mahmoud, Yuanqing Xia
November, 2017

PREFACE

In the last decades, the rapid development of communication, control and computer technologies has made a vital impact on the control system structure. In traditional control systems, the connections among the sensors, controllers and actuators are usually realized by port-to-port wiring, which may cause many problems such as difficulties in wiring, maintenance and low flexibility. Such drawbacks appear in many automation systems due to the increasing complexity of controlled plants. Incorporating a communication network in the control loop brings about Networked Control Systems (NCS). In this scenario, networked control systems have been attracting more and more attention. The utilization of a multipurpose shared network to connect spatially distributed elements results in flexible architectures and generally reduces installation and maintenance costs. Nowadays, NCSs have been extensively applied in many practical systems such as car automation, intelligent building, transportation networks, haptics collaboration over the Internet, and unmanned aerial vehicles. During the 1990s the advanced control theory has seen major computational advances and achieved advanced maturity, centered around the notion of convexity and convex analysis. This emphasis is twofold. On the one hand, the methods of convex programming have been introduced to the field and released a wave of computational methods which, interestingly, have impact beyond the study of control theory. Simultaneously a new understanding has developed on the computational complexity implications of uncertainty modeling; in particular, it has become clear that one must go beyond the time invariant structure to describe uncertainty in terms amenable to convex robustness analysis.

Thus the pedagogical objectives of the book are:

1. Introducing a coherent and unified framework for studying networked control systems;
2. Providing students with the control-theoretic background required to read and contribute to the research literature;
3. Presenting the main ideas and demonstrations of the major results of networked control theory;
4. Providing a modest coverage of cloud-based approaches to control systems, secure control methodologies to cyberphysical systems against to various types of malicious attacks.

- **Chapter 1** – An Overview
  This chapter provides a guided tour into the key ingredients of networked control systems and their prevailing features under normal operating environments and when subjected to cyberphysical attacks.
- **Chapter 2** – Networked Control Systems' Fundamentals
  The focus of this chapter is on the fundamental issues underlying the analysis and design of networked control systems. This chapter will demonstrate the operation of NCSs under normal conditions. The major imperfections that could affect NCSs are examined in-depth and their impact is critically examined.
- **Chapter 3** – Cloud Computing
  This chapter critically examines the principles of cloud computing.
- **Chapter 4** – Control From the Cloud
  This chapter introduces the paradigm of cloud control systems' design and discusses several approaches.
- **Chapter 5** – Secure Control Design Techniques
  This chapter examines available techniques of secure control design to networked control systems under cyberphysical attacks.
- **Chapter 6** – Case Studies
  This chapter introduces some typical practical case studies.
- **Chapter 7** – Smart Grid Infrastructures
  This chapter examines estimation and fault-tolerant diagnosis methods to analyze and counteract the impact of cyberphysical attacks in smart grids.
- **Chapter 8** – Secure Resilient Control Strategies
  This chapter deals with some typical strategies adopted for resilient and secure control of cyberphysical systems.
- **Chapter 9** – Cyberphysical Security Methods
  This chapter deals with game theory applicable to cyberphysical networked control systems facing various types of failure.

The book includes an appendix with basic lemmas and theorems needed.

CHAPTER 1

An Overview

1.1 INTRODUCTION

The accelerated integration and convergence of communications, computing, and control over the last decade has commanded researchers from different disciplines to become involved in the emerging field of networked control systems (NCSs). In general, an NCS consists of sensors, actuators, and controllers whose operations are distributed at different geographical locations and coordinated through information exchanged over communication networks. Some typical characteristics of those systems are reflected in their asynchronous operations, diversified functions, and complicated organizational structures. The widespread applications of the Internet have been one of the major driving forces for research and development of NCSs. More recently, the emergence of pervasive communication and computing has significantly intensified the effort of building such systems for control and management of various network-centric complex systems that have become more and more popular in process automation, computer integrated manufacturing, business operations, as well as public administration.

1.2 NETWORKED CONTROL SYSTEMS: CHALLENGES

Networking the control systems brings new challenges, which can be divided into six categories:

A. Data quantization;
B. Data packet loss;
C. Time-varying transmission intervals;
D. Time-varying transmission delays;
E. Communication constraints, which impose that not all of the sensor and actuator signals can be transmitted simultaneously; and
F. Network-induced external disturbance.

Combined with the performance constraints of the controlled object itself, such as actuator saturation, control algorithms should be pursued to handle these communication imperfections and constraints simultaneously.

Over the past four decades, digital computers became powerful tools in control system design, and microprocessors added a new dimension to the

Networked Control Systems
https://doi.org/10.1016/B978-0-12-816119-7.00009-5

capability of control systems. This is clearly manifested in the aerospace industry. With the introduction of a distributed control system (DCS), an advanced step in design evolution of process control was taken. Expanding needs of industrial applications pushed the limit of point-to-point control further and it became obvious that the networked control systems (NCSs) is the solution to achieve remote control operations. Same impact was mandatory in the case of teleoperation research to cope with safety and convenience issues in hazardous environments including space projects and nuclear reactor power plants.

Further development and research in NCSs were boosted by the tremendous increase in the deployments of wireless systems in the last few years. Today, NCSs are moving into distributed NCSs [1,2], which are multidisciplinary efforts whose aim is to produce a network structure and components that are capable of integrating distributed sensors, distributed actuators, and distributed control algorithms over a communication network in a manner that is suitable for real-time applications.

In a parallel direction, the field of control theory has been instrumental in developing a coherent foundation for system theory with sustained investigations into fundamental issues such as stability, estimation, optimality, adaptation, robustness and decentralization. These issues have been the major ingredients in many new proposed technologies, which are now within our collective purview. With the explosive advancements in microelectronics and information technology, it became increasingly evident recently that the *networked* nature of systems is drawing the major attention. From a technological standpoint, networked control systems (NCSs) comprise *the system to be controlled (plant), actuators, sensors, and controllers whose operation is orchestrated with the help of a shared band-limited digital communication network*. A contemporary layout of a networked control system is depicted in Fig. 1.1 where multiple plants, controllers, sensors, actuators and reference commands are connected through a network.

It is readily seen that the fundamental standing aspects of an NCS are:

- The component elements are spatially distributed;
- It may operate in an asynchronous manner;
- Its operation has to be coordinated to achieve some overall objective.

The intensification of this class of systems has addressed challenges to the communications, information processing, and control areas dealing with the relationship between operations of the network and the quality of the overall system's operation. The increasing interests in NCSs are due to their

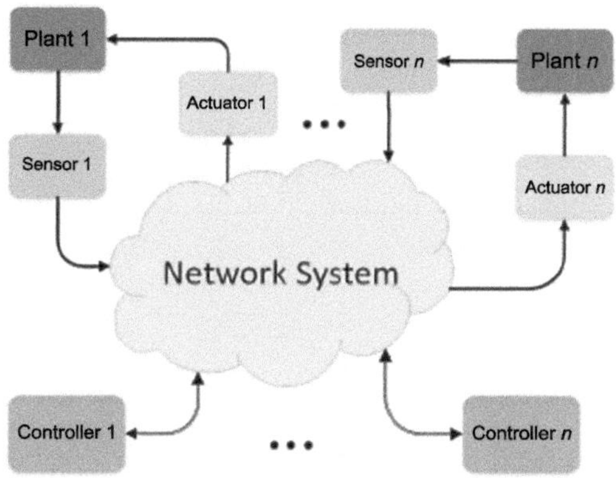

Figure 1.1 A typical networked control system

cost efficiency, lower weight and power requirements, simple installation and maintenance, and high reliability. Communication channels can lower the cost of wiring and power, ease the commissioning and maintenance of the whole system, and improve the reliability when compared with the point-to-point wiring system. Consequently, NCSs have been finding application in a broad range of areas such as mobile sensor networks [3], remote surgery [4], haptics collaboration over the Internet [5–7], and automated highway systems and unmanned aerial vehicles [8,9]. A modest coverage of numerous results and applications is found in [10]. In a parallel development, a study of the synchronization of the unified chaotic system via optimal linear feedback control and the potential use of chaos in cryptography through the presentation of a chaos-based algorithm for encryption are carried out in [11,12].

NCSs also allow remote monitoring and adjustment of plants over the communication channel, for example, the Internet in Internet-based control systems [13], which make the control systems benefit from the ways of retrieving data and reacting to plant actuation from anywhere around the world at any time.

Within network-based control systems there are three types of data, see Fig. 1.2:

- periodic data (or real-time synchronous data),
- sporadic data (or real-time asynchronous data), and
- messages (or non-real-time asynchronous data).

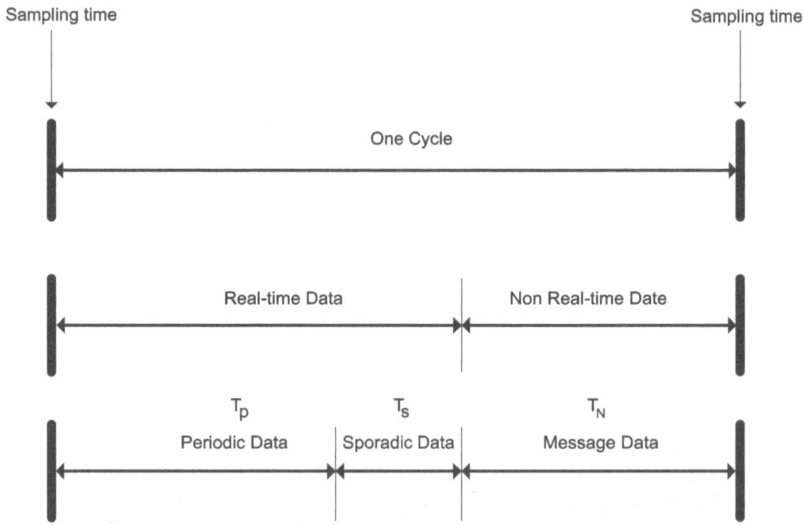

Figure 1.2 Definition of phases

1.2.1 Basic Classification

Evidently, the area of *Networks and Control* or *Networking and Control* emerges as a new discipline in the control circles. From this perspective, there are two essentially distinct directions of research:

- The first direction follows the interpretation, **control of networks**, and considers the control of communication networks, which falls into the broader field of information technology. This includes problems related to wireless networks and congestion control of asynchronous transfer mode (ATM) networks.
- The second area adopts the interpretation, **control through networks**, and deals with *networked control systems* (NCSs) which possess a defining characteristic in which one or more control loops are closed via a serial communication channel. The focus here is on stability/stabilization issues, application issues and several standard bus protocols.

This chapter belongs to the second area. Applications of *Networks and Control* include control and communications of active, intelligent, dynamic networks; distributed sensors systems; secure, reliable wireless communication; and cloud control systems.

1.2.2 Main Features

From a control systems' perspective, it turns out that the fundamental issues in networked control systems (NCSs) can be grouped into:

- *Time-varying transmission period.* When transmitting a continuous-time signal over a network, the signal must be sampled, encoded in a digital format, transmitted over the network, and finally the data must be decoded at the receiver side. This process is basically different from the conventional periodic sampling in digital control. The overall delay between sampling and decoding at the receiver can be highly variable because both the *network access delays* (the time it takes for a shared network to accept data) and the *transmission delays* (the time during which data are in transit inside the network) depend on highly variable network conditions such as congestion and channel quality.
- *Network schedulability.* To guarantee high performance, all periodic data have to be transmitted within the respective sampling period, while guaranteeing real-time transmission of sporadic data and minimum transmission of messages. This implies that transmissions of three types of data have to be allocated in the sampling period using a scheduling method for the network-based control system (NBCS).
- *Network-induced delays.* In the NBCS, network-induced delays arise due to sharing a common network medium. These delays depend on configurations of the network and the system under consideration. It is necessary to develop some methods to make these delays smaller and bounded.
- *Packet loss.* A significant difference between NCSs and standard digital control is the possibility that data may be lost while in transit through the network. Typically, packet loss results from transmission errors in physical network links or from buffer overflows due to congestion. Long transmission delays sometimes result in packet reordering, which essentially amounts to a packet dropout if the receiver discards "outdated" arrivals.

Typically in process industries, a *network* used at the lowest level of a process/factory communication hierarchy is called a *fieldbus*. Fieldbuses are intended to replace the traditional wiring between sensors, actuators, and controllers. In distributed control system applications, a feedback control loop is often closed through the network, which is called a network-based control system (NBCS). An example of the NBCS is shown in Fig. 1.3. In the NBCS, various delays with variable lengths occur due to shared common network medium, which are called *network-induced delays*. These delays

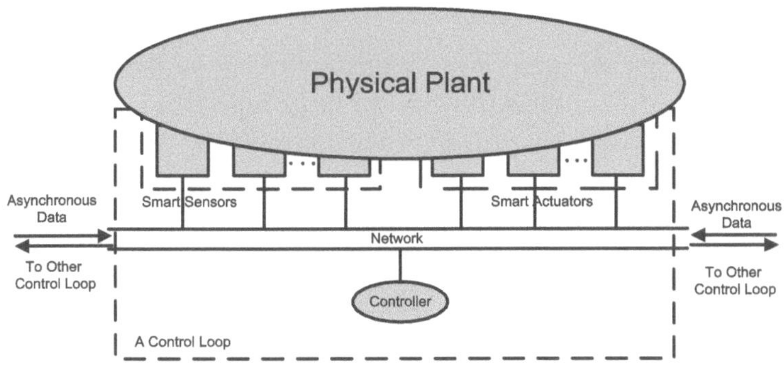

Figure 1.3 NCS with control loops

depend on configurations of the network and the given system, making the NBCS unstable.

In feedback control systems, it is significant that sampled data must be transmitted within a sampling period and stability of control systems should be guaranteed. While a shorter sampling period is preferable in most control systems, for some cases, it can be lengthened up to a certain bound within which stability of the system is guaranteed in spite of the performance degradation. This certain bound is called a *maximum allowable delay bound* (MADB). The MADB depends only on parameters and configurations of the given plant and the controller.

Remark 1.1. *It is noted that the MADB can be obtained from the plant model independent of network protocols, while the network-induced delays depend on network configurations.*

In addition, a faster sampling is said to be desirable in sampled-data systems because the performance of the discrete-time system controller can approximate that of the continuous-time system. But in NBCS, the high sampling rate can increase network load, which in turn results in longer delay of the signals. Thus finding a sampling rate that can both tolerate the network-induced delay and achieve desired system performance is of fundamental importance in the NBCS design.

Extending on Fig. 1.4, many control loops can be connected using a single network medium as depicted in Fig. 1.5. To simplify the analysis, the following notation is used:

- P is the total number of loops that use the same medium.
- N_α^i is the number of nodes in the ith loop; N, N_α, and N^i are the total number of nodes in the NBCS, the total number of α nodes in

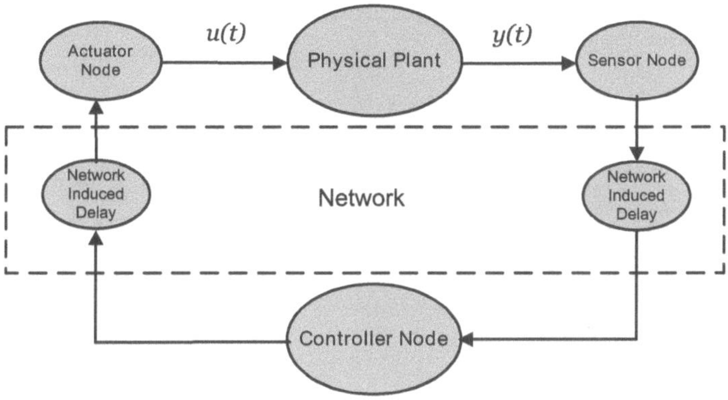

Figure 1.4 A feedback control loop with network-induced delays

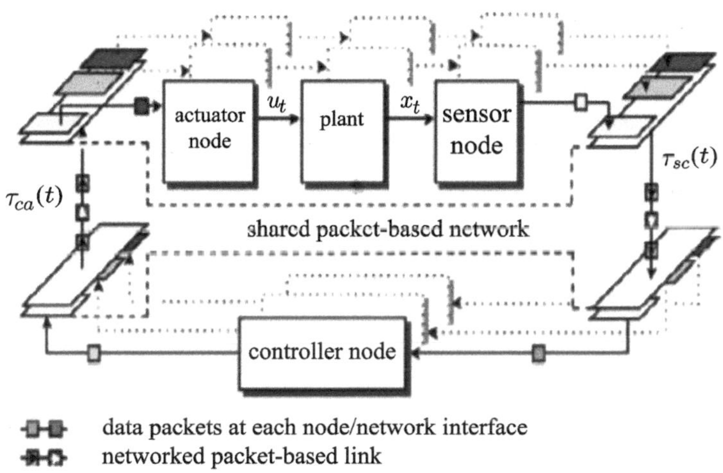

Figure 1.5 Schematic of NCS with several loops

the NBCS, and the total number of nodes in the *i*th loop, respectively. Hereinafter, α can be C (controller), A (actuator), or S (sensor).

- T^j is a sampling period of the *j*th loop.
- $T^j_{\alpha_i}$ is the data transmission time of periodic data in the *i*th α node in the *j*th loop.
- T_β is an interval for transmission of β data or messages. Hereinafter, β can be P (periodic data), S (sporadic data), or N (messages).

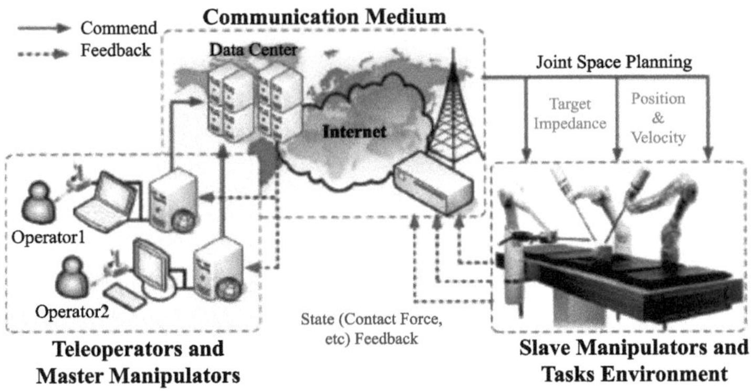

Figure 1.6 A schematic of Internet-based teleoperation

- N_S^M is the maximum number of sporadic data which arrived during a basic sampling period. The basic sampling period means the minimum sampling period in all loops.
- N_β^O is the maximum overhead time to transfer data or a message packet.
- N_D^j is the MADB in the jth loop.

In the foregoing definitions, N_S^M should be an integer and can be obtained from the maximum arrival rate of sporadic data in a basic sampling period. The maximum overhead time (N_β^O) can be time-varying in some network protocols (for example, token control); N_β^O consists of a message overhead time ($N_{O_\beta}^M$) and a protocol overhead time ($N_{P_\beta}^M$). The message overhead time occurs because of buffering, packetizing, and transmission of additional data frames such as addressing fields, control fields, or a frame check sequence. The protocol overhead time occurs because of various medium access control methods such as polling or token passing. In addition, each overhead time can have different values according to periodic data, sporadic data, and messages.

1.3 CURRENT KEY INGREDIENTS

As we learned, networked control systems (NCSs) have all the components, controllers, sensors and actuators, spatially distributed and connected via a certain digital communication network. The networks can be (see Fig. 1.6 and [19]):

- those incorporated for specialized real-time purposes including control area network (CAN), BACnet, Fieldbus, or

- the wired or wireless Ethernet, even Internet, for general-purpose data communication tasks.

Many infrastructures and service systems of the present day society can naturally be described as networks of a huge number of simple interacting units. Examples coming from the technological fields include *transportation networks*, *power grids*, *water distribution networks*, *telephone networks* and the *Internet*. Additional relevant examples can be found in other areas, that is, the *global financial network*, *genetic expression networks*, *ecological networks*, *social networks*, and so on.

Focusing on the engineering field, the reasons of the technological success of this design paradigm are manifold. The main winning features of networked structured systems come from their

- low cost, flexibility and easy reconfigurability,
- natural reliability and robustness to failure, and
- adaptation capability.

Some of these characteristics, though quite clearly recognizable from natural networked system examples listed above, are still not completely understood and so their engineering implementation is still to come. This is mainly due to the fact that it is still not completely clear how the rules describing the local interactions and the network structure influence the properties of the global emerging system. For this reason it is still a difficult task to design these local rules and the network architecture in order to obtain some prescribed global performance. However, it is clear both in the academia and industry that this paradigm will play a central role in the near future.

Currently engineers, and control engineers in particular, have to cope with many new problems arising from networked systems while designing complex networked systems. In fact, thanks to the low cost and flexibility of communication networks (Ethernet, Wi-Fi, the Internet, CAN, ...), these are widely used in industrial control systems (remote control, wireless sensors, collaborative systems, embedded systems, ...). For instance, the web technology on the Internet today appears as a natural, inexpensive way to ensure the communication link in remotely controlled systems. On the other hand, at the moment an engineer has few tools which allow him/her to exploit this architecture and these tools are mainly concerned with the negative impact caused by the very nature of the communication networks (delays, data loss, data quantization, asynchronous sampling, ...).

For these reasons we expect that the impact of networks in control is only going to increase in the future. Moreover, the problems arising in this context are rather novel, especially because of their interdisciplinary nature. In fact, their study requires both deep knowledge in automatic control, computer science and communication networks, and ability to capture the interplay between these disciplines.

Activities in the area of Networked Systems can be categorized into three major fields:

1. Control of networks, which is mainly concerned with providing a certain level of performance to a network data flow, while achieving efficient and fair utilization of network resources.
2. Control over networks, which deals with the design of feedback strategies adapted to control systems in which control data is exchanged through unreliable communication links.
3. Multi-agent systems, which focuses on the study of how network architecture and interactions between network components influence global control goals. More precisely, the problem here is to understand how local laws describing the behavior of the individual agents influence the global behavior of the networked system.

1.3.1 Control of Networks

Control-based approaches in communication networks is a very large research field. In order to be able to control the network performance, it is necessary to measure and modify the network parameters. The current layered communication architecture is not ideal for cross-layer designs, where information from the lower layers must be made available at the application layer and where the application layer must be able to modify the behavior of the lower layer protocols dynamically. Hence, new protocols and protocol models are needed.

In the area of control of networks, major problems of interest are call admission, scheduling, routing, flow control, power control, and various other resource allocation problems. The objective is to provide quality of service (QoS), while achieving efficient and fair utilization of network resources.

Wireless networks are especially interesting from a resource control point of view. Whereas the link capacities in wired networks are fixed, the capacities of wireless links can be adjusted by the allocation of communication resources, from transmission powers, bandwidths, or time slot fractions to different links. Adjusting the resource allocation changes the link capacities, influences the optimal routing of data flows, and alters the total utility of the network. Hence, optimal network operation can only be achieved by coordinating the operation across the networking stack. This is often referred to as cross-layer optimization.

1.3.2 Control Over Networks

A sensor and actuator network is a control architecture in which sensors, actuators and the computing units, where the control algorithms run, are remotely positioned and communicate with each other through a communication network. This scenario occurs in the field of embedded control systems, in large manufacturing systems, in automotive systems, etc. The advantages of this architecture have been mentioned above. However, there are also some drawbacks due to the characteristic features of the communication method adopted, namely there is the need to deal with the following problems:

1. Data transmission through communication network unavoidably introduces time delays in the control loops; see [15,16].
2. Data traffic congestion, data collision or interference can cause packet loss.
3. In networked control systems, communication occurs through digital channels with finite capacity.
4. In control over networks it is important to evaluate the influence of the QoS provided by the network on the performance of the control. The characteristics of the QoS are mainly determined by the delays in the message transfer, the message losses and the message integrity.
5. In case of energy consumption limitations, it may be convenient to adopt an event-based control strategy according to which the sensor does not communicate and the actuator does not act until some undesired event occurs.
6. While using wireless communication networks, much attention has to be devoted to the security issues; see [14].

1.3.3 Delay Characteristics

Considering the delays caused by communication networks, the mainstream research has been devoted to the design of control/observation algorithms that can stand the delay variations (jitter phenomenon) and their unpredictability.

An approach consists in combining robust control theory and network calculus theory. First, by network calculus theory it is possible to estimate a bound of the end-to-end delays by taking into account the whole network

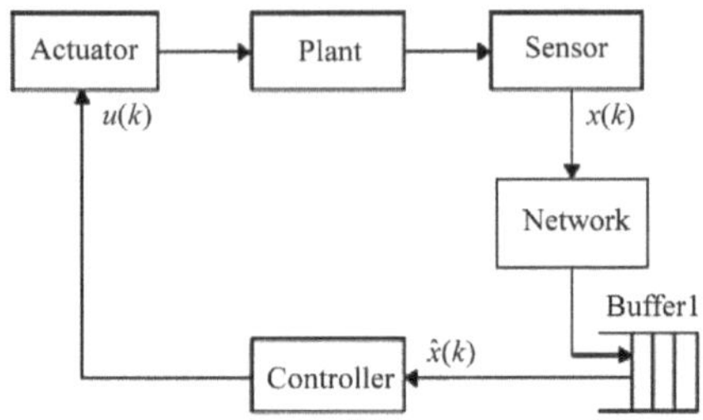

Figure 1.7 Structure of NCS with buffer

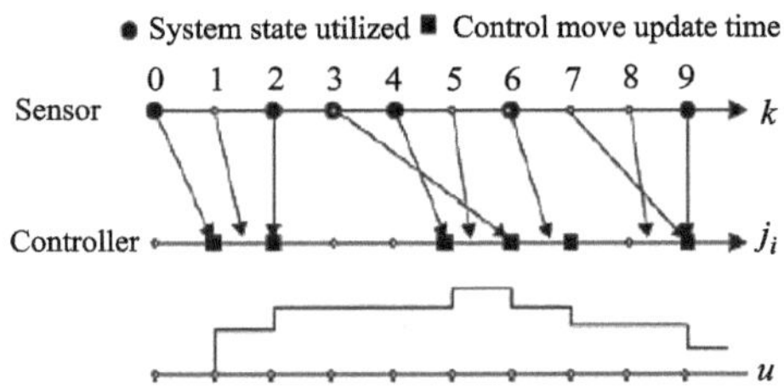

Figure 1.8 A representation of data packets transmitted and utilized

properties. These delays are then translated into uncertainties useful for robust control synthesis.

The characters of the delays in NCSs, once the specific networks are selected, can be summarized and categorized into three pairs:

1. constant versus time-varying delays;
2. deterministic versus stochastic delays;
3. delay smaller than one sampling interval and otherwise.

Such a classification can be justified in diverse applications using different networks. Consider the discrete linear time-invariant system in Fig. 1.7, where the data sent at time k is received at the buffer side at time $\hat{j}_k$; then, the received sequence is $\hat{\mathcal{J}} = \{\hat{j}_1; \hat{j}_2; \hat{j}_3; \ldots\}$. Due to packet disorder, it is not necessary true that $\hat{j}_1 \leq \hat{j}_2 \leq \hat{j}_3 \leq \cdots$. In Fig. 1.8, an example of data transmission is shown.

The delay characteristics on NCSs basically depend on the type of a network used, which are described as follows:

Cyclic service network. In local area network protocols with cyclic service such as IEEE 802.4, SAE token bus, PROFIBUS, IEEE 802.5, SAE token ring, MILSTD-1553B, and FIP, control and sensory signals are transmitted in a cyclic order with deterministic behaviors. Thus, the delays are periodic and can be simply modeled as a periodic function such that $\tau_k^{sc} = \tau_{k+1}^{sc}$ and $\tau_k^{ca} = \tau_{k+1}^{ca}$ where τ_k^{sc}, τ_k^{ca} are the sensor-to-controller delay and the controller-to-actuator delay at the sampling time period k. The models work perfectly in the ideal case. In practice, NCSs may experience small variations on periodic delays due to several reasons. For example, the discrepancies in clock generators on a controller and remote system may result in delay variations.

Random access network. Random access local area networks such as CAN and Ethernet involve more uncertain delays. The significant parts of random network delays are the waiting time delays due to queuing and frame collision on the networks. When an NCS operates across networks, several more factors can increase the randomness on network delays such as the queuing time delays at a switch or router, and the propagation time delays from different network paths. In addition, a cyclic service network connected to a random access network also results in random delays.

1.3.4 Data Loss

The state-of-the-art on design control systems has to take into account the effects of packet loss and delay in networked control systems. The observations on the difference of counting the data (packet) losses *online* or *offline* to develop control algorithms have led to categorizing the references into two frameworks. One framework for the analysis and design of NCSs with packet losses is the offline framework, where the controller is designed despite any situation of real packet losses.

The other framework is that the control input to plants is implemented online according to the situation whether a packet is lost or not, even though the controller gains are computed offline. To realize this, the intelligence of checking and comparing the indices of packets has to be equipped with the actuator, and the intelligence requires the running mode of receivers to be time driven. A typical approach in the framework is to extend the Bernoulli processes description of the packet losses and further model the correlation between packet loss and no packet loss as a two-dimensional Markov chain, and therefore the corresponding closed-loop system as a Markov jump system.

There generally exists a tradeoff between maximizing the number of consecutive packet losses, maximizing the allowable probability of packet losses, or lowering the transmission rate, and increasing the stability margins and/or system performance. The scheme of the approach is illustrated in

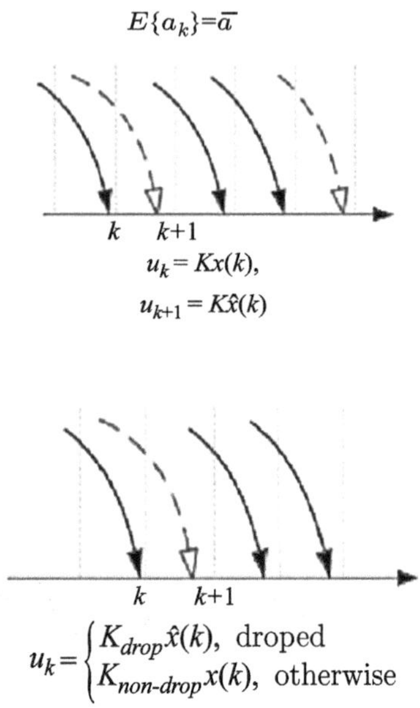

Figure 1.9 A typical implementation scheme to handle packet losses and disorder: scheme A (top), scheme B (bottom)

Fig. 1.9 (bottom). It can be seen that the controllers obtained offline are actually robust to the packet losses.

Although in the system the gains of the two controllers can be obtained offline, the control inputs will be implemented online depending on whether a packet is lost or not. As we commented in the last subsection, the approach of using the Markov jump systems can be replaced by non-deterministic switched systems. In this approach, the previous accepted measures or control inputs will be continuously used if the current packet is lost. The scheme is illustrated in Fig. 1.9 (bottom). Another typical approach in the framework is predictive control methodology. The natural property of predictive control is that it can predict a finite horizon of future control inputs at each time instant. Thus, if some consecutive packets (but within the horizon) after a successful packet transmission are lost, the control inputs to the plant can be successively replaced by the corresponding ones predicted in that successfully used packet. The idea necessitates a buffer with more than one storage unit (the length is the prediction horizon) and the actuator site to be equipped with selection intelligence. Fig. 1.10 (top) illustrates the approach.

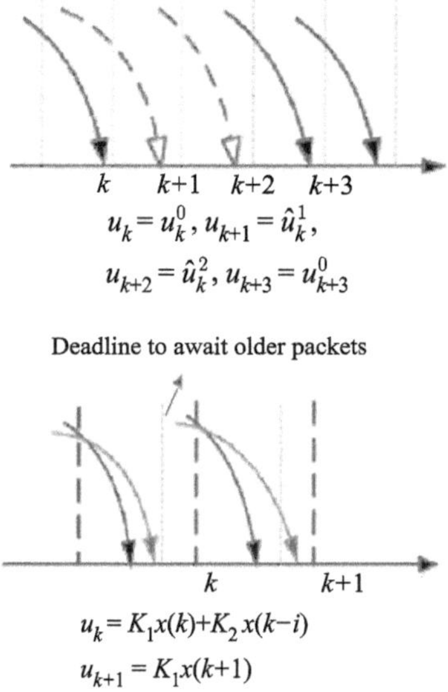

Figure 1.10 A typical implementation scheme to handle packet losses and disorder: scheme C (top), scheme D (bottom)

Note that although the predicted control inputs in a successfully transmitted packet are based on the plant in open-loop, studies have shown that implementing control inputs in such a way will in general result in a better performance than the approach where only the last used control input is kept and used in the presence of later packet losses. It can be concluded from the two above-reviewed approaches that the control input to the plant is finally selected at the actuator site according to whether the real packet transmission succeeded or not, and therefore it is online and can be viewed as an adaptation against packet losses. The studies on the packet disorder problem are sporadic since in most literature the so-called past packets rejection logic is commonly employed, which means that the older packets will be discarded if the most recent packet arrives at the receiver side. This logic can be easily realized by embedding some intelligence at the controllers and/or the actuators for comparison, checking, and selection. By this logic, the packet disorder can be viewed as packet losses in the analysis and design. However, it should be noted that discarding is a human made behavior at receiver sites, which essentially differs from the real packet losses. In the former, the information carried in the past packets still comes although it is not fresh. Therefore, a key point in the packet disorder problem should be whether the information is useful to the control design, and if the answer is positive, the past packets rejection logic is better if it is not used. Some works have shown that the incorporation of the past states will improve the system performance to some extent. Therefore, if, in some applications, some improvement in system performance is seriously required, the past system state at the controller site, or the past control inputs at the actuator site, are better if used. To enable this, an appropriate timeline within one control period should be available to mark the arrivals of the past packets at the receivers. An illustration is shown in Fig. 1.10 (bottom). If using the past information does not result in an appreciable increase in performance or if the clock-driven scheme is limited, the packet disorder problem has to be solved simply through the past packets rejection logic.

Considering different scenarios of both time delays and packet losses at the actuator nodes, the closed-loop system is modeled as a switched system. The controller gains are then optimized to ensure a certain performance of the system, and the control commands are selected online at the actuator node according to the real situation of delays and packet losses.

1.3.5 Quantization and Coding

The first problem considered in the literature is related to the minimum information rate which is needed for the stabilization of a linear system. It has been shown that this minimum rate depends on the unstable eigenvalues of the system. Several generalizations of this have been proposed. In particular, the results hold true for infinite memory quantizers. In particular, the quantized stabilization through finite memory and by memoryless quantizers is treated. In this scenario, the digital communication channel is noisy, the techniques proposed above do not work.

1.3.6 QoS and Control

If the controller is designed only considering the plant model and the specifications, without considering the problem of the implementation on a distributed system, then it is necessary to evaluate the quality of service (QoS) provided by the distributed computing system in order to characterize the robustness of the controller with respect to the provided QoS. In this way it is possible to compare different control strategies according to their sensitivity to the QoS level. Therefore, on the one hand, this method requires to identify all the mechanisms which govern the QoS, to model the QoS and to evaluate it, and, on the other hand, to evaluate its impact on the process control application. Moreover, it is worth mentioning methods in which the controller is designed by integrating the behavior of the distributed system.

1.3.7 Multiagent Systems

The research in this field can be categorized into two areas:

The first area deals with the design of distributed estimation techniques which can be applied in the framework of the sensor network technology. Since collecting measurements from distributed wireless sensors nodes at a single location for online data processing may not be feasible due to several reasons, there is a growing need for in-network data processing tools and algorithms that provide high performance in terms of online estimation. These algorithms have to reduce the communication load among all sensor nodes, to be robust to node failures and packet losses, and to be suitable for distributed control applications.

The second area deals with the control of mobile autonomous robots. Groups of autonomous agents with computing, communication, and mobility capabilities are expected to become economically feasible and perform a variety of spatially distributed sensing tasks, such as search and rescue, surveillance, environmental monitoring, and exploration. As examples of motion coordination problems, teams of mobile autonomous agents require the ability to cover a region of interest, to assume a specified pattern, to rendezvous at a common point, or to move in a synchronized manner as in flocking behaviors.

1.3.8 Internet-Based Control

The success of the Internet as a worldwide information network can be attributed to the feedback mechanisms that control the data transfer in the transport layer of the Internet Protocol (IP) stack. The tremendous complexity of the Internet makes it extremely difficult to be modeled and analyzed, and it has been questioned whether mathematical theory can offer any major improvements in this area. A block diagram of an Internet-based control system is depicted in Fig. 1.11. The total time of performing an operation (a control action) per cycle is $t_1 + t_2 + t_3 + t_4$ where the four types of time delay are:

t_1 – time delay in making control decision by a remote operator;

t_2 – time delay in transmitting a control command from the remote operator to the local system (the web server);

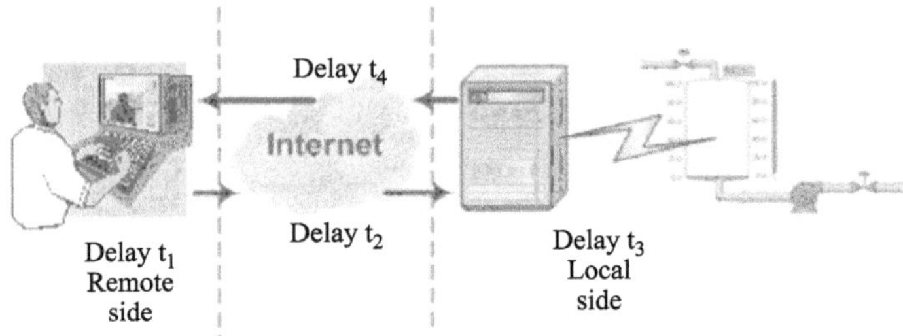

Figure 1.11 Internet-based control scheme

t_3 – execution time of the local system to perform the control action;
t_4 – time delay in transmitting the information from the local system to the remote operator.

If each of the four time delays is always a constant, the Internet-based control has a constant time delay.

Unfortunately this is not the case. The Internet time delays t_2 and t_4 increases with distance, but the delay depends also on the number of nodes traversed. Also the delay strongly depends on the Internet load. In effect, the Internet time delay is characterized by the processing speed of nodes, load of nodes, connection bandwidth, amount of data, transmission speed, etc. The Internet time delay $T_d(k)$ (t_2 and t_4) at instant k can be described in short as

$$T_d(k) = d_N + d_L(k)$$

where d_N is a term, which is independent of time, and $d_L(k)$ is a time-dependent term, the specification of both is given in [17].

It should be noted that computer-based control has been widely used in the process industries. Applications range from stand-alone computer-based control to local computer network-based control, such as a DCS. Fig. 1.12 shows a DCS linked with the Internet. Fig. 1.13 shows a typical computer-based process control system hierarchy comprising the following levels:

- plant-wide optimization,
- supervisory,
- regulatory, and
- protection.

The global database and plant data processing computer system are located at the top level where considerable computing power exists. The process

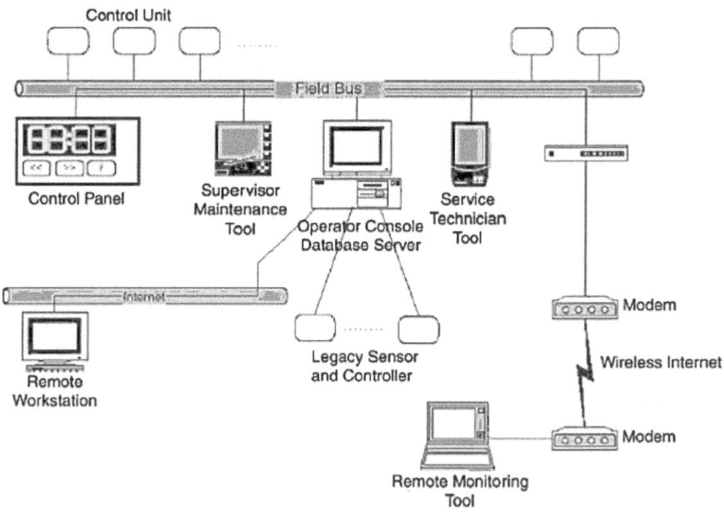

Figure 1.12 A distributed control scheme

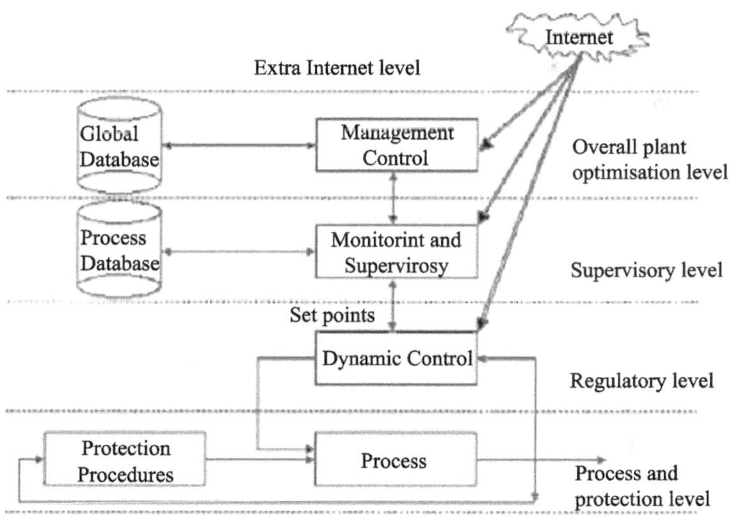

Figure 1.13 A process control system hierarchy

database and supervisory control are located at the second level in which many advanced control functions are implemented. DCS and process and protection are the lower two levels, respectively.

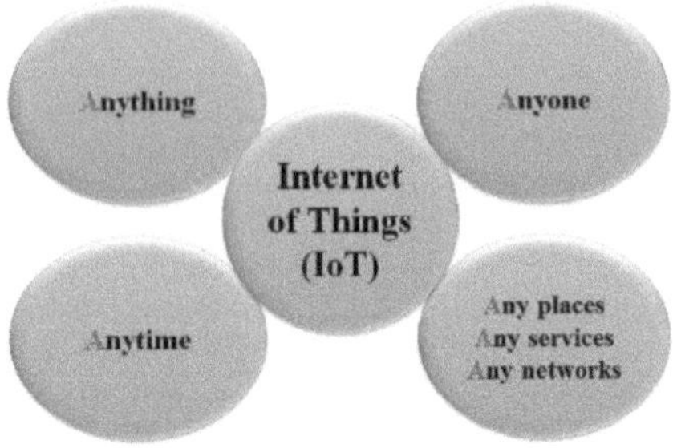

Figure 1.14 Concept of IoT

1.4 INTERNET-OF-THINGS

The term *Internet-of-Things (IoT)* was first introduced by Kevin Ashton to describe how IoT can be created by "adding radio frequency identification and other sensors to everyday objects" [18]. Over time, the term has evolved into one that describes the IoT as a network of entities that are connected through any form of sensor, enabling these entities, which we term as Internet-connected constituents, to be located, identified, and even operated upon.

Fig. 1.14 reviews that with the Internet-of-Things, anything will able to communicate to the Internet at any time from any place to provide any services by any network to anyone. This concept will create new applications, which can involve a smart vehicle and smart home, to provide many services such as notifications, security, energy saving, automation, communication, computers and entertainment.

For many years, the notions of *smart devices* like phones, cars, homes, cities, or even smart world have been adopted. Numerous research activities have been developed within the following prominent communities:

- Internet-of-Things (IoT),
- Mobile Computing (MC),
- Pervasive Computing (PC),
- Wireless Sensor Networks (WSN), and most recently,
- Cyberphysical Systems (CPS).

Due to the accelerated technological progress in each of these fields, there is an increasing overlap and merger of principles and research questions. In turn, close definitions of each of these fields are no longer appropriate. More importantly, the research endeavor in IoT, PC, MC, WSN and CPS often relies on underlying technologies such as

- Real-time computing,
- Machine learning,
- Security,
- Privacy,
- Signal processing,
- Big data, and others.

One vision of *smart world* lies in the degree of the density of sensing and actuation coverage which embedded in "things." These devices are incorporated in buildings in an attempt to save energy; in homes to realize full automation; in cars, taxis, and traffic lights to improve safety and transportation; in smartphones to run many useful applications and within healthcare services to support remote medicine and wellness. Despite the tremendous effort to use sensors/actuators, all of these are still at early stages of development. The steady increasing density of sensing and the sophistication of the associated processing will make for a significant qualitative change in how we work and live. We will truly have systems-of-systems that synergistically interact to form totally new and unpredictable services.

1.4.1 Architecture

Nowadays explosive number of things (objects) are connected to the Internet and hence it is necessary to have an adequate architecture that permits easy connectivity, control, communications, and useful applications. In some cases, things or sets of things must be disjoint and protected from other devices. In other cases, it makes sense to share devices and information. Various standards and automatic checks are made to ensure that an application can execute on a given platform.

While each application must solve its own problems, the sharing of a sensing and actuation utility across multiple simultaneously running applications can result in many systems-of-systems interference problems, especially with the actuators. Interferences arise from many issues, but primarily when the cyber depends on assumptions about the environment, hardware platform, requirements, naming, control and various device semantics.

1.4.2 Knowledge and Big Data

In an IoT world there will exist a vast amount of raw data being continuously collected. It will be necessary to develop techniques that convert this raw data into usable knowledge. For example, in the medical area, raw streams of sensor values must be converted into semantically meaningful activities performed by or about a person such as eating, poor respiration, or exhibiting signs of depression. Main challenges for data interpretation and the formation of knowledge include addressing noisy, physical world data and developing new inference techniques that do not suffer the limitations of Bayesian schemes. These limitations include the need to know a priori probabilities and the cost of computations. It can be expected that a very large number of real-time sensor data streams will exist, that it will be common for a given stream of data to be used in many different ways for many different inference purposes, that the data provenance and how it was processed will have to be known, and that privacy and security will have to be maintained. Data mining techniques are expected to provide the creation of important knowledge from all this data.

1.4.3 Robustness

Many IoT applications will be based on a deployed sensing, actuation, and communication platform (connecting a network of things). In these deployments it is common for the devices to know their locations, have synchronized clocks, know their neighbor devices when cooperating, and have a coherent set of parameter settings such as consistent sleep/wake-up schedules, appropriate power levels for communication, and pair-wise security keys. However, over time these conditions can deteriorate. The most common (and simple) example of this deterioration problem is with clock synchronization. Over time, clock drift causes nodes to have different enough times to result in application failures. While it is widely recognized that clock synchronization must reoccur, this principle is much more general. For example, some nodes may be physically moved unexpectedly. More and more nodes may become out of place over time. To make system-wide node locations coherent again, node relocalization needs to occur (albeit at a much slower rate than for clock sync). This issue can be considered a form of entropy where a system will deteriorate (tend towards disorder) unless energy in the form of rerunning protocols and other self-healing mechanisms is applied. Note that the control of actuators can also deteriorate due to their controlling software and protocols, but also due to physical wear and tear.

1.4.4 Standardizations

The Internet-of-Things (IoT) provides a technology to create the means of smart action for machines to communicate with one another and with

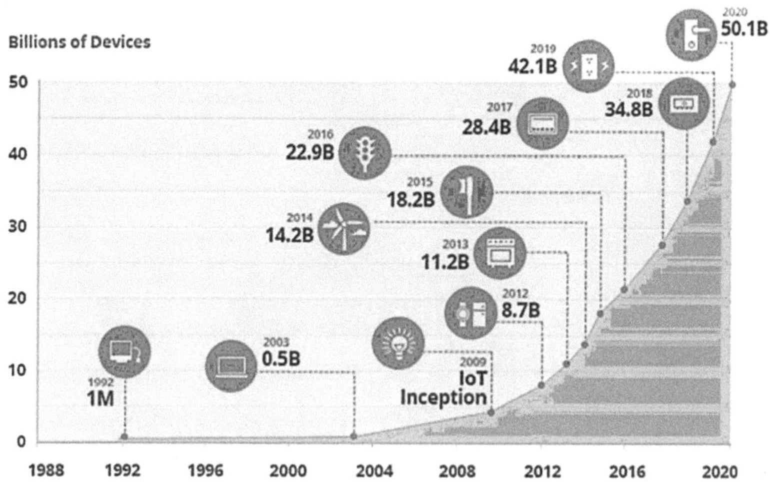

Figure 1.15 IoT growth [20]

many different types of information. By the year 2020 around 50 to 100 billion things will be connected electronically by the Internet. Fig. 1.15 shows the growth of the things connected to the Internet from 1988 to forecast 2020. The success of IoT depends on standardization, which provides interoperability, compatibility, reliability, and effective operations on a global scale. Today more than 60 companies for leading technology, in communications and energy, are working with standards, such as IETF, IEEE and ITU, to specify new IP based technologies for the Internet-of-Things.

1.5 CYBERPHYSICAL SYSTEMS

The term cyberphysical systems (CPS) emerged just over a decade ago as an attempt to unify the emerging application of embedded computer and communication technologies to a variety of physical domains, including aerospace, automotive, chemical production, civil infrastructure, energy, healthcare, manufacturing, materials, and transportation.

Fig. 1.16 illustrates a general architecture for CPS, where computation (top) interfaces through networks with physical processes (bottom). Security challenges arise when the computation is corrupted by false sensor information or when the control centers send malicious control actions to the physical process. Fig. 1.17 provides a pictorial layout of some of cyberphysical system applications.

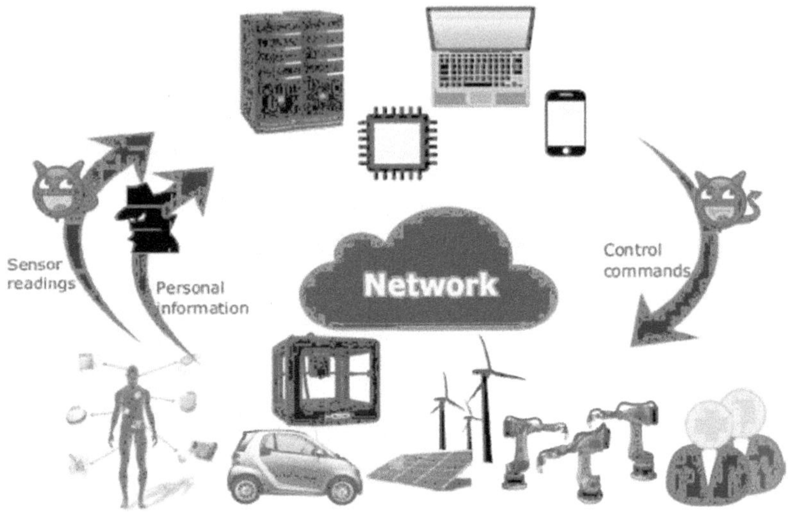

Figure 1.16 General representation of a CPS

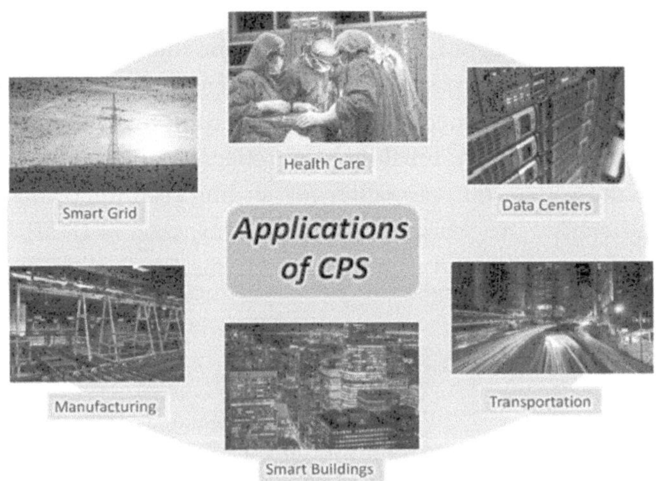

Figure 1.17 Applications of cyberphysical systems

1.5.1 Progress Ahead

Soon after the CPS term was coined, several research communities rallied to outline and understand how CPS security research is fundamentally different compared to conventional Information Technology (IT) systems. Because of the crosscutting nature of CPS, the background of early security position papers from 2006 to 2009 using the term CPS, ranged from real-time systems, embedded systems, control theory, and cyber-security.

Looking around and ahead, we see that cyberphysical systems bring advances in personalized health care, emergency response, traffic flow management, and electric power generation and delivery, as well as in many other areas now just being envisioned. Other phrases that you might hear when discussing these and related CPS technologies include:

- Internet-of-Things (IoT),
- Industrial Internet,
- Smart Cities,
- Smart Grid,
- "Smart" Anything (that is, cars, buildings, homes, manufacturing, hospitals, appliances).

1.5.2 Basic Features

Some of the major features of cyberphysical systems (CPS) are:

- In general, CPS has characteristics like distributed management and control, high degree of automation, real time performance requirements, reorganizing/reconfiguring dynamics, multiscale and system of systems control characteristics, networking at multiple scales, wide distribution geographically with components in location that lack physical security, integration at multiple temporal and spatial scales and take input and possible feedback from the physical environment. CPS objective is to monitor the behavior of physical processes and actuating actions to change its behavior in order to make the physical environment work correctly and in better way.
- As a new type of system that integrates computation with physical processes, components of a cyberphysical system (e.g., controllers, sensors, actuators, etc.) transmit the information to cyberspace through sensing a real world environment; also they reflect the policies of cyberspace back to the real world.
- CPS is a physical and engineered system whose operations are monitored, coordinated, controlled and integrated by a computing and communication core. This intimate coupling between the cyber and physical will be manifested from the nano-world to large-scale wide-area systems-of-systems. The Internet transformed how humans interact and communicate with one another, revolutionized how and where information is accessed, and even changed how people buy and sell products.

Similarly, CPS will transform how humans interact with and control the physical world around us.

- Cyberphysical systems may consist of many interconnected parts that must instantaneously exchange, parse and act upon heterogeneous data in a coordinated way. This creates two major challenges when designing cyberphysical systems: the amount of data available from various data sources that should be processed at any given time and the choice of process controls in response to the information obtained. An optimal balance needs to be attained between data availability and its quality in order to effectively control the underlying physical processes.

1.5.3 Cyberphysical Attacks

The Internet was established for the minimal processing and best-effort delivering of any packet, malicious or not. Recently, the Internet community was trying to cope with the series of distributed denial-of-service (DDoS) attacks that shut down some of the world's most high-profile and frequently visited Web sites, including *Yahoo* and *Amazon.com*, in the early 2000. For cyberattackers, motivated by revenge, prestige, politics, or money, the Internet architecture provided an unregulated network path to victims. Denial-of-service (DoS) attacks exploit this to target mission-critical services.

It is the nature of the attacks, as well as their scope, that has many people worried. In most of the cases, the perpetrators of the DDoS attacks planted *daemons* on perhaps thousands of intermediate computers, called *zombies*, which were then used as unwitting accomplices to flood and subsequently overwhelm defenseless Web servers with huge amounts of traffic.

The basic principle for all DDoS attacks is to use multiple intermediate machines to overwhelm other computers with traffic. A few prominent examples show the different types of DDoS techniques that hackers use:

UDP flood. Hackers have used UDP (User Datagram Protocol) technology to launch DDoS attacks. For example, by sending UDP packets with spoofed return addresses, a hacker links one system's UDP character-generating (chargen) service to another system's UDP echo service. As the chargen service keeps generating and sending characters to the other system, whose echo service keeps responding, UDP traffic bounces back and forth, preventing the systems from providing services, as shown in the figure on the previous page. Variations of this attack link two systems' chargen and echo services.

SYN flood. In TCP-based communications, an application sends an SYN synchronization packet to a receiving application, which acknowledges receipt of the packet with an SYN-ACK, to which the sending application responds with an ACK. At this point, the applications can begin communicating.

In an SYN flood attack, the hacker sends a large volume of SYN packets to a victim. However, the return addresses in the packets are spoofed with addresses that do not exist or are not functioning. Therefore, the victim queues up SYN-ACKs but cannot continue sending them because it never receives ACKs from the spoofed addresses. The victim then cannot provide services because its backlog queue fills up and cannot receive legitimate SYN requests.

Smurf. In this type of attack, a hacker broadcasts ICMP (Internet Control Message Protocol) echo (ping) requests, with the return addresses spoofed to show the ultimate victim's address, to a large group of hosts on a network. The hosts send their responses to the ultimate victim, whose system is overwhelmed and cannot provide services.

1.5.4 Detection Methods

For the interest of book objectives, we discuss here the methods for detecting DoS flooding attacks, *network-based attacks in which agents intentionally saturate system resources with increased network traffic.* As indicated earlier, in a distributed DoS (DDoS) attack, the assault is coordinated across many hijacked systems (*zombies*) by a single attacker *(master)*. It is worth mentioning that techniques that detect DoS also apply to DDoS [24].

The malicious workload in network-based DoS attacks comprises network datagrams or packets that consume network buffers, CPU processing cycles, and link bandwidth. When any of these resources form a bottleneck, system performance degrades or stops, impeding legitimate system use. Overloading a Web server with spurious requests, for example, slows its response to legitimate users. This specific DoS attack type does not breach the end (victim) system, either physically or administratively, and requires no other preexisting conditions except an Internet connection.

Although many high-profile DoS attacks have occurred, few have been empirically captured and analyzed. Given the potential for bad publicity, victims hesitate to share information regarding security incidents. It is therefore difficult for researchers to directly observe attacks and find their ubiquitous characteristics. In cases in which attack forensics are available, researchers can introduce classification systems, but attackers typically modify their techniques soon after discovery. As a result, the DoS attack-definition space is ever changing.

Based on the exploited weakness, the network-based DoS attack space can be divided into:

Vulnerability attacks in which malformed packets interact with some network protocol or application weakness present at the victim. This type of vulnerability typically originates in inadequate software assurance testing or negligent patching. The malformed attack packets interact with installed software, causing excessive memory consumption, extra CPU processing, system reboot, or general system slowing.

Flooding attacks in which the attacker sends the victim a large, occasionally continuous, amount of network traffic workload. As a result, legitimate workloads can become congested and lost at bottleneck locations near or removed from the victim. Such an attack requires no software vulnerability or other specific conditions.

1.5.5 Robustness, Resilience and Security

The notion of robustness often refers to a system's ability to withstand a known range of uncertain parameters or disturbances, whereas security describes the system's ability to withstand and be protected from malicious behaviors and unanticipated events. These two system properties are pre-event concepts, that is, the system is designed to be robust or secure offline before it is perturbed or attacked. Despite many engineering efforts toward designing robust and secure systems, it is costly and impractical, if not impossible, to achieve perfect robustness and security against all possible attacks and events. This fact, however, renders it essential to investigate the resilience aspect of a system, which refers to the system's ability to recover online after adversarial events occur. It is a post-event concept. Hence, to provide performance guarantees, control systems should be designed to be inherently resilient, allowing them to self-recover from unexpected attacks and failures.

The concept of resilience has appeared in various engineering fields, such as aviation, nuclear power, oil and gas, transportation, emergency health care, and communication networks [25–27]. It is suggested that the concept of resilient control systems, which emphasizes designing control systems for operation in an adversarial and uncertain environment. Resilient control systems are required to be capable of maintaining the state awareness of threats and anomalies and assuring an accepted level of operational normalcy in response to disturbances, including threats of an unexpected and malicious nature.

Metrics for robustness in control systems have been well studied in the literature [28,29]. A game-theoretic approach has been introduced to obtain the H_∞ optimal, disturbance-attenuating minimax controllers by viewing the controller as the cost minimizer and the disturbance as the maximizer. Likewise, cybersecurity problems have been studied using game theory [30], which provides a natural framework for capturing the conflict of goals between an attacker who seeks to maximize the damage inflicted on the system and a defender who aims to minimize it. Moreover, the design of security strategies is enabled by many existing analytical and computational tools [31]. Many metrics for the resilience of control systems have been proposed recently [32–37].

The design of resilient control systems pivots on the fundamental system tradeoffs between robustness, resilience, and security. Perfect security could be achieved by making the system unusable, and likewise perfect robustness could be attained by making the control performance completely inadequate. The need for resilience is due to the fact that no desirable control systems exhibit perfect robustness or security. Hence, it is imperative in the control design to know what type of uncertainties or malicious events need to be considered for enhancing robustness and security and what uncertainties or malicious events need to be considered for post-event resilience. Studying these tradeoffs requires extending the control system design problem to include the cyberlayers of the system and understand the cross-layer issues in cyberphysical systems.

Resilient control, however, poses new challenges, different from the ones encountered in robust control and security games. Resiliency should be considered together with robustness and security since the post-event resiliency relies on the pre-event designs. Resilience builds upon robustness and security frameworks and takes a cross-layer approach by considering post-event system features. Since game theory has been successfully applied to study robustness and security, it is natural to adopt it as the main tool to build an extended and integrated framework.

1.5.6 Multilayer Systems

A cross-layer approach is pivotal for designing resilient control systems. Integrating physical control systems with cyberinfrastructure to allow for new levels of HMI has been a growing trend in the past few decades. To manage the increasing complexity of cyberphysical systems, it is essential that

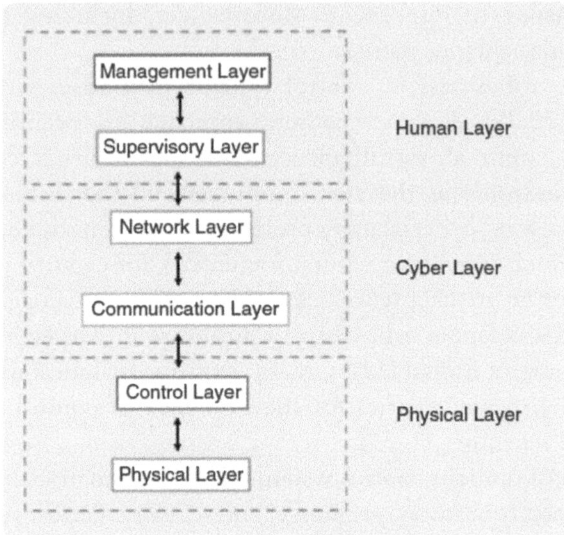

Figure 1.18 The multilayer structure of cyberphysical control systems

control designs exploit the hierarchical nature of such systems [36,37]. Depicted in Fig. 1.18, a cyberphysical control system can be conceptually divided into six layers:

- *Physical layer*, which deals with the physical devices or chemical processes, such as electric machines and transmission lines of power system infrastructure, and electric vehicles in transportation networks;
- *Control layer*, which monitors and controls the physical layer system for achieving desired system performance. It essentially consists of multiple control components, including observers/sensors, intrusion detection systems (IDSs), actuators, and other intelligent control components;
- *Communication layer*, which provides wired or wireless data communications that enable advanced monitoring and intelligent control;
- *Network layer*, which allocates network resources for routing and provides interconnections between system nodes;
- *Supervisory layer*, which is the executive brain of the entire system, provides human-machine interactions, and coordinates and manages lower layers through a centralized command and control, and
- *Management layer*, which resides at the highest echelon. It deals with social and economic issues, such as market regulation, pricing, and incentives.

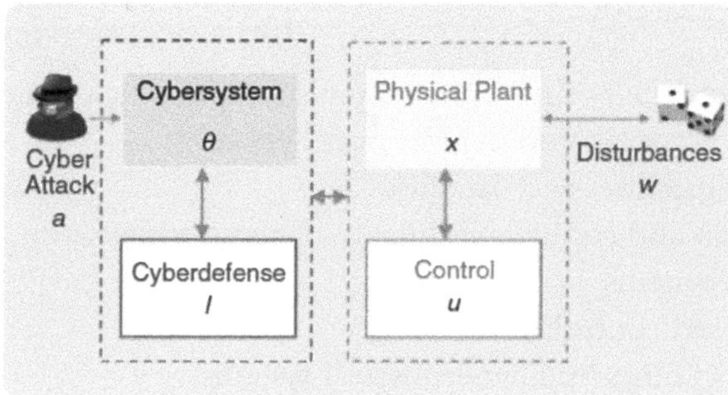

Figure 1.19 The interactions between the cyber and physical systems captured by their dynamics

One should observe that the physical layer, together with the control layer, can be viewed as the *physical world* of the system. On top of these two layers are the communication layer, which establishes physical layer wired or wireless communications, and the network layer that allocates resources and manages routing. The communication and network layers constitute the *cyberworld of the system*. Note that these two layers generally represent all the layers of open system interconnection (OSI) model [20], which can be incorporated into the cyberlayers of the system. The supervisory layer serves as the brain of the system, coordinating all lower layers by designing and sending appropriate commands. The management layer is a higher level decision-making engine, where the decision makers take an economic perspective towards the resource allocation problems in control systems. The supervisory and management layers are often interfaced with humans, and hence they contain human factor issues and HMIs.

The layered architecture can facilitate the understanding of the cross-layer interactions between the physical layers and the cyberlayers. In Fig. 1.19, $x(t)$ and $\theta(t)$ denote the continuous physical state and the discrete cyberstate of the system, which are governed by the laws f and Λ, respectively. The physical state $x(t)$ is subject to disturbances $x(t)$ and can be controlled by u. The cyberstate $\theta(t)$ is controlled by the defense mechanism l used by the network administrator as well as the attacker's action a. The hybrid nature of the cross-layer interaction leads to the adoption of a class of hybrid system models, as will be seen later.

Further developments will be carried out in Chapter 9.

1.6 NOTES

This chapter has briefly shed light on several topics, including

- Control of networks and control over networks,
- Delay characteristics and data loss,
- Quantization and coding and quality of service in control,
- Multiagent systems,
- Internet-based control,
- Internet-of-Things and cyberphysical systems.

On the one hand, IoT becomes a utility with increased sophistication in sensing, actuation, communications, control, and in creating knowledge from vast amounts of data. On the other hand, cloud computing is revolutionizing access to distributed information and computing resources that can facilitate future data and computation-intensive cloud control functions and improve dynamics reliability and safety. This has brought the security attacks which are problematic in networked control systems. The field of cyberphysical systems (CPSs), in a sense generalized networked control systems, is rapidly emerging due to a wide range of potential applications. However, there is a strong need for novel analysis and synthesis tools in control theory to guarantee safe and secure operation despite the presence of possible malicious attacks.

Recall that the communication networks in control systems are generally assumed to be deterministic and reliable. Real-time operating-system platforms rely on predetermined, static schedules for computation and communication. Some control is now occurring over the Internet, but at a supervisory level – for power-grid distribution stations, waste-water treatment plants, some commercial buildings, and other applications. These points have been remarked earlier in this chapter and will be explored further in the following chapters.

Along another research front, closed-loop automation and industrial process control, more often than not, require a dedicated, on-site end-to-end control system. Control in the IoT imposes control-theoretic challenges that we are unlikely to encounter in our usual application domains. More research is needed in many areas, including:

- Control over non-deterministic networks – due to unpredictability in sensor reading, packet delivery, or processing time, complicating closed-loop performance and stability.
- Latency and jitter – particularly relevant when considering control over the Internet and clouds, requiring much greater attention to latency

(the end-to-end delay from sensor reading to actuation) and jitter (the variance in the inter-sampling interval).

- Bandwidth – in modern applications, including mobile and/or feedback of video and other high-dimensional data, the issue of available bandwidth is critical. This requires that the involved sophisticated signal- and image-processing algorithms are best run on cloud platforms.

All of these topics and more will be further discussed in the subsequent chapters. The interested reader is advised to consult the papers/books in the list of references for additional demonstrations; in particular, we recommend [21–23].

REFERENCES

[1] N.N.P. Mahalik, K.K. Kim, A prototype for hardware-in-the-loop simulation of a distributed control architecture, IEEE Trans. Syst., Man, Cybern. C, Appl. Rev. 38 (2) (2008) 189–200.

[2] R.A. Gupta, Mo-Yuen Chow, Networked control system: overview and research trends, IEEE Trans. Ind. Electron. 57 (7) (2010).

[3] P. Ogren, E. Fiorelli, N.E. Leonard, Cooperative control of mobile sensor networks: adaptive gradient climbing in a distributed environment, IEEE Trans. Autom. Control 49 (8) (Aug. 2004) 1292–1302.

[4] C. Meng, T. Wang, W. Chou, S. Luan, Y. Zhang, Z. Tian, Remote surgery case: robot-assisted teleneurosurgery, in: IEEE Int. Conf. Robot. and Auto., vol. 1, ICRA'04, Apr. 2004, pp. 819–823.

[5] J.P. Hespanha, M.L. McLaughlin, G. Sukhatme, Haptic collaboration over the Internet, in: Proc. 5th Phantom Users Group Workshop, Oct. 2000.

[6] K. Hikichi, H. Morino, I. Arimoto, K. Sezaki, Y. Yasuda, The evaluation of delay jitter for haptics collaboration over the Internet, in: Proc. IEEE Global Telecomm. Conf., vol. 2, GLOBECOM, Nov. 2002, pp. 1492–1496.

[7] S. Shirmohammadi, N.H. Woo, Evaluating decorators for haptic collaboration over the Internet, in: Proc. 3rd IEEE Int. Workshop Haptic, Audio and Visual Env. Applic., Oct. 2004, pp. 105–109.

[8] P. Seiler, R. Sengupta, Analysis of communication losses in vehicle control problems, in: Proc. 2001 Amer. Contr. Conf., vol. 2, 2001, pp. 1491–1496.

[9] P. Seiler, R. Sengupta, An $\mathcal{H}_\infty$ approach to networked control, IEEE Trans. Autom. Control 50 (3) (2005) 356–364.

[10] M.S. Mahmoud, Control and Estimation Methods over Communication Networks, Springer-Verlag, UK, 2014.

[11] J.M.V. Grzybowski, M. Rafikov, J.M. Balthazar, Synchronization of the unified chaotic system and application in secure communication, Commun. Nonlinear Sci. Numer. Simul. 14 (6) (2009) 2793–2806.

[12] M. Rafikov, J.M. Balthazar, On control and synchronization in chaotic and hyperchaotic systems via linear feedback control, Commun. Nonlinear Sci. Numer. Simul. 14 (7) (2008) 1246–1255.

[13] M.S. Mahmoud, H.N. Nounou, Y. Xia, Robust dissipative control for Internet-based switching systems, J. Franklin Inst. 347 (1) (2010) 154–172.
[14] J.S. Baras, Security and trust for wireless autonomic networks: system and control methods, Eur. J. Control 13 (2007) 105–133.
[15] J.P. Richard, T. Divoux, Systemes commandes en reseau, in: Traites IC2, Information, Commande, Communication, Hermes Lavoisier, 2007.
[16] C. Canudas de Wit, Invited session on advances in networked controlled systems, in: Proc. the 25th IEEE American Control Conference, ACC'06, Minneapolis, Minnesota, USA, June 2006.
[17] S.H. Yang, Internet-Based Control Systems: Design and Applications, Springer, London, 2011.
[18] K. Ashton, That "internet of things" thing, RFID J. 22 (7) (2009) 97–114.
[19] Information Society, Mobile and wireless communications Enablers for the year Twenty-twenty (2020) [Online]. Available: https://www.metis2020.com.
[20] Z.K.A. Mohammed, E.S.A. Ahmed, Internet of things applications, challenges and related future technologies, World Sci. News 67 (2) (2017) 126–148.
[21] J. Deng, R. Han, S. Mishra, Secure code distribution in dynamically programmable wireless sensor networks, in: Proc. of ACM/IEEE IPSN, 2006, pp. 292–300.
[22] S. Munir, J. Stankovic, C. Liang, S. Lin, New cyber physical system challenges for human-in-the-loop control, in: 8th Int. Workshop on Feedback Computing, June 2013.
[23] G. Schirner, D. Erdogmus, K. Chowdhury, T. Padir, The future of human-in-the-loop cyberphysical systems, Computer 46 (1) (2013) 36–45.
[24] J. Mirkovic, J. Martin, P. Reiher, A taxonomy of DDoS attacks and DDoS defense mechanisms, ACM SIGCOMM Comput. Commun. Rev. 34 (2) (2004) 39–53.
[25] E. Hollnagel, J. Pariès, D.D. Woods, J. Wreathall, Resilience Engineering in Practice: A Guide Book, Ashgate Publishing, Farnham, UK, Jan. 2011.
[26] C. Rieger, D. Gertman, M. McQueen, Resilient control systems: next generation design research, in: Proc. 2nd Conf. Human System Interactions, 2009, pp. 632–636.
[27] C. Rieger, Notional examples and benchmark aspects of a resilient control system, in: Proc. 3rd Int. Symp. Resilient Control Systems, 2010, pp. 64–71.
[28] K. Zhou, J. Doyle, Essentials of Robust Control, Prentice-Hall, Englewood Cliffs, NJ, 1997.
[29] T. Basar, P. Bernhard, H_∞ Optimal Control and Related Minimax Design Problems: A Dynamic Game Approach, Birkhäuser, Switzerland, 1995.
[30] M. Manshaei, Q. Zhu, T. Alpcan, T. Basar, J.-P. Hubaux, Game theory meets network security and privacy, ACM Comput. Surv. 45 (3) (June 2013) 1–39.
[31] T. Alpcan, T. Basar, Network Security: A Decision and Game Theoretic Approach, Cambridge Univ. Press, Cambridge, UK, 2011.
[32] D. Wei, K. Ji, Resilient industrial control system: concepts, formulation, metrics, and insights, in: Proc. 3rd Int. Symp. Resilient Control Systems, 2010, pp. 15–22.
[33] W. Boyer, M. McQueen, Ideal based cyber security technical metrics for control systems, in: Proc. 2nd Int. Conf. Critical Information Infrastructures Security, Springer-Verlag, Berlin, Heidelberg, 2008, pp. 246–260.
[34] Q. Zhu, T. Basar, Robust and resilient control design for cyberphysical systems with an application to power systems, in: Proc. 50th IEEE Conf. Decision Control European Control, 2011, pp. 4066–4071.

[35] Q. Zhu, T. Basar, A dynamic game-theoretic approach to resilient control system design for cascading failures, in: Proc. 1st Conf. High Confidence Networked Systems, CPS Week, Beijing, China, Apr. 16, 2012, pp. 41–46.
[36] Q. Zhu, L. Bushnell, T. Basar, Resilient distributed control of multi-agent cyber-physical systems, in: Proc. Workshop Control Cyber-Physical Systems, Baltimore, MD, Mar. 20–21, 2013, pp. 301–316.
[37] Y. Yuan, Q. Zhu, F. Sun, Q. Wang, T. Basar, Resilient control of cyber-physical systems against denial-of-service attacks, in: Proc. 6th Int. Symp. Resilient Control Systems, 2013, pp. 54–59.

CHAPTER 2

Networked Control Systems' Fundamentals

A networked control system (NCS) is a system in which the traditional control loops are closed through a communication network such that signals of the system (control and feedback signals) can be exchanged among all components (sensors, controllers, and actuators) through a common network. Fig. 2.1 shows a typical structure of an NCS. In comparison with traditional control system, an NCS has several advantages including: less wiring, lower cost, and more flexibility and maintainability of the system. As a result, NCSs have been used widely in the last decades in many fields such as industry control, process control engineering systems, microgrids, aerospace systems, teleoperation, and intelligent systems. On the other hand, using a network system introduces new challenges to the system itself such as imperfections of the system which include quantization errors, varying delays, dropouts, etc. These imperfections could affect the behavior of the NCS by degrading the performance or causing instability. Therefore, it is essential to model the NCS correctly and design controllers that achieve stability under these circumstances.

The research of NCS can be classified into two main categories: (i) control of network, and (ii) control over or through network. The control

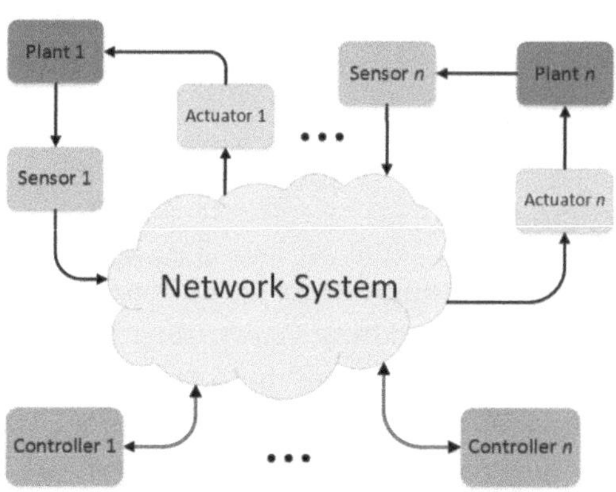

Figure 2.1 A typical networked control system

Networked Control Systems
https://doi.org/10.1016/B978-0-12-816119-7.00010-1

of network considers the problems of a communication network such as communication protocols, routing control, congestion control, and so on. While the control over or through network focuses on design and control of systems using a network as a transmission media to obtain the desired performance. The second topic "control over network", which will be the subject of this chapter, contains two main aspects: the quality of service (QoS) and the quality of control (QoC). Maintaining both QoS and QoC is a major objective of research of NCS. Measures of the network like transmission and error rates are subjected to QoS, whereas QoC is concerned with the system's stability subject to different conditions.

Several survey papers can be found in the literature summarizing the updated results on NCSs. The authors of [1] reviewed the stability of NCS in 2001. While, in 2006, [2] provided a general survey on NCS, in which the effect of NCS over the control methodologies of conventional large scale system was reviewed. In 2007, the challenges of control and communication in networked real-time system were presented in [3], and [4] provided an overview on estimation, analysis, and controller synthesis for NCS. Some of the research topics and trends of NCS were presented in 2010 [5]. A survey on network-induced constraints in NCS was proposed in 2013 [6]. In 2015, [7] discussed several aspects of NCS such as quantization, estimation, fault detection and networked predictive control. Also, it presented cloud control issues. Recently, an overview on the theoretical development of NCS was provided in [8] and [9]. An overview of the research investigations into the evolving area of NCS was provided in [10]. The interaction between control and computing theories was discussed in [11]. The authors of [12] proposed a review on event-based control and filtering of NCS. Finally, [13] focused on the distributed NCS. Besides, some results were presented in books, for example, coverage of analysis, stability and design of NCS can be found in [14–18].

2.1 MODELING OF NCS

As shown in Fig. 2.1, NCS components are connected via communication systems, and this connection addresses new imperfections and constraints that have to be considered in modeling the complete system. As presented in [19], the imperfections and constraints in NCS are classified into five types as follows: (i) quantization errors in the transmitted signals, due to the finite word length of the transmitted packets; (ii) packet dropouts, due to unreliable transmissions; (iii) variable sampling/transmission inter-

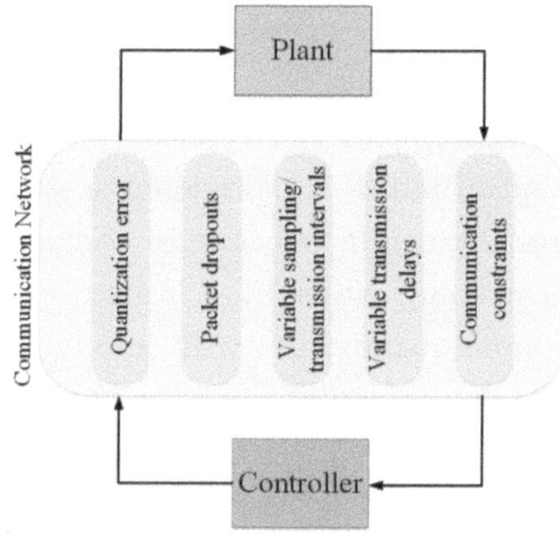

Figure 2.2 Types of imperfections and constraints in NCS

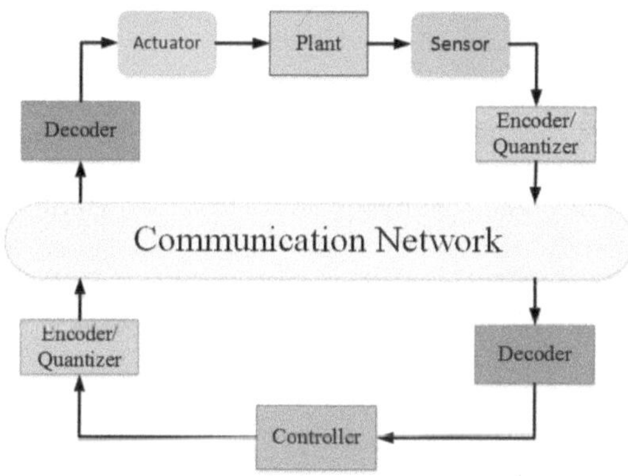

Figure 2.3 System configuration of NCS with quantizers

vals; (iv) variable transmission delays; (v) communication constraints, since not all sensor and actuator signals can be transmitted at the same time. These imperfections are summarized in Fig. 2.2.

2.1.1 Quantization Errors

Due to the existence of the communication network and its limited transmission capacity, signals have to be quantized before they are transmitted. The quantization occurs for both control signal and for plant output signal before they are sent to the network by implementing quantizers as shown in Fig. 2.3. A quantizer is a device that receives a real-valued signal and converts it to a piecewise constant one with a finite set of values.

In the literature, there are two common types of quantization, namely logarithmic and uniform quantization.

1. Logarithmic quantization
   The logarithmic quantization is considered as static quantization, its performance about the origin is better in comparison with uniform quantization, and it could either have infinite or finite quantization level. The logarithmic quantization with infinite quantization level is modeled as:

$$q(y) = \begin{cases} v_i, \text{ if } \frac{v_i}{1+\delta} < y < \frac{v_i}{1-\delta}, & y > 0, \\ 0, \quad \text{if } y = 0, \\ -q(-y), \quad \text{if } y < 0, \end{cases} \tag{2.1}$$

   where $U = \{\pm v_i \ : \ v_i = \rho^i v_0, \ i = \pm 1, \pm 2, \ldots\} \cup \{\pm v_0\} \cup \{0\}$, $0 < \rho < 1$, $v_0 > 0$ is the set of quantization values. And logarithmic quantization with finite quantization level is modeled as:

$$q(y) = \begin{cases} v_i, \text{ if } \frac{v_i}{1+\delta} < y < \frac{v_i}{1-\delta}, & 0 < i < N-1, \\ 0, \text{ if } 0 \le y \le \frac{v_{N-1}}{1+\delta}, \\ v_0, \text{ if } \quad y > \frac{v_0}{1+\delta}, \\ -q(-y), \text{ if } \quad y < 0, \end{cases} \tag{2.2}$$

   where $U = \{\pm v_i \ : \ v_i = \rho^i v_0, \ i = \pm 1, \pm 2, \ldots, \pm (N-1)\} \cup \{\pm v_0\} \cup \{0\}$, $0 < \rho < 1$, $v_0 > 0$ is the set of quantization values.

2. Uniform quantization
   A uniform quantizer is easier to be operated. The following are the conditions for uniform quantizers with an arbitrarily shaped quantitative area:
   a. If $\|y\|_2 \le M\mu$, then $\|q(y) - y\|_2 \le \Delta\mu$;
   b. If $\|y\|_2 > M\mu$, then $\|q(y)\|_2 > M\mu - \Delta\mu$;
   where M is the saturation value and Δ is the sensitivity. The upper bound of the quantization error when the quantization is not saturated is represented by the first condition, while the second condition gives the way of testing the saturation of the quantization. The rectangular

shaped quantitative area is modeled as:

$$q(y) = \begin{cases} 0, \text{ if } -\frac{1}{2} < y < \frac{1}{2}, \\ i, \text{ if } \frac{2i-1}{2} < y < \frac{2i+1}{2}, \; i = 1, 2, \ldots, K-1, \\ K, \text{ if } y \geq \frac{2K-1}{2}, \\ -q(-y), \text{ if } \; y \leq -\frac{1}{2}. \end{cases} \tag{2.3}$$

More details about the two types of quantization could be found in [7] and [20].

Remark 2.1. *The logarithmic quantization is mainly used with linear systems and with infinite quantization levels. Whereas, when uniform quantization is used, the zoom strategy is a useful control policy, which is divided into two stages: "zoom-in" and "zoom-out" [7].*

Remark 2.2. *The implementation of zoom-in and zoom-out stages was initially discussed in [21,22]. It was used to obtain a sufficient condition for the asymptotic stability for linear and nonlinear systems.*

As a result of the quantization, information loss will be introduced in the system. Therefore, a model of NCS has to take this into account. The quantization error is inversely proportional to the number of bits used for quantization, i.e., the small number of bits leads to a higher quantization error. Due to this fact, a significant research was directed to determine the minimum number of bits required to achieve the stability of the system, some examples could be found in [21–25].

Some researchers focused on controlling the quantization and its effects on the system. The authors of [26] proposed a sector bounded approach for dealing with the quantization errors, so their effects on NCS could be investigated using the procedures of robust analysis. Quantization and stochastic packet dropouts were considered in the study of the quadratic stability of NCS, and finite quantization was used for implementing the controller in [27]. The quantizer step size influence on NCS considering packet dropouts and finite-level quantization were studied in [28]. In [29] and [30], adaptable "center" and "zoom" parameters of the quantizers with finite values were considered, the input-to-state stability was obtained by applying a strategy of switching the controller continuously between "zoom-out" and "zoom-in". The same strategy of "zoom-out" and "zoom-in" was used to obtain parametrized input-to-state stability of NCS subjected to packet dropout and unknown disturbances but with random lengths of quantization regions based on the packet dropout process [31]. The authors of [32] have used a sector bound and convex combination property of quantizer to determine sufficient conditions needed to achieve the desired control of NCS subjected to several categories of asynchronous sampling and quantization. Some literature discussed quantization with the implementation of event triggering control, see, e.g., [33–37].

Remark 2.3. *There are two phenomena caused by quantization [7]:*

1. *Saturation, which occurs when the signal is larger than the quantization range, leading to a higher quantization error and causing instability in the closed-loop system.*
2. *Deterioration of the performance around the original point, which occurs near the origin when the signal is not exactly quantized due to the limitation of the accuracy of the quantizers, and this will prevent approaching the asymptotic stability of the closed-loop system.*

2.1.2 Packet Dropouts

Due to the use of the network for communication, the signals of the systems need to be grouped before being transmitted, each group of signals is called a "packet" and its size depends on the network used. The transmission of packets could be either single or multiple. In single packet transmission, all data are grouped from sensors or controller and transmitted together. On the other hand, in multiple packet transmission, the data are transmitted in several network packets, causing non-simultaneous arriving of data to the controller or actuator. The limited size of the network is not the only reason of using parallel transmission, but also the distribution of sensors and actuators practically over a large area make it difficult to lump data into one network packet, leading to the use of multiple transmission.

Failures or message collisions on nodes cause packet dropout. To avoid them, most protocols use transmission retry mechanisms. However, if the retransmission fails within a limited time, the packets are dropped. Since the communication network is the source of the losses, this type of dropout is called "network-induced packet dropout". Moreover, if a new packet sent earlier is available at the node later, it is more practical to discard it and use the recent one, and this type of packet dropout is called "active packet dropout". For tackling this issue, some techniques like logical ZOH mechanisms [38] and message rejection [39] were proposed.

One methodology of dealing with packet dropouts is to design the controller to withstand the upper bound of dropouts in the system; see [40–42].

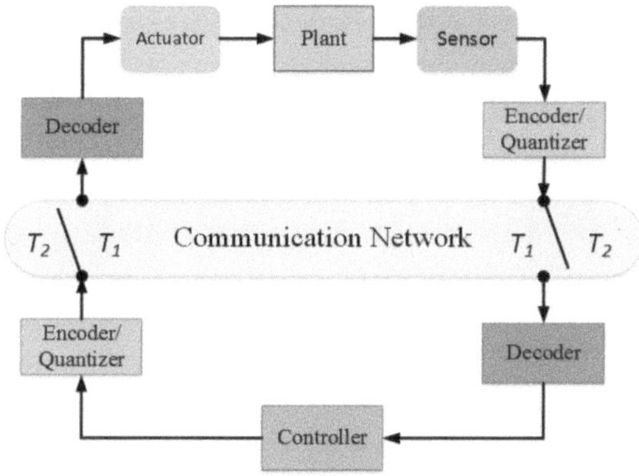

Figure 2.4 NCS with quantizers and packet dropout represented by a switch

Another famous approach is to represent the dropout in the system by a switch [43–45]. As shown in Fig. 2.4, when the switch is open (T_2), the dropout happens, while there is no dropout when it is closed (T_1). Then, the relation between the packet dropout rate and the H_∞ controller is derived to guarantee the exponential stability of the system. Consider the following discrete-time linear system model of NCS:

$$x(k+1) = A_{\sigma(k)}x(k) + B_{\sigma(k)}u(k); \tag{2.4}$$

and a hybrid state feedback controller is used such that

$$u(k) = -K_{\hat{\sigma}(k)}\hat{x}(k) \tag{2.5}$$

where $x(k)$ is the switched system state,

$$\sigma(k) : \mathbb{N}_0 = \{0, 1, 2, \ldots\} \rightarrow \Gamma = \{1, 2, \ldots, N\}$$

is the switching signal which is a piecewise constant function depending on time k and/or state $x(k)$. Subsystem i is activated when $\sigma(k) = i$, A_i and B_i are constant matrices; K_i is the controller gains to be designed. Also, subsystem i is activated at time k when $\sigma_i(k) = 1$, and $\sigma_i(k) = 0$ otherwise, so that $\sum_{i=1}^{N} \sigma_i(k) = 1$. Then, the discrete-time switched system is represented by

$$x_{k+1} = \sum_{i=1}^{N} \sigma_i(k)(A_i x_k + B_i u_k). \tag{2.6}$$

Let $\hat{x}_k$ be the switched system state and $\hat{\sigma}(k)$ be the switching signal received by the hybrid controllers over the network. Also, let s be the transmission indicator function defined as:

$$s = \begin{cases} 1, & \text{sample is transmitted,} \\ 2, & \text{sample is not transmitted.} \end{cases} \tag{2.7}$$

The dynamic of the switch system is described by:

$$\hat{x}(k) = \beta_s x(k) + (1 - \beta_s)\hat{x}(k-1), \tag{2.8}$$

$$\hat{\sigma}(k) = \beta_s \sigma(k) + (1 - \beta_s)\hat{\sigma}(k-1), \tag{2.9}$$

where β_s are switch variables such that $\beta_1 = 1$ and $\beta_2 = 0$. By combining Eqs. (2.4)–(2.6), the closed-loop networked switched control system with the hybrid state feedback controller is rewritten as

$$x(k+1) = \sum_{i=1}^{N} \sigma_i(k)(A_i x(k) - B_i(\sum_{i=1}^{N} \hat{\sigma}_i)K_i\hat{x}(k)). \tag{2.10}$$

Now, let $\zeta(k) = [x(k)^T \quad \hat{x}(k)^T]$ be the augmented state vector. Then the closed-loop networked switched control system with the network packet dropout effect is written as

$$\zeta(k+1) = \Phi_s \zeta(k), \tag{2.11}$$

and here we have two cases:

- If the switch is in T_1, $s = 1$, $\beta_1 = 1$, then

$$\hat{x}(k) = x(k),$$
$$\hat{\sigma}(k) = \sigma(k),$$
$$\Phi_1 = \begin{bmatrix} \sum_{i=1}^{N} \sigma_i(k)A_i & -\sum_{i=1}^{N} \sigma_i(k)B_iK_i \\ \sum_{i=1}^{N} \sigma_i(k)A_i & -\sum_{i=1}^{N} \sigma_i(k)B_iK_i \end{bmatrix};$$

- If the switch is in T_2, $s = 2$, $\beta_2 = 0$, then

$$\hat{x}(k) = x(k-1),$$
$$\hat{\sigma}(k) = \sigma(k-1),$$
$$\Phi_2 = \begin{bmatrix} \sum_{i=1}^{N} \sigma_i(k)A_i & -\sum_{i=1}^{N} \sigma_i(k)\sum_{l=1}^{N} \sigma_l(k-1)B_iK_l \\ 0 & I \end{bmatrix}.$$

Theorem 2.1 ([43]). *For system* (2.11)*, assume that the plant state and switching signal in a single packet are transmitted at a rate of r. If there exist symmetrical positive definite matrices* P_i*,* Q_i*,* $i \in \Upsilon$ *and scalars* $\alpha_1, \alpha_2 > 0$ *such that*

$$\alpha_1^r \alpha_2^{r-1} > 1, \tag{2.12}$$

$$\Phi_1^T P_j \Phi_1 + Q_i - \alpha_1^{-2} P_i \leq 0, \tag{2.13}$$

$$\Phi_2^T P_j \Phi_2 + Q_i - \alpha_2^{-2} P_i \leq 0 \tag{2.14}$$

hold, then system (2.11) *is exponentially stable.*

Proof. The proof could be derived using the following candidate switched Lyapunov–Krasovskii functional:

$$\begin{aligned} V(\zeta_k) = \zeta_k^T (\sum_{i=1}^{N} \sigma_i(k) P_i) \zeta_k \\ + \zeta_{k-1}^T (\sum_{i=1}^{N} \sigma_i(k-1) Q_i) \zeta_{k-1}. \end{aligned} \tag{2.15}$$

The detailed proof could be found in [43]. □

The other methodology is to consider the packet dropout as a random process, then model it as a Markov process as in [46] and [47], or as a Bernoulli distribution such as in [48] and [49]. In [50], the stability analysis and controller synthesis problems were presented for NCS with time-varying delays and affected by nonstationary packet dropouts. The plant is described by the following discrete-time linear time-invariant model as follows:

$$x_p(k+1) = Ax_p + Bu_p, \ \ y_p = Cx_p, \tag{2.16}$$

where $x_p(k) \in \mathbb{R}^n$ is the state vector of the plant and $u_p(k) \in \mathbb{R}^m$ and $y_p(k) \in \mathbb{R}^p$ are the control input and output vectors of the plant, respectively; A, B, and C are real matrices with appropriate dimensions. The measurement received by the controller is affected by a randomly varying communication delay and represented by

$$y_c(k) = \begin{cases} y_p(k - \tau_k^m), \ \delta(k) = 1, \\ y_p(k), \ \delta(k) = 0, \end{cases} \tag{2.17}$$

where τ_k^m is the "measurement delay" which follows the Bernoulli distribution, and $\delta(k)$ is Bernoulli distributed white noise sequence exhibiting the occurrence of packet dropouts in the NCS. Also, let

$$\mathrm{Prob}\{\delta(k) = 1\} = p_k$$

where p_k assumes discrete values. There are two classes to be considered [50]:

Class 1: p_k has the probability mass function where $q_r - q_{r-1} =$ constant for $r = 2, \ldots, n$. This covers a wide range of cases [50].

Class 2: $p_k = X/n$, $n > 0$ and $0 \leq X \leq n$ is a random variable that follows the binomial distribution $\mathbf{B}(q, n)$, $q > 0$, that is,

$$\mathrm{Prob}(p_k = (ax + b)/n) = \begin{pmatrix} n \\ x \end{pmatrix} q^x (1 - q)^{n-x}, \quad b > 0,$$
$$x = 0, 1, 2, \ldots, n, \quad an + b < n.$$

The following observer-based controller is required to be designed in case that the full state information is not available and the time delay occurs on the actuation side [51]:

$$\begin{aligned}
&\text{Observer}: \\
&\hat{x}(k+1) = A\hat{x} + Bu_p(k) + L(y_c(k) - \hat{y}_c(k)), \\
&\hat{y}_c(k) = \begin{cases} C\hat{x}(k), \ \delta(k) = 0, \\ C\hat{x}(k - \tau_k^m), \ \delta(k) = 1; \end{cases}
\end{aligned} \tag{2.18}$$

$$\begin{aligned}
&\text{Controller}: \\
&u_c(k) = K\hat{x}(k), \\
&u_p = \begin{cases} u_c(k), \ \alpha(k) = 0, \\ u_c(k - \tau_k^a), \ \alpha(k) = 1, \end{cases}
\end{aligned} \tag{2.19}$$

where $\hat{x}(k) \in \mathbb{R}^n$ is the estimate of the system (2.16), $\hat{y}_c(k) \in \mathbb{R}^p$ is the observer output, and $L \in \mathbb{R}^{n \times p}$ and $K \in \mathbb{R}^{m \times n}$ are the observer and controller gains, respectively, and τ_k^a is the *actuation delay*.

We assume that the "actuation delay" τ_k^a and the "measurement delay" τ_k^m are time-varying and satisfy the following inequalities:

$$\tau_m^- \leq \tau_k^m \leq \tau_m^+, \quad \tau_a^- \leq \tau_k^a \leq \tau_a^+. \tag{2.20}$$

Also, let the estimation error $e(k) = x_p(k) - \hat{x}(k)$. Then

$$x_p(k+1) = \begin{cases} Ax_p(k) + BKx_p(k-\tau_k^{\alpha}) & \\ -BKe(k-\tau_k^{\alpha}), & \alpha(k)=1, \\ (A+BK)x_p(k) - BKe(k), & \alpha(k)=0, \end{cases} \tag{2.21}$$

$$\begin{aligned} e(k+1) &= x_p(k+1) - \hat{x}(k+1) \\ &= \begin{cases} Ae(k) - LCe(k-\tau_k^{m}), & \delta(k)=1, \\ (A-LC)e(k), & \delta(k)=0. \end{cases} \end{aligned} \tag{2.22}$$

In terms of $\xi(k) = [x_p^T(k) \;\; e^T(k)]^T$, system (2.21)–(2.22) can be written in the following form:

$$\xi(k+1) = \mathsf{A}_j\xi(k) + \mathsf{B}_j\xi(k-\tau_k^{m}) + \mathsf{C}_j\xi(k-\tau_k^{a}), \tag{2.23}$$

where $\{\mathsf{A}_j,\ \mathsf{B}_j,\ \mathsf{C}_j,\ j=1,\ldots,4\}$ and j is an index identifying one of the following pairs $(\delta(k)=1,\ \alpha(k)=1)$, $(\delta(k)=1,\ \alpha(k)=0)$, $(\delta(k)=0,\ \alpha(k)=0)$ and $(\delta(k)=0,\ \alpha(k)=1)$ while

$$\begin{aligned} &\mathsf{A}_1 = \begin{bmatrix} A & 0 \\ 0 & A \end{bmatrix},\ \mathsf{A}_2 = \begin{bmatrix} A+BK & -BK \\ 0 & A \end{bmatrix}, \\ &\mathsf{A}_3 = \begin{bmatrix} A+BK & -BK \\ 0 & A-LC \end{bmatrix},\ \mathsf{A}_4 = \begin{bmatrix} A & 0 \\ 0 & A-LC \end{bmatrix}, \\ &\mathsf{B}_1 = \begin{bmatrix} BK & -BK \\ 0 & 0 \end{bmatrix},\ \mathsf{B}_2 = \begin{bmatrix} 0 & 0 \\ 0 & 0 \end{bmatrix}, \\ &\mathsf{B}_3 = \begin{bmatrix} 0 & 0 \\ 0 & 0 \end{bmatrix},\ \mathsf{B}_4 = \begin{bmatrix} BK & -BK \\ 0 & 0 \end{bmatrix}, \\ &\mathsf{C}_1 = \begin{bmatrix} 0 & 0 \\ 0 & -LC \end{bmatrix},\ \mathsf{C}_2 = \begin{bmatrix} 0 & 0 \\ 0 & -LC \end{bmatrix}, \\ &\mathsf{C}_3 = \begin{bmatrix} 0 & 0 \\ 0 & 0 \end{bmatrix},\ \mathsf{C}_4 = \begin{bmatrix} 0 & 0 \\ 0 & 0 \end{bmatrix}. \end{aligned} \tag{2.24}$$

Now, it is desired to design an observer-based feedback stabilizing controller in the form of (2.18) and (2.19) such that the closed loop system (2.23) is exponentially stable in the mean square sense. The switched time-delay systems based approach is used to solve this problem [50].

Theorem 2.2. *Let the controller and observer gain matrices K and L be given. The closed-loop system (2.23) is exponentially stable if there exist matrices* $0 < P$, $0 < Q_j^T = Q_j$, $j = 1, \ldots, 4$ *and matrices* R_i, S_i, *and* M_i, $i = 1, 2$, *such that the following matrix inequality holds:*

$$\Lambda_j = \begin{bmatrix} \Lambda_{1j} & \Lambda_{2j} \\ \bullet & \Lambda_{3j} \end{bmatrix} < 0, \tag{2.25}$$

$$\Lambda_{1j} = \begin{bmatrix} \Psi_j + \Phi_{j1} & -R_1 + S_1^T & -R_2 + S_2^T \\ \bullet & -S_1 - S_1^T - \hat{\sigma}_j Q_j & 0 \\ \bullet & \bullet & -S_2 - S_2^T - \hat{\sigma}_j Q_j \end{bmatrix},$$

$$\Lambda_{2j} = \begin{bmatrix} -R_1 + M_1^T - \Phi_{j2} & -R_2 + M_2^T - \Phi_{j3} \\ -S_1 - M_1^T & 0 \\ 0 & -S_2 - M_2^T \end{bmatrix},$$

$$\Lambda_{3j} = \begin{bmatrix} -M_1 - M_1^T + \Phi_{j4} & \Phi_{j5} \\ \bullet & -M_2 - M_2^T + \Phi_{j6} \end{bmatrix}, \tag{2.26}$$

where

$$\begin{aligned} \Psi_j &= -P + \hat{\sigma}_j(\tau_m^+ - \tau_m^- + \tau_a^+ - \tau_a^- + 2)Q_j \\ &\quad + R_1 + R_1^T + R_2 + R_2^T, \\ \Phi_{j1} &= (\mathsf{A}_j + \mathsf{B}_j + \mathsf{C}_j)^T \hat{\sigma}_j P (\mathsf{A}_j + \mathsf{B}_j + \mathsf{C}_j), \\ \Phi_{j2} &= (\mathsf{A}_j + \mathsf{B}_j + \mathsf{C}_j)^T \hat{\sigma}_j P \mathsf{B}_j, \\ \Phi_{j3} &= (\mathsf{A}_j + \mathsf{B}_j + \mathsf{C}_j)^T \hat{\sigma}_j P \mathsf{C}_j, \ \Phi_{j5} = \mathsf{B}_j^T P \mathsf{C}_j, \\ \Phi_{j4} &= \mathsf{B}_j^T \hat{\sigma}_j P \mathsf{B}_j, \ \Phi_{j6} = \mathsf{C}_j^T \hat{\sigma}_j P \mathsf{C}_j. \end{aligned}$$

The proof of this theorem could be found in [50].

2.1.3 Variable Sampling/Transmission Intervals

The signals in NCS need to be sampled before transmitting them through the network. The sampling periods are usually fixed in conventional systems due to their simplicity in design and analysis, and called "time-triggered sampling", "periodic sampling", and "uniform sampling". On the other hand, they are varying in the recent NCSs since they are waiting in a queue before the transmission process which will be based on the availability of the network and the protocol used. It was proved recently that sampling at varying time may have better performance than sampling at fixed intervals [52].

The other method of sampling is event-triggered sampling. It is also called "Lebesgue sampling," "level-crossing sampling," and "magnitude-driven sampling," etc., in this case, the sampling and transmission is triggered by an event such as changing one of the output signals to a specific value.

There are several approaches for modeling sampled/transmission intervals [52], but the most famous one is the input delay approach which uses a linear matrix inequality. By applying this approach, it is easy to determine the maximum upper bound of two consecutive samplings and design a proper controller for the NCS. Let the system with sampled signal be given by:

$$\begin{aligned} \dot{x} &= Ax(t) + Bu(t), \quad x(t_0) = x_0, \\ u(t) &= Kx(t_k), \quad t_k \le t < t_{k+1}, \end{aligned} \tag{2.27}$$

where $x(t) \in \mathbb{R}^n$ and $u(t) \in \mathbb{R}^m$ are the state vector of the system and control input vector, respectively; x_0 is the initial condition and $\{t_1, t_2, \ldots, t_k, \ldots\}$ is a sequence of sampling instances such that $t_k < t_{k+1}$, $\lim_{k\to\infty} t_k = \infty$ and $\sup_k\{t_{k+1} - t_k\} \le h_u$ for some known $h_u > 0$.

By applying the input delay approach, the above system is rewritten as [53]:

$$\begin{aligned} \dot{x} &= Ax(t) + BKx(t - \tau(t)), \quad t_k \le t < t_{k+1}, \\ x(\theta) &= x_0, \quad \text{for } \theta = t_0; \text{ and } 0 \text{ for } \theta < t_0, \end{aligned} \tag{2.28}$$

with piecewise time varying delay $\tau(t) := t - t_k$, $t_k \le t < t_{k+1}$ satisfying $0 \le \tau(t) \le h_u$ $\forall t \ge t_0$. Using this model, the Lyapunov–Krasovskii functional approach could be used to obtain the stability conditions and formulate the linear matrix inequalities for calculating the admissible upper bound of h_u and the corresponding controller gain K [53]. Other examples of the input delay approach can be found in [54–59].

2.1.4 Variable Transmission Delays

An NCS has two kinds of delay as shown in Fig. 2.5: (i) sensor-to-controller delay, which represents the time between sampling the signal from sensors and receiving it by the controller; and (ii) controller-to-actuator delay, which represents the time between generating the control signal and its availability on the actuator. Some of the sources of these delays are the limited data bandwidth, network traffic, and network protocols [60]. In this

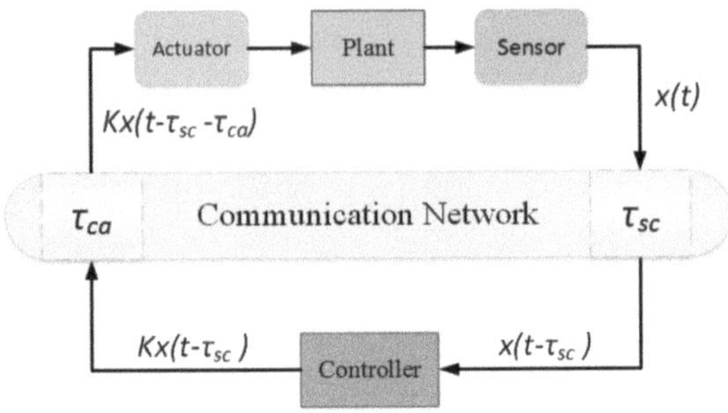

Figure 2.5 Network induced delay

way, only one of these delays was considered in the design of the controller, which is called one-mode controller. On the other hand, a two-mode controller is used to show that both of the aforementioned delays were considered in the model. One of the earlier results on a two-mode controller were discussed in [61–63], where both of the sensor-to-controller and controller-to-actuator random delays were considered and modeled as a Markov chain.

The networked induced delays are represented by

$$\tau(t_k) = \tau_{sc}(t_k) + \tau_{ca}(t_k), \tag{2.29}$$

where $\tau(t_k)$ is the total networked induced delay at sampling time t_k, τ_{sc} and τ_{ca} are the sensor-to-controller and controller-to-actuator delays, respectively. The timing diagram of the signals in the NCS is shown in Fig. 2.6.

In order to model all delays in the system, it is required to add the possible computational delays in the controller, actuator, and sensor nodes [64]. Then, the complete delay in the system is described by

$$\tau(t_k) = \tau_{sc}(t_k) + \tau_{ca}(t_k) + \tau_c(t_k) + \tau_a(t_k) + \tau_s(t_k), \tag{2.30}$$

where τ_c, τ_a, and τ_s are the computational delay in the controller, actuator, and sensor, respectively. Note that u as shown in Fig. 2.5 could be defined as

$$u(t) = Kx(t - \tau(t_k)), \tag{2.31}$$

where K represents the feedback control gain matrix.

Remark 2.4. *The induced delay τ could be extended to include the dropouts by representing it as a special case of time delay; thus,*

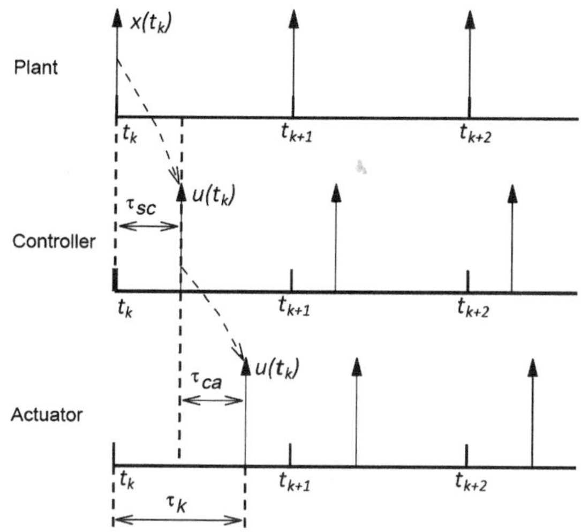

Figure 2.6 Signals in NCS with delay

$$\tau_d = \tau(t_k) + dh, \tag{2.32}$$

where $\tau_d(t_k)$ is the total delay including the dropouts delay, d is the number of dropouts, and h is the sampling period.

There are four main models for delays in an NCS [65]:

1. **Constant delay model.**
   Here an NCS is considered as a deterministic system with a constant time delay, normally equal to the maximum delay in the system similar to equations (2.29) and (2.30). It was used due its simplicity when the random delay is difficult to be characterized. In this model a receiver buffer is introduced at the controller (or actuator) node, and the buffer size is equal to the maximum delay (sensor to controller delay or controller to actuator delay); see [66] and [67]. And hence, the NCS can be treated as a deterministic system, and many deterministic control methods can be applied to this NCS.
2. **Mutually independent stochastic delay model.**
   When the probabilistic dependence is unknown, the constant delay model and the deterministic control strategies could hardly achieve the required performance of the system. The reason is due to the presence of many stochastic factors in networks such as load in network, competition between nodes, and network congestion, and these make the

network delay typically stochastic. The delay could be modeled either as mutually independent or probabilistically dependent.

3. **Markov chain model.**
   This type of model considers the special dependency relationships among the delays and includes a Markov chain. And the delay could be modeled in two ways:
   - One Markov chain to represent the sum of delays in the NCS (sensor-to-controller and controller-to-sensor), or
   - Two Markov chains for modeling both sensor-to-controller and controller-to-actuator delay, respectively.
4. **Hidden Markov model.**
   In this type of model, all of the stochastic factors such as load in the network, competition of nodes, and network congestion are grouped into a hidden variable and defined as the network state, and this network state governs the distribution of delays. Since the network state cannot be observed directly but rather it can be estimated through observing network delays, a hidden Markov model is used to represent the relation between the network state and the network delay.

2.1.5 Communication Constraints

In NCS, the communication network is normally shared with sensors and actuators by multiple nodes, and because of the limitation of data transmission, only one or some of these nodes are active at a time and have access to the network. This is the reason behind the communication constraints, sometimes called "medium access constraints", and hence, the network requires to have a protocol for allocating access of each node to the network. This protocol could be either deterministic or random [68]. And so, the model of constraints in NCS could be either deterministic or stochastic:

- Deterministic model of communication constraints
  Previously, the problem was to choose a periodic communication sequence and afterwards design a suitable controller for it [69]. But this method is an NP-hard problem as shown by [70]. So later researchers tried to design the controller first and then to find a suitable communication sequence either offline [71,72] or online [73–75]. Other examples of such attempts could be found in [76–80].
- Stochastic model of communication constraints
  In this case a random media access control (MAC) protocol is used. One example is that a node makes sure that there is no other traffic before

Table 2.1 References studying many imperfections.

References	Imperfections
[93]	(i)–(iii)
[94]	(i), (ii) and (v)
[95]	(i), (ii) and (iv)
[96]	(i), (iii) and (v)
[39]	(ii)–(iv)
[97]	(iii)–(v)
[98–102]	(i), (ii) and (iv)
[103]	(i), (iv), and (v)
[104]	(ii), (iv), and (v)
[105]	(i)–(iv)
[106,107]	(i), (iii), (iv) and (v)

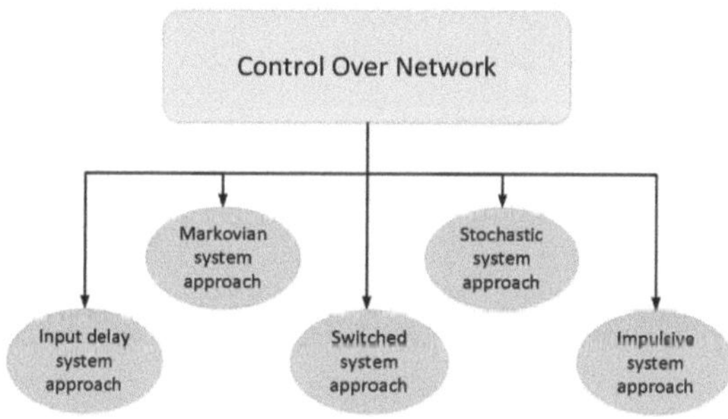

Figure 2.7 Theories of control over networks

transmitting its data [68]. Examples of this model could be found in [81–86].

2.2 CONTROL APPROACHES OVER NETWORKS

Several methods were developed for stabilizing NCS while considering one or more of the imperfections discussed in the previous section. Fig. 2.7 shows a list of these methods which include input delay, switched, Markovian, impulsive, stochastic system approaches, and predictive control approach. The following is a discussion on each of these approaches.

2.2.1 Input Delay System Approach

In this approach, the NCS is modeled as a system with time varying delay including the delay from sensor to controller, the delay from controller to actuator, and a representation of the dropout as a delay [1,108]. Additionally, the computational delay could be considered in the model as mentioned earlier in Eq. (2.32) [109]. Also, this approach was developed for considering the signal sampling [55], and applied for solving the synchronization problem of a complex network [110,111].

As mentioned in Section 2.1.2, some researchers have considered the disorder of data. In [38], a logical zeroth-order hold (ZOH) was designed to identify the newest control input signal that had arrived by making a comparison between time stamps and then used for controlling the process. The NCS with the sampler and the logical ZOH was described by the following discrete time system with input delay:

$$x(k+1) = Ax(k) + Bu(k - \tau(k)), \tag{2.33}$$

where $\tau(k)$ is the input delay which is bounded by τ_{max} such that $0 \leq \tau(k) \leq \tau_{max}$. Applying a state feedback controller $u = Kx$, the overall closed loop system is described by

$$x(k+1) = Ax(k) + BKx(k - \tau(k)), \quad K \in \mathbb{N}_0, \tag{2.34}$$

where K is to be selected and $\mathbb{N}_0$ is the set of nonnegative integers. A sufficient condition of stability of this NCS is derived based on a proper Lyapunov function and presented using the LMI method as follows [38]:

Theorem 2.3. *The NCS described in Eq.* (2.34) *is asymptotically stable if there exist matrices* $P \in \mathbb{S}^+$, $Z \in \mathbb{S}^+$, $T_1 \in \mathbb{R}^{n\times n}$, *and* $T_2 \in \mathbb{R}^{n\times n}$ *satisfying*

$$\begin{bmatrix} \Phi_{11} & \Phi_{12} & T_1 \\ \Phi_{12}^T & \Phi_{22} & T_2 \\ T_1^T & T_2^T & \frac{-1}{\tau_{\max} Z} \end{bmatrix} < 0, \tag{2.35}$$

where

$$\begin{aligned} \Phi_{11} &= A^T PA - P + \tau_{\max}(A^T - I)Z(A - I) + T_1 + T_1^T, \\ \Phi_{12} &= A^T PBK + \tau_{\max}(A^T - I)ZBK - T_1 + T_2^T, \\ \Phi_{22} &= K^T B^T PBK + \tau_{\max} K^T B^T ZBK - T_2 - T_2^T. \end{aligned}$$

The proof could be derived by using the following Lyapunov function:

$$V(x_k, k) \triangleq V_1(x_k, k) + V_2(x_k, k) + V_3(x_k, k), \tag{2.36}$$

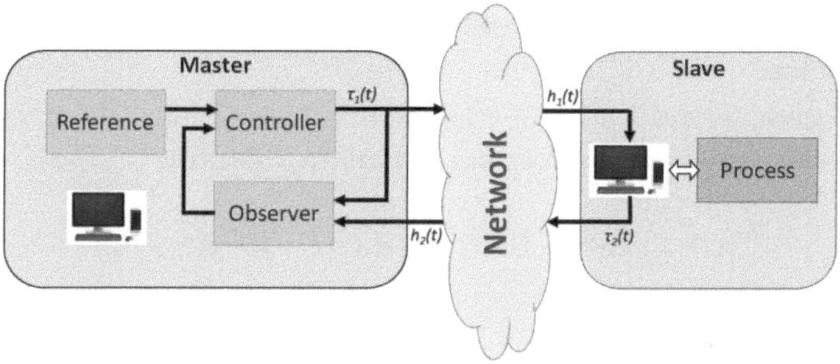

Figure 2.8 NCS with master–slave structure

where

$$V_1(x_k, k) = x^T(k)Px(k),$$
$$V_2(x_k, k) = \sum_{l=k-\tau(k)}^{k-1} x^T(l)Qx(l),$$
$$V_3(x_k, k) = \sum_{l=-\tau_{\max}+1}^{0} \sum_{h=k-1+l}^{k-1} \zeta^T(h)Z\zeta(h),$$

and

$$x_k \triangleq \left[x^T(k) \quad x^T(k-1) \ \ldots \ x^T(k-\tau_{\max})\right]^T,$$
$$\zeta(k) \triangleq x(k+1) - x(k),$$
$$\zeta(k) = (A - I)x(k) + BKx\left(k - \tau(k)\right),$$
$$\sum_{h=k-\tau(k)}^{k-1} \zeta(h) = x(k) - x\left(k - \tau(k)\right).$$

Further details of the proof can be found in [38].

Another representation of NCS using a structure of a master–slave system is discussed in [112] and [113]. As shown in Fig. 2.8, the control is performed in one PC, considered as a master, which communicates through a network with a slave that includes another PC and a mobile robot. Four delay sources are considered here: communication delay, data sampling, transmitting delay, and possible packet losses. The slave is considered to have the following linear form:

$$\dot{x}(t) = Ax(t) + Bu(t - \delta_1(t)),$$
$$y(t) = Cx(t), \tag{2.37}$$

where (A, B, C) is controllable and observable and $\delta_1(t)$ is the total master-to-slave delay.

For a given k and for any $t \in [t_{1,k} + h_{1m}, t_{1,k+1} + h_{1m}]$, there exists a k' such that the proposed observer has the following form:

$$\begin{aligned} \hat{x}(t) &= A\hat{x}(t) + Bu(t_{1,k}) - L(y(t_{2,k'}) - \hat{y}(t_{2,k'})), \\ \hat{y}(t) &= C\hat{x}(t), \end{aligned} \tag{2.38}$$

where k' corresponds to the newest output information received by the master. Now, given a signal $g(t)$, global delay $\delta(t)$ and packet loss delay $h(t_k)$ that the transmission line subjects to the packet containing the kth sample at time t_k, $g(t)$ can be written as:

$$\begin{aligned} g(t_k - h(t_k)) &= g(t - h(t_k) - (t - t_k)) \\ &= g(t - \delta(t)), \\ t_k \le t < t_{k+1}, \quad & \delta(r) \triangleq h(t_k) + t - t_k + d. \end{aligned} \tag{2.39}$$

Now, by using (2.39), Eq. (2.38) is rewritten as:

$$\begin{aligned} \hat{x}(t) &= A\hat{x}(t) + Bu(t - \delta_1(t)) - \\ &\quad L(y(t - \delta_2(t)) - \hat{y}(t - \delta_2(t))), \\ \hat{y}(t) &= C\hat{x}(t), \end{aligned} \tag{2.40}$$

where $\delta_1(t) \triangleq h_{1,k} + t - t_{1,k} + d_1$ and $\delta_2(t) \triangleq h_{2,k'} + t - t_{2,k'} + d_2$. The system features lead to $\delta_1(t) \le h_{1m} + T + d_1$ and $\delta_2(t) \le h_{2m} + T + d_2$. Also, the error vector between the estimated state $\hat{x}(t)$ and the real state $x(t)$ is defined as

$$e(t) = x(t) - \hat{x}(t). \tag{2.41}$$

So, the error is governed by:

$$\dot{e}(t) = Ae(t) - LCe(t - \delta_2(t)). \tag{2.42}$$

A delay-dependent state feedback control is designed by the master based on Lyapunov–Krasovskii functional and the approach of LMI, and using a remote observer for estimating the states of the salve. Thus, uniform stability is achieved for the system and characterized by the two following theorems [112]:

Theorem 2.4. *Suppose that, for some positive scalars α and ε, there exist $n \times n$ matrices $0 < P_1, P, S, Y_1, Y_2, Z_1, Z_2, Z_3, R, R_a$ and a matrix W with appropriate dimensions such that the following LMI conditions are satisfied for $j = 1, 2$:*

$$\begin{bmatrix} \Psi_2 & \begin{bmatrix} \beta_{2j} WC - Y_1 \\ \varepsilon_{2j} WC - Y_2 \end{bmatrix} & \mu_2 \beta_2 j \begin{bmatrix} WC \\ \varepsilon WC \end{bmatrix} \\ * & -S & 0 \\ * & * & -\mu_2 R_a \end{bmatrix} < 0,$$

$$\begin{bmatrix} R & Y \\ * & Z \end{bmatrix} \geq 0,$$

where β_{2j} are defined by:

$$\beta_{11} = e^{\alpha(\delta_1 - \mu_1)}, \; \beta_{12} = e^{\alpha(\delta_1 + \mu_1)},$$
$$\beta_{21} = e^{\alpha(\delta_2 - \mu_2)}, \; \beta_{22} = e^{\alpha}(\delta + \mu 2), \tag{2.43}$$

and the matrices Y, Z and Ψ_2 are given by:

$$Y = [Y_1 \quad Y_2], \quad Z = \begin{bmatrix} Z_1 & Z_2 \\ * & Z_3 \end{bmatrix}, \tag{2.44}$$

$$\Psi_2^{11} = P^{\mathrm{T}}(A_0 + \alpha I) + (A_0 + \alpha I)^{\mathrm{T}} P + S + \delta_2 Z_1 + Y_1 + Y_1^{\mathrm{T}},$$
$$\Psi_2^{12} = P_1 - P + \varepsilon P^{\mathrm{T}}(A_0 + \alpha I)^{\mathrm{T}} + \delta_2 Z_2 + Y_2,$$
$$\Psi_2^{22} = -\varepsilon(P + P^{\mathrm{T}}) + \delta_2 Z_3 + 2\,\mu_2 R_a + \delta_2 R.$$

Then, the gain

$$L = (P^T)^1 W \tag{2.45}$$

makes the error (2.42) of observer (2.40) exponentially converge to the solution $e(t) = 0$, with a decay rate α.

The solution of the LMI problem corresponding this theorem is written as

$$L = LMI_{obs}(\mu_2, \delta_2, \alpha). \tag{2.46}$$

For the control design, consider the controller $u = Kx$, $i = 1, 2$, i.e., the ideal situation $e(t) = 0$, $x(t) = \hat{x}(t)$, and

$$\dot{x}(t) = Ax(t) + BKx(t - \delta_1(t)). \tag{2.47}$$

Theorem 2.5. *Suppose that, for some positive scalars α and ε, there exist a positive definite matrix $\bar{P}_1$, $n \times n$ matrices $0 < \bar{P}, \bar{U}, \bar{Y}_1, \bar{Y}_2, \bar{Z}_1, \bar{Z}_2, \bar{Z}_3$ as in (2.44) and an $n \times m$ matrix W, such that the following LMI conditions hold:*

$$\Gamma_{3i}\begin{bmatrix} \Psi_3 & \begin{bmatrix} \beta_{1i}BW - \bar{Y}_1^T \\ \varepsilon_{1i}BW - \bar{Y}_2^T \end{bmatrix} & \mu_1 \begin{bmatrix} \beta_{1i}BW \\ \varepsilon\beta_{1i}BW \end{bmatrix} \\ * & -\bar{S} & 0 \\ * & * & -\mu_1\bar{R}_a \end{bmatrix} < 0, \quad \forall i = 1, 2,$$

$$\begin{bmatrix} \bar{R} & \bar{Y}_1 & \bar{Y}_2 \\ * & \bar{Z}_1 & \bar{Z}_2 \\ * & * & \bar{Z}_3 \end{bmatrix} \geq 0,$$

where β_{1i} for $i = 1, 2$ are defined by (2.43) *and:*

$$\Psi_3^{11} = (A_0 + \alpha I)\bar{P} + \bar{P}^T(A_0 + \alpha I)^T + \bar{S} + \delta_1\bar{Z}_1 + \bar{Y}_1 + \bar{Y}_1^T,$$
$$\Psi_3^{12} = \bar{P}_1 - \bar{P} + \varepsilon\bar{P}^T(A_0 + \alpha I)^T + \delta_1\bar{Z}_2 + \bar{Y}_2,$$
$$\Psi_3^{22} = -\varepsilon(\bar{P} + \bar{P}^T) + \delta_1\bar{Z}_3 + 2\,\mu_1\bar{R}_a + \delta_1\bar{R}.$$

Then, the gain

$$K = W\bar{P}^{-1} \tag{2.48}$$

exponentially stabilizes system (2.47) *with the decay rate α for all delays $\delta_1(t)$.*

The solution of the LMI problem corresponding to this theorem is written as

$$K = LMI_{con}(\mu_1, \delta_1, \alpha). \tag{2.49}$$

The input delay system approach was applied by other researchers with various protocols like round-robin (RR) protocol [114] and quadratic protocol (QP) [115], and the results of both of them were extended and applied on a discrete NCS with actuator constraints [116] and the results generated for system with more than two nodes [117].

2.2.2 Markovian System Approach

In this approach, a Markovian model is applied to represent a closed-loop NCS. In [46], the Markovian system approach was applied on a vehicle control problem that uses a wireless local network for communication in order to study the effect of packet dropout in the system and has the following description:

$$\begin{aligned} x(k+1) &= A_{\theta(k)}x(k) + B_{\theta(k)}u(k), \\ y(k) &= C_{\theta(k)}x(k), \\ x(0) &= x_0, \quad \theta(0) = \theta_0, \end{aligned} \tag{2.50}$$

where $\theta(k)$ is the time-varying dependence of the state matrices via the network packet loss parameters, and θ is either equal to 0 when the packet from sensor is dropped or equal to 1 when the packet is received. The state matrices are functions of a discrete-time Markov chain with values in a finite set $\mathcal{N} = \{1, \ldots, N\}$. The Markov chain has a transition probability matrix $\mathcal{P} = [p_{ij}]$ where $p_{ij} = \text{Prob}(\theta(k+1) = j|\theta(k) = i)$ subject to the restrictions $p_{ij} \geq 0$ and $\sum_{j=1}^{N} p_{ij} = 1$ for any $i \in \mathcal{N}$, which means that jumping probability is positive and that the Markov chain has to jump from mode i into some state that has a probability of one. Using the above model, the stochastic stability is derived for system (2.50) and represented in an LMI format. Using the above model, the stochastic stability is derived for system (2.50) and represented in an LMI format as follows:

Theorem 2.6. *System* (2.50) *is mean square stable (MSS) iff there exists* $G > 0$ *such that*

$$G - \sum_{j=1}^{N} p_j A_j^T G A_j. \tag{2.51}$$

Proof. The detailed proof of Theorem 2.6 could be found in [118]. □

Now, the objective is to design a dynamic output feedback controller that has the following form:

$$\begin{aligned} x_c(k+1) &= A_{c,\theta(k)}x_c(k) + B_{c,\theta(k)}y_c(k), \\ u(k) &= C_{c,\theta(k)}x_c(k), \end{aligned} \tag{2.52}$$

where $x_c(k) \in \mathbb{R}^n$ is the controller state and the subscript c denotes the controller matrices/states. Again, for $\theta(k) = \sigma \in \{0, 1\}$, $A_{c\sigma}$, $B_{c\sigma}$, and $C_{c\sigma}$ are used to denote the state space matrices of this two-mode controller. The closed-loop system is described by

$$A_{cl,\sigma} = \begin{bmatrix} \bar{A}_\sigma & \bar{B}_\sigma C_{c\sigma} \\ B_{c\sigma}\bar{C}_\sigma & A_{c\sigma} \end{bmatrix}, \tag{2.53}$$

where the subscript "*cl*" denotes the closed-loop matrices. By applying Theorem 2.6 and using Schur complements, system (2.53) is MSS for the following condition [46]:

$$\begin{bmatrix} Z & * & * \\ \sqrt{p}A_{cl,0}Z & Z & 0 \\ \sqrt{1-p}A_{cl,1}Z & 0 & Z \end{bmatrix} > 0, \tag{2.54}$$

where $Z = G^{-}1$.

In similar way, the model for an NCS with known packet loss was derived as a Markovian jumping system, and then an H_∞ controller was designed for that system [119]. In [120], a study of the stochastic stability of discrete-time NCS with random delays was presented. Two Markov chains were used to represent the sensor-to-controller and controller-to-actuator delays, and the obtained closed-loop systems are two-mode jump linear systems. In [121], the stability of the discrete time NCS contains polytopic uncertainty was considered, where a smart controller was updated with the buffered sensor information at stochastic intervals and the amount of the buffered data received by the controller under the buffer capacity constraint was also random. The exponential stability of generic switched NCS and the exponential mean-square stability of Markov-chain driven NCS were assured by establishing sufficient conditions.

Other examples from the literature that have used Markovian system approach are [122–127].

Remark 2.5. *A special class of hybrid and stochastic systems is called Markovian jump system. This system is applicable in many real situations such as manufacturing, power, chemical, economic, communication and control systems. A Markovian jump time-delay system model considering external disturbances of an event triggered NCS was presented in [128], the* H_∞ *control problem was solved and sufficient conditions to ensure stability of the closed-loop system were derived. More details about Markovian jump systems could be found in [129].*

2.2.3 Switched System Approach

In a switched system approach, an NCS is represented by a discrete-time switched system with a finite number of subsystems. A discrete-time switched system with arbitrary switching was formulated for an NCS while considering the effects of bounded uncertain access delay and packet losses.

Then both the asymptotic stability and L_∞ persistent disturbance attenuation issues were investigated [130]. The plant was represented by the following continuous-time linear time-invariant system:

$$\dot{x}(t) = A_c x(t) + B_c u(t) + E_c d(t),$$
$$z(t) = C_c x(t), \tag{2.55}$$

where $t \in \mathbb{R}^+$, $\mathbb{R}^+$ being the set of positive real numbers, $x(t) \in \mathbb{R}^n$ is the state variable, $u(t) \in \mathbb{R}^m$ is control input, and $z(t) \in \mathbb{R}^p$ is the controlled output. The disturbance input $d(t)$ is contained in $\mathcal{D} \subset \mathbb{R}^r$, and A_c, B_c, E_c, C_c are constant matrices. The control signal was assumed to be time-varying within a sampling period, and the sampling period was divided into a number of subintervals on which the controller reads its buffer at a higher frequency than the sampling frequency. The discrete time model of the system was described by:

$$\dot{x}(k+1) = Ax(k) + [B\ B\ \ldots\ B]_{nxN} \begin{bmatrix} u^1(k) \\ u^2(k) \\ \vdots \\ u^N(k) \end{bmatrix}$$
$$+ Ed(k), \tag{2.56}$$
$$z(t) = Cx(t), \tag{2.57}$$

where $A = e^{A_c T_s}$, $B = \int_0^{T_s/N} e^{A_c \eta} B_c d\eta$, $E = \int_0^{T_s} e^{A_c \eta} E_c d\eta$, and $C = C_c$. Then the time delay and dropouts effects were added to the model. There are three main scenarios that can occur during each sampling period.

1. If the delay is $\tau = h \times T$, where $T = \frac{T_s}{N}$, and $h = 1, 2, \ldots, d_{1_{max}}$, then $u_1[k] = u_2[k] = \cdots = u_h[k] = u[k-1]$, $u_{h+1}[k] = u_{h+2}[k] = \cdots = u_N[k] = u[k]$, and (2.56) can be written as:

$$x[k+1] = Ax[k] + [B\ B\ \ldots\ B] \begin{bmatrix} u[k-1] \\ \vdots \\ u[k-1] \\ u[k] \\ \vdots \\ u[k] \end{bmatrix} + Ed[k]$$
$$= Ax[k] + h \cdot Bu[k-1] + (N-h) \cdot Bu[k]$$
$$+ Ed[k]. \tag{2.58}$$

2. If a packet-dropout happens with delay less than τ_{max}, then the actuator will implement the previous control signal, i.e., $u_1[k] = u_2[k] = \cdots = u_N[k] = u[k-1]$. Therefore, the state transition equation (2.56) for this case can be written as:

$$\begin{aligned} x[k+1] &= Ax[k] + [B\ B \ldots\ B]\begin{bmatrix} u[k-1] \\ u[k-1] \\ \vdots \\ u[k-1] \end{bmatrix} + Ed[k] \\ &= Ax[k] + N \cdot Bu[k-1] + Ed[k]. \end{aligned} \tag{2.59}$$

3. The packets are dropped periodically, with period T_m which is an integer multiple of the sampling period T_s, i.e., $T_m = mT_s$. In case of $m = \frac{T_m}{T_s} \geq 2$, the first m_1 packets are dropped. Then, for these first m_1 steps, the previous control signal is used. Therefore

$$\begin{aligned} x(kT_m + T_s) &= Ax(kT_m) + NBu(kT_m - T_s) \\ &\quad + Ed(kT_m), \\ x(kT_m + 2T_s) &= A_2 x(kT_m) \\ &\quad + N \cdot (AB + B)u(kT_m - T_s) \\ &\quad + AEd(kT_m) + Ed(kT_m + T_s), \\ &\vdots \\ x(kT_m + (m-1)T_s) &= A_{m-1}x(kT_m) \\ &\quad + N\sum_{i=0}^{m-2} A_i Bu(kT_m - T_s) \\ &\quad + [A_{m-2}E, \cdots, E] \\ &\quad \cdot \begin{bmatrix} d(kT_m) \\ \vdots \\ d(kT_m + (m-2)T_s) \end{bmatrix}, \end{aligned}$$

where the integer $N = \frac{T_s}{T}$, and T is the period of the controller reading its receiving buffers.

During the period $t \in [kT_m + (m-1)T_s, (k+1)T_m)$, a new packet is transmitted successfully with some delay, so that $\tau = h\frac{T_s}{N}$, where $h = 0, 1, 2, \ldots, d_{max}$. Assume that $d(kT_m) = d(kT_m + 1) = \cdots = d(kT_m + m - 1)$, and that the controller uses just the time-invariant linear feedback control law, $u(t) = Kx(t)$. Then, we may obtain

$$\begin{aligned}x((k+1)T_m) = {} & [A_m + (N-h)BKA_{m-1}]x(kT_m) \\ & + [N\sum_{i=1}^{m-1} A_i + (N-h)NBK\sum_{i=0}^{m-2} A_i + h] \\ & \cdot BKx(kT_m - T_s) \\ & + [(N-h)BK\sum_{i=0}^{m-2} A_i + \sum_{i=0}^{m-1} A_i]Ed(kT_m).\end{aligned}$$

Now, by defining $\hat{x}[k] = \begin{bmatrix} x(kT_m - Ts) \\ x(kT_m) \end{bmatrix}$, the above equations can be written as

$$\begin{aligned}\hat{x}[k+1] &= \begin{bmatrix} x((k+1)T_m - T_s) \\ x((k+1)T_m) \end{bmatrix} \\ &= \Phi_{(m,h)} \begin{bmatrix} x(kT_m - T_s) \\ x(kT_m) \end{bmatrix} + E_m d(kT_m),\end{aligned}$$

where $\Phi_{(m,h)}$ equals to

$$\begin{bmatrix} N\sum_{i=0}^{m-2} A_iBK & A_{m-1} \\ N\sum_{i=0}^{m-2} A_iBK+ \\ (N-h)NBK\sum_{i-0}^{m-2} A_i + h)BK & A_m + (N-h)BKA_{m-1}A_{m-1} \end{bmatrix}$$

and

$$E_m = \begin{bmatrix} \sum_{i=0}^{m-2} A_iE \\ (N-h)BK\sum_{i=0}^{m-2} A_iE + \sum_{i=0}^{m-1} A_iE \end{bmatrix}$$

Here $m = \frac{T_m}{T_s} \geq 2$ and $h = 0, 1, \ldots, d_{max}$.

The aforementioned system could be represented in the following format:

$$\hat{x}[k+1] = \Phi_q\hat{x}[k] + E_q d[k]. \tag{2.60}$$

Theorem 2.7. *If the set* $\mathcal{P}^{(k)} \subset \text{int}\{X_0(\mu)\}$ *for some* k, *then the switched system* (2.60) *does not admit a positive disturbance invariant set under arbitrary* *switching in* $X_0(\mu)$. *In other words,* $\mu < \mu_\infty$, *where* $\mathcal{P}$ *is a positive disturbance invariant for the switched system* (2.60) *with arbitrary switching such that*

$$\mathcal{P}^{(k)} = \mathcal{P}^{(k-1)} \bigcap \underline{\text{pre}}(\mathcal{P}^{(k-1)}), \quad k = 0, 1, \ldots,$$
$$\mathcal{P}^{(0)} = X_0(\mu),$$

μ_∞ *is the* l_∞ *induced norm from* $d[k]$ *to* $z[k]$ *and defined as*

$$\mu_{inf} = \inf\{\mu : \ \|z[k]\|_{l^\infty} \leq \mu, \forall d[k], \|d[k]\|_{l^\infty} \leq 1\}.$$

The detailed stability analysis and L_∞ persistent disturbance attenuation for the above NCS could be found in [130]. Similar works are presented in [131–135].

In [136], a continuous-time NCS is modeled as an event based discrete-time model while allowing nonuniform sampling and varying delay larger than a sampling period. Then, the stability is achieved by solving a control problem for a switched polytopic system with an additive norm bounded uncertainty. A discrete-time switched linear uncertain system was used to model an NCS that includes time-varying transmission intervals, time-varying transmission delays, and communication constraints [137], at each single transmission, only one known node was allowed to communicate with the network for transmitting its data. Then, a convex overapproximation method in the form of a polytopic system with norm-bounded additive uncertainty was used to determine stability criteria for this NCS and represent it in an LMI format. The same procedure was followed, but with consideration of the quantization on the NCS [107], and the asymptotic stability was assured for the NCS with same imperfections but based on quantizer with a finite quantization level and suitably adjusted quantizer parameters [106].

Remark 2.6. *A switched system approach is implemented for NCS with network-induced delays by defining a switching function, the closed-loop NCS is represented by a time-delay switched system with two switching modes and each mode having a controller with a different gain [138]. Stability analysis is made based on both the time-delay switched system model and the average dwell time technique. Similarly, the exponential stability is achieved [139].*

Other examples of switched system approach could be found in [140–146].

2.2.4 Stochastic System Approach

The stochastic system approach is applied when network-induced delays and/or packet dropouts are random.

The authors of [147] discussed the stability of NCS with stochastic input delays, considering the following delayed NCS:

$$\dot{x}(t) = Ax(t) + Bu(t - \tau(t)), \tag{2.61}$$

where $\tau(t)$ refers to a piecewise continuous and bounded time delay including delay from sensor to actuator, delay from actuator to sensor, and the effect of dropout. Assume that the probability distribution of $\tau(t)$ is known a priori and takes values in $[0, \hat{\tau}_1]$ and $(\hat{\tau}_1, \hat{\tau}_2]$, and $0 \leq \hat{\tau}_1 \leq \hat{\tau}_2$, also, suppose that two random events $\mathcal{F}_1$, $\mathcal{F}_2$ are defined for the stochastic input delay $\tau(t)$ such that $\mathcal{F}_1 :\ \tau(t) \in [0, \hat{\tau}_1]$ and $\mathcal{F}_2 :\ \tau(t) \in (\hat{\tau}_1, \hat{\tau}_2]$, and define a random variable $\delta(t)$ such that $\delta(t) = 1$ when $\mathcal{F}_1$ occurs and $\delta(t) = 0$ when $\mathcal{F}_2$ occurs. Moreover, two functions are defined $\tau_1 : \mathbb{R}^+ \to [0, \hat{\tau}_1]$ and $\tau_2 : \mathbb{R}^+ \to [\hat{\tau}_1, \hat{\tau}_2]$ such that:

$$\begin{aligned}
\tau_1(t) &= \begin{cases} \tau(t), & \text{if } \delta(t) = 1, \\ \hat{\tau}_1, & \text{if } \delta(t) = 0, \end{cases} \\
\tau_2(t) &= \begin{cases} \hat{\tau}_1, & \text{if } \delta(t) = 1, \\ \tau(t), & \text{if } \delta(t) = 0. \end{cases}
\end{aligned} \tag{2.62}$$

Now, by using (2.62), Eq. (2.61) can be rewritten as

$$\begin{aligned}
\dot{x}(t) = Ax(t) &+ \delta(t)Bu(t - \tau_1(t)) \\
&+ (1 - \delta(t))Bu(t - \tau_2(t)),
\end{aligned} \tag{2.63}$$

where $u(t) = Kx(t)$, K is a constant matrix to be designed, and $\delta(t)$ is assumed to be a Bernoulli distribution sequence with $\text{Prob}\{\delta(t) = 1\} = E\{\delta(t)\} = \delta_0$, $\text{Prob}\{\delta(t) = 0\} = 1 - E\{\delta(t)\} = 1 - \delta_0$, and $0 \leq \delta_0 \leq 1$ being a constant. The closed-loop system (2.63) is described by the following formula:

$$\begin{aligned}
\dot{x}(t) = Ax(t) &+ \delta(t)BKx(t - \tau_1(t)) \\
&+ (1 - \delta(t))BKx(t - \tau_2(t)).
\end{aligned} \tag{2.64}$$

Theorem 2.8. *System (2.64) is exponentially stable in the mean-square sense if, for given constants $\hat{\tau}_1$, $\hat{\tau}_2$, δ_0 and matrix K, there exist matrices $P > 0$,*

$Q_i > 0,\ R_i > 0,\ Z_i > 0\ (i = 1, 2),\ N_j, M_j, T_j, W_j\ (j = 1, 2, 3, 4, 5, 6),$ *and* $S_l\ (l = 1, 2, 3, 4)$ *of appropriate dimensions such that the following LMIs hold:*

$$\begin{bmatrix} \Xi_{11} & * & * \\ \Xi_{21}^l & \Xi_{22} & * \\ \Xi_{31} & 0 & \Xi_{33} \end{bmatrix} < 0, \quad l = 1, 2, 3, 4, \tag{2.65}$$

where Ξ_{11}, Ξ_{21}^l, Ξ_{22}, Ξ_{31}, *and* Ξ_{33} *are functions of* $\hat{\tau}_1$, $\hat{\tau}_2$, δ_0, K, P, Q_i, R_i, Z_i, N_j, M_j, T_j, W_j *and* S_l *and defined in* [147].

Proof. The proof is obtained using the following candidate Lyapunov function:

$$V(x_t) = \sum_{i=1}^{7} V_i(x_t) \tag{2.66}$$

with

$$\begin{aligned}
V_1(x_t) &= x^T(t)Px(t), \\
V_2(x_t) &= \int_{t-\hat{\tau}_1}^{t} x^T(s)Q_1x(s)ds, \\
V_3(x_t) &= \int_{t-\hat{\tau}_2}^{t} x^T(s)Q_2x(s)ds, \\
V_4(x_t) &= \int_{t-\hat{\tau}_1}^{t} \int_{s}^{t} y^T(v)R_1y(v)dvds, \\
V_5(x_t) &= \int_{t-\hat{\tau}_2}^{t-\hat{\tau}_1} \int_{s}^{t} y^T(v)R_2y(v)dvds, \\
V_6(x_t) &= \delta_0(1-\delta_0)\int_{t-\hat{\tau}_1}^{t} \int_{s}^{t} \zeta^T(v)\beta^T Z_1\beta\zeta(v)dvds, \\
V_7(x_t) &= \delta_0(1-\delta_0)\int_{t-\hat{\tau}_2}^{t-\hat{\tau}_1} \int_{s}^{t} \zeta^T(v)\beta^T Z_2\beta\zeta(v)dvds.
\end{aligned}$$

The detailed proof of Theorem 2.8 can be found in [147]. □

A similar result can also be found in [148]. In [149], random communication delays from sensor to the controller and from controller to actuator through a limited bandwidth communication channel were represented by a linear function of the stochastic variable having a Bernoulli random binary distribution, and then the exponential stability is achieved by applying an H_∞ controller. Stability is discussed using a stochastic computational technique in NCS by representing transmission intervals and transmission delays by a sequence of continuous random variables [150]. Properties for input–output stability are derived for a general class of nonlinear NCS with exogenous disturbances using stochastic protocols subject to random network-induced delays and packet losses [151]. Other examples of using stochastic approach in NCS are found in [152–157].

2.2.5 Impulsive System Approach

In this approach, the NCS is represented by a hybrid discrete/continuous model or, in other words, impulsive system. An LTI system with uncertainties in the parameters of the process and intervals of sampling was modeled as a linear impulsive system described by the following equations [158]:

$$\begin{aligned} \dot{x}(t) &= f_k(x(t), t), \quad && t \neq s_k, \ \forall k \in \mathbb{N}, \\ x(s_k) &= g_k(x(s_k^-), s_k), \quad && t = s_k, \ \forall k \in \mathbb{N}, \end{aligned} \tag{2.67}$$

where f_k and g_k are locally Lipschitz functions from $\mathbb{R}^n \times \mathbb{R}$ to $\mathbb{R}^n$ such that $f_k(0, t) = 0$, $g_k(0, t) = 0$, $\forall t \geq 0$, and the impulse time sequence s_k forms a strictly increasing sequence in $[s_0, \infty)$ for some initial time $s_0 \geq 0$. Then, a discontinuous Lyapunov function at the impulse times was used for achieving the exponential stability of system (2.67). For an NCS with variable sampling time, the objective was to find the maximum allowable time interval τ_{MATI} for the following general LTI system [158]:

$$\dot{x}(t) = Ax(t) + Bu(t), \tag{2.68}$$

where x, u are the state and input of the process, respectively. At the sampling time s_k, $k \in \mathbb{N}$, the state of the system $x(s_k)$, is sent to the controller, and the control input $u = Kx(s_k)$ is sent back to the actuator. The resulting closed-loop system is modeled as an impulsive system with state $\xi(t) := [x'(t) \ z'(t)]$ where $z(t) := x(s_k)$, $t \in [s_k, s_{k+1})$. So, the impulsive system is described by:

$$\begin{aligned} \dot{\xi}(t) &= F\xi(t), \quad && t \neq s_k, \forall k \in \mathbb{N}, \\ \xi(s_k) &= \begin{bmatrix} x(s_k^-) \\ x(s_k^-) \end{bmatrix}, \quad && t = s_k, \forall k \in \mathbb{N}, \end{aligned} \tag{2.69}$$

where $F = \begin{bmatrix} A & BK \\ 0 & 0 \end{bmatrix}$.

Theorem 2.9. *System (2.69) is stable if there exist symmetric positive definite matrices P, R, X_1 and a slack matrix N such that*

$$M_1 + \tau_{MATI} M_2 < 0, \tag{2.70}$$

$$\begin{bmatrix} M_1 & \tau_{MATI} N \\ * & -\tau_{MATI} R \end{bmatrix} < 0, \tag{2.71}$$

where

$$\begin{aligned} M_1 = & \begin{bmatrix} P \\ 0 \end{bmatrix} [A \quad BK] + \begin{bmatrix} A^T \\ (BK)^T \end{bmatrix} [P \quad 0] \\ & - \begin{bmatrix} I \\ -I \end{bmatrix} X_1 [I \quad -I] - N[I \quad -I] - \begin{bmatrix} I \\ -I \end{bmatrix} N^T \\ & + \tau_{MATI} \overline{F}^T R \overline{F}, \\ M_2 = & \begin{bmatrix} I \\ -I \end{bmatrix} X_1 \overline{F} + \overline{F} X_1^T [I \quad -I], \quad \overline{F} = [A \quad BK]. \end{aligned}$$

Proof. The proof of Theorem 2.9 is found in [158] and can be obtained using the following candidate Lyapunov function:

$$V(\xi, \rho) := V_1(x) + V_2(\xi, \rho) + V_3(\xi, \rho), \tag{2.72}$$

where

$$\begin{aligned} V_1(x) &= x^T P x, \\ V_2(\xi, \rho) &= \xi^T \left(\int_{-\rho}^{0} (s + \tau_{MATI}) (F e^{Fs})^T \tilde{R} F e^{Fs} ds \right) \xi, \\ V_3(\xi, \rho) &= (\tau_{MATI} - \rho)(x - z)^T X_1 (x - z), \\ \tilde{R} &= \begin{bmatrix} R & 0 \\ 0 & 0 \end{bmatrix}, \\ \rho(t) &= t - s_k, \quad t \in [s_k, s_{k+1}), \forall k \in \mathbb{N}, \end{aligned}$$

with R, P, and X_1 being symmetric positive definite matrices. □

The Razumikhin technique and Lyapunov functions are applied to achieve exponential stability of an NCS subject to variable bounded delay by applying impulsive control [159]. A model of a threshold-error-dependent augmented impulsive system with an interval time-varying delay

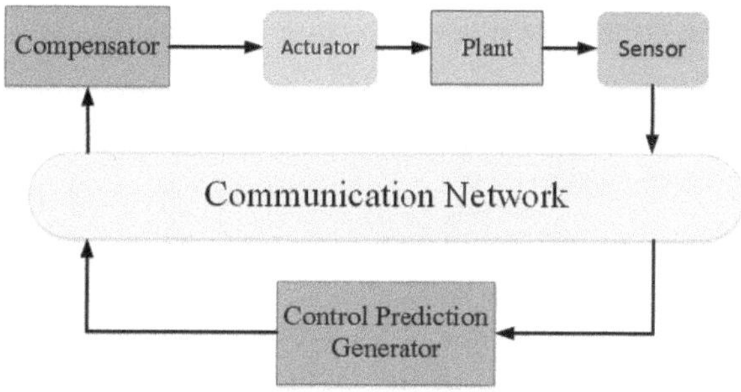

Figure 2.9 A network predictive controller scheme

is used to design a dissipative control for model-based NCS subject to event triggered communication in [160]. A delay scheduled impulsive (DSI) controller is presented to achieve robust stability of NCS subject to integral quadratic constraint and delays in [161]. Other examples of this approach could be found in [162–166].

2.2.6 Predictive Control Approach

In this approach, a network predictive controller (NPC) is designed in order to compensate for the effect of time delays and packet dropouts in the network. The NPC scheme consists of two parts as shown in Fig. 2.9, a control prediction generator and a compensator. A set of future control predictions is generated, packed, and transmitted to the plant side by the control prediction generator based on the signals received from the sensor. Using the most recent control value from the latest control prediction sequence, the compensator is designed to compensate for the delays and dropouts that occur from sensor to controller or from controller to actuator channels to achieve the desired performance.

The authors of [167] presented an NPC to compensate for the network-induced delay. For the following discrete-time system:

$$\begin{aligned} x_{k+1} &= Ax_k + Bu_k, \\ y_k &= Cx_k, \end{aligned} \tag{2.73}$$

where $x_k \in \mathbb{R}^n$ is the state vector, u_k and $y_k \in \mathbb{R}$ are the input and output vectors, respectively, while A, B, and C are system matrices. An output feedback controller is used such that $u_k = K_k y_k$. The NPC generates state predictions on the controller side up to time $t+\tau$ as follows:

$$\begin{aligned}\hat{x}_{k-d_k+1|k-d_k} &= Ax_{k-d_k} + B\hat{u}_{k-d_k},\\ &\vdots \\ \hat{x}_{k-d_k+i|k-d_k} &= Ax_{k-d_k+i-1} + B\hat{u}_{k-d_k+i-1},\end{aligned} \tag{2.74}$$

where $i = 2, 3, \ldots, d_k + \tau_k, \ldots, d_k + \tau$ and $\hat{x}_{k-d_k+1|k-d_k}$ is the state prediction for time $k - d_k + 1$ on the basis of the information at time $k - d_k$. And on the controller side, the following the control signals are generated:

$$U_{k|k-d_k} = \left[\hat{u}_{k|k-d_k}, \hat{u}_{k+1|k-d_k}, \ldots, \hat{u}_{k+\tau|k-d_k}\right], \tag{2.75}$$

where

$$\hat{u}_{k+i|k-d_k} = K_{i,d_k}\hat{y}_{k+i|k-d_k}, \tag{2.76}$$

with $\hat{u}_{k+i|k-d_k}$ being the control prediction and $\hat{y}_{k+i|k-d_k}$ the output prediction. The control input on the actuator side is considered as

$$u_k = \hat{u}_{k|k-\tau_k-d_k}. \tag{2.77}$$

The output feedback control input of system (2.73) after applying the NCS strategy is calculated as

$$\begin{aligned}u_k &= K_{d_k,\tau_k}y_{k-d_k-\tau_k}\\ &= K_{d_k,\tau_k}C\bar{U},\end{aligned} \tag{2.78}$$

where

$$\bar{U} = \left(A^{d_k+\tau_k}x_{k-d_k-\tau_k} + \sum_{i=1}^{d_k+\tau_k} A^{d_k+\tau_k-i}B\hat{u}_{k-d_k-\tau_k+i-1}\right).$$

Then, a Markovian jump system with discrete and distributed delays can be written as

$$\begin{aligned}x_{k+1} &= Ax_k + BK_{d_k,\tau_k}CA^{d_k+\tau_k}x_{k-d_k-\tau_k}\\ &\quad + BK_{d_k,\tau_k}\sum_{\iota=1}^{d_k+\tau_k}\mu_\iota\hat{u}_{k-\iota},\end{aligned} \tag{2.79}$$

where $\iota = d_k + \tau_k - i + 1$.

Theorem 2.10. *For the NCS in* (2.73) *with random delays d_k and τ_k in the feedback and forward channels, where d_k and τ_k are Markov processes, respectively, the closed-loop system in* (2.79) *with the predictive controller* (2.78) *is stochastically stable if there exist symmetric positive definite matrices $S_{i,j}$, R, and Q such that the following matrix inequality holds:*

$$\Omega_{i,j} + \Lambda_{i,j}^T \bar{S}_{i,j} \Lambda_{i,j} + \bar{\beta}_{i,j} \Gamma_{i,j}^T Q \Gamma_{i,j} < 0, \tag{2.80}$$

where

$$\Omega_{i,j} = U_1^T \left(\bar{\alpha} R - S_{i,j}\right) U_1 - U_2^T R U_2 - \frac{1}{\mu_{i+j}} U_3^T Q U_3,$$

$$\bar{\beta}_{i,j} = \bar{\mu}_{i+j} + \bar{\mu}_{d+\tau} \left(1 - p_{ii} \pi_{jj}\right),$$

and all other parameters are defined in [167] *for all i and j.*

Proof. The proof of Theorem 2.10 could be derived using the following Lyapunov–Krasovskii functional [167]:

$$V(t) = V_1(k) + V_2(k) + V_3(k) + V_4(k) + V_5(k), \tag{2.81}$$

with

$$V_1(k) = x_k^T S(d_k, \tau_k) x_k,$$

$$V_2(k) = \sum_{\nu=k-d_k-\tau_k}^{k-1} x_\nu^T R x_\nu,$$

$$V_3(k) = \sum_{\iota=0}^{d+\tau-1} \sum_{\nu=k-\iota}^{k-1} x_\nu^T \bar{R} x_\nu,$$

$$V_4(k) = \sum_{\iota=1}^{d_k+\tau_k} \mu_\iota \sum_{\nu=k-\iota}^{k-1} \hat{u}_\nu^T Q \hat{u}_\nu,$$

$$V_5(k) = \sum_{\iota=1}^{d+\tau} \mu_\iota \sum_{\omega=1}^{\iota-1} \sum_{\nu=k-\omega}^{k-1} \hat{u}_\nu^T \bar{Q} \hat{u}_\nu.$$

□

Other examples of implementing this approach can be found in [168–173].

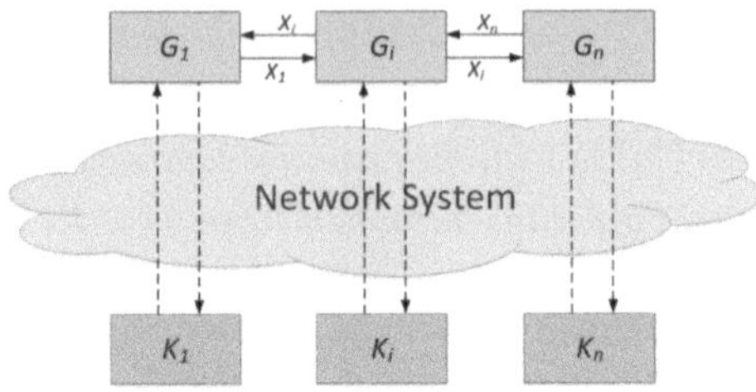

Figure 2.10 Models of decentralized NCS

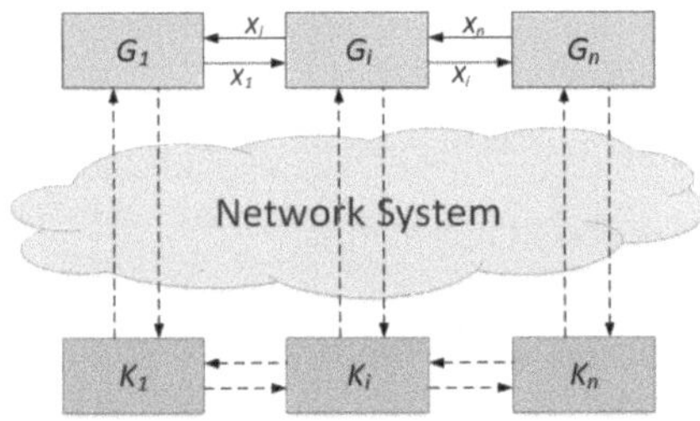

Figure 2.11 Models of distributed NCS

2.3 ADVANCED ISSUES IN NCS

2.3.1 Decentralized and Distributed NCS

In addition to the centralized configuration of NCS in which a single controller communicates with the system, there are two other configurations, namely decentralized and distributed configurations. Large-scale systems are usually modeled as a system-of-systems or system-with-subsystems with an interconnection among them, and several controllers are used for controlling the overall system; these controllers either communicate with each other as shown in Fig. 2.10 (this is called a decentralized configuration) or work separately as shown in Fig. 2.11 (this is called a distributed configuration).

The nodes of controllers in a decentralized configuration do not share information with nearby nodes, although the system may have one objective, this causes each controller to work locally. Due to lack of information, a suboptimal control performance may be achieved. Moreover, a deterioration in system performance and limitation in the application scope of the decentralized configuration may occur in wireless NCS due to the absence of communication and cooperation between decentralized controllers [13]. Examples of decentralized NCS are [174–178].

The main feature in a distributed configuration is that the exchange of information of each subsystem among components of the system and the plants, in order to achieve the objective of the system and which usually contains many interacting physical units, can be physically distributed and interconnected with others to coordinate their tasks, and this leads to the so-called cooperative control [13]. Since sharing of local information is allowed among the distributed controllers in this configuration, they are capable of coordination, which leads to modularity, scalability and robustness. Examples of distributed NCS are [179–185].

2.3.2 Event-Triggered Schemes

During the last decade, event-triggered control (ETC) has received increasing attention in real-time control systems. One conspicuous characteristic is that ETC provides a strategy under which the control task is executed only when necessary. Compared with traditional time-triggered control, ETC can efficiently reduce the number of execution of control tasks while preserving the desired closed-loop performance [186–189].

The first important issue on ETC is the minimum interevent time (the minimum time interval between two consecutive events). It is necessary to ensure that the minimum interevent time is larger than a certain positive constant; otherwise, an ETC system exhibits Zeno behavior, which is unexpected in the implementation of the control system. Under an event-triggering scheme, extra hardware is commonly used to monitor the instantaneous system state so that the next event time can be calculated. However, extra hardware incurs extra cost. In order to avoid using dedicated extra hardware to monitor the system state, self-triggered control (STC) is proposed in [190], in which the next event time can be derived by a predefined event-triggering condition related to the last measurement of the system state [191–194]. Nevertheless, under an ETC or STC scheme, although several sufficient conditions are obtained to calculate the minimum interevent time, the minimum interevent time may be quite small or may not exist for some real control systems. Recently, a periodic ETC (PETC) has been proposed in [195] and [196]. With the PETC, the system state is sampled with a proper period $h > 0$. The event-triggering condition in the PETC is verified only at the sampling instants, which means that the PETC directly provides a guaranteed minimum interevent time (at least $h > 0$) provided that the closed-loop stability can be ensured [196–198].

The second important issue on ETC is the feedback controller that computes the control signals. It is natural that the designed controller should reflect the event-triggering nature. However, some existing results follow a so-called emulation based approach [195], by which the controller is designed using standard period sampled-data controller design tools. This implies that the designed controller just reflects the nature of time-triggering rather than the one of event-triggering. Recently, under the PETC scheme, a co-design approach has been proposed in [196], where both event triggering parameters and state feedback controller can be designed simultaneously provided that a set of linear matrix inequalities (LMIs) are feasible. This idea is then employed to deal with $\mathcal{L}_2$ control for sampled-data systems [197], networked Takagi–Sugeno fuzzy systems [199,200], consensus of multiagents [201], output feedback control for networked control systems [202], and H_∞ filtering [198,203,204].

Among the results on ETC, STC, and PETC reported in the literature, it is found that most of them are about state feedback control, which requires the system state to be measurable. When the system state is not available, output feedback control is an alternative approach to the implementation of a control system. This important issue has been addressed by several researchers. For example, under an STC scheme, Almeida et al. [205] considered the output feedback control using a discrete-time observer. Employing a PETC scheme, Heemels et al. [195] analyzed the dynamical output-based control for a continuous-time linear system. Nevertheless, these results focused only on conditions for the existence of the minimum interevent time, while the design of desired output feedback controllers was not considered, which limits the application of the results.

2.3.3 Cloud Control System

Cloud computing is one of the recent essential tools in the industry since it provides customers with high computation power and reduces storage requirements. And this opens new windows to the control techniques. Cloud control systems have become one of the most promising directions [206]. The structure of cloud control systems is shown in Fig. 2.12. Cloud computing systems provide a medium of configurable resources including computation, software, data access, and storage services for practical systems while customers do not need to know the real location and configuration of the service provider while using them.

The requirements for computation and communication are increased in cloud systems due to the increase of the system scale. General speaking, most of complex systems cannot be controlled properly in the absence of powerful tools and adequate system information. But, a necessary platform

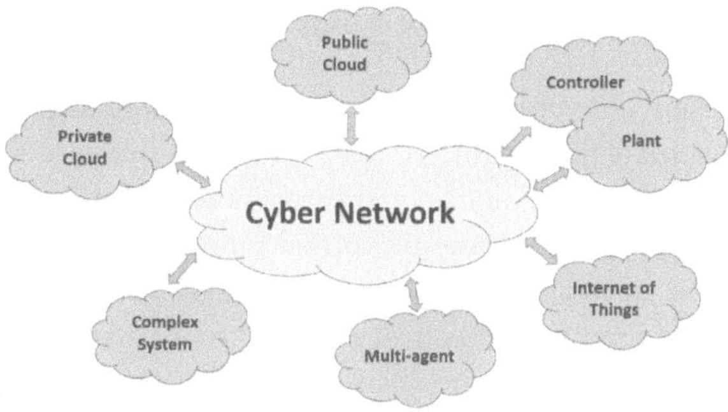

Figure 2.12 Cloud control systems

for computability is provided by the development of new technologies, including recent innovations in software and hardware. Additionally, big data is confronted with many challenges, including capturing, storage, search, sharing, transfer, analysis, visualization, etc. In the cloud control systems, big data will be sent to the centers of cloud computing to be processed first. Then, control signals such as scheduling schemes, predictive control sequences and any other useful information will be generated instantly for cloud control systems. So, cloud control systems will provide powerful tools to control the complex systems which were not available before.

Let us consider the following discrete dynamic system S with unknown process disturbances and measurement noises:

$$\begin{aligned} x(k+1) &= f(x(k), u(k), w(k)), \\ y(k) &= g(x(k), u(k), v(k)), \end{aligned} \tag{2.82}$$

where $x(k)$ is the system state, $u(k)$ and $y(k)$ are the system input and output, respectively; $f(x(k), u(k), w(k))$ and $g(x(k), u(k), v(k))$ are general models which can be linear or nonlinear; $w(k)$ and $v(k)$ are the unknown process disturbances and the unknown measurement noises, respectively. For NCS where there are network-induced time delay and data dropouts, it has been proven that a networked predictive method is a very effective method; see [16] and [206]. To estimate the state of system (2.82) and then produce the predictive states with finite horizon N_1, Kalman filter is adopted as follows:

$$\hat{x}(k|k) = KF(S, \hat{u}(k-1|k-1), y(k)), \tag{2.83}$$

$$\hat{x}(k+i|k) = KF(S, \hat{u}(k|k), y(k)), \quad i = 1, 2, \ldots, N_1, \tag{2.84}$$

$$\hat{u}(k+i|k) = K(k+i)\hat{x}(k+i|k), \quad i = 1, 2, \ldots, N_1, \tag{2.85}$$

where KF represents the compact form of Kalman filter, and $K(k+i)$ is the time-varying Kalman filter gain. A networked predictive control scheme consisting of a control prediction generator and a network delay compensator is proposed to overcome unknown random network transmission delays [206]. The network delay compensator considers the recent control value from the control prediction sequences available on the plant. So, when there is no time delay in the channel from sensor to controller and the time delay from controller to actuator is k_i, when the following predictive control sequences are received on the plant side:

$$\begin{aligned}
&\left[u^T_{t-k_1|t-k_1}, u^T_{t-k_1+1|t-k_1}, \ldots, u^T_{t|t-k_1}, \ldots, u^T_{t+N-k_1|t-k-1}\right]^T, \\
&\left[u^T_{t-k_2|t-k_2}, u^T_{t-k_2+1|t-k_2}, \ldots, u^T_{t|t-k_2}, \ldots, u^T_{t+N-k_2|t-k-2}\right]^T, \\
&\quad \vdots \\
&\left[u^T_{t-k_t|t-k_t}, u^T_{t-k_t+1|t-k_t}, \ldots, u^T_{t|t-k_t}, \ldots, u^T_{t+N-k_t|t-k_t}\right]^T,
\end{aligned} \tag{2.86}$$

where the control values $u_{t|t-k_i}$ for $i = 1, 2, \ldots, t$, are available to be chosen as the control input of the plant at time t, the output of the network delay compensator will be

$$u_t = u_{t|t-\min\{k_1, k_2, \ldots, k_t\}}. \tag{2.87}$$

The controller sends the following set of packets to the plant node:

$$\{u(k+i|k) \mid i = 0, 1, \ldots, N\}. \tag{2.88}$$

At each time instant k, the actuator selects a preferable control signal to be the actual input of the plant:

$$u(k) = u(k|k-i), \tag{2.89}$$

where $i = \arg\min_i\{u(k|k-i) \text{ is available}\}$.

In [207], a cloud predictive control scheme for networked multiagent systems (NMASs) via cloud computing was presented in order to achieve both consensus and stability simultaneously and for compensating actively the communication delays. More information about cloud control can be found in [208] and [209].

Remark 2.7. *The principle of cooperative cloud control system is used to describe the system in which two or more cloud controllers are used to achieve the control objectives in a form of cooperation, but reasonable allocation for control objectives is a difficult problem [208].*

2.3.4 Co-Design in NCS

The co-design of NCS deals with the interaction between control and computing theories. Normally, digital networks are used for connecting sensors, controllers, and actuators in NCS. The concept of centralized control was applied in the past, in which the sensors, controllers, and actuators are located in the area where the communication is peer-to-peer. As a result of the short distance, there is no time delay between the NCS parts and, also, no packet losses. After that, the concept of decentralized systems was applied, since manufactures tend to separate their large plants to subsystems with their own control system having indirect communication. Then, manufactures applied the hybrid systems which contain centralized and decentralized systems. As a result, Internet-of-Things (IoT) became a rising research area of both academia and industry, and it establishes the future interaction between the computing and communications [11,206].

Since IoT systems are complex, it is difficult to model them. Practically, a control system faces issues affecting the stability and overall performance of the system and may lead to instability such as critical time delay of data, irregular, time-varying, and packet dropout. Traffic congestion could cause loss of data, and unreliable nature of the link or protocol malfunctions [11]. On the other hand, the continuous growth of computing systems raises the capabilities to compute, store and process IoT data with high quality and reliable measurements [206].

The implementation of model estimation, optimization, and control approaches into the progressive data center control was presented in [210], and implementation of control theory in computing systems is discussed in [211]. Two methodologies for remote control systems were proposed, in which control system design is provided as a cloud service and hence time and cost are reduced and design of the plant-wide system becomes simpler [212]. In [213], both communication and control problems were solved simultaneously using co-design approaches. The authors of [214] proposed a solution for a co-design problem of a mixed event-triggering mechanism (ETM) and state feedback controller for discrete-time linear parameter-varying (LPV) systems in a network environment. A parameter-dependent co-design condition for event-triggered H_∞ control was written for the problem as a finite set of LMIs. The issue of performance limitation of networked systems was discussed by co-designing the controller and communication filter in [215]. It was shown that using the controller and communication filter co-design could enhance the performance limitation and revoke the effect of the channel noise.

2.4 NOTES

This chapter has demonstrated the operation of NCS under normal environment. As mentioned in the beginning of this chapter, there are five imperfections that could affect NCS. The stability analysis of NCS with two types of imperfections has seen the most effort in recent years, examples are [87–92]. Table 2.1 shows the references that discussed three or four imperfections. To the best of the authors' knowledge, no research has considered all of the five imperfections together and this calls for future investigations.

REFERENCES

[1] W. Zhang, M.S. Branicky, S.M. Phillips, Stability of networked control systems, IEEE Control Syst. 21 (1) (Feb. 2001) 84–99.
[2] T.C. Yang, Networked control system: a brief survey, IEE Proc., Control Theory Appl. 153 (4) (10 July 2006) 403–412.
[3] J. Baillieul, P.J. Antsaklis, Control and communication challenges in networked real-time systems, Proc. IEEE 95 (1) (Jan. 2007) 9–28.
[4] J. Hespanha, P. Naghshtabrizi, Y. Xu, A survey of recent results in networked control systems, Proc. IEEE 95 (1) (Jan. 2007) 138–162.
[5] R.A. Gupta, M.Y. Chow, Networked control system: overview and research trends, IEEE Trans. Ind. Electron. 57 (7) (July 2010) 2527–2535.
[6] L. Zhang, H. Gao, O. Kaynak, Network-induced constraints in networked control systems–a survey, IEEE Trans. Ind. Inform. 9 (1) (Feb. 2013) 403–416.
[7] Y. Xia, Y. Gao, L. Yan, M. Fu, Recent progress in networked control systems – a survey, Int. J. Autom. Comput. 12 (4) (Aug. 2015) 343–367.
[8] X.M. Zhang, Q.L. Han, Y.L. Wang, A brief survey of recent results on control and filtering for networked systems, in: 2016 12th World Congress on Intelligent Control and Automation, WCICA, Guilin, 2016, pp. 64–69.
[9] X.M. Zhang, Q.L. Han, X. Yu, Survey on recent advances in networked control systems, IEEE Trans. Ind. Inform. 12 (5) (Oct. 2016) 1740–1752.
[10] M.S. Mahmoud, Networked control systems analysis and design: an overview, Arab. J. Sci. Eng. 42 (3) (2016) 711–758.
[11] M.S. Mahmoud, Y.Q. Xia, The interaction between control and computing theories: new approaches, Int. J. Autom. Comput. 14 (3) (June 2017) 254–274.
[12] Z. Lei, W. Zidong, Z. Donghua, Event-based control and filtering of networked system: a survey, Int. J. Autom. Comput. 14 (3) (2017) 239–253.

[13] X. Ge, F. Yang, Q.L. Han, Distributed networked control systems: a brief overview, Inf. Sci. 380 (20 Feb. 2017) 117–131.
[14] F.Y. Wang, L. Derong, Networked Control Systems Theory and Applications, Springer, London, 2008.
[15] A. Bemporad, M. Heemels, M. Johansson, Networked Control Systems, Lect. Notes Control Inf. Sci., vol. 406, Springer, Berlin, 2010.
[16] Y.Q. Xia, M.Y. Fu, G.P. Liu, Analysis and Synthesis of Networked Control Systems, Springer, Berlin, 2014.
[17] M.S. Mahmoud, Control and Estimation Methods over Communication Networks, Springer, London, 2014.
[18] E. Garcia, M.J. McCourt, P.J. Antsaklis, Model-based event-triggered control of networked systems, in: Event-Based Control and Signal Processing, CRC Press, 2015, pp. 177–202.
[19] W.P.M.H. Heemels, D. Nešic, A.R. Teel, N. van de Wouw, Networked and quantized control systems with communication delays, in: Proceedings of the 48h IEEE Conference on Decision and Control, CDC, held jointly with 2009 28th Chinese Control Conference, Shanghai, 2009, pp. 7929–7935.
[20] H. Sun, N. Hovakimyan, T. Basar, $\mathcal{L}_1$ adaptive controller for quantized systems, in: Proceedings of the 2011 American Control Conference, San Francisco, CA, 2011, pp. 582–587.
[21] R.W. Brockett, D. Liberzon, Quantized feedback stabilization of linear systems, IEEE Trans. Autom. Control 45 (7) (July 2000) 1279–1289.
[22] D. Liberzon, Hybrid feedback stabilization of systems with quantized signals, Automatica 39 (9) (Sept. 2003) 1543–1554.
[23] N. Elia, S.K. Mitter, Stabilization of linear systems with limited information, IEEE Trans. Autom. Control 46 (9) (Sept. 2001) 1384–1400.
[24] M. Fu, L. Xie, The sector bound approach to quantized feedback control, IEEE Trans. Autom. Control 50 (11) (Nov. 2005) 1698–1711.
[25] H. Gao, T. Chen, A new approach to quantized feedback control systems, Automatica 44 (2) (2008) 534–542.
[26] M. Fu, L. Xie, Output feedback control of linear systems with input and output quantization, in: Proc. the 47th IEEE Conference on Decision and Control, Cancun, Mexico, Dec. 2008, pp. 4706–4711.
[27] K. Tsumura, H. Ishii, H. Hoshina, Tradeoffs between quantization and packet loss in networked control of linear systems, Automatica 45 (12) (Dec. 2009) 2963–2970.
[28] Y. Ishido, K. Takaba, D. Quevedo, Stability analysis of networked control systems subject to packet-dropouts and finite-level quantization, Syst. Control Lett. 60 (5) (May 2011) 325–332.
[29] Y. Sharon, D. Liberzon, Stabilization of linear systems under coarse quantization and time delays, IFAC Proc. Vol. 43 (19) (2010) 31–36.
[30] Y. Sharon, D. Liberzon, Input to state stabilizing controller for systems with coarse quantization, IEEE Trans. Autom. Control 57 (4) (Apr. 2012) 830–844.
[31] J. Yan, Y. Xia, Quantized control for networked control systems with packet dropout and unknown disturbances, Inf. Sci. 354 (1) (Aug. 2016) 86–100.
[32] T.H. Lee, J. Xia, J.H. Park, Networked control system with asynchronous samplings and quantizations in both transmission and receiving channels, Neurocomputing 237 (2017) 25–38.

[33] L. Bao, M. Skoglund, K.H. Johansson, Encoder-decoder design for event-triggered feedback control over band limited channels, in: Proceedings of the American Control Conference, IEEE, Minneapolis, USA, 2006, pp. 4183–4188.

[34] L. Li, X. Wang, M. Lemmon, Stabilizing bit-rates in quantized event triggered control systems, in: Proceedings of the 15th ACM International Conference on Hybrid Systems: Computation and Control, HSCC'12, Beijing, China, 2012, pp. 245–254.

[35] S. Hu, D. Yue, Event-triggered control design of linear networked systems with quantizations, ISA Trans. 51 (1) (Jan. 2012) 153–162.

[36] E. Garcia, P.J. Antsaklis, Model-based event-triggered control for systems with quantization and time-varying network delays, IEEE Trans. Autom. Control 58 (2) (2013) 422–434.

[37] H. Yan, S. Yan, H. Zhang, H. Shi, L_2 control design of event-triggered networked control systems with quantizations, J. Franklin Inst. 352 (1) (Jan. 2015) 332–345.

[38] J. Xiong, J. Lam, Stabilization of networked control systems with a logic ZOH, IEEE Trans. Autom. Control 54 (2) (Feb. 2009) 358–363.

[39] M. Cloosterman, L. Hetel, N. van de Wouw, W. Heemels, J. Daafouz, H. Nijmeijer, Controller synthesis for networked control systems, Automatica 46 (10) (Oct. 2010) 1584–1594.

[40] D. Yue, Q.L. Han, C. Peng, State feedback controller design of networked control systems, in: Proceedings of the 2004 IEEE International Conference on Control Applications, vol. 1, 2004, pp. 242–247.

[41] H. Gao, T. Chen, Network-based output tracking control, IEEE Trans. Autom. Control 53 (3) (2008) 655–667.

[42] X. Na, Y. Zhan, Y. Xia, J. Dong, Control of networked systems with packet loss and channel uncertainty, IET Control Theory Appl. 10 (17) (Nov. 21, 2016) 2251–2259.

[43] D. Ma, G.M. Dimirovski, J. Zhao, Hybrid state feedback controller design of networked switched control systems with packet dropout, in: Proceedings of the 2010 American Control Conference, Baltimore, MD, 2010, pp. 1368–1373.

[44] S. Yin, L. Yu, W.A. Zhang, A switched system approach to networked H_∞ filtering with packet losses, Circuits Syst. Signal Process. 30 (6) (2011) 1341–1354.

[45] Y. Sun, S. Qin, Stability of networked control systems with packet dropout: an average dwell time approach, IET Control Theory Appl. 5 (1) (Jan. 6, 2011) 47–53.

[46] P. Seiler, R. Sengupta, Analysis of communication losses in vehicle control problems, in: Proc. Am. Control Conf. 2, 2001, pp. 1491–1496.

[47] J. Xiong, J. Lam, Stabilization of linear systems over networks with bounded packet loss, Automatica 43 (1) (Jan. 2007) 80–87.

[48] Z. Wang, F. Yang, D.W.C. Ho, X. Liu, Robust H_∞ control for networked systems with random packet losses, IEEE Trans. Syst. Man Cybern., Part B, Cybern. 37 (4) (Apr. 2007) 916–924.

[49] M.S. Mahmoud, S.Z. Selim, P. Shi, Global exponential stability criteria for neural networks with probabilistic delays, IET Control Theory Appl. 4 (11) (2011) 2405–2415.

[50] M.S. Mahmoud, S.Z. Selim, P. Shi, M.H. Baig, New results on networked control systems with non-stationary packet dropouts, IET Control Theory Appl. 6 (15) (2012) 2442–2452.

[51] X. Luan, P. Shi, F. Liu, Stabilization of networked control systems with random delays, IEEE Trans. Ind. Electron. 58 (9) (2011) 4323–4330.

[52] D. Zhang, P. Shi, Q.G. Wang, L. Yu, Analysis and synthesis of networked control systems: a survey of recent advances and challenges, ISA Trans. 66 (2017) 376–392.

[53] E. Fridman, A. Seuret, J. Richard, Robust sampled-data stabilization of linear systems: an input delay approach, Automatica 40 (8) (Aug. 2004) 1441–1446.
[54] L. Mirkin, Some remarks on the use of time-varying delay to model sample-and-hold circuits, IEEE Trans. Autom. Control 52 (6) (June 2007) 1109–1112.
[55] E. Fridman, A refined input delay approach to sampled-data control, Automatica 46 (2) (2010) 421–427.
[56] K. Liu, E. Fridman, Wirtinger's inequality and Lyapunov-based sampled-data stabilization, Automatica 48 (1) (2012) 102–108.
[57] Z.G. Wu, P. Shi, H.Y. Su, J. Chu, Local synchronization of chaotic neural networks with sampled-data and saturating actuators, IEEE Trans. Cybern. 44 (12) (2014) 2635–2645.
[58] Z.G. Wu, P. Shi, H.Y. Su, J. Chu, Exponential stabilization for sampled-data neural-network-based control systems, IEEE Trans. Neural Netw. Learn. Syst. 25 (12) (2014) 2180–2190.
[59] Y. Liu, S.M. Lee, Stability and stabilization of Takagi-Sugeno fuzzy systems via sampled-data and state quantized controller, IEEE Trans. Fuzzy Syst. 24 (3) (2016) 635–644.
[60] Y.L. Wang, Q.L. Han, Modelling and controller design for discrete-time networked control systems with limited channels and data drift, Inf. Sci. 269 (10) (June 2014) 332–348.
[61] L. Zhang, Y. Shi, T. Chen, B. Huang, A new method for stabilization of networked control systems with random delays, IEEE Trans. Autom. Control 50 (8) (Aug. 2005) 1177–1181.
[62] Y. Shi, B. Yu, Output feedback stabilization of networked control systems with random delays modeled by Markov chains, IEEE Trans. Autom. Control 54 (7) (2009) 1668–1671.
[63] Y. Shi, B. Yu, Robust mixed H_2/H_∞ control of networked control systems with random time delays in both forward and backward communication links, Automatica 47 (4) (Apr. 2011) 754–760.
[64] A. Onat, T. Naskali, E. Parlakay, O. Mutluer, Control over imperfect networks: model-based predictive networked control systems, IEEE Trans. Ind. Electron. 58 (3) (Mar. 2011) 905–913.
[65] Y. Ge, Q.G. Chen, M. Jiang, Y.Q. Huang, Modeling of random delays in networked control systems, J. Control Sci. Eng. 2013 (2013) 383415.
[66] R. Luck, A. Ray, Delay compensation in integrated communication and control systems–I: conceptual development and analysis, in: Proceedings of American Control Conference, ACC'90, San Diego, CA, USA, May 1990, pp. 2045–2050.
[67] R. Luck, A. Ray, Delay compensation in integrated communication and control systems–II: implementation and verification, in: Proc. American Control Conference, ACC'90, San Diego, CA, USA, May 1990, pp. 2051–2055.
[68] A.L. Garcia, I. Widjaja, Communication Networks: Fundamental Concepts and Key Architectures, McGraw-Hill, 2001.
[69] R.W. Brockett, Stabilization of motor networks, in: Proc. the 34th IEEE Conference on Decision and Control, 1995, pp. 1484–1488.
[70] V. Blondell, J. Tsitsiklis, NP-hardness of some linear control design problem, SIAM J. Control Optim. 35 (6) (1997) 2118–2127.
[71] D.H. Varsakelis, Feedback control systems as users of a shared network: communication sequences that guarantee stability, in: Proc. the 40th IEEE Conference on Decision and Control, 2001, pp. 3631–3636.

[72] M.S. Branicky, S.M. Phillips, W. Zhang, Scheduling and feedback co-design for networked control systems, in: Proc. the 41st IEEE Conference on Decision and Control, 2002, pp. 1211–1217.
[73] L. Zhang, D.H. Varsakelis, Communication and control co-design for networked control systems, Automatica 42 (6) (2006) 953–958.
[74] W.J. Rugh, Linear System Theory, Prentice Hall, Upper Saddle River, NJ, 1996.
[75] Y.Q. Wang, H. Ye, S.X. Ding, G.Z. Wang, Fault detection of networked control systems subject to access constraints and random packet dropout, Acta Autom. Sin. 35 (9) (2009) 1235–1239.
[76] H.B. Song, W.A. Zhang, L. Yu, H_∞ filtering of network-based systems with communication constraints, IET Signal Process. 4 (1) (2010) 69–77.
[77] H.B. Song, L. Yu, W.A. Zhang, Networked H_∞ filtering for linear discrete-time systems, Inf. Sci. 181 (3) (2011) 686–696.
[78] G. Guo, A switching system approach to sensor and actuator assignment for stabilisation via limited multi-packet transmitting channels, Int. J. Control 84 (1) (2011) 78–93.
[79] D. Zhang, L. Yu, Q.G. Wang, Fault detection for a class of network-based nonlinear systems with communication constraints and random packet dropouts, Int. J. Adapt. Control Signal Process. 25 (10) (2011) 876–898.
[80] W.A. Zhang, L. Yu, G. Feng, Stabilization of linear discrete-time networked control systems via protocol and controller co-design, Int. J. Robust Nonlinear Control 25 (16) (2015) 3072–3085.
[81] P.D. Zhou, L. Yu, H.B. Song, L.L. Ou, H_∞ filtering for network-based systems with stochastic protocols, Control Theory Appl. 27 (12) (2010) 1711–1716.
[82] G. Guo, Z.B. Lu, Q.L. Han, Control with Markov sensors/actuators assignment, IEEE Trans. Autom. Control 57 (7) (2012) 1799–1804.
[83] C.Z. Zhang, G. Feng, J.B. Qiu, W.A. Zhang, T-S fuzzy-model-based piecewise H_∞ output feedback controller design for networked nonlinear systems with medium access constraint, Fuzzy Sets Syst. 248 (2014) 86–105.
[84] H. Zhang, Y. Tian, L.X. Gao, Stochastic observability of linear systems under access constraints, Asian J. Control 17 (1) (2015) 64–73.
[85] L. Zou, Z.D. Wang, H.J. Gao, Observer-based H_∞ control of networked systems with stochastic communication protocol: the finite-horizon case, Automatica 63 (2016) 366–373.
[86] D. Zhang, H. Song, L. Yu, Robust fuzzy-model-based filtering for nonlinear cyber-physical systems with multiple stochastic incomplete measurements, IEEE Trans. Syst. Man Cybern. Syst. 47 (8) (2017) 1826–1838.
[87] N. van de Wouw, P. Naghshtabrizi, M. Cloosterman, J. Hespanha, Tracking control for sampled-data systems with uncertain sampling intervals and delays, Int. J. Robust Nonlinear Control 20 (2010) 387–411.
[88] N.W. Bauer, P.J.H. Maas, W.P.M.H. Heemels, Stability analysis of networked control systems: a sum of squares approach, Automatica 48 (8) (2012) 1514–1524.
[89] M.S. Mahmoud, Estimator design for networked control systems with nonstationary packet dropouts, IMA J. Math. Control Inf. 30 (2012) 395–405.
[90] J. Yan, Y. Xia, Stabilization of fuzzy systems with quantization and packet dropout, Int. J. Robust Nonlinear Control 24 (2014) 1563–1583.
[91] M. Hussain, M. Rehan, Nonlinear time-delay anti-windup compensator synthesis for nonlinear time-delay systems: a delay-range-dependent approach, Neurocomputing 186 (2016) 54–65.

[92] L. Sheng, Z. Wang, W. Wang, F.E. Alsaadi, Output-feedback control for nonlinear stochastic systems with successive packet dropouts and uniform quantization effects, IEEE Trans. Syst. Man Cybern. Syst. 47 (2017) 1181–1191.
[93] R. Ling, D. Zhang, L. Yu, W.a. Zhang, H_∞ filtering for a class of networked systems with stochastic sampling – A Markovian system approach, in: Proc. the 32nd Chinese Control Conference, Xi'an, 2013, pp. 6616–6621.
[94] X. Jia, B. Tang, D. He, S. Peng, Fuzzy-model-based robust stability of nonlinear networked control systems with input missing, in: The 26th Chinese Control and Decision Conference, 2014 CCDC, Changsha, 2014, pp. 1995–2002.
[95] M.S. Mahmoud, A.-W. Al-Saif, Robust quantized approach to fuzzy networked control systems, IEEE J. Emerg. Sel. Top. Circuits Syst. 2 (2012) 71–81.
[96] D. Nesic, D. Liberzon, A unified approach to controller design for systems with quantization and time scheduling, in: 2007 46th IEEE Conference on Decision and Control, New Orleans, LA, 2007, pp. 3939–3944.
[97] M.C.F. Donkers, W.P.M.H. Heemels, N. van de Wouw, L. Hetel, Stability analysis of networked control systems using a switched linear systems approach, IEEE Trans. Autom. Control 56 (9) (2011) 2101–2115.
[98] Mu Li, Jian Sun, Lihua Dou, Stability of an improved dynamic quantized system with time-varying delay and packet losses, IET Control Theory Appl. 9 (6) (2014) 988–995.
[99] C. Zhang, G. Feng, J. Qiu, Y. Shen, Control synthesis for a class of linear network-based systems with communication constraints, IEEE Trans. Ind. Electron. 60 (8) (Aug. 2013) 3339–3348.
[100] P. Shi, R. Yang, H. Gao, State feedback control for networked systems with mixed delays subject to quantization and dropout compensation, in: Chinese Control and Decision Conference, CCDC, Mianyang, 2011, pp. 295–299.
[101] Faiz Rasool, Sing Kiong Nguang, Quantized robust H_∞ output feedback control of discrete-time systems with random communication delays, IET Control Theory Appl. 4 (11) (2010) 2252–2262.
[102] M.S. Mahmoud, N.B. Almutairi, Feedback fuzzy control for quantized networked systems with random delays, Appl. Math. Comput. 290 (2016) 80–97.
[103] G. Lai, Z. Liu, Y. Zhang, C.L.P. Chen, Adaptive fuzzy quantized control of time-delayed nonlinear systems with communication constraint, Fuzzy Sets Syst. 314 (2017) 61–78.
[104] M.S. Mahmoud, Improved networked-control systems approach with communication constraint, IMA J. Math. Control Inf. 29 (2011) 215–233.
[105] C. Jiang, D. Zou, Q. Zhang, S. Guo, Quantized dynamic output feedback control for networked control systems, J. Syst. Eng. Electron. 21 (6) (Dec. 2010) 1025–1032.
[106] J. Yan, Y. Xia, C. Wen, Quantized control for NCSs with communication constraints, Neurocomputing 267 (2017) 489–499.
[107] S.J.L.M. van Loon, M.C.F. Donkers, N. van de Wouw, W.P.M.H. Heemels, Stability analysis of networked and quantized linear control systems, Nonlinear Anal. Hybrid Syst. 10 (Nov. 2013) 111–125.
[108] H. Gao, T. Chen, J. Lam, A new delay system approach to network-based control, Automatica 44 (1) (2008) 39–52.
[109] D. Yue, Q.L. Han, J. Lam, Network-based robust H_∞ control of systems with uncertainty, Automatica 41 (6) (June 2005) 999–1007.

[110] Z.G. Wu, P. Shi, H. Su, J. Chu, Sampled-Data Exponential Synchronization of Complex Dynamical Networks With Time-Varying Coupling Delay, IEEE Trans. Neural Netw. Learn. Syst. 24 (8) (Aug. 2013) 1177–1187.
[111] C.D. Zheng, X. Zhang, Z. Wang, Mode-dependent stochastic stability criteria of fuzzy Markovian jumping neural networks with mixed delays., ISA Trans. 56 (May 2015) 8–17.
[112] J. Wenjuan, K. Alexandre, R. Jean-Pierre, T. Armand, A gain scheduling strategy for the control and estimation of a remote robot via Internet, in: 2008 27th Chinese Control Conference, Kunming, 2008, pp. 793–799.
[113] J. Wenjuan, K. Alexandre, R. Jean-Pierre, T. Armand, Networked control and observation for Master–Slave systems, in: Delay Differential Equations, Springer US, 2009, pp. 1–23.
[114] K. Liu, E. Fridman, L. Hetel, Stability and L_2-gain analysis of networked control systems under Round-Robin scheduling: a time-delay approach, Syst. Control Lett. 61 (5) (May 2012) 666–675.
[115] K. Liu, E. Fridman, L. Hetel, Network-based control via a novel analysis of hybrid systems with time-varying delays, in: 2012 IEEE 51st IEEE Conference on Decision and Control, CDC, Maui, HI, 2012, pp. 3886–3891.
[116] K. Liu, E. Fridman, Discrete-time network-based control under scheduling and actuator constraints, Int. J. Robust Nonlinear Control 25 (2012) 1816–1830.
[117] K. Liu, E. Fridman, L. Hetel, Networked control systems in the presence of scheduling protocols and communication delays, SIAM J. Control Optim. 53 (4) (2015) 1768–1788.
[118] O.L.V. Costa, M.D. Fragoso, Stability results for discrete-time linear systems with Markovian jumping parameters, J. Math. Anal. Appl. 179 (1) (1993) 154–178.
[119] P. Seiler, R. Sengupta, An H_∞ approach to networked control, IEEE Trans. Autom. Control 50 (3) (Mar. 2005) 356–364.
[120] M.S. Mahmoud, Stabilizing control for a class of uncertain interconnected systems, IEEE Trans. Autom. Control 39 (12) (Dec. 1994) 2484–2488.
[121] D. Wu, J. Wu, S. Chen, Stability of networked control systems with polytopic uncertainty and buffer constraint, IEEE Trans. Autom. Control 55 (5) (2010) 1202–1208.
[122] L. Xiao, A. Hassibi, J.P. How, Control with random communication delays via a discrete-time jump system approach, in: American Control Conference, 2000. Proceedings of the 2000, vol. 3, IEEE, 2000, pp. 2199–2204.
[123] L. Zhang, E.K. Boukas, Stability and stabilization of Markovian jump linear systems with partly unknown transition probabilities, Automatica 45 (2) (2009) 463–468.
[124] R. Ling, L. Yu, D. Zhang, W.A. Zhang, A Markovian system approach to distributed H_∞ filtering for sensor networks with stochastic sampling, J. Franklin Inst. 351 (11) (2014) 4998–5014.
[125] Z. Li, F. Alsaadi, T. Hayat, H. Gao, New results on stability analysis and stabilisation of networked control system, IET Control Theory Appl. 8 (16) (Nov. 6, 2014) 1707–1715.
[126] R. Ling, J. Chen, W.A. Zhang, D. Zhang, Energy-efficient H_∞ filtering over wireless networked systems – a Markovian system approach, Signal Process. 120 (2016) 495–502.
[127] F.L. Qu, B. Hu, Z.H. Guan, Y.H. Wu, D.X. He, D.F. Zheng, Quantized stabilization of wireless networked control systems with packet losses, ISA Trans. 64 (Sept. 2016) 92–97.

[128] L. Zha, J. Fang, X. Li, J. Liu, Event-triggered output feedback control for networked Markovian jump systems with quantizations, Nonlinear Anal. Hybrid Syst. 24 (May 2017) 146–158.
[129] P. Shi, F. Li, A survey on Markovian jump systems: modeling and design, Int. J. Control. Autom. Syst. 13 (1) (2015) 1–16.
[130] H. Lin, P.J. Antsaklis, Persistent disturbance attenuation properties for networked control systems, in: 2004 43rd IEEE Conference on Decision and Control, vol. 1, CDC, Nassau, 2004, pp. 953–958.
[131] H. Lin, P. Antsaklis, Stability and persistent disturbance attenuation properties for a class of networked control systems: switched system approach, Int. J. Control 78 (18) (Dec. 2005) 1447–1458.
[132] H. Lin, G. Zhai, P.J. Antsaklis, Asymptotic stability and disturbance attenuation properties for a class of networked control systems, J. Control Theory Appl. 4 (1) (2006) 76–85.
[133] Y.L. Wang, G.H. Yang, H_∞ control of networked control systems with time delay and packet disordering, IET Control Theory Appl. 1 (5) (2007) 1344–1354.
[134] W.A. Zhang, L. Yu, Modelling and control of networked control systems with both network-induced delay and packet-dropout, Automatica 44 (12) (Dec. 2008) 3206–3210.
[135] W.A. Zhang, L. Yu, New approach to stabilisation of networked control systems with time-varying delays, IET Control Theory Appl. 2 (12) (2008) 1094–1104.
[136] L. Hetel, J. Daafouz, C. Lung, Analysis and control of LTI and switched systems in digital loops via an event-based modelling, Int. J. Control 81 (7) (Jul. 2008) 1125–1138.
[137] M.S. Mahmoud, A.W.A Saif, P. Shi, Stabilization of linear switched delay systems: H_2 and H_∞ methods, J. Optim. Theory Appl. 142 (3) (Sept. 2009) 583–601.
[138] X.M. Sun, G.P. Liu, W. Wang, R. David, Stability analysis for networked control systems based on average dwell time method, Int. J. Robust Nonlinear Control 20 (15) (Oct. 2010) 1774–1784.
[139] A. Kruszewski, W.J. Jiang, E. Fridman, J.P. Richard, A. Toguyeni, A switched system approach to exponential stabilization through communication network, IEEE Trans. Control Syst. Technol. 20 (4) (July 2012) 887–900.
[140] W.A. Zhang, L. Yu, Output feedback stabilization of networked control systems with packet dropouts, IEEE Trans. Autom. Control 52 (9) (Sept. 2007) 1705–1710.
[141] R. Wang, G.P. Liu, W. Wang, D. Rees, Y.B. Zhao, H_∞ control for networked predictive control systems based on the switched Lyapunov function method, IEEE Trans. Ind. Electron. 57 (10) (Oct. 2010) 3565–3571.
[142] W.A. Zhang, L. Yu, S. Yin, A switched system approach to H_∞ control of networked control systems with time-varying delays, J. Franklin Inst. 348 (2) (2011) 165–178.
[143] X. Zhao, L. Zhang, P. Shi, M. Liu, Stability and stabilization of switched linear systems with mode-dependent average dwell time, IEEE Trans. Autom. Control 57 (7) (2012) 1809–1815.
[144] D. Zhang, L. Yu, W.A. Zhang, Exponential H_∞ filtering for nonlinear discrete-time switched stochastic systems with mixed time delays and random missing measurements, Asian J. Control 14 (3) (2012) 807–816.
[145] W.A. Zhang, H. Dong, G. Guo, L. Yu, Distributed sampled-data H_∞ filtering for sensor networks with nonuniform sampling periods, IEEE Trans. Ind. Inform. 10 (2) (2014) 871–881.

[146] W.A. Zhang, A. Liu, K. Xing, Stability analysis and stabilization of aperiodic sampled-data systems based on a switched system approach, J. Franklin Inst. 353 (4) (2016) 955–970.

[147] D. Yue, E. Tian, Z. Wang, J. Lam, Stabilization of systems with probabilistic interval input delays and its applications to networked control systems, IEEE Trans. Syst. Man Cybern., Part A, Syst. Hum. 39 (4) (2009) 939–945.

[148] C. Peng, D. Yue, E. Tian, Z. Gu, A delay distribution based stability analysis and synthesis approach for networked control systems, J. Franklin Inst. 346 (4) (2009) 349–365.

[149] F. Yang, Z. Wang, Y.S. Hung, M. Gani, H_∞ control for networked systems with random communication delays, IEEE Trans. Autom. Control 51 (3) (Mar. 2006) 511–518.

[150] M. Donkers, W. Heemels, D. Bernardini, A. Bemporad, V. Shneer, Stability analysis of stochastic networked control systems, Automatica 48 (5) (May 2012) 917–925.

[151] M. Tabbara, D. Nesic, Input-output stability of networked control systems with stochastic protocols and channels, IEEE Trans. Autom. Control 53 (5) (June 2008) 1160–1175.

[152] H. Gao, X. Meng, T. Chen, Stabilization of networked control systems with a new delay characterization, IEEE Trans. Autom. Control 53 (9) (Sept. 2008) 2142–2148.

[153] H. Dong, Z. Wang, H. Gao, Robust H_∞ filtering for a class of nonlinear networked systems with multiple stochastic communication delays and packet dropouts, IEEE Trans. Signal Process. 58 (4) (Apr. 2010) 1957–1966.

[154] J.G. Li, J.Q. Yuan, J.G. Lu, Observer-based H_∞ control for networked nonlinear systems with random packet losses, ISA Trans. 49 (1) (2010) 39–46.

[155] Y. Wang, C. Li, X. Liu, Consensus-based filter designing for wireless sensor networks with packet loss, ISA Trans. 53 (2) (2014) 578–583.

[156] H. Yan, F. Qian, F. Yang, H. Shi, H_∞ filtering for nonlinear networked systems with randomly occurring distributed delays, missing measurements and sensor saturation, Inf. Sci. 370 (2016) 772–782.

[157] Q. Xu, Y. Zhang, W. He, S. Xiao, Event-triggered networked H_∞ control of discrete-time nonlinear singular systems, Appl. Math. Comput. 298 (1) (Apr. 2017) 368–382.

[158] P. Naghshtabrizi, J. Hespanha, A. Teel, Exponential stability of impulsive systems with application to uncertain sampled-data systems, Syst. Control Lett. 57 (5) (May 2008) 378–385.

[159] L. Tongtao, L. Lixiong, F. Minrui, Exponential stability of a classic networked control systems with variable and bounded delay based on impulsive control theory, in: Intelligent Computation Technology and Automation, 2009, ICICTA'09, Second International Conference on, vol. 1, IEEE, 2009, pp. 785–789.

[160] D. Zhang, Y. Wang, X. Jia, Event-triggered dissipative control for model-based networked control systems: an impulsive system approach, in: 2015 Chinese Automation Congress, CAC, Wuhan, 2015, pp. 2007–2012.

[161] C. Yuan, F. Wu, Delay scheduled impulsive control for networked control systems, IEEE Trans. Control Netw. Syst. 4 (3) (Sept. 2017) 587–597.

[162] S. Xu, T. Chen, Robust H_∞ filtering for uncertain impulsive stochastic systems under sampled measurements, Automatica 39 (3) (Mar. 2003) 509–516.

[163] N. van de Wouw, P. Naghshtabrizi, M. Cloosterman, J. Hespanha, Tracking control for networked control systems, in: Proc. 46th IEEE Conf. Decis. Control, 2007, pp. 4441–4446.

[164] W.H. Chen, W.X. Zheng, Input-to-state stability for networked control systems via an improved impulsive system approach, Automatica 47 (4) (Apr. 2011) 789–796.
[165] C. Briat, A. Seuret, Convex dwell-time characterizations for uncertain linear impulsive systems, IEEE Trans. Autom. Control 57 (12) (2012) 3241–3246.
[166] C. Briat, Convex conditions for robust stability analysis and stabilization of linear aperiodic impulsive and sampled-data systems under dwell-time constraints, Automatica 49 (11) (Nov. 2013) 3449–3457.
[167] R. Yang, G.P. Liu, P. Shi, C. Thomas, M.V. Basin, Predictive output feedback control for networked control systems, IEEE Trans. Ind. Electron. 61 (1) (Jan. 2014) 512–520.
[168] Y.B. Zhao, G.P. Liu, D. Rees, A predictive control-based approach to networked Hammerstein systems: design and stability analysis, IEEE Trans. Syst. Man Cybern., Part B, Cybern. 38 (3) (June 2008) 700–708.
[169] G.P. Liu, Predictive controller design of networked systems with communication delays and data loss, IEEE Trans. Circuits Syst. II, Express Briefs 57 (6) (June 2010) 481–485.
[170] J. Zhang, Y. Xia, P. Shi, Design and stability analysis of networked predictive control systems, IEEE Trans. Control Syst. Technol. 21 (4) (July 2013) 1495–1501.
[171] Z.H. Pang, G.P. Liu, D. Zhou, Design and performance analysis of incremental networked predictive control systems, IEEE Trans. Cybern. 46 (6) (June 2016) 1400–1410.
[172] G.P. Liu, Predictive control of networked multiagent systems via cloud computing, IEEE Trans. Cybern. 47 (8) (Aug. 2017) 1852–1859.
[173] Z. Razavinasab, M.M. Farsangi, M. Barkhordari, State estimation-based distributed model predictive control of large-scale networked systems with communication delays, IET Control Theory Appl. 11 (15) (Oct. 13, 2017) 2497–2505.
[174] F. Abdollahi, K. Khorasani, A decentralized Markovian jump H_∞ control routing strategy for mobile multi-agent networked systems, IEEE Trans. Control Syst. Technol. 19 (2) (Mar. 2011) 269–283.
[175] N.W. Bauer, M.C.F. Donkers, Nathan van de Wouw, W.P.M.H. Heemels, Decentralized observer-based control via networked communication, Automatica 49 (7) (2013) 2074–2086.
[176] W.P.M.H. Heemels, Output-Based Event-Triggered Control With Guaranteed-Gain and Improved and Decentralized Event-Triggering, 2012.
[177] M. Mazo, M. Cao, Asynchronous decentralized event-triggered control, Automatica 50 (12) (2014) 3197–3203.
[178] F. Zhou, Z. Huang, Y. Yang, J. Wang, L. Li, J. Peng, Decentralized event-triggered cooperative control for multi-agent systems with uncertain dynamics using local estimators, Neurocomputing 237 (10 May 2017) 388–396.
[179] X. Wang, M.D. Lemmon, Event-triggering in distributed networked control systems, IEEE Trans. Autom. Control 56 (3) (Mar. 2011) 586–601.
[180] M. Guinaldo, D.V. Dimarogonas, K.H. Johansson, J. Sánchez, S. Dormido, Distributed event-based control strategies for interconnected linear systems, IET Control Theory Appl. 7 (6) (2013) 877–886.
[181] M. Andersson, D.V. Dimarogonas, H. Sandberg, K.H. Johansson, Distributed control of networked dynamical systems: static feedback, integral action and consensus, IEEE Trans. Autom. Control 59 (7) (July 2014) 1750–1764.
[182] M. Guinaldo, D. Lehmann, J. Sánchez, S. Dormido, K.H. Johansson, Distributed event-triggered control for non-reliable networks, J. Franklin Inst. 351 (12) (2014) 5250–5273.

[183] G. Guo, L. Ding, Q.L. Han, A distributed event-triggered transmission strategy for sampled-data consensus of multi-agent systems, Automatica 50 (5) (2014) 1489–1496.
[184] M.S. Mahmoud, M. Sabih, Experimental investigations for distributed networked control systems, IEEE Syst. J. 8 (3) (Sept. 2014) 717–725.
[185] M.S. Mahmoud, M. Sabih, M. Elshafei, Event-triggered output feedback control for distributed networked systems, ISA Trans. 60 (2016) 294–302.
[186] E. Hendricks, M. Jensen, A. Chevalier, T. Vesterholm, Problems in event based engine control, in: Proc. Amer. Control Conf., vol. 2, Baltimore, MD, USA, 1994, pp. 1585–1587.
[187] K.J. Åström, B.M. Bernhardsson, Comparison of periodic and event based sampling for first-order stochastic systems, in: Proc. IFAC World Congr., Beijing, China, 1999, pp. 301–306.
[188] K.-E. Arzén, A simple event-based PID controller, in: Proc. IFAC World Congr., vol. 18, Beijing, China, 1999, pp. 423–428.
[189] M.S. Mahmoud, A. Ismail, Role of delay in networked control systems, in: Proc. the 10th IEEE International Conference on Electronics, Circuits and Systems-UoS, UAE, Dec. 2003, pp. 40–43.
[190] M. Velasco, J.M. Fuertes, P. Marti, The self triggered task model for real-time control systems, in: Proc. Real-Time Syst. Symp., RTSS, Cancun, Mexico, 2003, pp. 67–70.
[191] M. Lemmon, T. Chantem, X. Hu, M. Zyskowski, On self-triggered full-information H_∞ controllers, in: Hybrid Systems: Computation and Control, Springer, Berlin, Germany, 2007, pp. 371–384.
[192] X. Wang, M. Lemmon, Self-triggered feedback control systems with finite-gain L_2 stability, IEEE Trans. Autom. Control 54 (3) (Mar. 2009) 452–467.
[193] C. Fiter, L. Hetel, W. Perruquetti, J.-P. Richard, A state dependent sampling for linear state feedback, Automatica 48 (8) (2012) 1860–1867.
[194] A. Anta, P. Tabuada, To sample or not to sample: self-triggered control for nonlinear systems, IEEE Trans. Autom. Control 55 (9) (Sept. 2010) 2030–2042.
[195] W.P.M.H. Heemels, M.C.F. Donkers, A.R. Teel, Periodic event-triggered control for linear systems, IEEE Trans. Autom. Control 58 (4) (Apr. 2013) 847–861.
[196] D. Yue, E. Tian, Q.-L. Han, A delay system method for designing event-triggered controllers of networked control systems, IEEE Trans. Autom. Control 58 (2) (Feb. 2013) 475–481.
[197] C. Peng, Q.-L. Han, A novel event-triggered transmission scheme and L_2 control co-design for sampled-data control systems, IEEE Trans. Autom. Control 58 (10) (Oct. 2013) 2620–2626.
[198] L. Zou, Z. Wang, H. Gao, X. Liu, Event-triggered state estimation for complex networks with mixed time delays via sampled data information: the continuous-time case, IEEE Trans. Cybern. 45 (12) (2015) 2804–2815.
[199] C. Peng, Q.-L. Han, D. Yue, To transmit or not to transmit: a discrete event-triggered communication scheme for networked Takagi–Sugeno fuzzy systems, IEEE Trans. Fuzzy Syst. 21 (1) (Feb. 2013) 164–170.
[200] D. Zhang, Q.-L. Han, X. Jia, Network-based output tracking control for T-S fuzzy systems using an event-triggered communication scheme, Fuzzy Sets Syst. 273 (Aug. 2015) 26–48.
[201] W. Zhang, Y. Tang, T.W. Huang, J. Kurths, Sampled-data consensus of linear multi-agent systems with packet losses, IEEE Trans. Neural Netw. Learn. Syst. 28 (11) (Nov. 2017) 2516–2527.

[202] X.-M. Zhang, Q.-L. Han, Event-triggered dynamic output feedback control for networked control systems, IET Control Theory Appl. 8 (4) (Mar. 2014) 226–234.

[203] X.-M. Zhang, Q.-L. Han, Event-based H_∞ filtering for sampled data systems, Automatica 51 (Jan. 2015) 55–69.

[204] X. Ge, Q.-L. Han, Distributed event-triggered H_∞ filtering over sensor networks with communication delays, Inf. Sci. 291 (Jan. 2015) 128–142.

[205] J. Almeida, C. Silvestre, A.M. Pascoal, Self-triggered output feedback control of linear plants, in: Proc. Amer. Control Conf., San Francisco, CA, USA, 2011, pp. 2831–2836.

[206] Y.Q. Xia, From networked control systems to cloud control systems, in: Proceedings of the 31st Chinese Control Conference, Hefei, China, 2012, pp. 5878–5883.

[207] A. Adaldo, D. Liuzza, D.V. Dimarogonas, K.H. Johansson, Coordination of multi-agent systems with intermittent access to a cloud repository, in: T.I. Fossen, K.Y. Pettersen, H. Nijmeijer (Eds.), Sensing and Control for Autonomous Vehicles, Springer-Verlag, Aug. 2017, pp. 352–369.

[208] Y.Q. Xia, Cloud control systems, IEEE/CAA J. Autom. Sin. 2 (2) (2015) 134–142.

[209] Y.Q. Xia, Y. Qin, D.H. Zhai, S. Chai, Further results on cloud control systems, Sci. China Inf. Sci. 59 (7) (2016) 073201.

[210] J.M. Luna, C.T. Abdallah, Control in computing systems: Part I, in: IEEE Int. Symposium on Computer-Aided Control System Design, CACSD, Sept. 2011, pp. 25–31.

[211] J.M. Luna, C.T. Abdallah, Control in computing systems: Part II, in: IEEE Int. Symposium on Computer-Aided Control System Design, CACSD, Sept. 2011, pp. 32–36.

[212] M.S. Mahmoud, Approaches to remote control systems, in: IECON 2016 – 42nd Annual Conference of the IEEE Industrial Electronics Society, Florence, 2016, pp. 4607–4612.

[213] D. Sauter, M.A. Sid, S. Aberkane, D. Maquin, Co-design of safe networked control systems, Annu. Rev. Control 37 (2) (Dec. 2013) 321–332.

[214] S. Li, D. Sauter, B. Xu, Co-design of event-triggered control for discrete-time linear parameter-varying systems with network-induced delays, J. Franklin Inst. 352 (5) (2015) 1867–1892.

[215] J. Wu, X.S. Zhan, X.H. Zhang, T. Han, H.L. Gao, Performance limitation of networked systems with controller and communication filter co-design, Trans. Inst. Meas. Control 40 (4) (Febr. 1, 2018) 1167–1176.

CHAPTER 3

Cloud Computing

3.1 INTRODUCTION

With the rapid development of processing and storage technologies and the success of the Internet, computing resources have become cheaper, more powerful and more ubiquitously available than ever before. This technological trend has enabled the realization of a new computing model called cloud computing, in which resources (e.g., CPU and storage) are provided as general utilities that can be leased and released by users through the Internet in an on-demand fashion. In a cloud computing environment, the traditional role of service provider is divided into two: *the infrastructure providers* who manage cloud platforms and lease resources according to a usage-based pricing model, and *service providers* who rent resources from one or many infrastructure providers to serve the end users. The emergence of cloud computing has made a tremendous impact on the Information Technology (IT) industry over the past few years, where large companies such as Google, Amazon and Microsoft strive to provide more powerful, reliable and cost-efficient cloud platforms, and business enterprises seek to reshape their business models to gain benefit from this new paradigm. Indeed, cloud computing provides several compelling features that make it attractive to business owners, as shown below:

No up-front investment. Cloud computing uses a pay-as-you-go pricing model. A service provider does not need to invest in the infrastructure to start gaining benefit from cloud computing. It simply rents resources from the cloud according to its own needs and pay for the usage.

Lowering operating cost. Resources in a cloud environment can be rapidly allocated and de-allocated on demand. Hence, a service provider no longer needs to provision capacities according to the peak load. This provides huge savings since resources can be released to save on operating costs when service demand is low.

Highly scalable. Infrastructure providers pool large amount of resources from data centers and make them easily accessible. A service provider can easily expand its service to large scales in order to handle rapid increase in service demands (e.g., flash-crowd effect). This model is sometimes called surge computing [5].

Networked Control Systems
https://doi.org/10.1016/B978-0-12-816119-7.00011-3

Easy access. Services hosted in the cloud are generally web-based. Therefore, they are easily accessible through a variety of devices with Internet connections. These devices not only include desktop and laptop computers, but also cell phones and PDAs.

Reducing business risks and maintenance expenses. By outsourcing the service infrastructure to the clouds, a service provider shifts its business risks (such as hardware failures) to infrastructure providers, who often have better expertise and are better equipped for managing these risks. In addition, a service provider can cut down the hardware maintenance and the staff training costs.

However, although cloud computing has shown considerable opportunities to the IT industry, it also brings many unique challenges that need to be carefully addressed. In this paper, we present a survey of cloud computing, highlighting its key concepts, architectural principles, state-of-the-art implementations, as well as research challenges. Our aim is to provide a better understanding of the design challenges of cloud computing and identify important research directions in this fascinating topic.

3.2 OVERVIEW OF CLOUD COMPUTING

This section presents a general overview of cloud computing, including its definition and a comparison with related concepts.

3.2.1 Definitions

The main idea behind cloud computing is not a new one. John McCarthy in the 1960s already envisioned that computing facilities will be provided to the general public like a utility [39]. The term *cloud* has also been used in various contexts such as describing large ATM networks in the 1990s. However, it was after Google's CEO Eric Schmidt used the word to describe the business model of providing services across the Internet in 2006 that the term really started to gain popularity. Since then, the term cloud computing has been used mainly as a marketing term in a variety of contexts to represent many different ideas. Certainly, the lack of a standard definition of cloud computing has generated not only market hypes, but also a fair amount of skepticism and confusion. For this reason, recently there has been work on standardizing the definition of cloud computing. As an example, the work in [49] compared over 20 different definitions from a variety of sources to confirm a standard definition. In this paper, we adopt the definition of cloud computing provided by The National Institute of Standards and Technology (NIST) [36], as it covers, in our opinion, all the essential aspects of cloud computing:

NIST definition of cloud computing. *Cloud computing is a model for enabling convenient, on-demand network access to a shared pool of configurable computing resources (e.g., networks, servers, storage, applications, and services) that can be rapidly provisioned and released with minimal management effort or service provider interaction.*

The main reason for the existence of different perceptions of cloud computing is that cloud computing, unlike other technical terms, is not a new technology, but rather a new operations model that brings together a set of existing technologies to run business in a different way. Indeed, most of the technologies used by cloud computing, such as virtualization and utility-based pricing, are not new. Instead, cloud computing leverages these existing technologies to meet the technological and economic requirements of today's demand for information technology.

3.2.2 Related Technologies

Cloud computing is often compared to the following technologies, each of which shares certain aspects with cloud computing:

Grid Computing is a distributed computing paradigm that coordinates networked resources to achieve a common computational objective. The development of grid computing was originally driven by scientific applications which are usually computation-intensive. Cloud computing is similar to grid computing in that it also employs distributed resources to achieve application-level objectives. However, cloud computing takes one step further by leveraging virtualization technologies at multiple levels (hardware and application platform) to realize resource sharing and dynamic resource provisioning.

Utility Computing represents the model of providing resources on-demand and charging customers based on usage rather than a flat rate. Cloud computing can be perceived as a realization of utility computing. It adopts a utility-based pricing scheme entirely for economic reasons. With on-demand resource provisioning and utility-based pricing, service providers can truly maximize resource utilization and minimize their operating costs.

Virtualization is a technology that abstracts away the details of physical hardware and provides virtualized resources for high-level applications.

A virtualized server is commonly called a virtual machine (VM). Virtualization forms the foundation of cloud computing, as it provides the capability of pooling computing resources from clusters of servers and dynamically assigning or reassigning virtual resources to applications on-demand.

Autonomic Computing was originally coined by IBM in 2001 and aims at building computing systems capable of self-management, that is, reacting to internal and external observations without human intervention. The goal of autonomic computing is to overcome the management complexity of today's computer systems. Although cloud computing exhibits certain autonomic features such as automatic resource provisioning, its objective is to lower the resource cost rather than to reduce system complexity. In summary, cloud computing leverages virtualization technology to achieve the goal of providing computing resources as a utility. It shares certain aspects with grid computing and autonomic computing but differs from them in other aspects. Therefore, it offers unique benefits and imposes distinctive challenges to meet its requirements.

3.3 CLOUD COMPUTING ARCHITECTURE

This section describes the architectural, business and various operation models of cloud computing.

3.3.1 A Layered Model of Cloud Computing

Generally speaking, the architecture of a cloud computing environment can be divided into 4 layers: the hardware/data center layer, the infrastructure layer, the platform layer and the application layer, as shown in Fig. 3.1. We describe each of them in detail.

The hardware layer is responsible for managing the physical resources of the cloud, including physical servers, routers, switches, power and cooling systems. In practice, the hardware layer is typically implemented in data centers. A data center usually contains thousands of servers that are organized in racks and interconnected through switches, routers or other fabrics. Typical issues at the hardware layer include hardware configuration, fault tolerance, traffic management, power and cooling resource management.

The infrastructure layer, also known as the virtualization layer, creates a pool of storage and computing resources by partitioning the physical resources using virtualization technologies such as Xen [55], KVM [30], and VMware [52]. The infrastructure layer is an essential component of cloud

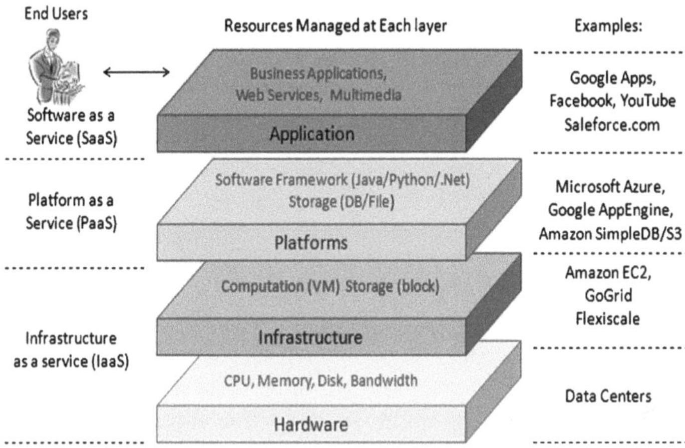

Figure 3.1 Cloud computing architecture

computing, since many key features, such as dynamic resource assignment, are only made available through virtualization technologies.

The platform layer is built on top of the infrastructure layer and consists of operating systems and application frameworks. The purpose of the platform layer is to minimize the burden of deploying applications directly into VM containers. For example, Google App Engine operates at the platform layer to provide API support for implementing storage, database and business logic of typical web applications.

The application layer sits at the highest level of the hierarchy and consists of the actual cloud applications. Different from traditional applications, cloud applications can leverage the automatic-scaling feature to achieve better performance, availability and lower operating cost.

Compared to traditional service hosting environments such as dedicated server farms, the architecture of cloud computing is more modular. Each layer is loosely coupled with the layers above and below, allowing each layer to evolve separately. This is similar to the design of the OSI model for network protocols. The architectural modularity allows cloud computing to support a wide range of application requirements while reducing management and maintenance overhead.

3.3.2 Business Model

Cloud computing employs a service-driven business model. In other words, hardware and platform-level resources are provided as services on an on-

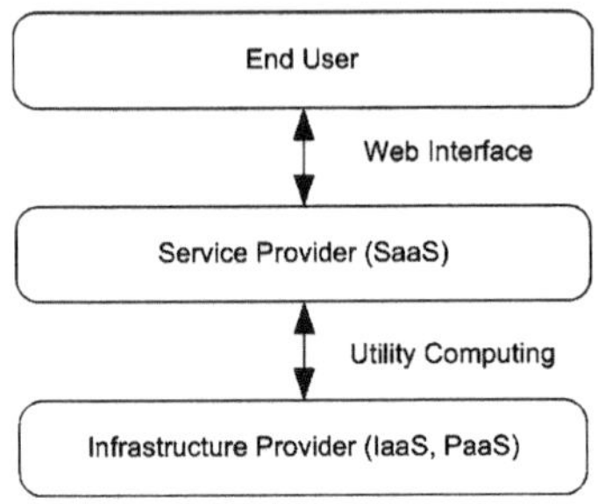

Figure 3.2 Business model of cloud computing

demand basis. Conceptually, every layer of the architecture described in the previous section can be implemented as a service to the layer above. Conversely, every layer can be perceived as a customer of the layer below. However, in practice, clouds offer services that can be grouped into three categories: software-as-a-service (SaaS), platform-as-a-service (PaaS), and infrastructure-as-a-service (IaaS).

1. Infrastructure-as-a-Service refers to on-demand provisioning of infrastructural resources, usually in terms of VMs. The cloud owner who offers IaaS is called an IaaS provider. Examples of IaaS providers include Amazon EC2 [2], GoGrid [15], and Flexiscale [18].
2. Platform-as-a-Service refers to providing platform layer resources, including operating system support and software development frameworks. Examples of PaaS providers include Google App Engine [20], Microsoft Windows Azure [53], and Force.com [41].
3. Software-as-a-Service refers to providing on demand applications over the Internet. Examples of SaaS providers include Salesforce.com [41], Rackspace [17], and SAP Business By Design [44].

The business model of cloud computing is depicted by Fig. 3.2. According to the layered architecture of cloud computing, it is entirely possible that a PaaS provider runs its cloud on top of an IaaS provider's cloud. However, in the current practice, IaaS and PaaS providers are often parts of the same organization (e.g., Google and Salesforce). This is why PaaS and IaaS providers are often called *the infrastructure providers or cloud providers* [5].

3.3.3 Types of Clouds

There are many issues to consider when moving an enterprise application to the cloud environment. For example, some service providers are mostly interested in lowering operation cost, while others may prefer high reliability and security. Accordingly, there are different types of clouds, each with its own benefits and drawbacks:

Public clouds are clouds in which service providers offer their resources as services to the general public. Public clouds offer several key benefits to service providers, including no initial capital investment on infrastructure and shifting of risks to infrastructure providers. However, public clouds lack fine-grained control over data, network and security settings, which hampers their effectiveness in many business scenarios.

Private clouds are also known as internal clouds and are designed for exclusive use by a single organization. A private cloud may be built and managed by the organization or by external providers. A private cloud offers the highest degree of control over performance, reliability and security. However, they are often criticized for being similar to traditional proprietary server farms and do not provide benefits such as no up-front capital costs.

Hybrid clouds are a combination of public and private cloud models that try to address the limitations of each approach. In a hybrid cloud, part of the service infrastructure runs in private clouds while the remaining part runs in public clouds. Hybrid clouds offer more flexibility than both public and private clouds. Specifically, they provide tighter control and security over application data compared to public clouds, while still facilitating on-demand service expansion and contraction. On the down side, designing a hybrid cloud requires carefully determining the best split between public and private cloud components.

Virtual Private Cloud (VCP) is an alternative solution to addressing the limitations of both public and private clouds. A VPC is essentially a platform running on top of public clouds. The main difference is that a VPC leverages virtual private network (VPN) technology that allows service providers to design their own topology and security settings such as firewall rules. VPC is essentially a more holistic design since it not only virtualizes servers and applications, but also the underlying communication network as well. Additionally, for most companies, VPC provides seamless transition from a proprietary service infrastructure to a cloud-based infrastructure, owing to the virtualized network layer.

For most service providers, selecting the right cloud model is dependent on the business scenario. For example, computation-intensive scientific applications are best deployed on public clouds for cost-effectiveness. Arguably, certain types of clouds will be more popular than others.

In particular, it was predicted that hybrid clouds will be the dominant type for most organizations [14]. However, virtual private clouds have started to gain more popularity since their inception in 2009.

3.4 INTEGRATING CPS WITH THE CLOUD

There are three main reasons for integrating CPS with the cloud:

1. First, CPS devices, such as embedded sensor nodes or mobile robots, are typically resource-constrained devices with limited on-board processing and storage capacities. Even though some CPSs might have appropriate resources, it induces a much higher cost and makes them unaffordable for large-scale uses. This results in the need for offloading intensive computations from the CPS devices to much more powerful machines in the cloud, where very high computing capacity is available at low cost.
2. Second, the amount of data generated and used by CPSs is inherently very large as CPS devices are in continuous and direct interaction with the physical world getting instantaneous data to feed the computing system. Analyzing and extracting useful information from such large volumes of data require powerful computing resources. In addition to the need of offloading computation, abundant storage resources becomes a major requirement to cope with big data manipulation.
3. Third, cyberphysical systems are heterogeneous in nature, making interoperability a serious challenge. Accessing these heterogeneous systems from the Internet is not straightforward without having standard and common interfaces. To overcome this issue, there is a need to virtualize their resources and expose them as services to facilitate their integration.

Considering the aforementioned challenges, it is clear that CPSs have three main requirements namely:

i. offloading intensive computation,
ii. storing and analyzing large amount of data,
iii. enabling seamless access through virtual interfaces.

3.5 CLOUD COMPUTING CHARACTERISTICS

Cloud computing provides several salient features that are different from traditional service computing, which we summarize below:

Multi-tenancy. In a cloud environment, services owned by multiple providers are co-located in a single data center. The performance and management issues of these services are shared among service providers and the infrastructure provider. The layered architecture of cloud computing provides a natural division of responsibilities: the owner of each layer only needs to focus on the specific objectives associated with this layer. However, multitenancy also introduces difficulties in understanding and managing the interactions among various stakeholders.

Shared resource pooling. The infrastructure provider offers a pool of computing resources that can be dynamically assigned to multiple resource consumers. Such dynamic resource assignment capability provides much flexibility to infrastructure providers for managing their own resource usage and operating costs. For instance, an IaaS provider can leverage VM migration technology to attain a high degree of server consolidation, hence maximizing resource utilization while minimizing cost such as power consumption and cooling.

Geo-distribution and ubiquitous network access. Clouds are generally accessible through the Internet and use the Internet as a service delivery network. Hence any device with Internet connectivity, be it a mobile phone, a PDA or a laptop, is able to access cloud services. Additionally, to achieve high network performance and localization, many of today's clouds consist of data centers located at many locations around the globe. A service provider can easily leverage geodiversity to achieve maximum service utility.

Service oriented. As mentioned previously, cloud computing adopts a service-driven operating model. Hence it places a strong emphasis on service management. In a cloud, each IaaS, PaaS and SaaS provider offers its service according to the Service Level Agreement (SLA) negotiated with its customers. SLA assurance is therefore a critical objective of every provider.

Dynamic resource provisioning. One of the key features of cloud computing is that computing resources can be obtained and released on the fly. Compared to the traditional model that provisions resources according to peak demand, dynamic resource provisioning allows service providers to acquire resources based on the current demand, which can considerably lower the operating cost.

Self-organizing. Since resources can be allocated or deallocated on-demand, service providers are empowered to manage their resource consumption according to their own needs. Furthermore, the automated resource management feature yields high agility that enables service providers to respond quickly to rapid changes in service demand such as the flash crowd effect.

Utility-based pricing. Cloud computing employs a payer-use pricing model. The exact pricing scheme may vary from service to service. For example, an SaaS provider may rent a virtual machine from an IaaS provider on a per-hour basis. On the other hand, an SaaS provider that provides on-demand customer relationship management (CRM) may charge its customers based on the number of clients it serves (e.g., Salesforce). Utility-based pricing lowers service operating cost as it charges customers on a per-use basis. However, it also introduces complexities in controlling the operating cost. In this perspective, companies like VKernel [51] provide software to help cloud customers understand, analyze and cut down the unnecessary cost on resource consumption.

3.5.1 State-of-the-Art

In this section, we present the state-of-the-art implementations of cloud computing. We first describe the key technologies currently used for cloud computing. Then, we survey the popular cloud computing products.

3.5.2 Cloud Computing Technologies

This section provides a review of technologies used in cloud computing environments.

3.5.3 Architectural Design of Data Centers

A data center, which is home to the computation power and storage, is central to cloud computing and contains thousands of devices like servers, switches and routers. Proper planning of this network architecture is critical, as it will heavily influence applications performance and throughput in such a distributed computing environment. Further, scalability and resiliency features need to be carefully considered. Currently, a layered approach is the basic foundation of the network architecture design, which has been tested in some of the largest deployed data centers. The basic layers of a data center consist of the core, aggregation, and access layers, as shown in Fig. 3.3. The access layer is where the servers in racks physically connect to the network. There are typically 20 to 40 servers per rack, each connected to an access switch with a 1 Gbps link. Access switches usually connect to two aggregation switches for redundancy with 10 Gbps links. The aggregation layer usually provides important functions, such as do-

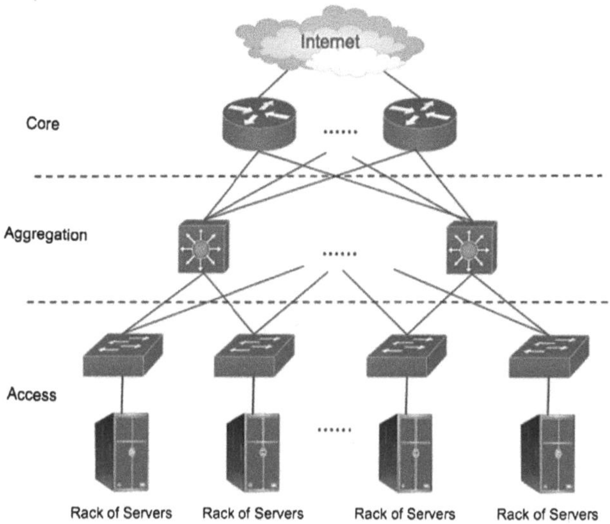

Figure 3.3 Basic layered design of data center network infrastructure

main service, location service, server load balancing, and more. The core layer provides connectivity to multiple aggregation switches and provides a resilient routed fabric with no single point of failure. The core routers manage traffic into and out of the data center. A popular practice is to leverage commodity Ethernet switches and routers to build the network infrastructure. In different business solutions, the layered network infrastructure can be elaborated to meet specific business challenges. Basically, the design of a data center network architecture should meet the following objectives [1, 21–23,35]:

Uniform high capacity. The maximum rate of a server-to-server traffic flow should be limited only by the available capacity on the network-interface cards of the sending and receiving servers, and assigning servers to a service should be independent of the network topology. It should be possible for an arbitrary host in the data center to communicate with any other host in the network at the full bandwidth of its local network interface.

Free VM migration. Virtualization allows the entire VM state to be transmitted across the network to migrate a VM from one physical machine to another. A cloud computing hosting service may migrate VMs for statistical multiplexing or dynamically changing communication patterns to achieve high bandwidth for tightly coupled hosts or to achieve variable heat distribution and power availability in the data center. The communication topology should be designed so as to support rapid virtual machine migration.

Resiliency. Failures will be common at scale. The network infrastructure must be fault-tolerant against various types of server failures, link outages, or server-rack failures. Existing unicast and multicast communications should not be affected to the extent allowed by the underlying physical connectivity.

Scalability. The network infrastructure must be able to scale to a large number of servers and allow for incremental expansion.

Backward compatibility. The network infrastructure should be backward compatible with switches and routers running Ethernet and IP. Because existing data centers have commonly leveraged commodity Ethernet and IP based devices, they should also be used in the new architecture without major modifications.

Another area of rapid innovation in the industry is the design and deployment of shipping-container based, modular data center (MDC). In an MDC, normally up to a few thousands of servers, are interconnected via switches to form the network infrastructure. Highly interactive applications, which are sensitive to response time, are suitable for geodiverse MDC placed close to major population areas. The MDC also helps with redundancy because not all areas are likely to lose power, experience an earthquake, or suffer riots at the same time. Rather than the three-layered approach discussed above, Guo et al. [22,23] proposed server-centric, recursively defined network structures of MDC.

3.5.4 Distributed File System Over Clouds

Google File System (GFS) [19] is a proprietary distributed file system developed by Google and specially designed to provide efficient, reliable access to data using large clusters of commodity servers. Files are divided into chunks of 64 MB, and are usually appended to or read and only extremely rarely overwritten or shrunk. Compared with traditional file systems, GFS is designed and optimized to run on data centers to provide extremely high data throughputs, low latency and survive individual server failures.

Inspired by GFS, the open source Hadoop Distributed File System (HDFS) [24] stores large files across multiple machines. It achieves reliability by replicating the data across multiple servers. Similarly to GFS, data is stored on multiple geodiverse nodes. The file system is built from a cluster of data nodes, each of which serves blocks of data over the network using a block protocol specific to HDFS. Data is also provided over HTTP, allowing access to all content from a web browser or other types of clients. Data nodes can talk to each other to rebalance data distribution, to move copies around, and to keep the replication of data high.

3.5.5 Distributed Application Framework Over Clouds

HTTP-based applications usually conform to some web application framework such as Java EE. In modern data center environments, clusters of servers are also used for computation and data-intensive jobs such as financial trend analysis, or film animation on clusters of computers. MapReduce consists of one master, to which client applications submit MapReduce jobs. The master pushes work out to available task nodes in the data center, striving to keep the tasks as close to the data as possible. The master knows which node contains the data, and which other hosts are nearby. If the task cannot be hosted on the node where the data is stored, priority is given to nodes in the same rack. In this way, network traffic on the main backbone is reduced, which also helps to improve throughput, as the backbone is usually the bottleneck. If a task fails or times out, it is rescheduled. If the master fails, all ongoing tasks are lost. The master records what it is up to in the file system. When it starts up, it looks for any such data, so that it can restart work from where it left off.

The open source Hadoop MapReduce project [25] is inspired by Google's work. Currently, many organizations are using Hadoop MapReduce to run large data-intensive computations. MapReduce [16] is a software framework introduced by Google to support distributed computing on large data sets.

3.5.6 Commercial Products

In this section, we provide a survey of some of the dominant cloud computing products.

3.5.6.1 Amazon EC2

Amazon Web Services (AWS) [3] is a set of cloud services, providing cloud-based computation, storage and other functionality that enable organizations and individuals to deploy applications and services on an on-demand basis and at commodity prices. Amazon Web Services offerings are accessible over HTTP, using REST and SOAP protocols.

Amazon Elastic Compute Cloud (Amazon EC2) enables cloud users to launch and manage server instances in data centers using APIs or available tools and utilities. EC2 instances are virtual machines running on top of the Xen virtualization engine [55]. After creating and starting an instance, users can upload software and make changes to it. When changes are finished, they can be bundled as a new machine image. An identical copy can then be launched at any time. Users have nearly full control of the entire software stack on the EC2 instances that look like hardware to them. On the other hand, this feature makes it inherently difficult for Amazon to offer automatic scaling of resources.

EC2 provides the ability to place instances in multiple locations, which are composed of Regions and Availability Zones. Regions consist of one or more Availability Zones and are geographically dispersed. Availability Zones are distinct locations that are engineered to be insulated from failures in other Availability Zones and provide inexpensive, low latency network connectivity to other Availability Zones in the same Region.

EC2 machine images are stored in and retrieved from Amazon Simple Storage Service (Amazon S3). S3 stores data as "objects" that are grouped in "buckets". Each object contains from 1 byte to 5 gigabytes of data. Object names are essentially URI [6] path names. Buckets must be explicitly created before they can be used. A bucket can be stored in one of several Regions. Users can choose a Region to optimize latency, minimize costs, or address regulatory requirements.

Amazon Virtual Private Cloud (VPC) is a secure and seamless bridge between a company's existing IT infrastructure and the AWS cloud. Amazon VPC enables enterprises to connect their existing infrastructure to a set of isolated AWS compute resources via a Virtual Private Network (VPN) connection, and to extend their existing management capabilities such as security services, firewalls, and intrusion detection systems to include their AWS resources.

For cloud users, Amazon CloudWatch is a useful management tool which collects raw data from partnered AWS services such as Amazon EC2 and then processes the information into readable, near real-time metrics. The metrics about EC2 include, for example, CPU utilization, network in/out bytes, disk read/write operations, etc.

3.5.7 Microsoft Windows Azure Platform

Microsoft's Windows Azure platform [53] consists of three components and each of them provides a specific set of services to cloud users. Windows Azure provides a Windows-based environment for running applications and storing data on servers in data centers; SQL Azure provides data services in the cloud based on SQL Server; and .NET Services offer distributed infrastructure services to cloud-based and local applications. Windows Azure platform can be used both by applications running in the cloud and by applications running on local systems.

Windows Azure also supports applications built on the .NET Framework and other ordinary languages supported in Windows systems, like C#, Visual Basic, C++, and others. Windows Azure supports general-purpose programs, rather than a single class of computing. Developers can create web applications using technologies such as ASP.NET and Windows Communication Foundation (WCF), applications that run as independent background processes, or applications that combine the two. Windows Azure allows storing data in blobs, tables, and queues, all accessed in a RESTful style via HTTP or HTTPS.

SQL Azure components are SQL Azure Database and "Huron" Data Sync. SQL Azure Database is built on Microsoft SQL Server, providing a database management system (DBMS) in the cloud. The data can be accessed using ADO.NET and other Windows data access interfaces. Users can also use on-premises software to work with this cloud–based information. "Huron" Data Sync synchronizes relational data across various on-premises DBMSs. The NET Services facilitate the creation of distributed applications. The Access Control component provides a cloud-based implementation of single identity verification across applications and companies. The Service Bus helps an application expose web services endpoints that can be accessed by other applications, whether on-premises or in the cloud. Each exposed endpoint is assigned a URI, which clients can use to locate and access a service.

All of the physical resources, VMs and applications in the data center are monitored by software called the fabric controller. With each application, the users upload a configuration file that provides an XML-based description of what the application needs. Based on this file, the fabric controller decides where new applications should run, choosing physical servers to optimize hardware utilization.

3.5.7.1 Google App Engine

Google App Engine [20] is a platform for traditional web applications in Google-managed data centers. Currently, the supported programming languages are Python and Java. Web frameworks that run on the Google App Engine include Django, CherryPy, Pylons, and web2py, as well as a custom Google-written web application framework similar to JSP or

Table 3.1 A comparison of representative commercial products.

Cloud Provider	Amazon EC2	Windows Azure	Google App Engine
Classes of Utility Computing	Infrastructure service	Platform service	Platform service
Target Applications	General-purpose applications	General-purpose	Windows applications
Traditional web applications with supported framework	Computation OS Level on a Xen Virtual Machine	Microsoft Common Language Runtime (CLR) VM; Predefined roles of app. instances	Predefined web application frameworks
Storage	Elastic Block Store; Amazon Simple Storage Service (S3); Amazon SimpleDB	Azure storage service and SQL Data Services	BigTable and MegaStore
Auto Scaling	Automatically changing the number of instances based on parameters that users specify	Automatic scaling based on application roles and a configuration file specified by users	Automatic Scaling which is transparent to users

ASP.NET. Google handles deploying code to a cluster, monitoring, failover, and launching application instances as necessary. Current APIs support features such as storing and retrieving data from a BigTable [10] non-relational database, making HTTP requests and caching. Developers have read-only access to the file-system on App Engine.

Table 3.1 summarizes the three examples of popular cloud offerings in terms of the classes of utility computing, target types of application, and more importantly their models of computation, storage and auto-scaling. Apparently, these cloud offerings are based on different levels of abstraction and management of the resources. Users can choose one type or combinations of several types of cloud offerings to satisfy specific business requirements.

3.6 ADDRESSED CHALLENGES

Although cloud computing has been widely adopted by the industry, the research on cloud computing is still at an early stage. Many existing issues have not been fully addressed, while new challenges keep emerging from industry applications. In this section, we summarize some of the challenging research issues in cloud computing.

3.6.1 Automated Service Provisioning

One of the key features of cloud computing is the capability of acquiring and releasing resources on-demand. The objective of a service provider in this case is to allocate and deallocate resources from the cloud to satisfy its service level objectives (SLOs), while minimizing its operational cost. However, it is not obvious how a service provider can achieve this objective. In particular, it is not easy to determine how to map SLOs such as QoS requirements to low-level resource requirement such as CPU and memory requirements. Furthermore, to achieve high agility and respond to rapid demand fluctuations such as in flash crowd effect, the resource provisioning decisions must be made online.

Automated service provisioning is not a new problem. Dynamic resource provisioning for Internet applications has been studied extensively in the past [47,57]. These approaches typically involve: (i) constructing an application performance model that predicts the number of application instances required to handle demand at each particular level, in order to satisfy QoS requirements; (ii) periodically predicting future demand and determining resource requirements using the performance model; and (iii) automatically allocating resources using the predicted resource requirements. Application performance model can be constructed using various techniques, including Queuing theory [47], Control theory [28], and Statistical Machine Learning [7].

Additionally, there is a distinction between proactive and reactive resource control. The proactive approach uses predicted demand to periodically allocate resources before they are needed. The reactive approach reacts to immediate demand fluctuations before periodic demand prediction is available. Both approaches are important and necessary for effective resource control in dynamic operating environments.

3.6.2 Virtual Machine Migration

Virtualization can provide significant benefits in cloud computing by enabling virtual machine migration to balance load across the data center. In addition, virtual machine migration enables robust and highly responsive provisioning in data centers.

Virtual machine migration has evolved from process migration techniques [37]. More recently, Xen [55] and VMWare [52] have implemented "live" migration of VMs that involves extremely short downtimes ranging from tens of milliseconds to a second. Clark et al. [13] pointed out that migrating an entire OS and all of its applications as one unit allows to avoid many of the difficulties faced by process level migration approaches, and analyzed the benefits of live migration of VMs.

The major benefits of VM migration is to avoid hotspots; however, this is not straightforward. Currently, detecting workload hotspots and initiating a migration lacks the agility to respond to sudden workload changes. Moreover, the in-memory state should be transferred consistently and efficiently, with integrated consideration of resources for applications and physical servers.

3.6.3 Server Consolidation

Server consolidation is an effective approach to maximize resource utilization while minimizing energy consumption in a cloud computing environment. Live VM migration technology is often used to consolidate VMs residing on multiple under-utilized servers onto a single server, so that the remaining servers can be set to an energy-saving state. The problem of optimally consolidating servers in a data center is often formulated as a variant of the vector bin-packing problem [11], which is an NP-hard optimization problem. Various heuristics have been proposed for this problem [33,46]. Additionally, dependencies among VMs, such as communication requirements, have also been considered recently [34].

However, server consolidation activities should not hurt application performance. It is known that the resource usage (also known as the footprint [45]) of individual VMs may vary over time [54]. For server resources that are shared among VMs, such as bandwidth, memory cache and disk I/O, maximally consolidating a server may result in resource congestion when a VM changes its footprint on the server [38]. Hence, it is sometimes important to observe the fluctuations of VM footprints and use this information for effective server consolidation. Finally, the system must quickly react to resource congestions when they occur [54].

3.6.4 Energy Management

Improving energy efficiency is another major issue in cloud computing. It has been estimated that the cost of powering and cooling accounts for 53% of the total operational expenditure of data centers [26]. In 2006, data centers in the US consumed more than 1.5% of the total energy generated in that year, and the percentage is projected to grow 18% annually [33]. Hence infrastructure providers are under enormous pressure to reduce energy consumption. The goal is not only to cut down energy cost in data centers, but also to meet government regulations and environmental standards.

Designing energy-efficient data centers has recently received considerable attention. This problem can be approached from several directions. For example, energy efficient hardware architecture that enables slowing down CPU speeds and turning off partial hardware components [8] has become commonplace. Energy-aware job scheduling [50] and server consolidation [46] are two other ways to reduce power consumption by turning off unused machines. Recent research has also begun to study energy-efficient network protocols and infrastructures [27]. A key challenge in all the above methods is to achieve a good trade-off between energy savings and application performance. In this respect, few researchers have recently started to investigate coordinated solutions for performance and power management in a dynamic cloud environment [32].

3.6.5 Traffic Management and Analysis

Analysis of data traffic is important for today's data centers. For example, many web applications rely on analysis of traffic data to optimize customer experiences. Network operators also need to know how traffic flows through the network in order to make many of the management and planning decisions.

However, there are several challenges for existing traffic measurement and analysis methods in Internet Service Providers (ISPs) networks and enterprise to extend to data centers. Firstly, the density of links is much higher than that in ISPs or enterprise networks, which makes the worst case scenario for existing methods. Secondly, most existing methods can compute traffic matrices between a few hundred end hosts, but even a modular data center can have several thousand servers. Finally, existing methods usually assume some flow patterns that are reasonable in Internet and enterprises networks, but the applications deployed on data centers, such as MapReduce jobs, significantly change the traffic pattern. Further, there is tighter coupling in application's use of network, computing, and storage resources, than what is seen in other settings.

Currently, there is not much work on measurement and analysis of data center traffic. Greenberg et al. [21] report data center traffic characteristics on flow sizes and concurrent flows, and use these to guide network infrastructure design. Benson et al. [16] perform a complementary study of traffic at the edges of a data center by examining SNMP traces from routers.

3.6.6 Data Security

Data security is another important research topic in cloud computing. Since service providers typically do not have access to the physical security system of data centers, they must rely on the infrastructure provider to achieve full data security. Even for a virtual private cloud, the service provider can only specify the security setting remotely, without knowing whether it is fully implemented. The infrastructure provider, in this context, must achieve the following objectives: (i) *confidentiality*, for secure data access and transfer, and (ii) *auditability*, for attesting whether security setting of applications has been tampered or not. Confidentiality is usually achieved using cryptographic protocols, whereas auditability can be achieved using remote attestation techniques. Remote attestation typically requires a trusted platform module (TPM) to generate unforgeable system summary (i.e., system state encrypted using TPM's private key) as the proof of system security. However, in a virtualized environment like the clouds, VMs can dynamically migrate from one location to another, hence directly using remote attestation is not sufficient. In this case, it is critical to build trust mechanisms at every architectural layer of the cloud. Firstly, the hardware layer must be trusted using hardware TPM. Secondly, the virtualization platform must be trusted using secure virtual machine monitors [43]. VM migration should only be allowed if both source and destination servers are trusted. Recent work has been devoted to designing efficient protocols for trust establishment and management [31,43].

3.6.7 Software Frameworks

Cloud computing provides a compelling platform for hosting large-scale data-intensive applications. Typically, these applications leverage MapReduce frameworks such as Hadoop for scalable and fault-tolerant data processing. Recent work has shown that the performance and resource consumption of a MapReduce job is highly dependent on the type of the application [29,42,56]. For instance, Hadoop tasks such as sort is I/O intensive, whereas grep requires significant CPU resources. Furthermore, the VM allocated to each Hadoop node may have heterogeneous characteristics. For example, the bandwidth available to a VM is dependent on other VMs collocated on the same server. Hence, it is possible to optimize the performance and cost of a MapReduce application by carefully selecting its configuration parameter values [29] and designing more efficient scheduling algorithms [42,56]. By mitigating the bottleneck resources, execution time of applications can be significantly improved. The key challenges include performance modeling of Hadoop jobs (either online or offline), and adaptive scheduling in dynamic conditions.

Another related approach argues for making MapReduce frameworks energy-aware [50]. The essential idea of this approach is to turn Hadoop node into sleep mode when it has finished its job while waiting for new assignments. To do so, both Hadoop and HDFS must be made energy-aware. Furthermore, there is often a trade-off between performance and energy-awareness. Depending on the objective, finding a desirable trade-off point is still an unexplored research topic.

3.6.8 Storage Technologies and Data Management

Software frameworks such as MapReduce and its various implementations such as Hadoop and Dryad are designed for distributed processing of data-intensive tasks. As mentioned previously, these frameworks typically operate on Internet-scale file systems such as GFS and HDFS. These file systems are different from traditional distributed file systems in their storage structure, access pattern and application programming interface. In particular, they do not implement the standard POSIX interface, and therefore introduce compatibility issues with legacy file systems and applications. Several research efforts have studied this problem [4,40]. For instance, the work in [4] proposed a method for supporting the MapReduce framework using cluster file systems such as IBM's GPFS. Patil et al. [40] proposed new API primitives for scalable and concurrent data access.

3.6.9 Novel Cloud Architectures

Currently, most of the commercial clouds are implemented in large data centers and operated in a centralized fashion. Although this design achieves economy-of-scale and high manageability, it also comes with its limitations such high energy expense and high initial investment for constructing data centers. Recent work [12,48] suggests that small size data centers can be more advantageous than big data centers in many cases: a small data center does not consume so much power, hence it does not require a powerful and yet expensive cooling system; small data centers are cheaper to build and better geographically distributed than large data centers. Geo-diversity is often desirable for response time-critical services such as content delivery and interactive gaming. For example, Valancius et al. [48] studied the feasibility of hosting video-streaming services using application gateways (a.k.a. nano-data centers).

Another related research trend is on using voluntary resources (i.e., resources donated by end-users) for hosting cloud applications [9]. Clouds built using voluntary resources, or a mixture of voluntary and dedicated resources are much cheaper to operate and more suitable for non-profit applications such as scientific computing. However, this architecture also imposes challenges such managing heterogeneous resources and frequent churn events. Also, devising incentive schemes for such architectures is an open research problem.

3.7 PROGRESS OF CLOUD COMPUTING

Cloud computing can be classified as a new paradigm for the dynamic provisioning of computing services supported by state-of-the-art data centers that usually employ Virtual Machine (VM) technologies for consolidation and environment isolation purposes [30]. Cloud computing delivers an infrastructure, platform, and software (applications) as services that are made available to consumers in a *pay-as-you-go* model. In industry these services are referred to as Infrastructure-as-a-Service (IaaS), Platform-as-a-Service (PaaS), and Software-as-a-Service (SaaS), respectively. Many computing service providers including Google, Microsoft, Yahoo, and IBM are rapidly deploying data centers in various locations around the world to deliver Cloud computing services.

A recent Berkeley report [31] stated: "Cloud computing, the long-held dream of computing as a utility, has the potential to transform a large part of the IT industry, making software even more attractive as a service." Cloud offers significant benefit to IT companies by relieving them from the necessity in setting up basic hardware and software infrastructures, and thus enabling more focus on innovation and creating business value for their services. Moreover, developers with innovative ideas for new Internet services

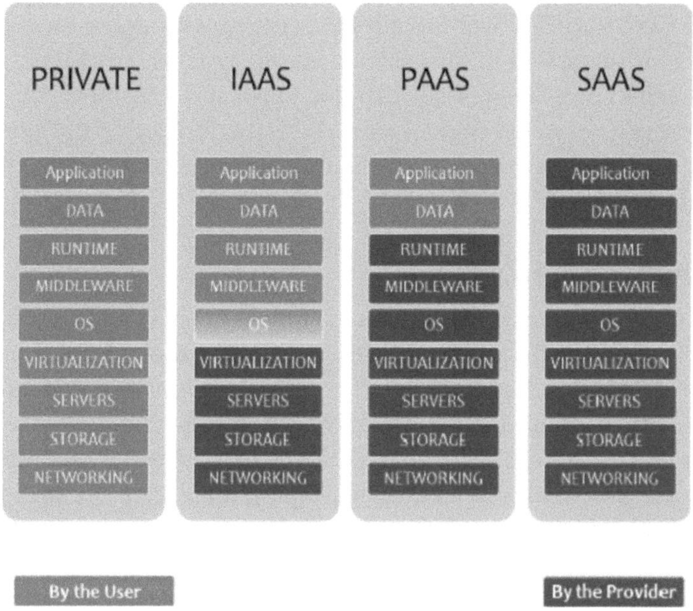

Figure 3.4 Models of service in the cloud

Source: NIST

no longer require large capital outlays in hardware to deploy their service or human expenses to operate it [31].

Well, there was no revolutionary technological innovation, no new patent, and no ingenuity not ever seen before. The trend to move servers out of the enterprise to an external provider was made possible by incremental improvements in communications infrastructures and by declining network traffic costs over the past two decades.

Several business models for engaging cloud computing providers exist, the most widespread of which enable enterprises to buy service packages ranging from basic infrastructures, to operating systems, databases and even a whole application provided as a service.

Fig. 3.4 illustrates the differences between the basic business models for cloud computing systems that exist. In the left column, all the computing and software services are provided by the enterprise itself; in the rightmost column, SaaS (Software-as-a-Service), all the elements of the solution are furnished by the service provider. Enterprises currently tend to use the models in the middle – IaaS (Infrastructure-as-a-Service) – a basic model in which only the hardware is provided as a service: the servers, storage, network services and virtualization systems. Another wider model

PaaS (Platform-as-a-Service) provides the entire platform, i.e., not only the infrastructure services, but also the operating systems, databases, and the supplementary software services that the application requires. Under this model, all that is required of the enterprise is to take care of the application and the data for it; all the other computing infrastructures are furnished by the cloud provider.

3.7.1 The Major Providers

It is not every day that one encounters a player that has garnered 90% of the market share in the United States. Amazon is the biggest provider of cloud computing services in the world; analysts expect it to end 2013 with a turnover of $3.8 billion in this area alone, with growth projections to $8.8 billion in 2015. Later on, Microsoft announced a competing service, Windows Azure, which aims to eat at Amazon's market share. Amazon's cloud computing services, AWS, offer a user-friendly interface; computing infrastructures can be set up at the press of a button, with starting costs of just few dollars a day for a Windows 2012 Server. Amazon's S3 storage services guarantee 99.99% up-time, with a 1 GB storage cost of just 5 to 10 cents a month (depending on the amount of usage). The services are provided in a simple and straightforward manner via a user-friendly interface; setting up an array of new servers is a simple task that does not require IT experts, and customers can also define more advanced services such as DRP, CLUSTER and VPN, etc. Gartner, the leading research and advisory company on IT issues, published an article in March 2013 stating that the traditional service model will disappear by 2015: the software, infrastructure and platform will all be provided as a service.

3.7.2 Control in the Cloud

In the light of the huge momentum in implementing computing systems in the cloud, the question arises: how are these trends expected to impact automation and control, the areas we specialize in? Control and automation systems are responsible for controlling machinery and production line processes. The basic requirements of control systems are maximum up-time, reliability, quality and speed in order to make optimal use of the enterprise's production resources, while at the same time preserving operating security and the confidentiality of sensitive information liable to exist in those systems. For more than three decades such controls have been implemented by means of dedicated Programmable Logic Controllers (PLCs), which provide an excellent solution to the unique requirements of enterprises in the control business and are geared to work in noisy industrial surroundings. What differentiates and sets control systems apart from computer embedded systems is the issue of reliability. No enterprise is going to want to rely on a server or PC computer if the operating system and the software are less stable and their up-time inferior to that of PLCs. Stopping a production facility can have huge financial costs, which is why dedicated hardware systems were developed and the operating logic of the production systems implemented on them. Though control engineers instinctively reject outright the possibility of a meeting point between cloud computing and the world of traditional control, we questioned whether these two tracks will indeed remain separate, never to converge.

We would like to present a different line of thinking in this section. Cloud computing providers are assessed by the uptime of their systems, and they invest extensively in the security of the computing infrastructure; they have far greater financial strength to make these investments than any individual enterprise does on its own, and as a result, their computing systems are becoming ever more stable and more secure too, not less so. Transmission lines are expanding all the time and the communications providers are constantly improving the up-time of their communication lines. Hence, the forces that drove enterprises to seek dedicated hardware solutions – PLCs – are effectively the same forces driving cloud computing providers to improve their systems and to offer higher levels of service and survivability. In our estimation, therefore, implementation of control systems based on a controller in the cloud will become viable in the not-too-distant future.

3.7.3 PLC as a Service

For many years now, the leading makers of controls, such as Rockwell Automation, have enabled installation of their control system on a computer or a server, with all the same programming tools and functions that can be implemented by means of their standard hardware-based controller. The computer-based programmable controller, SoftLogix, never came into its own despite the fact that its use the powerful RSLogix 5000 programming software, the same tool used to develop software for the hardware-based Logix series of controllers. Contel has never supplied a solution to a customer that uses this "soft" controller. Once again, this is due to the lower reliability of computer systems compared to the high up-time of the traditional PLC. If, at a future time, cloud computing providers install SoftLogix on computing and communications infrastructures that feature greater reli-

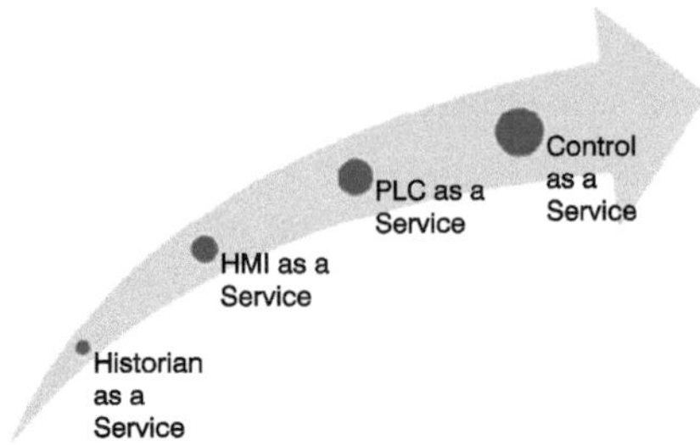

Figure 3.5 Control in the cloud-milestones

ability and greater uptime, we will be able to look at the option of using a controller in the cloud.

3.7.4 Historian as a Service

While a controller in the cloud sounds like a distant vision, the control system can be expected to migrate to the cloud in stages. First to go up will be the monitoring components, which have no real impact on the production resources of the enterprise. Thus, for example, the data collection systems – Plant Historians – are the first systems that can be expected to migrate to the cloud. A study published by American institute ARC entitled the Plant Historian as a Cloud Application is the first significant publication to examine and recommend that the major global makers of industrial controls adopt a strategy in this direction. The 26-page study, published in November 2012, reviewed steps already taken by manufacturers towards solutions in this direction.

3.7.5 HMI as a Service

Several control systems and Human–Machine Interfaces (HMIs) based on cloud computing are already in evidence today. Israeli company RealiteQ, for example, provides monitoring services for control systems by means of an application installed in the cloud. While the scan speed of PLCs constitutes a barrier to their entry to the cloud, the software – the HMI – providing humans with the operating interfaces to the control and production systems do not require the rapid response times of the actual controllers themselves, and are therefore likely to join the cloud before the controllers do. See Fig. 3.5.

3.7.6 Control as a Service

A conventional manufacturing system is composed of a variety of devices installed in the field (motors, sensors, control valves, pumps, etc.). The devices are wired up to PLCs by means of electrical signals (I/Os) and by means of communications networks specific to the production floor. With growing use of Ethernet-based industrial communications networks, such as the Ethernet Industrial Protocol (Ethernet IP), the assumption is that the technical barriers will be eliminated in the not-too-distant future, and that all the device arrays will be controllable via IP communication, wireless or landline, interfacing with the control system. Taking the long view, the need for control cabinets, IO cabinets, and wiring of control points from the devices to the control system will be done away with. Also in the long term, possibly two decades down the road, all the devices will be installed at the production site, each equipped with its own individual IP address. The control system in the cloud will identify every device, and all that will be left for us to do as control engineers will be to write the operating logic.

The lecture at the Israchem Exhibition, in the Future Trends in Instrumentation & Process Control session had a mixed reception. Most of the participants we spoke to expressed an interest in the subject, but in the same breath said, when it comes to traditional control, these processes will take many years, if they happen at all. Contel has set up a new cloud-based test lab using the computing services of Amazon Ireland. A SoftLogix controller – the "soft" controller from Rockwell Automation – was installed on a virtual server there and linked up to remote IO units in the Point IO Series installed at our HQ offices in Israel. We are currently examining the performance of the controller in Ireland vis-à-vis the IO units in the company's in Tel-Aviv area. The aim of the lab is to test the controller scan speed as a function of the communications infrastructures, and to create a feasibility coefficient, as shown in Fig. 3.6.

3.8 CLOUD-BASED MANUFACTURING

Data center infrastructure design has recently been receiving significant research interest both from academia and industry, in no small part due to the growing importance of data centers in supporting and sustaining the rapidly growing Internet-based applications including search (e.g., Google, Bing), video content hosting and distribution (e.g., YouTube, NetFlix), social networking (e.g., Facebook, Twitter), and large-scale computations (e.g., data mining, bioinformatics, indexing). For example, the Microsoft Live online services are supported by a Chicago-based data center, which is one of the largest data centers ever built, spanning more than 700,000 square feet. In particular, cloud computing is characterized as the culmination of the integration of computing and data infrastructures to provide a scalable, agile and cost-effective approach to support the ever-growing critical IT needs (in terms of computation, storage, and applications) of both enterprises and the general public [58,59].

3.8.1 Cloud-Based Services

Large-scale data centers form the core infrastructure support for the ever expanding cloud based services. Thus the performance and dependability characteristics of data centers will have significant impact on the scalability of these services. In particular, the data center network needs to be agile and reconfigurable in order to respond quickly to ever changing application demands and service requirements. Significant research work has been done on designing the data center network topologies in order to improve the performance of data centers.

Massive data centers providing storage form the core of the infrastructure for the cloud [59]. It is thus imperative that the data center infrastructure, including the data center networking is well designed so that both the deployment and maintenance of the infrastructure is cost-effective. With data availability and security at stake, the role of the data center is more critical than ever.

The topology of the network interconnecting the servers has a significant impact on the agility and reconfigurability of the data center infrastructure to respond to changing application demands and service requirements. Today, data center networks primarily use top-of-rack (ToR) switches that are interconnected through end-of-rack (EoR) switches, which are in turn connected via core switches. This approach leads to significant bandwidth over subscription on the links in the network core [58], and prompted several researchers to suggest alternate approaches for scalable cost-effective network architectures. According to the reconfigurability of the topology after the deployment of the DCN, there are fixed architectures and flexible architectures. Fixed architectures can be further classified into two categories: the tree-based architectures such as fat tree [60] and Clos Network [61], and recursive topologies such as DCell [62] and BCube [63]. Flexible architectures such as c-Through [64], Helios [65] and OSA [66] enable reconfigurability of their network topology. Every approach is characterized by its unique network architecture, routing algorithms, fault tolerance and fault recovery approaches.

3.8.2 Conceptual Framework

The main requirements and challenges of a cloud manufacturing environment can be summarized in Big Data support, real-time operation, configurability and agility, security, CPS, social interaction, and finally quality-of-service (QoS). These requirements drive the development of new architecture for cloud manufacturing.

Cyberphysical systems are a major characteristic in the conceptual architecture, capable of processing a large amount of data and performing real-time operations. CPSs have already made significant evolutionary steps and are moving toward an infrastructure that increasingly depends on monitoring the real world, on timely evaluation of data acquired and on timely applicability of management and control. Moreover, the combination of cloud and cyberphysical technologies will lead to a more secure design with a higher QoS. The prevalence of the cloud and its benefits enable the manufacturing sector to expand the cyber-part of the CPS and distribute it on-device, and in Cloud.

More specifically, modern industrial shop floors will introduce a new common information flow through incorporating the novel conceptual architecture. Smart sensor networks, along with real-time simulation, will introduce a new generation of CPSs that will enable quick and accurate information flow, thus providing feedback to other systems and finally to the end-users. The fusion of the cyberphysical system with the cloud will introduce a new way of manufacturing as well as a new information driven infrastructure. The current automated manufacturing systems are structured in hierarchical levels based on the specifications of the standard enterprise architecture. With the proposed infrastructure, the functionalities of each subsystem in a production system will be delivered as an aware and self-adaptive service based on the captured information and the new information flow. The conceptual framework, addressing all the aforementioned needs, is illustrated in Fig. 3.6.

In [70], the authors surveyed existing tools and platforms for gathering and transmitting large volumes of sensor data to clouds such as MicroStrains, TempoDB, SensaTrack Monitor and Ostia Portus platforms. They identified a set of research challenges, namely, communication, data exchange formats, security and interoperability. To address these challenges, the authors proposed a cloud-based sensor monitoring platform to collect,

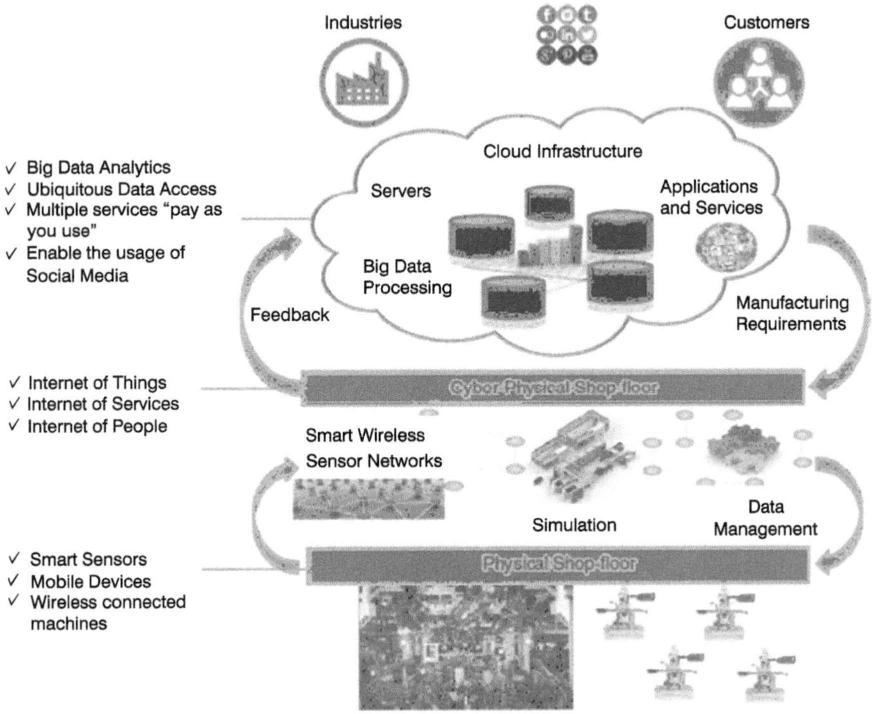

Figure 3.6 Conceptual framework

store, analyze and process big sensor data, and then send control information to actuators. The proposed architecture depicted in Fig. 3.7 uses the generic cloud interface as an intermediate layer between sensor servers and the cloud to assure:

i. secure data access from and transfer to the cloud,
ii. portability and interoperability, and
iii. efficient data transport by formatting the sensor data into platform-neutral data exchange format, namely XML and JSON.

3.9 NOTES

Cloud computing has recently emerged as a compelling paradigm for managing and delivering services over the Internet. The rise of cloud computing is rapidly changing the landscape of information technology, and ultimately turning the long-held promise of utility computing into a reality.

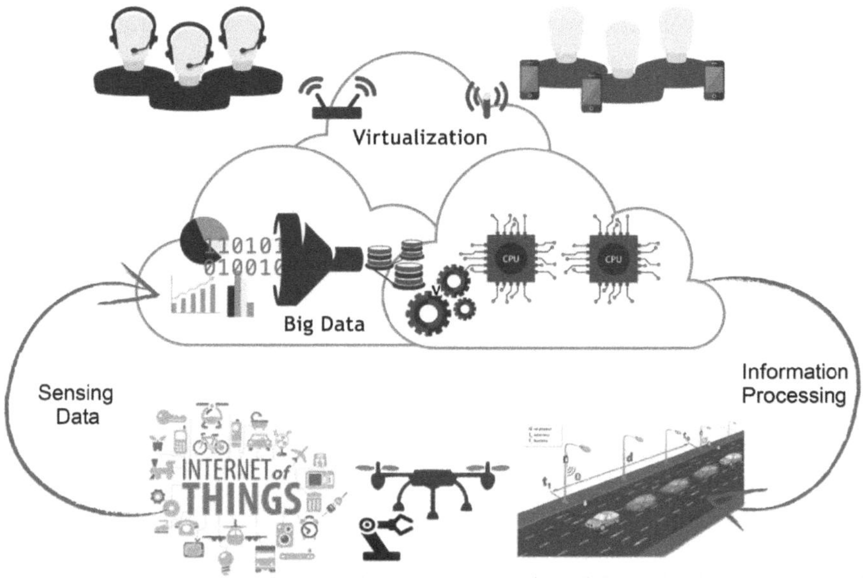

Figure 3.7 Conceptual framework

However, despite the significant benefits offered by cloud computing, the current technologies are not matured enough to realize its full potential. Many key challenges in this domain, including automatic resource provisioning, power management and security management, are only starting to receive attention from the research community. Therefore, we believe there is still a tremendous opportunity for researchers to make groundbreaking contributions in this field, and bring significant impact to their development in the industry.

In this section, we have surveyed the state-of-the-art of cloud computing, covering its essential concepts, architectural designs, prominent characteristics, key technologies as well as research directions. As the development of cloud computing technology is still at an early stage, we hope our work will provide a better understanding of the design challenges of cloud computing, and pave the way for further research in this area.

In the next section, we investigated the problem of stabilizing distributed systems under Denial-of-Service, characterizing DoS frequency and duration under which stability can be preserved. In order to save communication resources, we also consider a hybrid communication strategy. It turns out that the hybrid transmission strategy can reduce communication load effectively and prevent Zeno behavior while preserving the same robustness as pure round-robin protocol.

An interesting research direction is the stabilization problem of networked distributed systems, where only a fraction of subsystems, possibly time-varying are under DoS. It is also interesting to investigate the problem where DoS attacks imposing on systems are asynchronous with different frequencies and durations. Finally, in the hybrid transmission strategy, the effect of event-triggered control with communication collision can be an interesting direction from a practical viewpoint.

REFERENCES

[1] M. Al-Fares, et al., A scalable commodity data center network architecture, in: Proc SIGCOMM, 2008.
[2] Amazon Elastic Computing Cloud, aws.amazon.com/ec2.
[3] Amazon Web Services, aws.amazon.com.
[4] R. Ananthanarayanan, K. Gupta, et al., Cloud analytics: do we really need to reinvent the storage stack?, in: Proc. of HotCloud, 2009.
[5] M. Armbrust, et al., Above the Clouds: A Berkeley View of Cloud Computing, UC Berkeley Technical Report, 2009.
[6] T. Berners-Lee, R. Fielding, L. Masinter, RFC 3986: Uniform Resource Identifier (URI): Generic Syntax, Jan. 2005.
[7] P. Bodik, et al., Statistical machine learning makes automatic control practical for Internet data centers, in: Proc. HotCloud, 2009.
[8] D. Brooks, et al., Power-aware microarchitecture: design and modeling challenges for the next-generation microprocessors, IEEE MICRO 20 (6) (2000) 26–44.
[9] A. Chandra, et al., Nebulas: using distributed voluntary resources to build clouds, in: Proc. of HotCloud, 2009.
[10] F. Chang, et al., Bigtable: a distributed storage system for structured data, in: Proc. of OSDI, 2006.
[11] C. Chekuri, S. Khanna, On multi-dimensional packing problems, SIAM J. Comput. 33 (4) (2004) 837–851.
[12] K. Church, et al., On delivering embarrassingly distributed cloud services, in: Proc. of HotNets, 2008.
[13] C. Clark, K. Fraser, S. Hand, J.G. Hansen, E. Jul, C. Limpach, I. Pratt, A. Warfield, Live migration of virtual machines, in: Proc. of NSDI, 2005.
[14] Cloud Computing on Wikipedia, en.wikipedia.org/wiki/Cloudcomputing, 20 Dec. 2009.
[15] Cloud Hosting, Cloud Computing and Hybrid Infrastructure from GoGrid, http://www.gogrid.com.
[16] J. Dean, S. Ghemawat, MapReduce: simplified data processing on large clusters, in: Proc. of OSDI, 2004.
[17] Dedicated Server, Managed Hosting, Web Hosting by Rackspace Hosting, http://www.rackspace.com.
[18] FlexiScale Cloud Comp and Hosting, www.flexiscale.com.

[19] S. Ghemawat, H. Gobioff, S-T. Leung, The Google file system, in: Proc. of SOSP, October 2003.
[20] Google Application Engine, http://code.google.com/appengine.
[21] A. Greenberg, et al., VL2: a scalable and flexible data center network, in: Proc. SIGCOMM, 2009.
[22] C. Guo, et al., DCell: a scalable and fault-tolerant network structure for data centers, in: Proc. SIGCOMM, 2008.
[23] C. Guo, et al., BCube: a high performance, server-centric network architecture for modular data centers, in: Proc. SIGCOMM, 2009.
[24] Hadoop Distributed File System, hadoop.apache.org/hdfs.
[25] Hadoop MapReduce, hadoop.apache.org/mapreduce.
[26] J. Hamilton, Cooperative expendable micro-slice servers (CEMS): low cost, low power servers for Internet-scale services, in: Proc. of CIDR, 2009.
[27] IEEE P802.3az Energy Efficient Ethernet Task Force, www.ieee802.org/3/az.
[28] E. Kalyvianaki, et al., Self-adaptive and self-configured CPU resource provisioning for virtualized servers using Kalman filters, in: Proc. Int. Conference Autonomic Computing, 2009.
[29] K. Kambatla, et al., Towards optimizing Hadoop provisioning in the cloud, in: Proc. of HotCloud, 2009.
[30] Kernal Based Virtual Machine, www.linux-kvm.org/page/MainPage.
[31] F.J. Krautheim, Private virtual infrastructure for cloud computing, in: Proc. of HotCloud, 2009.
[32] S. Kumar, et al., Manage: loosely coupled platform and virtualization management in data centers, in: Proc. Int. Conference on Cloud Computing, 2009.
[33] B. Li, et al., EnaCloud: an energy-saving application live placement approach for cloud computing environments, in: Proc. Int. Conference on Cloud Computing, 2009.
[34] X. Meng, et al., Improving the scalability of data center networks with traffic-aware virtual machine placement, in: Proc. INFOCOM, 2010.
[35] R. Mysore, et al., Portland: a scalable fault-tolerant layer 2 data center network fabric, in: Proc. SIGCOMM, 2009.
[36] NIST Definition of Cloud Computing, v15, csrc.nist.gov/groups/SNS/cloud-computing/cloud-def-v15.doc.
[37] S. Osman, et al., The design and implementation of ZAP: a system for migrating computing environments, in: Proc. of OSDI, 2002.
[38] P. Padala, et al., Automated control of multiple virtualized resources, in: Proc. of EuroSys, 2009.
[39] D. Parkhill, The Challenge of the Computer Utility, Addison Wesley, Reading, 1966.
[40] S. Patil, et al., In search of an API for scalable file systems: under the table or above it?, in: Proc. HotCloud, 2009.
[41] Salesforce CRM, http://www.salesforce.com/platform.
[42] T. Sandholm, K. Lai, MapReduce optimization using regulated dynamic prioritization, in: Proc. of SIGMETRICS/Performance, 2009.
[43] N. Santos, K. Gummadi, R. Rodrigues, Towards trusted cloud computing, in: Proc. of HotCloud, 2009.
[44] SAP Business ByDesign, www.sap.com/sme/solutions/businessmanagement/businessbydesign/index.epx.
[45] J. Sonnek, et al., Virtual putty: reshaping the physical footprint of virtual machines, in: Proc. of HotCloud, 2009.

[46] S. Srikantaiah, et al., Energy aware consolidation for cloud computing, in: Proc. of HotPower, 2008.

[47] B. Urgaonkar, et al., Dynamic provisioning of multi-tier Internet applications, in: Proc. of ICAC, 2005.

[48] V. Valancius, et al., Greening the Internet with nano data centers, in: Proc. of CoNext, 2009.

[49] L. Vaquero, L. Rodero-Merino, J. Caceres, M. Lindner, A break in the clouds: towards a cloud definition, in: ACM SIGCOMM Computer Communications Review, 2009.

[50] N. Vasic, et al., Making cluster applications energy-aware, in: Proc. Automated Control for Data Centers and Clouds, 2009.

[51] Virtualization Resource Charge Back, www.vkernel.com/products/EnterpriseChargebackVirtualAppliance.

[52] VMWare ESX Server, www.vmware.com/products/esx.

[53] Windows Azure, www.microsoft.com/azure.

[54] T. Wood, et al., Black-box and gray-box strategies for virtual machine migration, in: Proc. of NSDI, 2007.

[55] XenSource Inc, Xen, www.xensource.com.

[56] M. Zaharia, et al., Improving MapReduce performance in heterogeneous environments, in: Proc. of HotCloud, 2009.

[57] Q. Zhang, et al., A regression-based analytic model for dynamic resource provisioning of multi-tier applications, in: Proc. ICAC, 2007.

[58] K.C. Joshi, Cloud computing: in respect to grid and cloud approaches, Int. J. Mod. Eng. Res. 2 (3) (May-June 2012) 902–905.

[59] A. Greenberg, J. Hamilton, D.A. Maltz, P. Patel, The cost of a cloud: research problems in data center networks, SIGCOMM Comput. Commun. Rev. 39 (1) (2009) 68–73.

[60] R.N. Mysore, A. Pamboris, . Farrington, N. Huang, P. Miri, S. Radhakrishnan, V. Subramanya, A. Vahdat, PortLand: a scalable fault-tolerant layer 2 data center network fabric, in: Proceedings of the ACM SIGCOMM 2009 Conference on Data Communication, ACM, 2009, pp. 39–50.

[61] M. Hajibaba, S. Gorgin, A review on modern distributed computing paradigms: cloud computing, jungle computing and fog computing, J. Comput. Inf. Technol. 22 (2) (2014) 69–84.

[62] D.K. Viswanath, S. Kusuma, S.K. Gupta, Cloud computing issues and benefits modern education, Glob. J. Comput. Sci. Technol. Cloud Distrib. 12 (10) (July 2012) 271–277.

[63] C. Guo, G. Lu, D. Li, H. Wu, X. Zhang, Y. Shi, C. Tian, Y. Zhang, S. Lu, Bcube: a high performance, server-centric network architecture for modular data centers, ACM SIGCOMM Comput. Commun. Rev. 39 (4) (2009) 63–74.

[64] G. Wang, D. Andersen, M. Kaminsky, K. Papagiannaki, T. Ng, M. Kozuch, M. Ryan, c-Through: part-time optics in data centers, ACM SIGCOMM Comput. Commun. Rev. 40 (4) (2010) 327–338.

[65] N. Farrington, G. Porter, S. Radhakrishnan, H. Bazzaz, V. Subramanya, Y. Fainman, G. Papen, A. Vahdat, Helios: a hybrid electrical/optical switch architecture for modular data centers, ACM SIGCOMM Comput. Commun. Rev. 40 (4) (2010) 339–350.

[66] K. Chen, A. Singla, A. Singh, K. Ramachandran, L. Xu, Y. Zhang, X. Wen, Y. Chen, Osa: an optical switching architecture for data center networks with unprecedented flexibility, in: Proceedings of the 9th USENIX Conference on Networked Systems Design and Implementation, USENIX Association, 2012, p. 18.

[67] K. Chen, C. Hu, X. Zhang, K. Zheng, Y. Chen, A. Vasilakos, Survey on routing in data centers: insights and future directions, IEEE Netw. 25 (4) (2011) 6–10.
[68] M. Bari, R. Boutaba, R. Esteves, L. Granville, M. Podlesny, M. Rabbani, Q. Zhang, M. Zhani, Data center network virtualization: a survey, IEEE Commun. Surv. Tutor. PP (99) (2012) 1–20.
[69] C. Kachris, I. Tomkos, A survey on optical interconnects for data centers, IEEE Commun. Surv. Tutor. 14 (4) (2012) 1021–1036.
[70] V.C. Emeakaroha, K. Fatema, Philip Healy, J.P. Morrison, Towards a generic cloud-based sensor data management platform: a survey and conceptual architecture, in: The Eighth Int. Conference on Sensor Technologies and Applications, 2014, pp. 88–96.

CHAPTER 4

Control From the Cloud

4.1 INTRODUCTION

And so, in the weeks leading up to a conference on the subject of this chapter, one of us found himself delving deeper and deeper into the fascinating literature and articles on the web about a growing trend among enterprises and institutions to relocate their computing resources outside the organization and enter into service agreements with some of the biggest providers of computing resources in the world.

One begins to wonder whether the world of production control and production continuity, which we are specializing in, could benefit from the advantages that cloud computing services have to offer. It turns out that there are huge quantities of literature describing a whole myriad of benefits that numerous enterprises had gained by placing their data centers in the professional hands of strangers, outside their organizations. Cloud computing companies, by delivering computing services to numerous enterprises on a vast scale, are able to invest in and allocate technological resources to their customers' computer systems of a magnitude that no individual enterprise could on its own. Today, enterprises that take advantage of cloud computing services generally enjoy higher levels of up time, redundancy, information security, and quality than they did in the past. Cloud service providers are obliged to maintain high standards of service in order to ensure their continued prosperity. They are assessed by their standards, and it is by their standards that they differentiate themselves from one another.

Another advantage is the *Pay-As-You-Go* business model that many enterprises prefer, a fixed monthly installment tailored to the actual computing needs of the enterprise: payment is based on the quantities of computing resources actually consumed, which can be adjusted at any time according to the enterprise's changing needs.

A recent extension of networked computer systems is the application of virtual operating systems and applications to create a single system. This means that cloud computing can be envisioned as a *System-of-Systems (SoS)*. Essentially, this system is created out of many individual hardware and software components that were historically used separately. From this angle, cloud computing is:

Networked Control Systems
https://doi.org/10.1016/B978-0-12-816119-7.00012-5

- Internet-based development and use of computer technology,
- Dynamically scalable, virtualized resources provided "as a service",
- "Cloud" is a metaphor for the Internet, based on how it is depicted in computer network diagrams and an abstraction for the complex infrastructure it conceals.

The involved technologies are:

A. Software-as-a-Service (SaaS) – online applications available over the web,

B. Hardware-as-a-Service (HaaS) – computer processing power purchased over the web,

C. Grid Computing (GC), Utility Computing (UC) and Autonomic Computing (AC).

An appropriate layout of cloud computing network is depicted in Fig. 4.1.

4.2 TOWARDS CONTROLLING FROM THE CLOUD

The field of control theory has been instrumental in developing a coherent foundation for systems theory with sustained investigations into fundamental issues such as stability, estimation, optimality, adaptation, robustness and decentralization. These issues have been the major ingredients in many new proposed technologies, which are now within our collective purview. On the implementation side, a centralized control system architecture was frequently employed (see Fig. 4.2), where the computational complexity is high and the distribution of sensors cover vast geographical region leading to long delays and loss of data.

In a parallel direction, digital computers became powerful tools in control system design, and microprocessors added a new dimension to the capability of control systems. This is clearly manifested in the aerospace industry. The introduction of a distributed control system (DCS) provides an advanced step in design evolution of process control. Expanding needs of industrial applications pushed the limit of point-to-point control further and it became obvious that remote control operations require some sort of effective coupling media.

4.2.1 Overview

With the explosive advancements in microelectronics and information technology, it became increasingly evident recently that it is the *networked* nature of systems which is drawing the major attention. From a technological standpoint, networked control systems (NCS) comprise *the system to be*

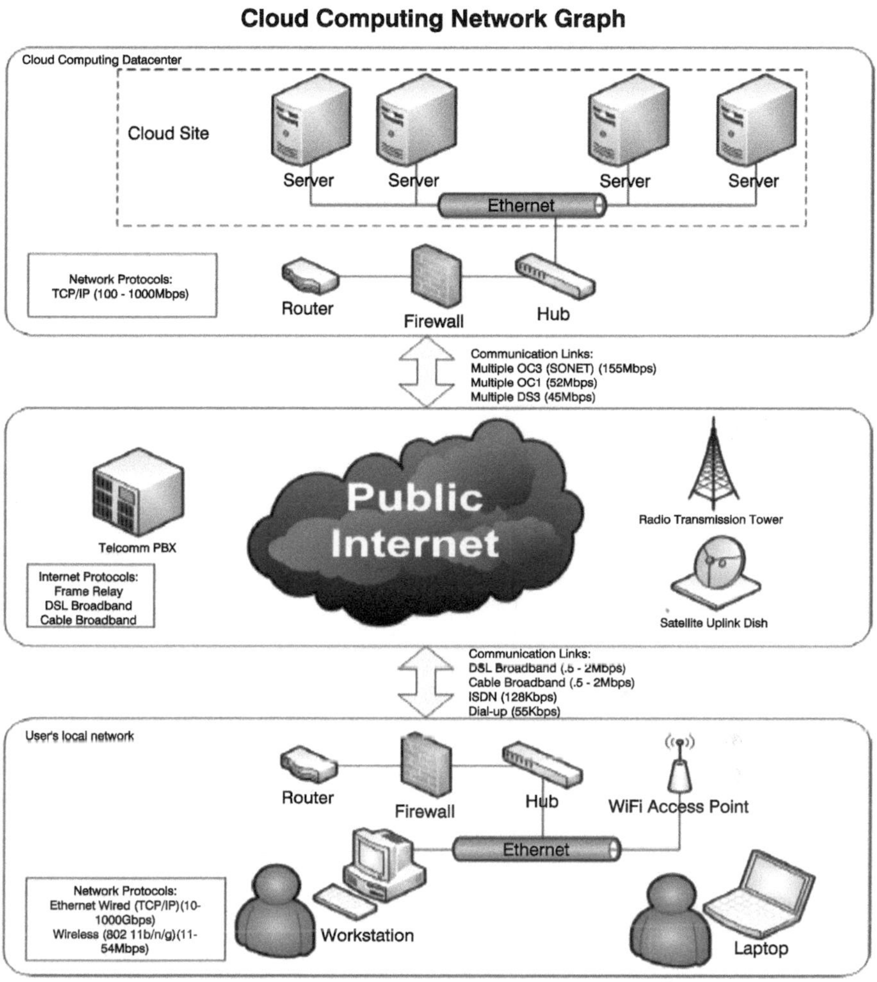

Figure 4.1 Cloud computing network graph

controlled (plant), actuators, sensors, and controllers whose operation is orchestrated with the help of a shared band-limited digital communication network. A contemporary layout of a networked control system is depicted in Fig. 4.3, where multiple plants, controllers, sensors, actuators and reference commands are connected through a network.

The intensification of this class of systems has addressed challenges to the communications, information processing, and control areas dealing with the relationship between operations of the network and the quality of the

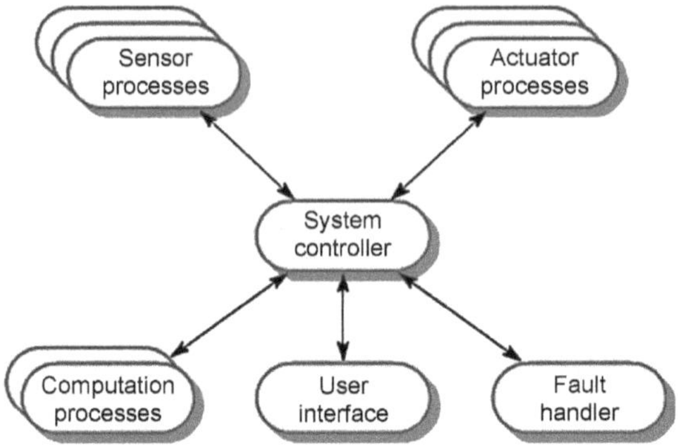

Figure 4.2 Centralized control system (CCS)

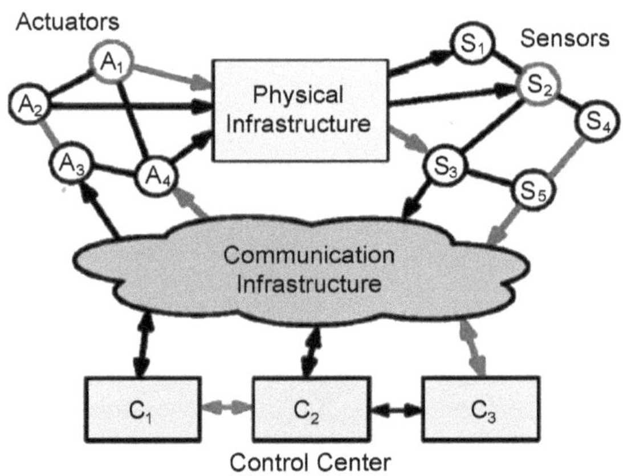

Figure 4.3 A typical networked control system structure

overall system's operation. The increasing interests in NCSs are due to their cost efficiency, lower weight and power requirements, simple installation and maintenance, and high reliability. Communication channels can lower the cost of wiring and power, ease the commissioning and maintenance of the whole system, and improve the reliability when compared with the point-to-point wiring system; see Fig. 4.4 where decentralized/distributed control systems are envisioned. The apparent advantages are scalability, robustness, adaptability, and computational efficiency.

Consequently, NCSs have been finding application in a broad range of areas such as mobile sensor networks [1], remote surgery [2], haptics collaboration over the Internet [3–5], and automated highway systems and unmanned aerial vehicles [6,7]. A modest coverage of numerous results

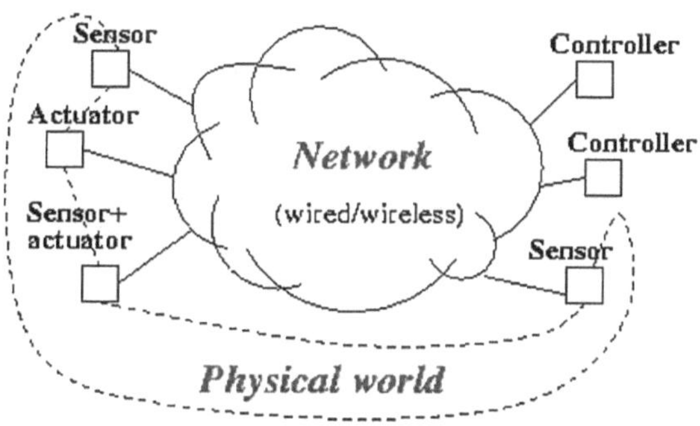

Figure 4.4 A schematic of wireless networked control system (WNCS)

and applications is found in [8]. In a parallel development, a study of the synchronization of the unified chaotic system via optimal linear feedback control and the potential use of chaos in cryptography, through the presentation of a chaos-based algorithm for encryption, are carried out in [9, 10].

4.2.2 Wireless Control Systems

Further development and research in NCSs were boosted by the tremendous increase in the deployments of wireless systems in the last few years. Today, NCSs are moving into distributed NCSs [11,12], which are multidisciplinary efforts aiming to produce a network structure and components that are capable of integrating distributed sensors, actuators, and control algorithms over a communication network in a manner that is suitable for real-time applications.

Essentially, the standing aspects of an NCS are:

- The component elements are spatially distributed,
- It may operate in an asynchronous manner,
- Its operation is coordinated to achieve some overall objective.

NCSs also allow remote monitoring and adjustment of plants over the communication channel, for example, the Internet in Internet-based control systems [13,14], which make the control systems benefit from the ways of retrieving data and reacting to plant actuation from anywhere around the world at any time.

The application of network technology in practical applications gave rise to industrial networked control paradigm (INCP), whereby at least one of the sensor–controller and controller–actuator links is completed using a communication network as shown in Fig. 4.5, where sensor-coder/decoder

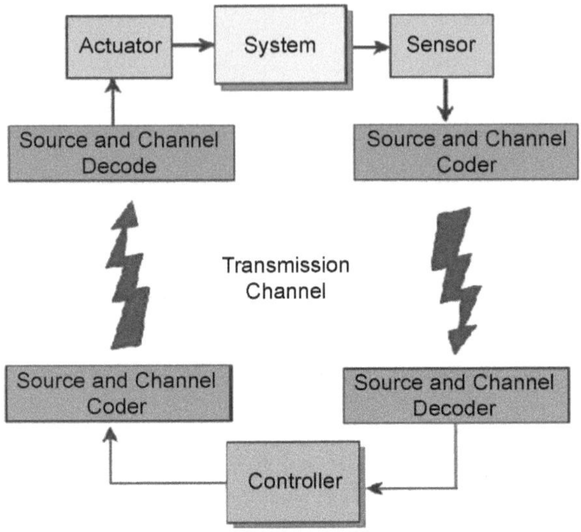

Figure 4.5 Components in industrial networked control paradigm (INCP)

devices and coder/decoder-actuator devices are located in the backward channel. The idea of introducing a communication network is to be able to remotely monitor and control the plant parameters. A huge amount of literature exists which studies such control systems in detail; see, for example, [15] and references therein.

4.2.3 Basic Classification

Since then, the area of *Networks and Control* or *Networking and Control* emerges as a new discipline in the control circles. From this perspective, there are two essentially distinct areas of research:

- The first area follows the interpretation, **control of networks**, and considers the control of communication networks, which falls into the broader field of information technology. This includes problems related to wireless networks and congestion control of asynchronous transfer mode (ATM) networks.
- The second area adopts the interpretation, **control through networks**, and deals with *networked control systems* (NCSs) which possesses a defining characteristic in which one or more control loops are closed via a serial communication channel. The focus here is on stability/stabilization issues, application issues and several standard bus protocols.

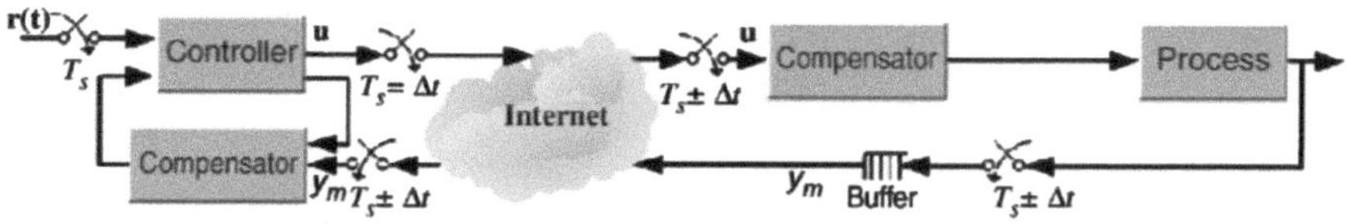

Figure 4.6 General structure of an internet-based control system

This chapter belongs to the second area where the objective is to provide an overview of the main results of networked control systems analysis and design. We adopt the approach of focusing on the developments of the different models and mathematical treatment while stating the main results in terms of theorems/lemmas without formal proofs. We further add numerous remarks for the purpose of illustration.

4.2.4 Remote Control Systems

In such applications, where the communication network at hand is the Internet and the controller is implemented on a remote server or a computing platform, then the resulting scenario is called a *remote control system (RCS)* [14], where concepts of switched control theory were deployed in the design stage; refer to Fig. 4.6.

In a parallel development, an extension of the networked control system (NCS) concept was coined as a *cloud control system (CCS)* in [16]. This new paradigm, which sees control as a service residing on a cloud computing platform, received attention from both the industry and academia in the past few years.

The motivating factors for utilizing a cloud service to implement control algorithm(s) are:

- The availability of internet-enabled devices (IoT), which generate such large amounts of data that companies have to maintain terabytes of data banks. This data can be used for quality check and/or control purposes;
- Powerful computational platforms found even in the hand held devices like smart phones;
- Increased complexity and geographical distribution of the plant or facility. This involved structure and dynamics of the systems translates from the requirement of increased efficiency, security and production demands resulting from an ever increasing size of the consumer market.

Remark 4.1. *The control of plants distributed along multiple geographical locations is feasible only with the introduction of a communication channel for the*

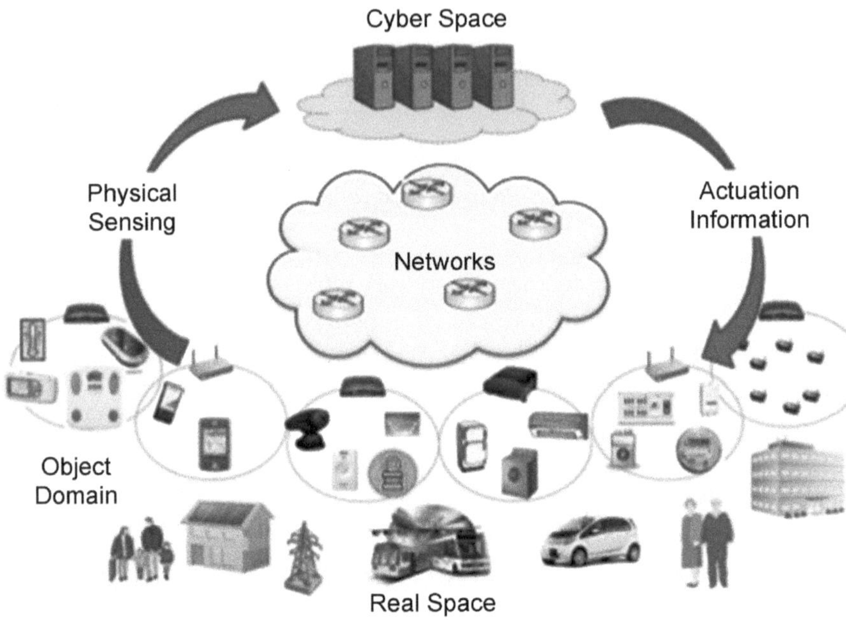

Figure 4.7 Cloud control system perspectives

transmission of data, or more precisely, the plant information and corresponding control command. The ubiquitous communication infrastructure of the internet, which has connected virtually every part of the world, is instrumental to this control structure. The control algorithm can be located in a remote server, the knowledge of whose geographical location is, in fact, unnecessary for the plant's operation! Every plant has its own network manager or a gateway that connects it with the internet. After finding a suitable cloud server, the link is established and the feedback loop is closed. This signifies the phrase "Control from the Cloud".

It is becoming increasingly evident that cloud control systems span multiple fields of study, such as control theory, information science, communication engineering, software and computer science. With such an amalgam of different research and study areas come increased complexities. Also, from our point of view, the terminologies used for the NCSs by professionals in different areas of study can be visualized in Fig. 4.7. The areas of applications include, but are not limited to, traffic networks, distributed energy resources and microgrids, mobile control applications and water transportation networks. The motivation behind pointing this out is to allow the researchers to clearly define different paradigms, which apparently differ but are extensions or applications of a single concept. Another factor is that it will allow people from different academic areas to explore the area collectively and be able to use and consider the approaches and mathematical tools prevailing in a different research area.

4.2.5 Some Prevailing Challenges

In what follows, we list the major issues and challenges pertaining to the development of rigorous track to "cloud control system":

A. The main challenge is the high dynamic nature of the cloud control system, which can jeopardize real-time operation of the overall control system. The imperfections introduced by communication networks include:

- delays,
- dropouts,
- back-offs due to unavailability of network resources,
- quantization effects,
- limited packet length,
- sharing of resources among several applications, and
- constraints on the bandwidth usage multiply the challenges for cloud control systems.

Other related issues are:

- large amounts of data communicated over the Internet by plants and controllers,
- derivation of data-driven models for the plants due to the asynchronous transmission of data packets which can be corrupted,
- huge amounts of data generated by the devices which becomes difficult to manage, and lastly,
- security of the data.

Standing alone, the cost of cloud service is another frontier which can constrain the usage per unit of time.

B. An excellent work which practically implements industrial automation service was reported in [17], where an adaptive delay compensator is proposed and implemented to mitigate the effects of delays. In addition, distributed fault tolerant scheme is proposed to tackle the faults. The overall scheme was implemented on commercially available clouds to control a solar power plant and the results showed comparable performance to the local controllers. Another work which deals with the time delays and data loss for process control systems over the

Internet was presented in [13,14]. It was stated based on the existing works that the performance of the control system associated with delay and data loss shows large temporal and spatial variations. It was also reported that the traditional periodic control schemes cannot be implemented on cloud control systems due to a large variation in the time-delay patterns.

C. Analyzing other existing works in the literature suggests that the cloud control paradigm is not developed rigorously when it comes to theoretical background. This is the main area where this newly suggested control scheme lacks.

1. In the authors' opinion, the main shortcoming comes from the fact that mathematical modeling of the cloud control system is still immature, at least to some extent, and this should result in interesting research areas in analyzing stability of cloud based systems. Treating the cloud as a black or grey box will always result in conservative results. Therefore, a comprehensive model is necessary. For this purpose, some works which focus on detailed modeling of industrial networks such as Wireless HART and ZigBee [18–21] can be consulted. This will provide a start towards modeling of more complex and relatively undeterministic characteristics of the Internet.
2. Another important aspect to explore is the available simulation tools to create scenarios representative of actual cloud control systems. One such platform is *Contiki based Cooja simulator*, which allows introducing nodes with physically defined distances. One can create servers as well as hook up the software with Matlab-based Simulink; this allows simulating the physical plant along with the Internet architecture. Nevertheless, other software tools for packet sniffing such as Wireshark can also be used when the simulation results have to be validated experimentally.
3. Typical application would be the class of large-scale networked systems, see Fig. 4.8, where theoretical analysis is warranted for cloud-control system implementations.

4.2.6 Reflections on Industrial Automation

With increasing usage of information technologies such as the Internet-of-Things (IoT), Service Oriented Architectures (SOA) and mobile computing in industrial automation, there is a further need for a platform to provide computing resources, information integration and a data repository

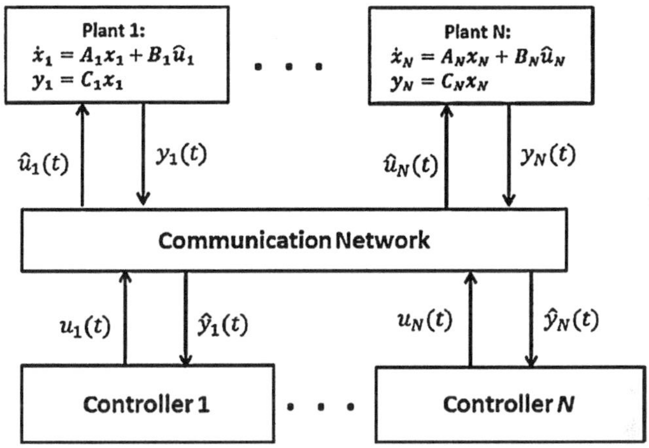

Figure 4.8 Large-scale networked system

for the connected devices (things). Cloud computing could be a possible solution for this reason. On the other hand, this technology could be assumed as a solution for implementation of newly proposed architectures for automation. This would be possible with migration of current automation functionalities to the cloud. However, application requirements determine potential for each function for moving to the cloud or staying in the floor as a physical device.

Looking ahead, control levels must be further investigated in order to address reliability and real-time issues. To satisfy industrial companies for migrating to cloud-based systems, new approaches should be applied based on vendor independent standards to ensure their interoperability with special consideration of security issues.

In the preparation phase of design, the scenario unfolds as follows: a successful cloud-based control approach should aim at offering control-as-a-service for an industrial automation. Based on application requirements for each specific module, there are possible avenues:

- Local PID controllers could remain in the shop floor as a physical device traditionally to gain more reliable control functions.
- Alternatively, PID can be implemented as a virtual entity and be delivered to the field as a service from a cyberphysical system via the network.

The cloud-based control system can then be realized as a private cloud platform with capability to offer virtual PID appliances using, for example, *VMware's vCloud suite* solution [22] which is connected to the *Profinet* real-time Ethernet system (RT) via physical interface. This turns out to increase

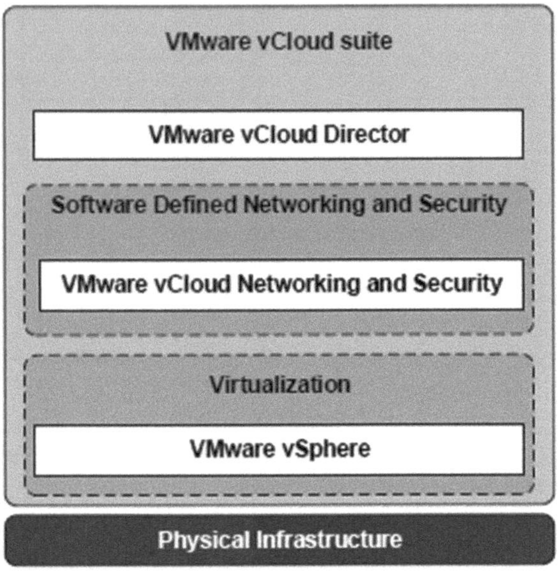

Figure 4.9 VMware's vCloud suite architecture [23]

bandwidth and the number of physical interfaces could be increased by connecting more network adapters to the hypervisor computer. There will be a virtual switch (vSwitch) directly managed by the vCloud Networking and Security component. Network rules and policies such as access permissions for virtual machines defined for the cloud will be applied by vSwitch. For enabling industrial communications such as Profinet on vSwitch, Virtual Machine Communication Interface (VMCI) [23] is enabled for the adapter driver of VMs. VMCI is a high-speed interface which boosts the communication between VMs and the host computer [23].

A virtual PID appliance is connected to each virtual interface. Technically, each appliance includes a virtualized operating system which runs a PC-based software PID. See Fig. 4.9.

Another issue to observe is watching the task of reading the input of the I/O device generated by a signal generator periodically at a fixed interval of T_I and writing the inverted value to the output. This task repeats cyclically in determined time intervals, and the output will be measured and recorded by an appropriate display (an oscilloscope). The ensuing output is labeled T_O. Typically, outputs will be influenced by some delays, and a variation for output called ΔT will be observed for each output given by

$$T_O = T_I + \Delta T. \tag{4.1}$$

Observe that the network is not fully synchronized, and therefore the cycle update time should be considered when frames start to be transmitted on the network. This time value is defined as T_{CD} corresponding to the Profinet cycle update delay. In fact, this delay would be $2T_{CD}$ in the worst case where a frame waits for the whole cycle time at its upstream and downstream times. The T_{CD} of devices frequently has a maximum of 1 ms in the worst case. Let T_{CON} be the control task interval time plus execution time performed by the PID to process each control task. The minimum supported value for the used PIDs is 2 ms. Due to simple inverting operation, execution time expected as a small value compared to the entire T_{CON} value. Denote by T_N the network delay for a frame to send over the network. Consider that each *Profinet* frame is 64 bytes and its T_N is roughly 7 μs which is a small value. The sum of these delays will be defined as *End-to-End (E2E) delay*

$$T_{E2E} = 2T_{CD} + T_{CON} + 2T_N. \tag{4.2}$$

This delay will cause the mentioned variance ΔT for each output. Suppose that N experiments were run. Then the mean value of ΔT defined by

$$\overline{\Delta T} = \frac{1}{N}\sum_{j=1}^{N} \Delta T_j \tag{4.3}$$

is considered as *a metric to evaluate the performance of the system.*

4.2.7 Quality-of-Service (QoS)

The problem of Quality-of-Service (QoS) provisioning has been an extremely active area of research for many years and was mainly focused on the Internet. From the earlier Integrated Services (IntServ) architecture [25] to the more recent Differentiated Services (DiffServ) architecture [26], many QoS control mechanisms, especially in the areas of packet scheduling and queue management algorithms, have been proposed. Elegant theories such as network calculus and effective bandwidths have also been developed. Several books have been written on the subject, some focus more on architectural and other practical issues [27,28], while others on theoretical aspects of QoS provisioning [29].

Network QoS can be defined in a variety of ways and include a diverse set of service requirements such as performance, availability, reliability, security, etc. All these service requirements are important aspects of a comprehensive network QoS service offering. However, in this paper we will take a more performance-centric view of network QoS and focus primarily on the issues in providing performance guarantees. Typical performance metrics used in defining network QoS are bandwidth, delay/delay jitter, and packet loss rate. Using these performance metrics, network performance guarantees can be specified in various forms, such as absolute (or deterministic), that is, a network connection is guaranteed with 10 Mbps bandwidth all the time; probabilistic (or stochastic), e.g., network delay is guaranteed to be no more than 100 ms for 95% of the packets; time average, that is, packet loss rate is less than 10^{-5} measured over a month. The guarantee feature of network QoS is what differentiates it from the best-effort network services.

4.2.8 Preliminary Control Models

Experience has indicated that the round-trip delay between the cloud controllers and controlled processes varies with time [63]. In the sequel, we present some models to investigate the effect of delays in cloud-based control systems.

A. Model With Forward and Backward Delays

For the purpose of illustration, consider a plant represented by a linear shift-invariant system

$$\begin{aligned} x(k+1) &= Ax(k) + Bu(k - d_s - d_a) + \Gamma\omega(k), \\ z(k) &= Cx(k) + \Lambda\omega(k), \end{aligned} \tag{4.4}$$

where $x(k) \in \mathbb{R}^n$ is the state vector, $u(k) \in \mathbb{R}^m$ is the local control vector, $z(k) \in \mathbb{R}^q$ is the output observation vector and $\omega(k) \in \mathbb{R}^q$ is the exogenous vector which is assumed to belong to $\ell_2[0, \infty)$. The matrices $A \in \mathbb{R}^{n\times n}$, $B \in \mathbb{R}^{n\times m}$, $C \in \mathbb{R}^{q\times n}$, $\Gamma \in \mathbb{R}^{n\times q}$, $\Lambda \in \mathbb{R}^{q\times q}$ are constants.

It can be seen that there are two delays: d_s is used to represent the forward communication delay from a sensor to the cloud and d_a is used to represent backward communication delay from the cloud to the actuator. A natural assumption on the time delay components d_s and d_a can be given by

$$0 < \alpha_m \leq d_s \leq \alpha_M, \quad 0 < \tau_m \leq d_a \leq \tau_M, \tag{4.5}$$

where α_m, α_M, τ_m, τ_M are constants designating delay bounds with α_m, τ_m reflecting a finite delay irrespective of the technology level of the communication links and α_M, τ_M reflecting the maximum allowable bound beyond which the networked system will become unstable.

The following theorem provides a sufficient asymptotic stability condition for system (4.4):

Theorem 4.1. *System (4.4) with $u(k) \equiv 0$, $\omega(k) \equiv 0$ is asymptotically stable if there exist matrices $\mathcal{P}_1 > 0$, $\mathcal{R}_j > 0$, $\mathcal{S}_j > 0$, $j = 1, 2$, $Q > 0$, $L > 0$, P_2, P_3, P_4, P_5, P_6 and P_7 satisfying*

$$\Omega = \begin{bmatrix} \Omega_{11} & \bullet & \bullet & \bullet & \bullet \\ \Omega_{21} & \Omega_{22} & \bullet & \bullet & \bullet \\ \Omega_{31} & \Omega_{32} & \Omega_{33} & \bullet & \bullet \\ \Omega_{41} & \Omega_{42} & \Omega_{43} & \Omega_{44} & \bullet \\ \Omega_{51} & 0 & \Omega_{53} & 0 & \Omega_{55} \end{bmatrix} < 0, \tag{4.6}$$

where

$$\begin{aligned}
\Omega_{11} &= P_2^T(A-I) + (A-I)^T P_2 + P_4 + P_4^t \tau - \alpha_M^{-1}\mathcal{R}_1 + \mathcal{S}_1 + Q, \\
\Omega_{21} &= P_3^T(A-I) + P_3^T + P_1 - P_2, \\
\Omega_{22} &= -P_3 - P_3^T + P_1 + \alpha_M \mathcal{R}_1 + \tau_M \mathcal{R}_2 + (\alpha_M + \tau_M - \alpha_m - \tau_m)L, \\
\Omega_{31} &= P_6^T + B^T P_2 - P_4, \quad \Omega_{32} = B^T P_3 - P_5, \\
\Omega_{33} &= -P_6 - P_6^T - \tau_M^{-1}\mathcal{R}_2 - \mathcal{S}_2 - Q - (\alpha_M + \tau_M - \alpha_m \quad \tau_m)^{-1}L, \\
\Omega_{41} &= P_7^T - P_4, \quad \Omega_{42} = -P_5, \\
\Omega_{43} &= -P_7^T - P_6^T, \quad \Omega_{44} = -P_7^T - P_7, \quad \Omega_{51} = \alpha_M^{-1}\mathcal{R}_1, \\
\Omega_{53} &= \tau_M^{-1}\mathcal{R}_1 + (\alpha_M + \tau_M - \alpha_m - \tau_m)^{-1}L, \\
\Omega_{55} &= -\alpha_M^{-1}\mathcal{R}_1 - \tau_M^{-1}\mathcal{R}_2 - \mathcal{S}_1 + \mathcal{S}_2 - (\alpha_M + \tau_M - \alpha_m - \tau_m)^{-1}L.
\end{aligned} \tag{4.7}$$

The proof of Theorem 4.1 hinges on Lyapunov–Krasovskii stability theory [64]. We construct the following Lyapunov–Krasovskii function candidate:

$$\begin{aligned}
V(k) &= V_1(k) + V_2(k) + V_3(k) + V_4(k), \\
V_1(k) &= x^t(k)\mathcal{P}_1 x(k), \\
V_2(k) &= \sum_{\theta=-d_s}^{-1} \sum_{\ell=k+\theta}^{k-1} \eta^T(\ell)\mathcal{R}_1\eta(\ell) + \sum_{\theta=-d_s-d_a}^{-d_s-1} \sum_{\ell=k+\theta}^{k-1} \eta^T(\ell)\mathcal{R}_2\eta(\ell),
\end{aligned}$$

$$V_3(k) = \sum_{\ell=k-d_s}^{k-1} x^T(\ell)S_1x(\ell) + \sum_{\ell=k-d_s-d_a}^{k-d_s-1} x^T(\ell)S_2x(\ell),$$

$$V_4(k) = \sum_{\ell=k-d_s-d_a}^{k-1} x^T(\ell)Qx(\ell) + \sum_{\ell=-d_s-d_a}^{-\alpha_M-\tau_M 1} \eta^T(\ell)L\eta(\ell), \tag{4.8}$$

where $\eta(k) = x(k+1) - x(k)$, $V(k) > 0$. We then proceed to compute $\Delta V(k) = V(k+1) - V(k)$. A straightforward computation shows that $\Delta V(k) \leq \xi^T\Omega\xi$ for a given $\xi(k) \neq 0$. It is clear that if (4.6) holds, then $\Delta V(k) < 0$, which completes the proof. □

Remark 4.2. *Theorem 4.1 can be readily extended to establish stabilization of system (4.4) under linear controllers including state feedback $u(k) = K_s x(k)$, or output feedback $u(k) = K_o z(k)$. Moreover, robust controllers could be designed with prescribed performance criteria including $\mathcal{H}_2$, $\mathcal{H}_\infty$, or mixed $\mathcal{H}_2/\mathcal{H}_\infty$.*

B. A Predictor Feedback Controller

Here, consider a plant described by a linear shift-invariant system with multiple input delays

$$x(k+1) = \mathsf{A}x(k) + \sum_{j=0}^{p} B_j u(k-d_j), \tag{4.9}$$

where $x(k) \in \mathbb{R}^n$ is the state vector, $u(k) \in \mathbb{R}^m$ is the local control vector. The matrices $\mathsf{A} \in \mathbb{R}^{n\times n}$, $B \in \mathbb{R}^{n\times m}$ are constants and d_j is the delay in control channel j. Adopting the predictor feedback approach [64] for controlling system (4.9), we introduce

$$\sigma(k) = x(k) + \sum_{j=1}^{p}\sum_{m=1}^{d_j} \mathsf{A}^{m-d_j-1}B_j u(k-m). \tag{4.10}$$

Algebraic manipulation of (4.10) using (4.9) yields

$$\sigma(k+1) = \mathsf{A}\sigma(k) + \sum_{j=1}^{p} \beta_j(k) + B_0 u(k), \tag{4.11}$$

$$\beta_j(k) = \mathsf{A}^{-d_j} B_j u(k). \tag{4.12}$$

Further simplification results in the delay-free model

$$\sigma(k+1) = \mathsf{A}\sigma(k) + \mathsf{B}u(k), \tag{4.13}$$

$$\mathsf{B} = \mathsf{A}^{d_0} B_0 + \mathsf{A}^{d_1} B_1 + \cdots + \mathsf{A}^{d_p} B_p. \tag{4.14}$$

Now, seeking a state-feedback stabilization design we define

$$\begin{aligned} u(k) &= \mathsf{K}_s \sigma(k) \\ &= \mathsf{K}_s x(k) + u_d(k), \end{aligned} \tag{4.15}$$

where $\mathsf{K}_s \triangleq \mathsf{K}_s(d_0, \ldots, d_p)$ such that $(\mathsf{A} + \mathsf{BK}_s)$ is Schur stable [64]. From (4.14) and (4.15), we obtain the closed loop system

$$\sigma(k+1) = (\mathsf{A} + \mathsf{BK}_s)\sigma(k). \tag{4.16}$$

Schur stability of system (4.16) means that $\lim_{k \to 0} ||\sigma(k)|| = 0$ which in turn implies that $\lim_{k \to 0} ||u(k)|| = 0$. This takes us to

$$\begin{aligned} \lim_{k \to 0} ||x(k)|| \leq & \lim_{k \to 0} ||\sigma(k)|| \\ & + \lim_{k \to 0} \sum_{j=1}^{p} \sum_{m=1}^{d_j} ||\mathsf{A}^{m-d_j-1} B_j|| \, ||u(k-m)|| \\ = & \, 0, \end{aligned} \tag{4.17}$$

from which we conclude that system (4.9) is stabilizable by control (4.15) as desired. □

Remark 4.3. *It is noted that the control law (4.15) is referred to as predictor feedback because it effectively compensates the delays by a predictor-like form [64]. Since this control involves the history of the control signal, it is considered as a full-delayed state feedback. This feature does not compound the computational load as the previous required are stored in the cloud.*

4.3 CLOUD CONTROL SYSTEMS

During the last decade, network technology has been dramatically developed. Recently, more and more network technologies have been applied to control systems [32,33]. This kind of control systems in which a control loop is closed via communication channel is called networked control systems. Now, networked control is a new area in control systems. Particularly, Internet-based control systems allow remote monitoring and adjustment of plants over the Internet, which make the control systems benefit from the

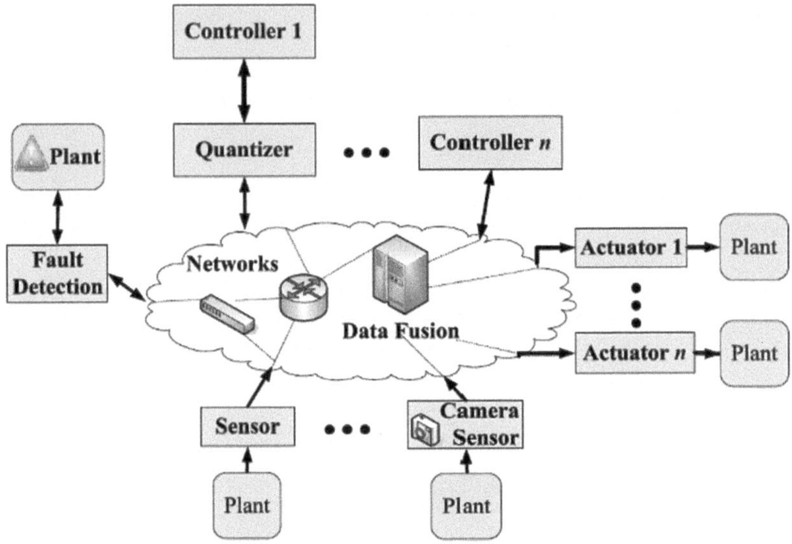

Figure 4.10 Networked control system schematic

ways of retrieving data and reacting to plant fluctuations from anywhere around the world at any time. Most attention in this area has been paid to the design and analysis of NCSs [34–38]. Many applications have been carried out in practice. In NCSs, the plant, controller, sensor, actuator and reference command are connected through a network. A typical diagram of an NCS is shown in Fig. 4.10.

4.3.1 Introduction

In recent years, the techniques of IoT are developed rapidly, the NCSs research plays a key role in the Internet-of-Things which will draw on the functionality offered by all of these technologies to realize the vision of a fully interactive and responsive network environment. Fig. 4.11 shows the future applications of the Internet-of-Things. While researching IoT, we find that data collection and processing are very important. First, it is difficult, and even impossible, for us to get an accurate physical model of every object in IoT. The only way we know the objects in IoT is through data we can get. Secondly, due to the sensor technology, we can detect changes in the physical status of things. Thus, big data about things are collected and stored. The advances in computer science, especially in the aspects of computing ability and storage, together with high quality and reliable mea-

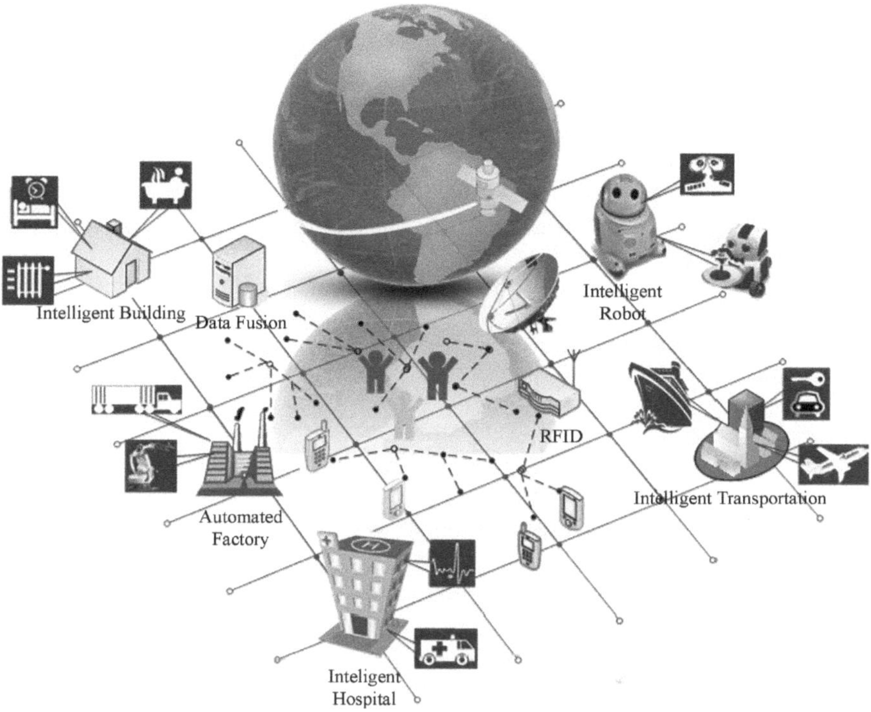

Figure 4.11 Internet-of-Things

surements from process instruments, make it possible to collect and process the data efficiently. Finally, all the objects and devices are connected to large databases and networks and indeed to the network of networks (the Internet). Information and commands are transmitted through the network. However, the use of the network will lead to intermittent losses or delays, bandwidth constraints, asynchronization and other unpredictable factors of the communicated information. These losses could deteriorate the performance and may even cause the system to go unstable. Many effective control theories and applications have been proposed with the development of IoT.

4.3.2 Model Based Networked Control Systems

First, fruitful research results on NCSs are model based, especially, for linear time invariant systems. More generally, consider a discrete dynamical system S with unknown process disturbances and measurement noises, which is

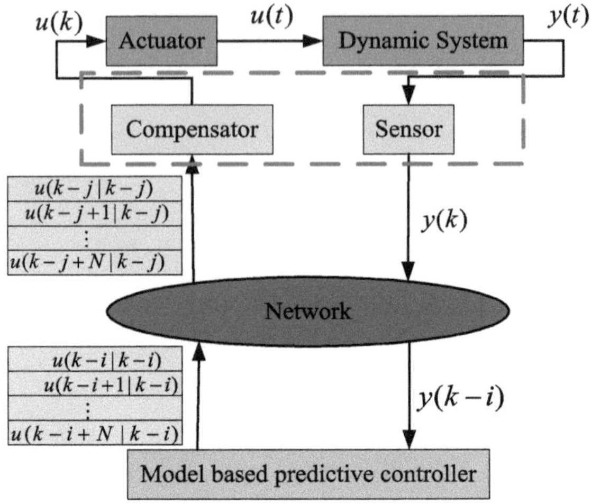

Figure 4.12 Model based networked control system

described in state-space form as:

$$x(k+1) = f(x(k), u(k), w(k)), \tag{4.18}$$

$$y(k) = g(x(k), u(k), v(k)), \tag{4.19}$$

where $x(k)$ is the system state, $u(k)$ is the system input, $y(k)$ is the system output, $x(k)$, $u(k)$ and $y(k)$ have suitable finite dimensions; $f(x(k), u(k), w(k))$ and $g(x(k), u(k), v(k))$ are models, which can be linear or nonlinear; $w(k)$ is the unknown process disturbance and $v(k)$ is the unknown measurement noise.

Many methods have been proposed to solve problems related to networked control systems. It has been proven that the networked predictive method is a very effective method for networked control systems where there are network induced time delay and data dropouts [32,33]. The design of a feedback control scheme for system (4.18)–(4.19) is shown in Fig. 4.12.

Here, a buffer is set at the controller node to assure that the measurements are processed in a sequence. Kalman filter is adopted to estimate the state and then produce the predictive states with finite horizon N_1:

$$\begin{aligned} \hat{x}(k|k) &= KF(S, \hat{u}(k-1|k-1), y(k)), \\ \hat{x}(k+i|k) &= KF(S, \hat{u}(k|k), y(k)), \quad i = 1, 2, \ldots, N_1, \\ \hat{u}(k+i|k) &= K(k+i)\hat{x}(k+i|k), \quad i = 1, 2, \ldots, N_1, \end{aligned} \tag{4.20}$$

where KF represents the compact form of Kalman filter expression and $K(k+i)$ is the time-varying Kalman filter gain [33]. To overcome unknown network transmission delays, a networked predictive control scheme is proposed. It mainly consists of a control prediction generator and a network delay compensator. The control prediction generator is designed to generate a set of future control predictions. The network delay compensator is used to compensate for the unknown random network delays. The network delays are in the forward (from controller to actuator) channel (CAC) and feedback (from sensor to controller) channel (SCC). A very important characteristic of the network is that it can transmit a set of data at the same time. Thus, it is assumed that predictive control sequence at time k is packed and sent to the plant side through a network. The network delay compensator chooses the latest control value from the control prediction sequences available on the plant side.

For example, when there is no time delay in the channel from a sensor to a controller and the time delay from the controller to an actuator is k_i, and if the following predictive control sequences are received on the plant side:

$$\begin{aligned}
&[u^T_{t-k_1|t-k_1}, u^T_{t-k_1+1|t-k_1}, \ldots, u^T_{t|t-k_1}, \ldots, u^T_{t+N-k_1|t-k_1}]^T,\\
&[u^T_{t-k_2|t-k_2}, u^T_{t-k_2+1|t-k_2}, \ldots, u^T_{t|t-k_2}, \ldots, u^T_{t+N-k_2|t-k_2}]^T,\\
&\vdots\\
&[u^T_{t-k_t|t-k_t}, u^T_{t-k_t+1|t-k_t}, \ldots, u^T_{t|t-k_t}, \ldots, u^T_{t+N-k_t|t-k_t}]^T,
\end{aligned} \tag{4.21}$$

where the control values $u_{t|t-k_i}$ for $i=1, 2, \ldots, t$, are available to be chosen as the control input of the plant at time t, the output of the network delay compensator, i.e., the input to the actuator, will be

$$u_t = u_{t|t-\min\{k_1, k_2, \ldots, k_t\}}. \tag{4.22}$$

In fact, by using the networked predictive control scheme presented in this section, the control performance of the closed-loop system with network delay is very similar to that of the closed-loop system without network delay.

The controller sends packets to the plant node:

$$\{u(k+i|k)|i=0, 1, \ldots, N\} \tag{4.23}$$

At time instant k, the actuator chooses a preferable control signal as the actual input of the controlled dynamic system:

$$u(k) = u(k|k-i) \tag{4.24}$$

where $i = \arg\min_i \{u(k|k-i) \text{ is available}\}$.

The stability proof and idea can be found in [32,33].

4.3.3 Data Driven Networked Control Systems Design

Under traditional design frameworks, the data from the plant are used to build the model, since dynamic models are the prerequisite of control and monitoring. Once the design of a controller or a monitor is completed, the model often ceases to exist. However, the use of models also introduces unavoidable modeling error and complexity in building the model. With the emergence of IoT, big data has to be processed. The data-driven subspace approach has been proposed for industrial process control field. The data-driven control method also can be developed for complex systems. In particular, the data-driven method is especially suitable for NCSs since only digital data can be transferred through the network and received by a controller and actuator.

In this section, data-driven networked control systems are introduced. The subspace projection method is applied to generate the predictive control signals. The only difference between data-driven networked control systems and model-based control systems is the controller, which is shown in Fig. 4.13.

When we get data which includes past input and output from the sensors over network, we apply the data-driven predictive control algorithm described as above. Thus, we can obtain a sequence of predictive control inputs, which will be transmitted to a buffer over network. The compensator will choose the right control input according to the predictive networked control scheme described in Eqs. (4.19)–(4.22). From the description we can see that the proposed data driven networked control scheme is different from the model based networked control scheme in [39], by which the control input can be obtained directly without modeling. Certainly, there are still many questions about data-driven control method: for example, how to distinguish linear and nonlinear systems under the data-driven control scheme? How to evaluate the performance? As for fast controlled plants, one can wonder how to generate the initial control signals. If the data drop

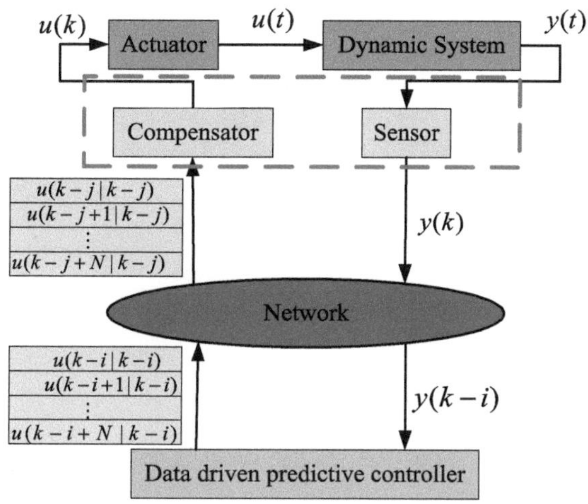

Figure 4.13 Data driven networked control systems

occurs, how to compute the subspace objection with intermittent observations? How to analyze the stability of data-driven based nonlinear systems? More details about data-driven predictive networked control systems can be found in our recent publication [40].

4.3.4 Networked Multiagent Systems

At the beginning of research on NCSs, more attention was paid on single plant through network. Recently, fruitful research results on multiplant, especially, on multiagent networked control systems, have been obtained, which focused on the design of more general models, where each agent has its own distinct nonlinear dynamics that are unknown for the other agents [41]. Agents in networks update their states based on information exchange among them. There are many interesting problems for networked multiagent systems, for example, aggregation, clustering, coordination, consensus, formation, synchronization, evolution and swarm, as is shown in Fig. 4.14.

If each agent runs in a common state, then, the multiagent system achieves consensus. Consensus problems have a long history in computer science and form the foundation of the field of distributed computing and control [42]. Consensus protocols are distributed control policies based on neighbor's state feedback that allows coordination of multiagent systems. According to the usual meaning of consensus, the system state components

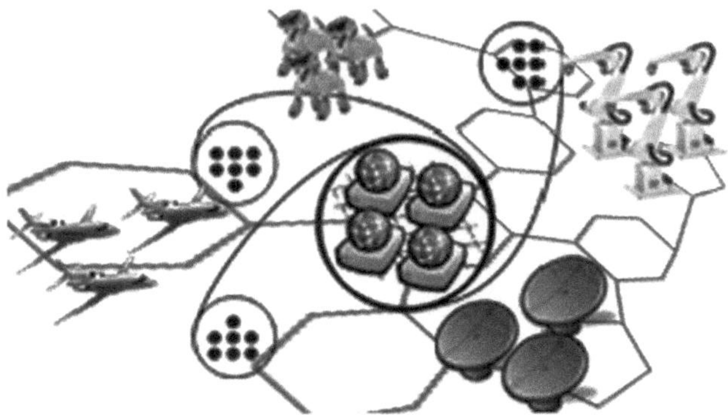

Figure 4.14 Control of multiagent systems

must converge, in finite time or asymptotically, to an equilibrium point where they all have the same value lying somewhere between the minimum and maximum of their initial values. In recent years, there have been a tremendous amount of research on the consensus problem, and many interesting results have been obtained [43].

On the other hand, the formation control problem for a group of agents is a popular research topic in decentralized control [44]. Formation control of multiple robots, spacecrafts and unmanned aerial vehicles, mobile autonomous agents are often treated as rigid bodies or mass points. Formation control of multiple autonomous vehicles has recently received increasing interest in the control community. This interest is motivated by their potential applications in areas such as search and rescue missions, reconnaissance operations, forest fire detection, surveillance and multimissile attack [45]. Work in this area is generally inspired by the recent results in the coordinated control of multiagent systems. The leader–follower method uses several agents as leaders and others as followers; one also employs the behavior based approach and the virtual structure approach.

Coordination control is an active research area presently and has more practical meaning [46]. For example, when we eat food, our eyes help in locating the food, our nose senses the food, our hand brings the food to our mouth, and our jaw muscles help the teeth to chew the food. All these activities occur in a coordinated manner, and if any of these activities misses or does not occur in time then the body will not get nutrition. Now, more researchers are devoted to multirobot systems which can be used to increase the system effectiveness. That is, with respect to a single autonomous robot or to a team of noncooperating robots, multirobot systems can better per-

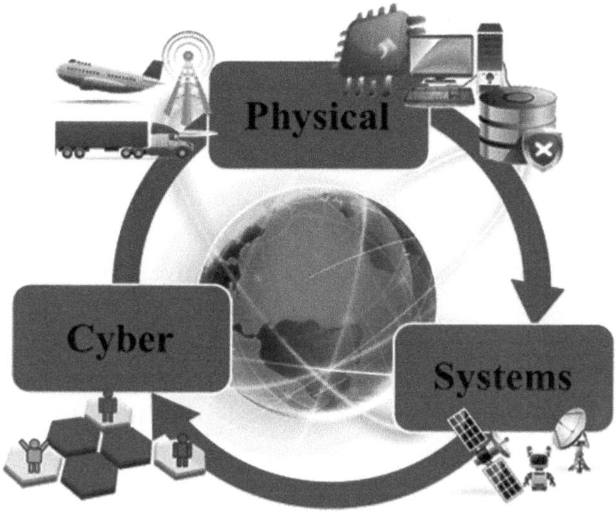

Figure 4.15 Cyberphysical system

form a mission in terms of time and quality, can achieve tasks not executable by a single robot (e.g., moving a large object) or can take advantages of distributed sensing and actuation.

With the development of software, hardware and other advance techniques in multiagent systems, a system of collaborating computational elements controlling physical entities has come into practice, that is, cyber-physical systems (CPSs), which interface physics-based and digital world models as shown in Fig. 4.15. Currently, CPS research aims to integrate physical and computational models in a manner that outperforms a system in which the two models are kept separate [47]. In the context of the feedback control system, objectives of the physical system (e.g., disturbance rejection, tracking accuracy, etc.) are translated to computing actuator commands that minimize errors between reference and actual trajectories through physical space.

4.3.5 Control of Complex Systems

Up to now, there is no unified understanding for complex systems both in terms of the precise definition of complexity and the basic intuition behind the concept. From different points of view, there are many definitions of complex systems, for example, a complex system refers to a system composed of interconnected parts that as a whole exhibit one or more properties (behavior among the possible properties) not obvious from the properties of the individual parts [48]. In [49], complex systems are considered as natural or social systems which are composed of a large number of nonlinear modules. Loosely speaking, the complex systems require large, messy models that are difficult to formulate and are uncertain in their status, which is one of the main features of a complex system. Other features also include nonlinear and complexity of components itself, strong coupling between components, coexistence of positive and negative feedback, etc. Complex systems may be open. They may interchange information or mass with the environment, and during the process, they can modify their internal structure and patterns of activity in a way of self-organization, which causes such systems to be flexible and easily adapt to variable external conditions. However, for complex systems, the most significant feature is that one cannot derive or predict solely from the knowledge of such systems' structure and the interactions between their individual elements. Thus, for the same system, one often needs to provide the parallel descriptions on different levels of its organization. A few of the most characteristic properties of the structure and behavior of the systems, which are commonly referred to as complex, are reviewed in the same paper, i.e., power law, self-organization, coexistence of collective effects and noise, variability and adaptability, hierarchical structure, scale invariance, self-organized criticality, and highly optimized tolerance.

Fig. 4.16 shows the composition of complex systems. As shown in the figure, many practical systems can be modeled as complex systems, for example, climate, ant colonies, human economies and social structures, and even living things, including human beings, as well as modern energy or telecommunication infrastructures, etc. In addition, multiagent networked control systems can also be viewed as one simple kind of complex systems, each agent is an element of a complex system.

However, at the current stage, it is still a huge challenge to apply directly existing stability theory to present an analysis of stability and instability for such complex systems. Generally speaking, the existing methods are limited to feedback control which has been implemented to regulate local and global behavior, and it is awkward to apply them directly to deal with complex systems, especially the interconnected systems. It is well known that Lyapunov theory based stability theory and its many extensions have been successfully applied to many systems like continuous, discrete, impulsive, hybrid and time-delay systems. The emphasis has been on so-called closed systems. An alternative view has been taken in engineering where the em-

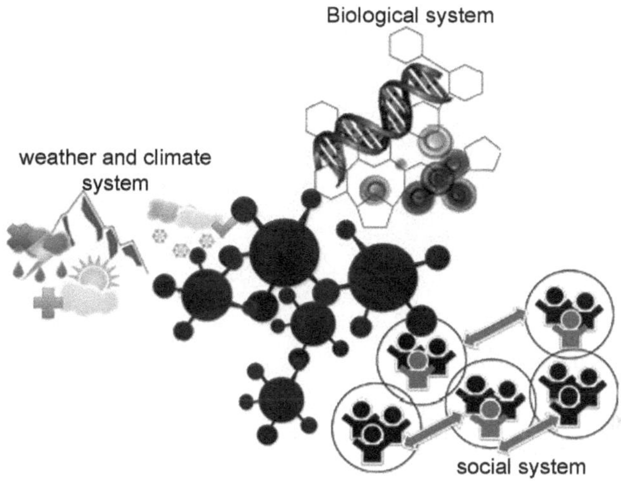

Figure 4.16 Control of complex systems

phasis is on responses to the influence of many externalities such as inputs, disturbances and interconnections. This view treats systems as open with inputs and outputs. Lyapunov theory is also included in this approach, but functional analysis is needed since all the inputs, outputs and other signals live in extended function spaces. It has been reviewed in the systems and control community that there are unfulfilled theoretical opportunities in the theory of open systems for mathematicians and physicists [50].

The techniques of stability analysis related to Lyapunov theory, so-called energy-function methods, have been well developed and can give regions of synchronization after disturbances, and so change the control rules, and transfer limits thereby influencing economic performance [51]. This kind of development appears to be needed for larger classes of complex systems. A further approach, which developed in electrical engineering as an extension of operator methods in circuit theory, uses functional analysis methods to make statements about input–output behavior [52]. The more rigorous recent results on complex systems mainly use Lyapunov theory (for example, see [53,54]). However, there are also a lot of phenomena in complex systems which cannot be analyzed or proven by Lyapunov theory.

4.3.6 Cloud Control System Concepts

Initially, the concept of cloud control systems was used by an alternative rock band originating from the Blue Mountains near Sydney, Australia.

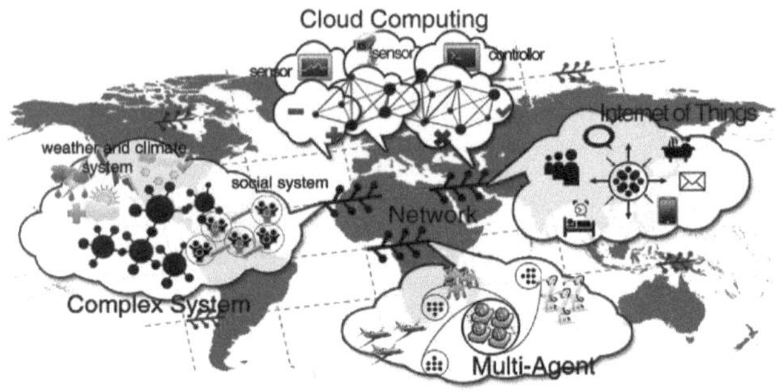

Figure 4.17 Cloud control systems

We also adopt this name for a new kind of control perspective [55]. As we know, most of complex systems cannot be controlled properly, since we know less about them and lack powerful tools to control this kind of systems. The development of new technologies, especially, magical innovations in software and hardware, provides necessary conditions to speed-up computability and distribute computing. Together with the development of networks, cloud computing has come into our life. Nowadays, cloud computing has exceeded the original product concept, and more often, it means a service. General speaking, it is a byproduct of the ease-of-access to remote computing sites provided by the Internet [58] which describes a new supplement, consumption, and delivery model for IT services based on Internet protocols. It often includes the provisioning of dynamically scalable and often virtualized resources [56,57]. In practical systems, the cloud control system provides a shared pool of configurable resources, which include computation, software, data access, and storage services, etc., wherein end-users consume power without requiring to know the physical location and configuration of the service provider [60]. The computers and other devices in such a system are as a utility over the network which can share resources, software and information, etc., so that the users can access and use them through a web browser as if the programs were installed locally on their own computers [59].

Since cloud control systems combine the merits of cloud computing, advanced theory of NCSs and other recently developed related results, they will exhibit unbelievable potential applications in the industry and other related areas. Fig. 4.17 shows the structure of cloud control systems.

In a practical cloud control system, due to the scale of the system, which includes ubiquitous information-sensing mobile devices, aerial sen-

Figure 4.18 Big Data

sory technologies (remote sensing), software logs, cameras, microphones, radio-frequency identification readers, wireless sensor networks are increasingly being gathered [61] and the captured data usually grows in size. To name this kind of data, a new concept – "Big data" – has emerged, which is a collection of data sets [62]. However, the set is so large and complex that it is difficult to be processed with the help of any on-hand traditional database management or processing tools. From the stochastic point of view, when there are sufficient data, some useful deterministic conclusions can be achieved based on the law of large numbers, while single or a few data are quite random. However, many challenges still exist, which include capturing, storage, search, sharing, transfer, analysis, and visualization of big data as shown in Fig. 4.18.

In cloud control systems, big data will be sent out to the systems of cloud computing, and after the data is processed, control signals, such as schedule schemes, predictive control sequences and other useful information, will be generated instantly for the cloud control systems. Cloud control systems will provide us powerful tools to control the complex systems which we could not imagine before.

4.3.7 A Rudiment of Cloud Control Systems

In this section, we present a rudiment of classical cloud control systems to supply a plethora for researchers who are interested in this topic. The basic assumptions are as follows:

Assumption 4.1. A broadcast domain is involved, in which all nodes can reach each other by broadcasting at the data link layer.

Assumption 4.2. All nodes in the broadcast domain are intelligent enough to undertake the cloud control task, but their computation abilities are assumed to be equal, and the available computation resources change unpredictably.

Assumption 4.3. The network delivery is not ideal, bounded time delays and data delivery dropout could occur during any transmission.

Assumption 4.4. The network delivery time delays and dropout statistic between any two particular nodes could be obtained by the same nodes in some ways.

Assumption 4.5. The controlled object, that is, some physical plant, is located at a node P. The cloud control task starts from a controller located at a node CT, and in the meantime the node CT is also the cloud control task management node.

Though the controlled plant P could be any type of equipment, it is assumed to be a linear discrete-time dynamical system to make the stability of the proposed cloud control system easily obtainable by employing the existing NCS knowledge. The actuator and sensor work in time-driven mode with same step time T. The proposed rudiment of the cloud control systems is divided into two phases: initial (NCS) and cloud control phases.

The cloud control task starts from the initial phase when the control system is initialized as an NCS, which involves the controller CT and the controlled plant P. The networked control scheme applied by controller CT could be arbitrary for this cloud control rudiment, but it is assumed here to be the same as mentioned earlier, since this method is easily extendable from NCSs to cloud control systems and the stability remains unchanged after the extension. The controller CT receives measured outputs from the plant P, and generates the manipulated variable sets according to the model based predictive control algorithm. A compensator is set at the plant node P to cancel the network-induced time delays. During the initial phase, the cloud control system only involves two nodes in the predefined broadcast domain, and it is actually an NCS.

After the initial phase is well maintained, the system switches to the second phase, the cloud control phase. In this phase, the node CT is not only a controller but also a task management node. Node CT starts broadcasting a requirement over the domain at a predefined frequency. All nodes in this broadcast domain can receive this requirement. The requirement has to include, but is not limited to, the following information:

1. The IP address of the plant node P;
2. The control algorithm applied and its corresponding parameters;
3. The mathematical model of the plant;
4. Computation burden estimated.

During the initial period of the cloud control phase, the CT node undertakes two tasks, one of them is to apply the predefined control algorithm, generate the manipulated variables, and hence send encapsulated predictive control signals to the node P, while the other task is to keep broadcasting the requirement over the domain. It should be noticed that the first task (control task, in short) is not permanent for node CT, the reason of broadcasting the requirement for node CT is to find "suitable successors" to undertake the control task instead of itself.

Once a node, for example, let it be C_i, has enough computation resource or it is powerful enough to undertake both its current local computation and the potential cloud control task after it receives the requirement from the node CT, it responds to the node CT with an acknowledgment. In a similar fashion, the acknowledgment includes, but is not limit to, the following information:

1. The network delivery time delays and data dropout statistic between node C_i and node P (bounded time delay N_i and the maximum consecutive data package dropout D_i, for example);
2. Computation capability available (may be quantized to positive numbers CCA_i).

After the acknowledgment of C_i arrives at the node CT, the node CT will evaluate the superiority of node C_i. The superiority mentioned here is assumed to be a weighed sum function of the delivery time delay statistic N_i, the maximum consecutive data package dropout D_i, and computation capability available CCA_i, that is,

$$S_i = \alpha f(N_i) + \beta g(D_i) + \gamma h(CCA_i), \tag{4.25}$$

where S_i denotes the superiority of node C_i, functions $f(\cdot)$ and $g(\cdot)$ are monotonically decreasing, and function $h(\cdot)$ is monotonically increasing. The positive weight coefficients α, β and γ should be chosen according to the engineering practice. The greater the superiority, the more suitable the node C_i. In the meantime, the node CT also has knowledge of its own superiority, S_{CT}. The node CT maintains a list of willing nodes whose

Table 4.1 The list of willing nodes.

Node	IP Address	Superiority	Ranking
C_{i1}	Add_{i1}	S_{i1}	1
C_{i2}	Add_{i2}	S_{i2}	2
C_{i3}	Add_{i3}	S_{i3}	3
C_{i4}	Add_{i4}	S_{i4}	4
⋮	⋮	⋮	⋮
CT	Add_{CT}	S_{CT}	k_{CT}

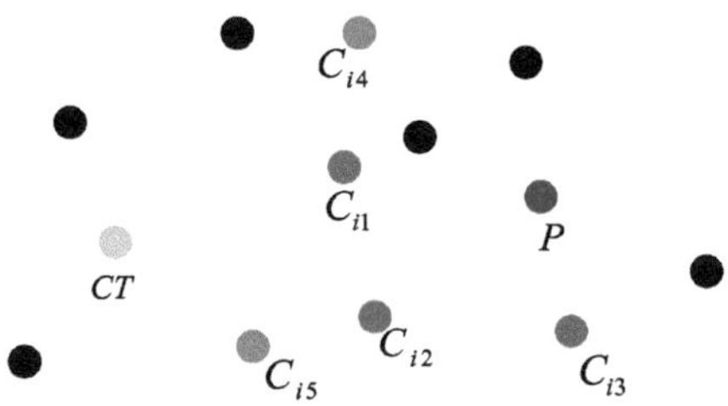

Figure 4.19 The schematic diagram of a cloud control system

superiorities are greater than its own, i.e., any node j that sent acknowledgment to node CT if $S_j > S_{CT}$. Those nodes whose superiorities are not greater S_{CT} are ignored. The list of willing nodes is shown in Table 4.1, where the length of the list, k_{CT}, is dynamic, it depends on how many available willing nodes exist. All willing nodes are ranked in the list, and the node CT is the last since its superiority is the smallest among all willing nodes. If there are no willing nodes available, node CT could be the first one, and the length of the list is one. It is advised to set a maximum list length, k_{MAX}, to avoid unnecessary memory space cost and computation burden. If there are too many willing nodes available, node CT only contains the first $k_{MAX} - 1$ nodes, and abandons the others.

The next step for node CT is to pick up some successors from the candidates listed. The number of successors, l_{MAX}, is predefined. Node CT chooses the first l_{MAX} nodes as the cloud controllers for node P. If the number of the listed candidates is smaller than l_{MAX}, node CT can use all available willing nodes as the cloud controllers. As shown in Fig. 4.19, the yellow node is node CT and the blue one is node P. And in this snap, there are five available willing nodes, C_{i1}–C_{i5}, and the l_{MAX} is defined as 3, so we use red color to denote the cloud control nodes C_{i1}–C_{i3}, and green color to denote the inactive willing nodes C_{i4}–C_{i5}.

After the cloud controllers are confirmed, node CT delivers a control statement to the cloud controllers, this statement includes the controller states in greater detail as follows:

1. The mathematical model of the plant;
2. The estimate of plant states and manipulated variables up to the current time instant;
3. The controller parameters.

As mentioned above, the Kalman filter based predictive control method introduced earlier is applied in this rudiment, and the second item in the statement could be as follows:

$$\{\hat{x}_0, \hat{x}_1, \ldots, \hat{x}_t\}, \tag{4.26}$$

$$\{u_0, u_1, \ldots, u_t\}, \tag{4.27}$$

where t denotes the current time instant. At the same time, node CT sends a copy the current cloud control nodes' list to the plant node P, and after node P receives the list, it starts sending its measurements and also the historical measurements to the cloud control nodes. Once any cloud control node receives both the statement from node CT and the measurements from node P, it can apply the appointed control method and send the manipulated variable packets to node P.

To maintain the cloud control system working well, all active cloud control nodes send feedback to node CT at every sampling time instant, if node CT has not received the feedback of a particular cloud control node for a predefined duration, this cloud control node should be removed from the list, and node CT will indicate the first node among all inactive willing nodes to substitute the position of the removed node. At the same time, node P is informed about this substitution. The management of the proposed cloud control system is a dynamic process, node CT keeps on seeking willing nodes, removing inactive nodes, and exchanging information of current cloud control nodes to node P. Node P can receive multiple manipulated variables packets from different cloud control nodes, and the compensator picks up the newest one as the actual input for the controlled object. The classical control flow diagram of cloud control system is shown in Fig. 4.20, in which C_2, C_3 and C_6 are the cloud control nodes, while C_6 is the active cloud control node at the current stage.

4.3.8 Cooperative Cloud Control

In a practical cloud control system, most of the big cloud infrastructure/services are often provided by the enterprises, which include Amazon, Sales-

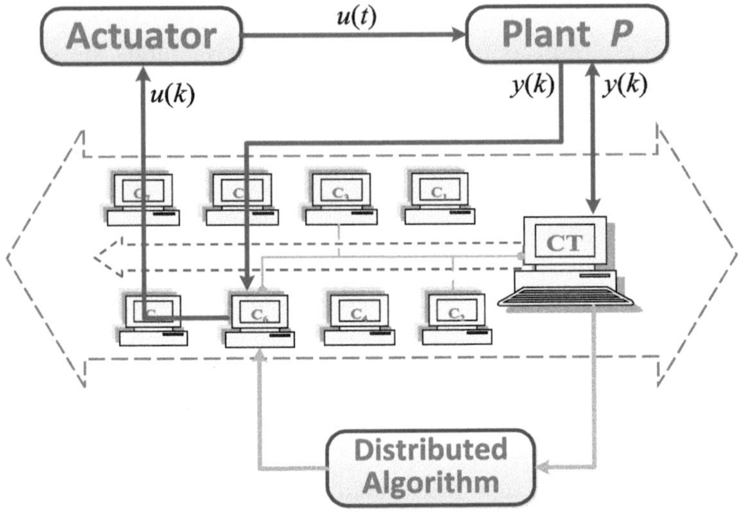

Figure 4.20 A representative control flow diagram of a cloud control system

force, Google, Microsoft, etc. In general, the end-users have to pay a lot of money for that. Most end-users are more inclined to get low-cost services, which is one of the original intentions of the cloud control industry. However, in such systems, a single cloud controller lacks enough computing resources or power due to limited finances. This thus calls for the cooperative cloud control theory, which retains the advantages of the cloud control with a lower cost.

The principle of a cooperative cloud control system is similar to that of the classic cloud control, which, however, has the following difference: the control task will be completed by several (two or more) cloud controllers in a form of cooperation, as shown in Fig. 4.21. Generally speaking, CT is not only a controller but also a task management server, while C_1–C_8 are the cloud controllers with the same step time, all of which have the same definition as in the cloud control system in Fig. 4.20. At the beginning of the task, the CT node has to select several suitable cloud controllers from the candidates listed according to the scale of the task, for example, C_2, C_3, C_6, and then, using the distributed algorithm, CT will assign part of the total task for each cloud controller based on its current computation resource. At the same time, node CT will also send a copy of the current cloud control nodes list to the plant node P. Then, the plant P starts to

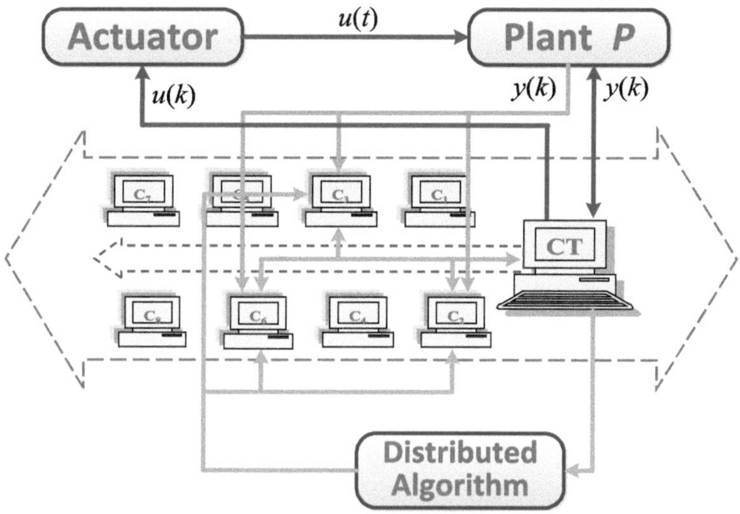

Figure 4.21 A diagram of cooperative cloud control system

send its and also historical measurements to the cloud control nodes, i.e., C_2, C_3, C_6. After that, at each step time, every cloud controller will send feedback result to CT, and at the same time, CT will give the newest control signal according to the current task assignment algorithm, and then send to the actuator. It's worth noting that, at every sampling instant, both the active cloud controllers and candidates need to send their status to server CT, which includes the current computation resources. Then, CT will make a new candidate list. And to make sure the cloud control system works well, the CT will also reassign the task according to the newest state of the cloud nodes' resources in the next sampling time. For the cooperative cloud control system, other technical details are similar to those of a typical cloud control system as described in the above subsection, and here we omit them.

The rudiment of the cloud control systems proposed above supplies a simple platform, in which researchers can theoretically develop or test their new algorithms for cloud control systems. Although we did not consider many potential features of cloud control systems in this rudiment, it indeed involves the most basic principle of the cloud control systems, taking advantage of the available computation resources and the superiorities in topology of all possible nodes to fulfill the given control task.

4.4 NOTES

A brief overview of NCSs has been given, and new trends in NCSs have also been pointed out, i.e., with the development of cloud computing and the processing techniques of big data, the dawn of cloud control systems will soon emerge. A preliminary structure and algorithm are proposed. Some new results on cloud control systems will be found in our future publications. We believe that there will be more interesting and important results, which will be produced in this new research area.

REFERENCES

[1] P. Ogren, E. Fiorelli, N.E. Leonard, Cooperative control of mobile sensor networks: adaptive gradient climbing in a distributed environment, IEEE Trans. Autom. Control 49 (8) (Aug. 2004) 1292–1302.
[2] C. Meng, T. Wang, W. Chou, S. Luan, Y. Zhang, Z. Tian, Remote surgery case: robot-assisted teleneurosurgery, in: IEEE Int. Conf. Robot. and Auto., vol. 1, ICRA'04, Apr. 2004, pp. 819–823.
[3] J.P. Hespanha, M.L. McLaughlin, G. Sukhatme, Haptic collaboration over the Internet, in: Proc. 5th Phantom Users Group Workshop, Oct. 2000.
[4] K. Hikichi, H. Morino, I. Arimoto, K. Sezaki, Y. Yasuda, The evaluation of delay jitter for haptics collaboration over the Internet, in: Proc. IEEE Global Telecomm. Conf., vol. 2, GLOBECOM, Nov. 2002, pp. 1492–1496.
[5] S. Shirmohammadi, N.H. Woo, Evaluating decorators for haptic collaboration over the Internet, in: Proc. 3rd IEEE Int. Workshop Haptic, Audio and Visual Env. Applic., Oct. 2004, pp. 105–109.
[6] P. Seiler, R. Sengupta, Analysis of communication losses in vehicle control problems, in: Proc. 2001 Amer. Contr. Conf., vol. 2, 2001, pp. 1491–1496.
[7] P. Seiler, R. Sengupta, An $\mathcal{H}_\infty$ approach to networked control, IEEE Trans. Autom. Control 50 (3) (2005) 356–364.
[8] M.S. Mahmoud, Control and Estimation Methods over Communication Networks, Springer-Verlag, UK, 2014.
[9] J.M.V. Grzybowski, M. Rafikov, J.M. Balthazar, Synchronization of the unified chaotic system and application in secure communication, Commun. Nonlinear Sci. Numer. Simul. 14 (6) (2009) 2793–2806.
[10] M. Rafikov, J.M. Balthazar, On control and synchronization in chaotic and hyperchaotic systems via linear feedback control, Commun. Nonlinear Sci. Numer. Simul. 14 (7) (2008) 1246–1255.
[11] N.N.P. Mahalik, K.K. Kim, A prototype for hardware-in-the-loop simulation of a distributed control architecture, IEEE Trans. Syst. Man Cybern., Part C, Appl. Rev. 38 (2) (2008) 189–200.
[12] R.A. Gupta, M.Y. Chow, Networked control system: overview and research trends, IEEE Trans. Ind. Electron. 57 (7) (2010).
[13] S.H. Yang, X. Chen, L.S. Tan, L. Yang, Time delay and data loss compensation for internet-based process control systems, Trans. Inst. Meas. Control 27 (2) (2005) 103–118.

[14] M.S. Mahmoud, H.N. Nounou, Y. Xia, Robust dissipative control for Internet-based switching systems, J. Franklin Inst. 347 (1) (2010) 154–172.
[15] M.S. Mahmoud, A.M. Memon, Aperiodic triggering mechanisms for networked control systems, Inf. Sci. 296 (2015) 282–306.
[16] Y. Xia, Cloud control systems, IEEE/CAA J. Autom. Sin. 2 (2) (Apr. 2015).
[17] T. Hegazy, M. Hafeeda, Industrial automation as a cloud service, IEEE Trans. Parallel Distrib. Syst. 26 (10) (Oct. 2015).
[18] P. Park, J. Araujo, K.H. Johansson, Wireless networked control system co-design, in: Proc. of 2011 IEEE Int. Conf. on Networking, Sensing and Control, ICNSC, Delft, the Netherlands, 2011, pp. 486–491.
[19] P. Park, P.D. Marco, C. Fischione, K.H. Johansson, Delay distribution analysis of wireless personal area networks, in: Proc. of IEEE Conf. on Decision and Control, Maui, HI, 2012, pp. 5864–5869.
[20] P. Park, P.D. Marco, C. Fischione, K.H. Johansson, Modeling and optimization of the IEEE 802.15.4 protocol for reliable and timely communications, IEEE Trans. Parallel Distrib. Syst. 24 (Mar. 2013) 550–564.
[21] P. Park, C. Fischione, K.H. Johansson, Modeling and stability analysis of hybrid multiple access in the IEEE 802.15.4 protocol, ACM Trans. Sens. Netw. 9 (2) (2013).
[22] S. Gallagher, VMware Private Cloud Computing with VCloud Director, John Wiley & Sons, 2013.
[23] VMware, Getting Started with VMCI, 2007.
[24] VMware, VMware vCloud Suite Datasheet, 2012.
[25] R. Braden, D. Clark, S. Shenker, Integrated services in the internet architecture: an overview, in: RFC1633, June 1994.
[26] S. Blake, et al., An architecture for differentiated services, in: RFC2475, Dec. 1998.
[27] P. Ferguson, G. Houston, Quality of Service: Delivering QoS on the Internet and in Corporate Networks, John Wiley & Sons, 1998.
[28] Z. Wang, Internet QoS: Architectures and Mechanisms for Quality of Service, Morgan Kaufmann, 2001.
[29] C.S. Chang, Performance Guarantees in Communication Networks, Springer-Verlag, New York, 2000.
[30] P. Barham, B. Dragovic, K. Fraser, S. Hand, T. Harris, A. Ho, R. Neugebauer, I. Pratt, A. Warfield, Xen and the art of virtualization, in: Proceedings of the 19th ACM Symposium on Operating Systems Principles, SOSP 2003, Bolton Landing, NY, USA, 2003, p. 177.
[31] M. Armbrust, A. Fox, R. Griffith, A.D. Joseph, R. Katz, A. Konwinski, G. Lee, D. Patterson, A. Rabkin, I. Stoica, M. Zaharia, A view of cloud computing, Commun. ACM 53 (4) (2009) 50–58.
[32] Y. Xia, M. Fu, P. Shi, Analysis and Synthesis of Dynamic Systems with Time-Delays, Springer, 2009.
[33] Y. Xia, M. Fu, G.-P. Liu, Analysis and Synthesis of Networked Control Systems, Springer, 2011.
[34] H.S. Park, Y.H. Kim, D.S. Kim, W.H. Kwon, A scheduling method for network based control systems, IEEE Trans. Control Syst. Technol. 10 (3) (2002) 318–330.
[35] P.V. Zhivoglyadov, R.H. Middleton, Networked control design for linear systems, Automatica 39 (4) (2003) 743–750.
[36] D. Yue, Q.L. Han, C. Peng, State feedback controller design of networked control systems, IEEE Trans. Circuits Syst. II, Express Briefs 51 (11) (2004) 640–644.

[37] H. Gao, T. Chen, A new approach to quantized feedback control systems, Automatica 44 (2) (2008) 534–542.
[38] H. Gao, T. Chen, Network-based H_∞ output tracking control, IEEE Trans. Autom. Control 53 (3) (2008) 655–667.
[39] Y. Xia, G.P. Liu, M. Fu, D. Rees, Predictive control of networked systems with random delay and data dropout, IET Control Theory Appl. 3 (11) (2008) 1476–1486.
[40] Y. Xia, W. Xie, B. Liu, X. Wang, Data-driven predictive control for networked control systems, Inf. Sci. 235 (20) (2013) 45–54.
[41] R.O. Saber, J.A. Fax, R.M. Murray, Consensus and cooperation in networked multi-agent systems, Proc. IEEE 95 (1) (2007) 215–233.
[42] G. Yan, L. Wang, G. Xie, B. Wu, Consensus of multi-agent systems based on sampled-data control, Int. J. Control 82 (12) (2009) 2193–2205.
[43] D. Meng, Y. Jia, Finite-time consensus for multi-agent systems via terminal feedback iterative learning, IET Control Theory Appl. 5 (8) (2011) 2098–2110.
[44] F. Xiao, L. Wang, J. Chen, Y. Gao, Finite-time formation control for multi-agent systems, Automatica 45 (11) (2011) 2605–2611.
[45] Y. Xia, M. Fu, Compound Control Methodology for Flight Vehicles, Springer, 2013.
[46] D.J. Park, P.D. Lelima, G.J. Toussaint, G. York, Cooperative control of UAVs for localization of intermittently emitting mobile targets, IEEE Trans. Syst. Man Cybern. 39 (4) (2009) 959–970.
[47] J.M. Bradley, E.M. Atkins, Toward continuous state–space regulation of coupled cyber-physical systems, Proc. IEEE 100 (1) (2012) 60–74.
[48] C. Joslyn, L. Rocha, Towards semiotic agent-based models of socio-technical organizations, in: Proc. AI, Simulation and Planning in High Autonomy Systems Conference, AIS 2000, Tucson, Arizona, 2000, pp. 70–79.
[49] J. Kwapień, S. Drożdż, Physical approach to complex systems, Phys. Rep. 515 (3–4) (2012) 115–226.
[50] J.C. Willems, In control, almost from the beginning until the day after tomorrow, Eur. J. Control 13 (1) (2007) 71–81.
[51] D.J. Hill, Advances in stability theory for complex systems and networks, in: Proceedings of the 27th Chinese Control Conference, Kunming, China, 2008, pp. 13–17.
[52] C.A. Desoer, M. Vidyasagar, Feedback Systems: Input–Output Properties, Academic Press, New York, 1975.
[53] X. Wang, G. Chen, Synchronization in scale-free dynamical networks: robustness and fragility, IEEE Trans. Circuits Syst. I, Fundam. Theory Appl. 49 (1) (2002) 54–62.
[54] C.W. Wu, Synchronization in Complex Networks of Nonlinear Dynamical Systems, World Scientific, New Jersey, 2007.
[55] Y. Xia, From networked control systems to cloud control systems, in: Proceedings of the 31th Chinese Control Conference, Hefei, China, 2012, pp. 5878–5883.
[56] G. Says, Cloud computing will be as influential as e-business, http://www.gartner.com/newsroom/id/707508, 22 Aug. 2010.
[57] G. Galen, What cloud computing really means, InfoWorld, http://www.infoworld.com/d/cloud-computing/what-cloud-computing-really-means-031, 6 June 2009.
[58] The economist, cloud computing: clash of the clouds, http://www.economist.com/node/14637206, 3 Nov. 2009.
[59] Cloud computing, http://www.cloudcomputingdefined.com/, 17 July 2010.
[60] Cloud computing, http://en.wikipedia.org/wiki/Cloud_computing, 27 Nov. 2011.

[61] T. Segaran, J. Hammerbacher, Beautiful Data: The Stories Behind Elegant Data Solutions, O'Reilly Media, 2009.
[62] T. White, Hadoop: The Definitive Guide, O'Reilly Media, 2012.
[63] S. Yang, X. Chen, Dealing with time delay and data loss for internet-based control systems, IFAC Proc. 36 (19) (Sept. 2003) 117–122.
[64] M.S. Mahmoud, Y. Xia, Applied Control Systems Design, Springer, London, 2012.

CHAPTER 5

Secure Control Design Techniques

5.1 INTRODUCTION

Cyberphysical systems (CPS) integrate computing and communication capabilities with monitoring and control of entities in the physical world. These systems are usually composed by a set of networked agents, including: sensors, actuators, control processing units, and communication devices; see Fig. 5.1.

While some forms of CPS are already in use, the widespread growth of wireless embedded sensors and actuators is creating several new applications – in areas such as medical devices, autonomous vehicles, and smart structures – and increasing the role of existing ones, such as Supervisory Control and Data Acquisition (SCADA) systems. Many of these applications are *safety-critical*: their failure can cause irreparable harm to the physical system being controlled and to people who depend on it. SCADA systems, in particular, perform vital functions in national critical infrastructures, such as electric power distribution, oil and natural gas industry, water and waste-water distribution systems, and transportation systems. The disruption of these control systems could have a significant impact on public health, safety, and lead to large economic losses. While most of the effort

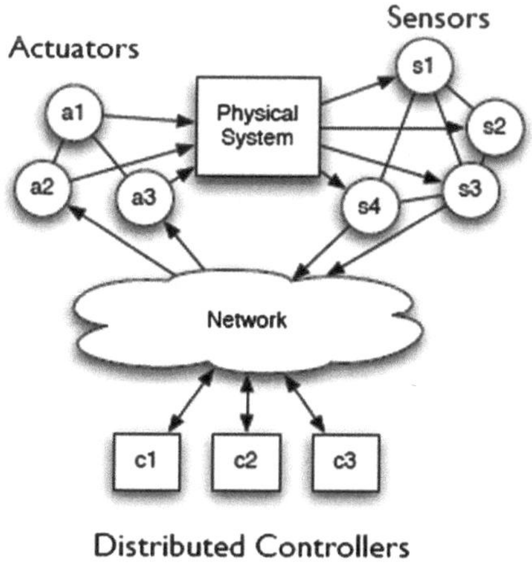

Figure 5.1 An architecture of cyberphysical systems

Networked Control Systems
https://doi.org/10.1016/B978-0-12-816119-7.00013-7

for protecting CPS systems (and SCADA, in particular) has been done in reliability (protection against random failures), there is an urgent growing concern for the protection against malicious cyberattacks [1,2].

In this chapter we study the problem of secure control. We first characterize the properties required from a secure control system and the possible threats. Then we analyze what elements from

1. information security,
2. sensor network security, and
3. control theory

can be used to solve our problems. It is concluded that while these fields can give necessary mechanisms for the security of control systems, they alone are not sufficient for the security of CPS.

Before proceeding further, the following definition is provided:

Definition 5.1. The **secure control** problem refers to any of the algorithms and architectures designed to **survive** deception and denial-of-service (DoS) attacks against CPS under a well-defined adversary model and trust assumptions.

We recall that computer and sensor network security has focused on prevention mechanisms, but not on addressing how a control system can continue to function under attack. Control systems, on the other hand, have strong results on robust and fault-tolerant algorithms against well-defined uncertainties or faults, but there is very little work accounting for faults caused by a malicious adversary.

5.1.1 Security Goals

In this section we study how the traditional security goals of integrity, availability, and confidentiality can be interpreted for CPS.

Integrity refers to the trustworthiness of data or resources [3]. A lack of integrity results in **deception**, i.e., when an authorized party receives false data and believes it to be true [4]. Integrity in CPS can therefore be viewed as the ability to maintain the operational goals by preventing, detecting, or surviving deception attacks in the information sent and received by the sensors, controllers, and actuators.

Availability refers to the ability of a system be accessible and usable upon demand [4]. Lack of availability results in **denial of service (DoS)** [5]. While in most computer systems a temporary DoS attack may not compromise their services (a system may operate normally when it becomes available again), the strong real-time constraints of many cyber-physical systems introduce new challenges. For example, if a critical physical process is unstable in open loop, a DoS attack on the sensor measurements may render the controller unable to prevent irreparable damages to the system and entities around it.

The goal of availability in CPS is therefore to maintain the operational goals by preventing or surviving DoS attacks to the information collected by the sensor networks, the commands given by the controllers, and the physical actions taken by the actuators.

Confidentiality refers to the ability of keeping information secret from unauthorized users. A lack of confidentiality results in disclosure, a circumstance or event whereby an entity gains access to data for which it is not authorized [4].

The use of CPS in commercial applications has the potential risk of violating a users' **privacy**: even apparently innocuous information such as humidity measurements may reveal sensitive personal information [6]. Additionally, CPS used for medical systems must abide with federal regulations such as the Health Insurance Portability and Accountability Act (HIPAA), which mandates the protection of a patient's data.

Confidentiality in CPS must prevent an adversary from inferring the state of the physical system by eavesdropping on the communication channels between the sensors and the controller, and between the controller and the actuator.

While confidentiality is an important property in CPS, we believe that the inclusion of a physical system and real-time automated decision making does not affect current research in mechanisms for enforcing confidentiality. Therefore, in the remaining of this chapter we focus only on deception and DoS attacks.

5.1.2 Workflow Within CPS

A general workflow of CPS can be categorized into four main steps:

Monitoring of physical processes and environment is a fundamental function of CPS. It is also used to give feedback on any past actions which are taken by the CPS and ensure correct operations in the future. The physical process is to achieve the original physical goal of the CPS.

Networking step deals with the data aggregation diffusion. There can be much more than one sensor in a CPS. These sensors can generate a lot of data in real-time, which is to be aggregated or diffused for

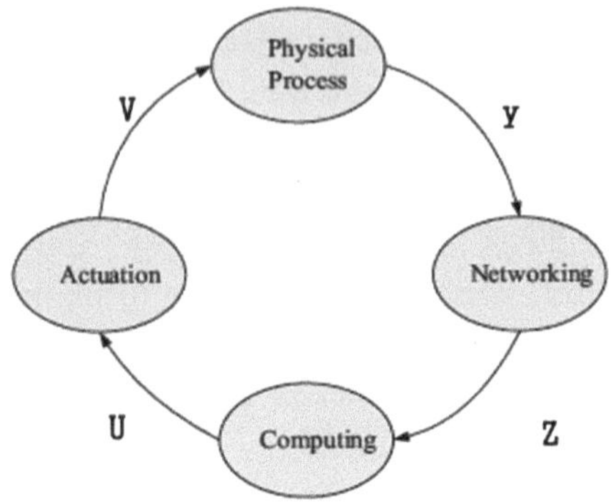

Figure 5.2 Schematic of CPS

analyzers to process further. At the same time, different applications need to be interacted via network communication.

Computing step is for reasoning and analyzing the data collected during monitoring to check whether the physical process satisfies certain predefined criteria. If the criteria are not satisfied, corrective actions are proposed to be executed in order to ensure criteria satisfaction. For instance, a datacenter CPS can have a model to predict the temperature rise with respect to various scheduling algorithms, which can be used to determine future operations.

Actuation step executes the actions determined during the computing phase. It can actuate various forms of actions such as correcting the cyberbehavior of the CPS, changing the physical process.

Fig. 5.2 shows a general workflow of a CPS. Let y represent the data acquisition from sensors, z the physical data aggregation in the network, u the valid computed result of the physical system states, which could advise controller to select valid commands, and v the control commands sent to the actuators. A controller can usually be divided in two components: an *estimation algorithm* to track the state of the physical system given y, and the *control algorithm* which selects a control command u given the current estimate.

5.1.3 Summary of Attacks

As Fig. 5.3 shows, the types of attacks to CPS are summarized as follows:

(1) Eavesdropping. It refers to the attack that adversary can intercept any information communicated by the system [13]. It is called a passive attack since the attacker does not interfere with the working of the system and simply observes its operation. CPS is particularly sus-

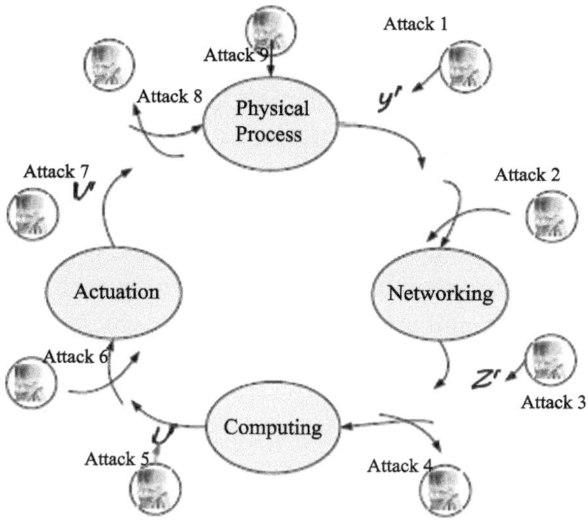

Figure 5.3 Possible attacks in CPS

ceptible to eavesdropping through traffic analysis such as intercepting the monitoring data transferred in sensor networks collected through monitoring. Eavesdropping also violates user's privacy such as a patient's personal health status data transferred in the system. In Fig. 5.3, attack 4 can represent the eavesdropping attacks on data aggregation processes; attack 8 can represent the tapping on controller demands.

(2) Compromised-Key Attack. A key is a secret code which is necessary to interpret secure information. Once an attacker obtains a key, this key is considered compromised [14]. An attacker can gain access to a secured communication without the perception of the sender or receiver by using the compromised key. The attacker can decrypt or modify data by using the compromised key, and try to use this key to compute additional keys, which could allow the attacker access to other secured communications or resources. Actually, it is possible for an attacker to obtain a key, even though the process may be difficult and resource intensive. For example, the attacker could capture the sensors to execute a reverse engineering job in order to figure out the keys inside, which could be represented in attack 9 shown in Fig. 5.3, or the attacker could pretend to be a valid sensor node to cheat to agree on keys with other sensors.

(3) Man-in-the-Middle Attack. In such an attack, false messages are sent to the operator, and can take the form of a false negative or a false positive. This may cause the operator to take an action, such as flipping a breaker, when it is not required, or it may cause the operator to think everything is fine and not take an action when an action is required. For example, in Fig. 5.3, attack 7 shows that the adversary sends V' to indicate a system change, however, V' is not the real actuation command. When the operator follows normal procedures and attempts to correct the problem, the operator's action could cause an undesirable event. There are numerous variations of the modification and replay of control data which could impact the operations of the system. Attacks 1, 3 and 5 can also represent this kind of attack.

(4) Denial-of-Service Attack. Such an attack is one of the network attacks that prevent the legitimate traffic or requests for network resources from being processed or responded by the system. This type of attack usually transmits a huge amount of data to the network to overwhelm handling the data so that normal services cannot be provided. The denial-of-service attack prevents normal work or use of the system. After gaining access to the network of cyberphysical systems, the attacker can always do any of the following:

- *Flood a controller or the entire sensor network* with traffic until a shutdown occurs due to the overload by sending invalid data to controller or system networks, which causes abnormal termination or behavior of the services.
- *Block traffic*, which results in a loss of access to network resources by authorized elements in the system.

For instance, in Fig. 5.3, attack 2 can represent a situation where adversaries flood the entire sensor network with a large amount of jamming data to block the normal network traffic, while attack 6 can represent adversaries sending a huge amount of invalid data to actuators to cause abnormal termination of actuation process.

5.1.4 Robust Networked Control Systems

We begin by considering a scenario where the system and remote estimator communicate over a communication network. The estimator's goal is to generate recursive state estimates based on measurements sent by the sensor. Under perfect communication, there is no loss of data and packets arrive at the estimator instantaneously. Under perfect integrity, the measured data is not compromised. For this ideal case, Kalman filter is the optimal estimator.

Let us recall the basic Kalman filter in systems' theory for a discrete-time linear dynamical system

$$\begin{aligned} x(k+1) &= Ax(k) + Bu(k) + w(k), \\ y(k) &= Cx(k) + v(k), \end{aligned} \tag{5.1}$$

where $x(k) \in \mathbb{R}^n$, $u(k) \in \mathbb{R}^p$, $y(k) \in \mathbb{R}^m$ are the state, input and measurement vectors with $w(k) \in \mathbb{R}^n$ and $v(k) \in \mathbb{R}^m$ denoting the state and measurement noise, respectively. Here, x_0 is the initial state, a Gaussian random vector with zero mean and covariance Σ_0, while $w(k)$ and $v(k)$ are independent Gaussian random vectors with zero mean and covariances $Q \geq 0$ and $R > 0$, respectively. It is known that under the assumption that $(A,\ C)$ is detectable and $(A,\ Q)$ is stabilizable, the estimation error covariance of the Kalman filter converges to a unique steady state value from any initial condition.

The optimal estimate of $x(k) \in \mathbb{N}$ and the error covariance matrix given the past measurements $\mathbb{Y}(k) = \{y_0, \ldots, y_{k-1}\}$ are denoted by $\hat{x}(k|k-1) = \mathbb{E}[x(k)|\mathbb{Y}_k$ and $P_{k|k-1} = \mathbb{E}[(x(k) - \hat{x}_{k|k-1})(x(k) - \hat{x}_{k|k-1})^T$, respectively. Starting with $x(0|-1) = 0$ and $P(0|-1) = \Sigma_0$, the update equations for the basic Kalman filter can be computed as

$$\begin{aligned} \hat{x}(k+1|k) &= A\hat{x}(k|k), \\ P(k+1|k+1) &= AP(k|k)A^T + Q, \\ \hat{x}(k+1|k+1) &= \hat{x}(k|k) + K(k+1)[y(k+1) \\ &\quad - C\hat{x}(k+1|k)], \end{aligned} \tag{5.2}$$

$$P(k+1|k+1) = (I - K(k+1)C)P(k+1|k), \tag{5.3}$$

where $K(k+1) = P(k+1|k)C^T[CP(k+1|k)C^T + R]^{-1}$ is the Kalman gain matrix.

5.1.5 Resilient Networked Systems Under Attacks

In what follows, we formulate and analyze the problem of secure control for discrete-time linear dynamical systems. Our work is based on two ideas:

1. The introduction of safety-constraints as one of the top security requirements of a control system, and
2. The introduction of new adversary models.

The goal in our model is to minimize a performance function such that a safety specification is satisfied with high probability and power limitations are obeyed in expectation when the sensor and control packets can be dropped by a random or a resource-constrained attacker.

We consider a linear time invariant stochastic system with measurement and control packets subject to DoS attacks $(\gamma(k), \sigma(k))$ over $k = 0, \dots, N-1$:

$$x(k+1) = Ax(k) + Bu^a(k) + w(k), \tag{5.4}$$

$$u^a(k) = \sigma(k)u(k), \quad \sigma(k) \in \{0, 1\}, \tag{5.5}$$

$$y^a(k) = \gamma(k)Cx(k) + v(k), \quad \gamma(k) \in \{0, 1\}, \tag{5.6}$$

where $x(k) \in \mathbb{R}^n$, $u(k) \in \mathbb{R}^m$ are the state and control input, respectively, $w(k) \in \mathbb{R}^n$ is independent, Gaussian distributed noise with zero mean and covariance W (denoted as $w(k) \approx \mathcal{N}(0, W)$), $v(k) \in \mathbb{R}^p$ is independent, Gaussian distributed noise with zero mean and covariance V (denoted as $v(k) \approx \mathcal{N}(0, V)$) $x(0) \approx \mathcal{N}(\bar{x}, P(0))$ is the initial state and $\{\gamma(k)\}$ (resp. $\{\sigma(k)\}$) is the sensor (resp. the actuator) attack sequence. Also, $x(0)$, $v(k)$ and $w(k)$ are uncorrelated. The available output (resp. available control input) is denoted by $y^a(k)$ (resp. $u^a(k)$) after a DoS attack on the measurement (resp. control) packet.

Employing an acknowledgment based communication protocol such as TCP, the information set available at time k is

$$\mathcal{I}_k = \{y^a(0), \dots, y^a(k), \gamma_o(k), \sigma_o(k-1)\}$$

where $\gamma_i(j) = (\gamma_i, \dots, \gamma_j)$, $\sigma_i(j) = (\sigma_i, \dots, \sigma_j)$ and we denote $u_o^{N-1} = (u(0), \dots, u(N-1)$.

Remark 5.1. *In view of (5.4), we observe that the controller receives perfect output information $y(k)$ corrupted by noise $v(k)$ when $\gamma = 1$. However, when $\gamma = 0$, no information is received. In addition, there are no restrictions on the DoS attack actions except that $\gamma(k) \in \{0, 1\}$, $\sigma(k) \in \{0, 1\}$, $k = 0, \dots, N-1$.*

Given the above constraints, our objective is to construct a feedback control law $u(k) = \eta(k, \mathcal{I}_k)$ such that for system (5.4)–(5.6), the following finite-horizon objective functional is minimized

$$\begin{aligned} J(N, \bar{x}, P(0), u_o^{N-1}) = \mathbb{E}\Bigg[& y^T(N) Q_y y(N) \\ & + \sum_{m=0}^{N-1} \begin{pmatrix} y(m) \\ u(m) \end{pmatrix}^T \begin{pmatrix} I_p & 0 \\ 0 & \gamma(m) I_m \end{pmatrix} Q \begin{pmatrix} y(m) \\ u(m) \end{pmatrix} \\ & \mid \bar{x}, P(0), u_o^{N-1} \Bigg], \end{aligned} \tag{5.7}$$

whenever $Q_y > 0$:

$$0 \leq Q = \begin{pmatrix} Q_y & 0 \\ 0 & Q_u \end{pmatrix} \in \mathbb{R}^{(p+m)\times(p+m)}$$

subject to the following inequalities:

(C1) Constraints on both the output and input in an expected sense

$$\mathbb{E}\left[\begin{pmatrix} y(m) \\ u(m) \end{pmatrix}^T \begin{pmatrix} I_p & 0 \\ 0 & \gamma(m) I_m \end{pmatrix} H_j \begin{pmatrix} y(m) \\ u(m) \end{pmatrix} \middle| \bar{x},\ P(0),\ u_o^{N-1}\right] \leq \delta_j,$$
$$\text{for}\, j = 1, \ldots, L,\quad k = 0, \ldots, N-1, \tag{5.8}$$

with $H_j \geq 0$

(C2) Scalar constraints on both the output and input in a probabilistic sense

$$\mathbb{P}\left[g_j^T \begin{pmatrix} I_p & 0 \\ 0 & \gamma(m) I_m \end{pmatrix} \begin{pmatrix} y(m) \\ u(m) \end{pmatrix} \leq \theta_j\right] \leq (1-\varepsilon),$$
$$\text{for}\, j = 1, \ldots, T,\quad k = 0, \ldots, N-1, \tag{5.9}$$

where $g_j \in \mathbb{R}^{p+m}$.

Remark 5.2. *It is important to note that constraints (5.8) can be viewed as "power constraints" that limit the energy of output and control inputs at each time step. In addition, constraint (5.9) can be interpreted as a "safety specification" stipulating that the output and input remain within the hyperplanes specified by g_j and θ_j with a sufficiently high probability, $(1-\varepsilon)$, $k = 0, \ldots, N-1$. Together (5.8) and (5.9) are to be interpreted as conditioned on the initial state, that is, $\mathbb{E}[\cdot] := \mathbb{E}[\cdot|x(0)]$ and $\mathbb{P}[\cdot] := \mathbb{P}[\cdot|x(0)]$.*

5.2 TIME-DELAY SWITCH ATTACK

Time delays are ubiquitous in nature. They occur in a wide variety of natural and man-made control systems. Time delays can impact the stability of a system and degrade its performance. Time delays exist in power systems, specifically in the sensing and control loops. The traditional controller of power systems is designed based on the availability of current information and ignored time delays. However, power grids are being enhanced by introducing new telecommunication technologies for monitoring to improve the efficiency, reliability and sustainability of supply and distribution. For example, the introduction of a wide-area measurement system (WAMS) provides synchronized, near real-time measurements in phase measurement units (PMUs). WAMSs are used for stability analysis of power systems and can be used for efficient controller designs. Nevertheless, time delays are present in PMUs measurements in natural transmission lines.

Furthermore, modern power grids rely on computers and multipurpose networks, which make these type of grids vulnerable to cyberattacks [28], resulting in major negative impacts on lives and the economy. Investigating the methods of attacks on industrial control systems of sensitive infrastructures and devising countermeasures and security control protocols have attracted the attention of academia, industries, and governments.

5.2.1 Introduction

Few studies [28–30] considered the control of power systems with time delays introduced by adversary. In this section, we will focus on time delays injected by a hacker in a control system, with the purpose of destabilizing the system. This kind of attack was named a "time-delay-switch attack" or "TDS" for short. To circumvent a TDS attack, system controllers must be redesigned in a way to be robust, or if possible, be able to estimate variable random time delays. An easily implementable and effective method will be described to address time-delay switch attacks, that is, a TDS attack on the observed states of a controlled system. The proposed method utilizes a system state estimator, a time-delay estimator, a buffer to store the history of controller commands, and a PID or optimal controller to stabilize the plant that tracks a reference signal. For now, only the linear time-invariant systems in state feedback will be discussed.

5.2.2 Model Setup

A two-area power plant with automatic gain control under attack is considered in Fig. 5.4. The load frequency controller sends control signals to the plant and obtains state feedback through the communication channels from the turbines and measurements for remote terminal units (RTUs). The communication channels are wireless networks. Attacks can be launched by jamming the communication channels (i.e., DOS attack [31]), by distorting feedback signals (e.g., FDI attack [32]), or by injecting delays (i.e., TDS attack [29,30]) in data from telemetry measurements.

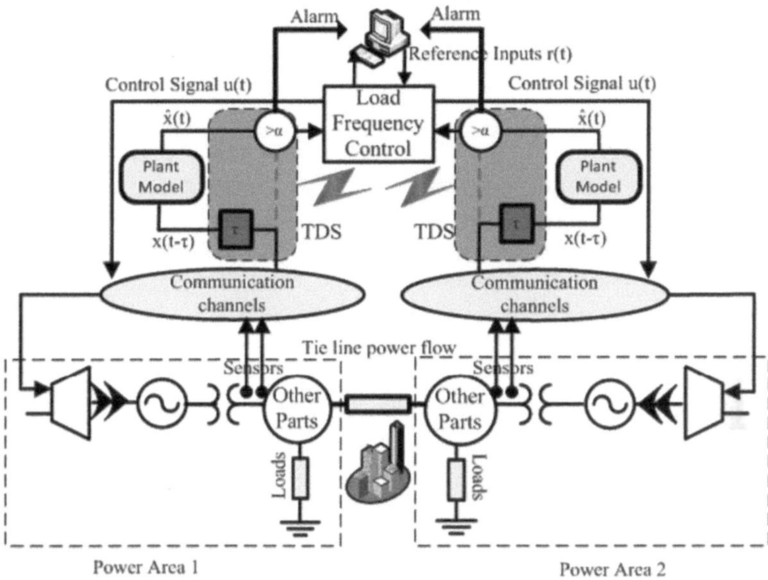

Figure 5.4 An adaptive load frequency control for a two-area power system

LFC is usually designed as an optimal feedback controller, but to operate optimally it requires power state estimations to be telemetered in real time. If an adversary introduces significant time delays in the telemetered control signals or measured states, the LFC will deviate from its optimality and in most cases, the system will break down. In [29], an LFC power system under TDS attack was modeled as a hybrid system, specifically in a switch action, "Off/Delay-by-τ," where τ is some delay time of the sensed system state or the control signals.

The LFC multiarea interlock power system is briefly discussed in [33] and [34]. More details can be found in [35] and [36]. The LFC dynamic model for the ith area is given by

$$\begin{cases} \dot{x}^i &= A_{ii}x^i(t) + B_i u^i(t) - h(x^j(t), \Delta P_l^i), \\ x^i(0) = x_0^i, \end{cases} \tag{5.10}$$

where $x \in \mathbb{R}^5$ and $u \in \mathbb{R}^5$ are the state and control vectors, respectively. The model of the ith area is influenced by the jth power area. Matrices A_{ii} and B_i are constant matrices with suitable dimensions, and ΔP_l^i is the power deviation of the load. The initial state vector is denoted by x_0^i for the ith power area. Then the state vector is defined as

$$x^i(t) = [\Delta f^i(t) \quad \Delta P^i_g(t) \quad \Delta P^i_{tu}(t) \quad \Delta P^i_{pf}(t) \quad \Lambda^i(t)]^T, \tag{5.11}$$

where $\Delta f^i(t)$, $\Delta P^i_g(t)$, $\Delta P^i_{tu}(t)$, $\Delta P^i_{pf}(t)$, and $\Lambda^i(t)$ are frequency and power deviations of the generator, position value of the turbine, tie-line power flow and control error on the ith power area, respectively [29]. The control error of the ith power area is expressed as

$$\Lambda^i(t) = \int_0^t \beta_i \Delta f^i(s) ds, \tag{5.12}$$

where β_i denotes the frequency bias factor. In the dynamic model of the LFC, A_{ii}, B_i, and $h(x^j(t), \Delta P^i_l)$ are represented by

$$A_{ii} = \begin{bmatrix} -\frac{\mu_i}{j} & \frac{1}{j} & 0 & -\frac{1}{j} & 0 \\ 0 & -\frac{1}{T_{tui}} & \frac{1}{T_{tu_i}} & 0 & 0 \\ -\frac{1}{\omega_i T_{gi}} & 0 & -\frac{1}{T_{gi}} & 0 & 0 \\ \sum_{i\neq j, j=1}^N 2\pi T_{ij} & 0 & 0 & 0 & 0 \\ \beta_i & 0 & 0 & 0 & 1 \end{bmatrix}, \tag{5.13}$$

$$B_i = \begin{bmatrix} 0 & 0 & \frac{1}{T_{gi}} & 0 & 0 \end{bmatrix}^T,$$

$$h(x^j(t), \Delta P^i_l) = \sum_{i\neq j, j=1}^N A_{ii} x^j(t) + D_i \Delta P^i_l, \tag{5.14}$$

where N is the total number of power areas, J_i, ω_i, μ_i, T_{gi} and T_{tu_i} are the generator moment of inertia, the speed-droop coefficient, generator damping coefficient, the governor time constant, the turbine time constant in the ith power area, and T_{ij} is the stiffness constant between the ith and the jth power areas, respectively. Also,

$$A_{ii} = \begin{bmatrix} 0 & 0 & 0 & 0 & 0 \\ 0 & 0 & 0 & 0 & 0 \\ 0 & 0 & 0 & 0 & 0 \\ -2\pi T_{ij} & 0 & 0 & 0 & 0 \\ 0 & 0 & 0 & 0 & 0 \end{bmatrix} \tag{5.15}$$

and

$$D_i = \begin{bmatrix} -\frac{1}{j} & 0 & 0 & 0 & 0 \end{bmatrix}^T. \tag{5.16}$$

Note that (5.17) gives the extension of the dynamic model (5.10) to the multiarea power system with the attack model using (5.13)–(5.16):

$$\begin{cases} \dot{X}(t) = AX(t) + BU(t) + D\Delta P_l, \\ X(0) = X_0, \end{cases} \tag{5.17}$$

where

$$X(t) = \begin{bmatrix} x^1(t)^T & x^2(t)^T & \dots & x^N(t)^T \end{bmatrix} \tag{5.18}$$

$$A = \begin{bmatrix} A_{11} & A_{12} & A_{13} & \dots & A_{1N} \\ A_{21} & A_{22} & A_{23} & \dots & A_{2N} \\ A_{31} & A_{32} & A_{33} & \dots & A_{3N} \\ \vdots & \vdots & \vdots & \ddots & \vdots \\ A_{N1} & A_{N2} & A_{N3} & \dots & A_{NN} \end{bmatrix} \tag{5.19}$$

$$B = \text{diag}\{\begin{bmatrix} B_1^T & B_2^T & B_3^T & \dots & B_N^T \end{bmatrix}^T\}, \tag{5.20}$$

$$D = \text{diag}\{\begin{bmatrix} D_1^T & D_2^T & D_3^T & \dots & D_N^T \end{bmatrix}^T\} \tag{5.21}$$

where B and D are $5N \times 5N$ matrices.

The optimal feedback controller is given by

$$U = -KX, \tag{5.22}$$

where optimal gain K is a $5N \times 5N$ matrix, and the control and state signals are $5N \times 1$ matrices. The design of the optimal controller for the LFC system in the normal operation (i.e., with no attack) involves minimizing a cost function described by

$$J = \frac{1}{2}\int_0^{t_f} \{X^T(t)QX(t) + U^T(t)RU(t)dt\}, \tag{5.23}$$

where the matrix $Q \in \mathbb{R}^{5N \times 5N}$ is positive semi-definite and $R \in \mathbb{R}^{5N \times 5N}$ is positive definite. Then, the optimal control problem is to obtain the optimal control signal $U(t)$ that minimizes the performance index (5.23), subject to the dynamic of the system with no time delay in its state.

The system with the optimal controller is described by the following equation:

$$\dot{X}(t) = (A - BK)X(t) + D\Delta P_l, \quad X(0) = X_0. \tag{5.24}$$

With the time-delay attack, the control signal will be modified by

$$U = -K\bar{X} \tag{5.25}$$

and the new state after the attack can be modeled by

$$\bar{X} = \begin{bmatrix} \bar{x}_1 \\ \bar{x}_2 \\ \vdots \\ \bar{x}_N \end{bmatrix} = \begin{bmatrix} x_1(t - t_{d1}) \\ x_2(t - t_{d2}) \\ \vdots \\ x_N(t - t_{dN}) \end{bmatrix}. \tag{5.26}$$

In (5.26), $t_{d1}, t_{d2}, \ldots, t_{dN}$ can be different/random time delays and take positive values. When $t_{d1}, t_{d2}, \ldots, t_{dN}$ are all zero, the system is in its normal operation. An adversary can gain access to the communication link and inject a delay attack on the line to direct the system to abnormal operations.

Remark 5.3. *In this section, ΔP_l is considered constant. This is a reasonable case because the stability of power system will not be influenced for an appropriate period following a step load change [33].*

In [29] and [30], it was mathematically proved and visually shown that a TDS attack can sabotage and disable the networked control system, and in particular the LFC system.

To circumvent the detrimental effect of time-delay attacks, control strategies must be developed that are resistant to time delays, and must be able to detect and track the TDS attack and manage a response strategy. A strategy is proposed with the modified method to estimate time-delay attacks from the history of sensed signals. This method can control the system under TDS attack.

5.2.3 Control Methodology

A method is developed to control an LTI system with natural delays or a system under TDS attack which consists of the plant model, a time-delay estimator, and a PID or optimal controller. The control scheme will detect and track time delays introduced by a hacker and guide the plant to track the reference signal to guarantee the stability for the system. Fig. 5.5 shows a diagram of the proposed time-delay estimator and controller.

Remark 5.4. *There has been much effort in the study of time delayed systems [37–40]. For example, two different time-varying time-delay estimation methods were proposed in [37] using a neural network for a class of nonlinear systems with*

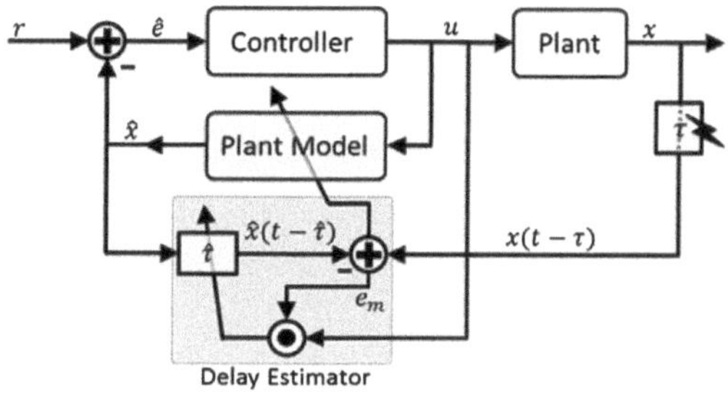

Figure 5.5 A block diagram of the control technique

time-varying time delays. An indirect time delay estimator procedure in the first method using nonlinear programming, and a direct time-delay estimation scheme is utilized in the second method that uses a neural network to construct a time-delay estimator. For a class of networked control systems, an adaptive control algorithm to guess random time delays is developed in [38]. The algorithm updates delay estimation using the gradient descent method, and discovers plant parameters by an improved recursive least squares.

Authors asserted that the method is superior to the typical networked predictive control. However, the method is complex for even the simplest linear system. Furthermore, the authors did not show results for time-delay control compensation, nor any time tracking of variable delays. Another example of time-delay estimation is the method of variable sampling to compensate for the time delay in a networked control system. A multilayer perceptron (MLP) neural network was used to learn the time delay offline and predict its value during the online control operation [39]. This method assumes that time delay is constant, so it cannot be applied to a system under time-delay attack. All control methods developed in the past to compensate for time delays either rely on a controller that is strong enough to resist a maximum time delay, offline estimates of time delays, or approximation of time-delayed signals.

In this section, we look at a general method for the control of continuous linear time-invariant systems under TDS attack.

Suppose the system being dealt with is given by or can be approximated in a region of interest by the LTI system

$$\dot{x}(t) = Ax(t) + Bu(t) \tag{5.27}$$

with its solution is given by

$$x(t) = e^{At}x_0 + \int_0^t e^{A(t-s)} Bu(s)ds. \tag{5.28}$$

With time delay τ, due to a time-delay switch attack or any natural delay, the solution becomes

$$x(t-\tau) = e^{A(t-\tau)}x_0 + \int_0^{t-\tau} e^{A(t-\tau-s)} Bu(s)ds. \tag{5.29}$$

The solution $x(t)$ at time t as a function of time-delayed signal $x(t-\tau)$ is

$$\begin{aligned} x(t) &= e^{At}x_0 + e^{A\tau} \int_0^{t-\tau} e^{A(t-s)} e^{A\tau} Bu(s)ds + \int_{t-\tau}^{t} e^{A(t-s)} Bu(s)ds \\ &= eAtx_0 + e^{A\tau}\left[x(t-\tau) - e^{A(t-\tau)}x_0\right] + \int_{t-\tau}^{t} e^{A(t-s)} Bu(s)ds. \end{aligned} \tag{5.30}$$

In general, the time delay τ is an unknown variable. If we assume that τ is a constant value, and $\hat{\tau}$ is the estimate of the time delay τ, then $\epsilon = \hat{\tau} - \tau$ is the error. The state estimate $\hat{x}(t)$ of the system based on the plant model and the estimate of time delay $\hat{\tau}$ can be calculated as

$$\hat{x}(t) = e^{At}x_0 + eA\hat{\tau}\left[\hat{x}(t-\hat{\tau}) - e^{A(t-\hat{\tau})}x_0\right] + \int_{t-\hat{\tau}}^{t} eA(t-s)Bu(s)ds, \tag{5.31}$$

where $\hat{x}(t-\hat{\tau})$ is the delayed estimate of the state given the estimate of the delay $\hat{\tau}$ (i.e., a simulated signal).

It should be noted that $x(t-\tau)$ is what is actually measured and delivered to the plant model. So, at every time instance, variables

$$\hat{x}(t),\ \hat{x}(t-\hat{\tau}),\ u(t),\ A,\ B,\ x(t-\tau)$$

are known to the controller and plant model. On the other hand, the current state $x(t)$ and the time delay τ are unknown. It is essential that the plant model estimates $x(t)$ correctly. Because of the delay, an accurate estimation of $x(t)$ requires a good estimate of the delay τ. The process will be shown how to estimate delay τ, the state $x(t)$, control the system using a PID controller, and extend the method to an optimal controller.

The modeled error signals in states can be described by $e_m(t) = x(t) - \hat{x}(t)$ and

$$e_m(t; \tau, \hat{\tau}) = x(t-\tau) - \hat{x}(t-\hat{\tau}). \tag{5.32}$$

The idea is to estimate $\hat{\tau}$ in time as quickly as possible to minimize the modeling error $e_m(t; \tau, \hat{\tau})$. To do so, let $\nu = 1/2e_m^2$. Using the gradient descent method,

$$\frac{d\hat{\tau}}{dt} = -\eta \frac{\delta \nu}{\delta \hat{\tau}}, \tag{5.33}$$

where η is the learning parameter, which is set to guarantee convergence of the time-delay estimate to the appropriate time delay as quickly as possible without causing system instability. Algebraic manipulation of (5.33) using (5.31) yields

$$\begin{aligned} \frac{d\hat{\tau}}{dt} &= -\eta e_m \frac{\delta e_m}{\delta \hat{\tau}} \\ &= -\eta e_m \left[Bu(t-\hat{\tau}) - e^{A(t-\hat{\tau})} Bu(0) - Ae^{A(t-\hat{\tau})} x_0 \right]. \end{aligned} \tag{5.34}$$

Letting $u(0) = 0$ as an appropriate initial point gives

$$\frac{d\hat{\tau}}{dt} = -\eta e_m [Bu(t-\hat{\tau}) - Ae^{A(t-\hat{\tau})} x_0], \quad 0 \leq \tau \leq t. \tag{5.35}$$

Remark 5.5. *Note that (5.35) will be used to estimate the time delay τ. However, there are practical issues that must be considered. Computing machines have finite memory and temporal resolution. Therefore, (5.35) cannot be implemented without discrete approximation and boundedness assumptions. To guarantee the stability of calculations and limit memory usage, the following condition must be added: $\tau < \tau_{max}$. This condition will allow for the construction of a finite buffer that will store the history of $u(t)$ from t to $t - \tau_{max}$. Also, this will prevent runaway condition on $\hat{\tau}$. It should be noticed if the delay injected by an adversary is more than τ_{max}, a trap condition signal will be sent to the supervisory control and data acquisition (SCADA) center, and the controller will then switch to an open loop control to stabilize the system. This switch is robust since our controller is equipped with a plant model that can predict the next state.*

We now proceed to include the combination of plant model and controller. Let the performance error be $e(t) = r(t) - x(t)$ and the estimate of the performance error be $\hat{e}(t) = r(t) - \hat{x}(t)$. The PID controller is defined in terms of the estimated error as

$$u(t) = K_P \hat{e}(t) + K_D \frac{d\hat{e}(t)}{dt} + K_I \int_0^t \hat{e}(s) ds \tag{5.36}$$

and the optimal feedback controller as

$$u(t) = K\hat{e}(t), \tag{5.37}$$

where K_P, K_D, K_I and K are proportional, derivative, integral and optimal gain, respectively.

Remark 5.6. *Depending on the error $\hat{e}(t)$ that results from the estimate $\hat{x}(t)$, the controller can be adjusted accordingly. It turns out that, when the estimate $\hat{x}(t)$ converges to $x(t)$, $\hat{e}(t)$ converges to $e(t)$ and is minimized by the controller such that the system $x(t)$ converges to $r(t)$. The crucial factor is the presence of $x(t-\tau)$ in estimating $\hat{x}(t)$. This is an important ingredient in finding a stable controller.*

To address the foregoing concern, the following argument must be considered. Let the plant model estimation be given by

$$\dot{\hat{x}}(t) = A\hat{x}(t) + Bu(t). \tag{5.38}$$

Then the delayed state model becomes

$$\dot{x}(t-\tau) = Ax(t-\tau) + Bu(t-\tau). \tag{5.39}$$

Note that $x(t-\tau)$ and $\dot{x}(t-\tau)$ are measured by the plant model, and $u(t-\tau)$ is unknown since τ is not known. Likewise, we have

$$\dot{\hat{x}}(t-\tau) = A\hat{x}(t-\tau) + Bu(t-\tau), \tag{5.40}$$

where the elements of (5.40) are unknown because τ is unknown. From (5.39) and (5.40), we have for a constant gain matrix $C > 0$ that

$$\dot{\hat{x}}(t) = A\hat{x}(t) + C\dot{\hat{x}}(t-\tau) - CA\hat{x}(t-\tau) - CBu(t-\tau) + Bu(t). \tag{5.41}$$

Further manipulations of (5.41) yield

$$\begin{aligned}
\dot{\hat{x}}(t) &= A\hat{x}(t) + Bu(t) + C\dot{\hat{x}}(t-\tau) - CA\hat{x}(t-\tau) - C\dot{x}(t-\tau) + CAx(t-\tau) \\
&= A\hat{x}(t) + Bu(t) - C\left[\dot{x}(t-\tau) - \dot{\hat{x}}(t-\tau)\right] + CA\left[x(t-\tau) - \hat{x}(t-\tau)\right] \\
&= A\hat{x}(t) + Bu(t) - C\left[\dot{e}_m(t'; \tau, \tau) - Ae_m(t; \tau, \tau)\right].
\end{aligned} \tag{5.42}$$

Remark 5.7. *Considering the term $e_m(\cdot; \cdot, \cdot)$ in (5.42), it turns out the first τ is the actual delay signal that is associated with the delayed signal $x(t-\tau)$ itself which* *is usually read via the communication channel. This τ is not accessible to the control system. The second τ is associated with the plant estimate of the delay signal $\hat{\tau}$ and delayed states $\hat{x}$. Hence, (5.42) is presented assuming the fact that $\hat{\tau} = \tau$.*

By replacing $e_m(t; \tau, \tau)$ by $e_m(t; \tau, \hat{\tau})$ of (5.32), we obtain

$$\dot{\hat{x}}(t) = A\hat{x}(t) + Bu(t) - C[\dot{e}_m(t; \tau, \hat{\tau}) - Ae_m(t; \tau, \hat{\tau})]. \tag{5.43}$$

It must be emphasized that the above replacement makes the current predicted estimate of the plant state $\hat{x}(t)$ dependent on the estimate of the time delay $\hat{\tau}$. This guarantees that an accurate estimate of the state depends on an accurate estimate of the time delay.

Remark 5.8. *When $e_m(t; \tau, \hat{\tau}) - Ae_m(t; \tau, \hat{\tau})$ in (5.43) equals zero as a result of $\hat{\tau}$ converging to τ, $\hat{x}(t)$ will converge to $\hat{x}(t)$. This means that the modeling error e_m should be exponentially damped, that is, $\dot{e}_m(t; \tau, \hat{\tau}) = Ae_m(t; \tau, \hat{\tau})$. It is significant to observe that the plant estimate is constructed to depend on the measured states of the plant, $x(t - \tau)$, and the estimate of the state given the estimated time delay, $\hat{x}(t - \hat{\tau})$. The difference $x(t - \tau) - \hat{x}(t - \hat{\tau})$ is obviously the modeling error signal $e_m(t; \tau, \hat{\tau})$.*

Remark 5.9. *The choice of the gain matrix C in (5.43) is dictated by balancing the requirements for having the plant model state depend on errors in time-delays estimation and guarantees that the control system remains stable.*

This method has been implemented in MATLAB, and its performance has been verified using a single input, single output system and an LFC control of two-area distributed power systems. The next section presents and discusses the simulation results.

5.2.4 Procedure for Controller Design

Step 1: Initialize time-delay estimate $\hat{\tau}$, plant model stat estimate $\hat{x}$ and model error e_m. Then, set the learning parameter η to a suitable value. Also, set the matrix C.

Step 2: Obtain a plant state measurement (that is, the sensed states of the plant $x(t - \tau)$), which could be time delayed by $\tau(t)$.

Step 3: Compute the current state estimate, $\hat{x}(t)$, using (5.43). In the discrete form, (5.43) can be approximated in terms of the sampling period Γ as:

$$\hat{x}(k+1) = \hat{x}(k) + A\Gamma\hat{x}(k) + B\Gamma u(k) - C[e_m(k) - e_m(k-1) - A\Gamma e_m(k)].$$

Step 4: Compute the delayed plant state estimate $\hat{x}(t-\hat{\tau})$ based on the model equation and the estimate of the performance error $\hat{e}(t) = r(t) - \hat{x}(t)$ and model error $e_m(t;\tau,\hat{\tau}) = x(t-\tau) - \hat{x}(t-\hat{\tau})$.

Step 5: Compute the time-delay estimate $\hat{\tau}$, from Eq. (5.35). The discrete approximation of (5.35) is described below:

$$\hat{\tau}(t+1) = \hat{\tau}(k) - \eta\Gamma e_m(k)(Bu(k-\omega) - Ae^{A\Gamma(k-\omega)}x(0)),$$

where $\omega = \text{round}(\hat{\tau}(k)/\Gamma)$ is the nearest integer to $\hat{\tau}(k)/\Gamma$.

Step 6: Compute the control signal $u(t)$. For example, u can be set by using (5.36) or (5.37).

Step 7: To prevent runaway conditions, bound the control signal by $\pm u_{max}$, time-delay estimate by τ_{max} and plant model by $\pm x_{max}$.

Step 8: Repeat Steps 2–7 until the estimate of the performance error $\hat{e} < \epsilon$. In the case of time-delay tracking and tracking of a reference trajectory r, continuously repeat Steps 2–7.

5.2.5 Simulation Example 5.1

In the sequel, a simple single-input, single-output (SISO) system under variable time-delay attack with a variable reference signal is considered. This test is conceived to demonstrate the usability and efficiency of the proposed method. The simulation model is given by

$$\begin{aligned}
\dot{x}(t) &= -0.7x(t) + 2u(t), \\
r(t) &= \begin{cases} 1 - e^{-\gamma 2t}, & \text{if } t \le T_a, \\ 4e^{-\gamma t(t-T_b)}(\sin(2\pi\omega_a t) + 1), & \text{otherwise,} \end{cases} \\
\tau(t) &= T_1 e^{-\lambda_1 t}(\sin(2\pi\omega_1 t) + 1) + T_2(1 - e^{-\lambda_2 t}) + 1,
\end{aligned} \tag{5.44}$$

where the total simulation time $T_{final} = 500$ s; $Ta = T_{final}/2$, $T_b = T_{final}/2.2$, $T_1 = T_2 = T/10$, $\lambda_1 = 0.005$, $\lambda_2 = 0.0005$, $\gamma_1 = 0.07$, $\gamma_2 = 0.004$, $\omega_a = 0.06$, $\omega_1 = 0.005$ and the sampling time is 0.01 s. The proposed PID controller was applied to track the reference signal under a TDS attack. Fig. 5.6A shows the state of the plant given in Eq. (5.40), tracking the desired trajectory $r(t)$. The tracking is almost perfect, even though the time delay varies by $\tau(t)$. Fig. 5.6B shows the TDS attack detection and its tracking; the estimated time delay $\hat{\tau}(t)$ tracks the time-varying time delay $\tau(t)$ that can either be injected by an adversary or occurs naturally. Note that in the first 80 seconds of simulation, the plant's system operation does not track the reference signal because the time variable t is less than the time

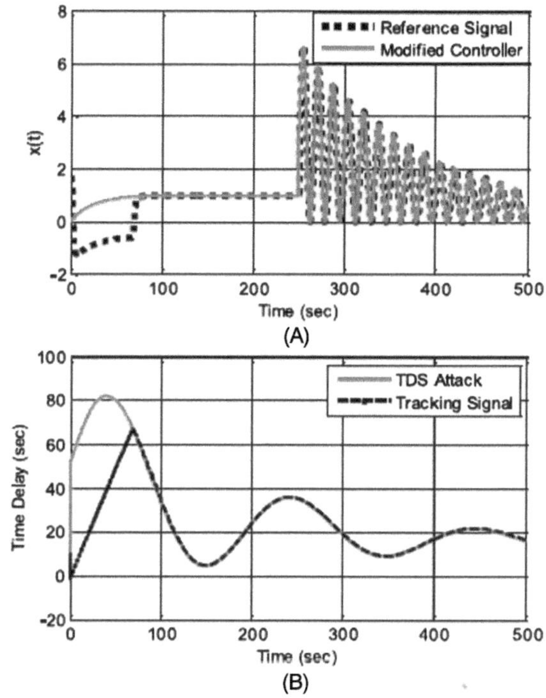

Figure 5.6 (A) Tracking performance under TDS attack. The dotted black line is the state of the system while the solid green line is the reference signal. (B) Time-delay tracking. The dotted black line is the time-delay estimate while the solid green line is the TDS attack

delay $\tau(t)$. In this plant simulation, a PID controller with the parameters $K_P = 5$, $K_I = 2$ and $K_D = 1.5$ was used. A time-delay estimator learning rate $\eta = 0.32$ was used, as well as a plant model teacher forcing effort parameter, $C = 2$.

The simple modified model based on control and time-delay estimation has shown that it works for simple single-input, single-output systems under a complex variable time-delay attack.

5.2.6 Simulation Example 5.2

The model described in (5.44) was used for the purpose of comparison with other methods, with some minor modifications in the sampling time, simulation time and TDS attack model. This modification was made for visualization purposes, as well as to make the attack model simpler. The methods selected were those with better performance.

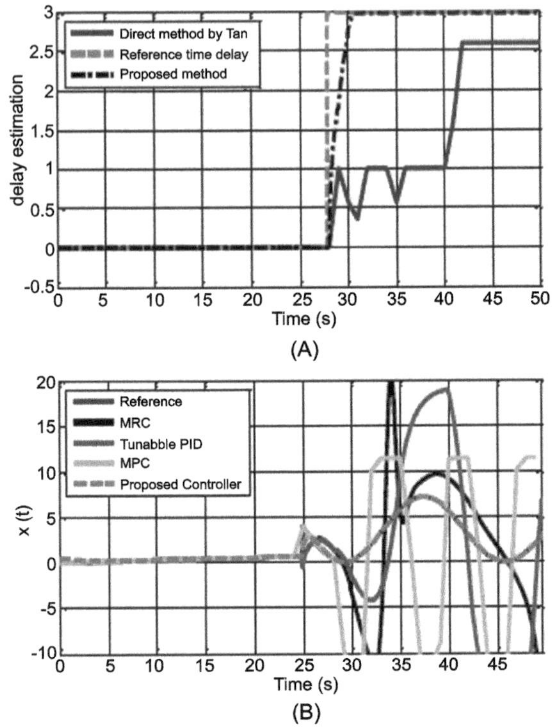

Figure 5.7 (A) Behavior of time-delay estimation under time-delay switch attack. (B) Performance comparison with other time-delay control methods

The new TDS attack can be modeled as

$$\tau(t) = \begin{cases} 0, & t \leq 28 \text{ s}, \\ 3, & t > 28 \text{ s}. \end{cases} \tag{5.45}$$

The total simulation time was set to $T_{final} = 50$ s, and the sampling time to 0.01 s.

The direct method time-delay estimation proposed in [37] was applied to the proposed controller to show the performance of the proposed time-delay estimation method. The results in Fig. 5.7A show that the proposed method tracks the step function TDS attack accurately in a shorter amount of time when compared with other methods.

5.2.7 Simulation Example 5.3

We applied other control methods to control the system described in Eq. (5.44) under TDS attack model described in Eq. (5.45). These methods

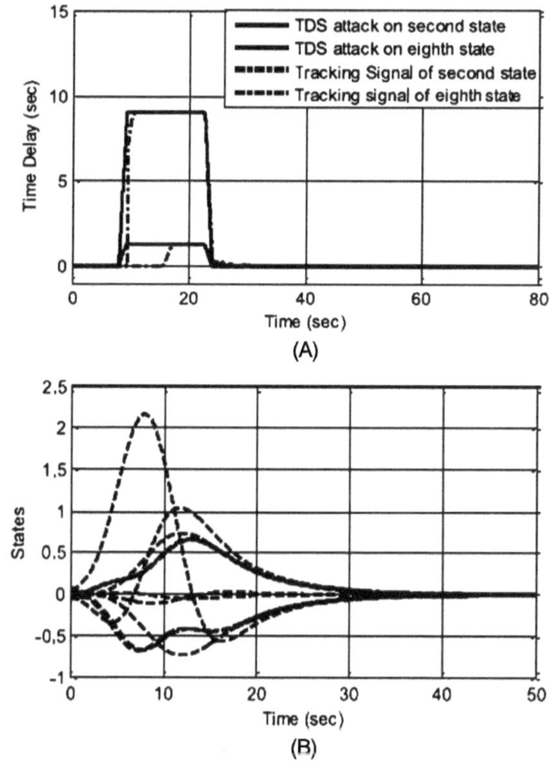

Figure 5.8 (A) TDS attack tracking. (B) All states for the two-area LFC power system under TDS attack using the modified optimal controller. Initially, delays throw the system away from the stable point, however, shortly after time delays are estimated and the controller recovers, the regulating power states back to zero

are the model referenced control (MRC), tunable PID controller (TPID), and model predictive controller (MPC). The methods used are part of the MATLAB Simulink package.

It should be noted that all of the considered controllers were designed in the optimal way for possible TDS attacks.

The TPID, MRC and the proposed controller were used with the following parameters: $K_P = 5$, $K_I = 2$ and $K_D = 1.5$.

The TDS attack of 0.1 s was applied from the starting time of 28 s to the system, and all of the controllers responded appropriately, with some methods responding better than others. Fig. 5.8B shows the simulation results.

Table 5.1 Parameter values for a two-area power system controller design.

Parameter	Value	Parameter	Value
J_1	10	ω_1	0.05
μ_1	1.5	T_{g1}	0.12 s
T_{tu1}	0.2 s	T_{tu2}	0.45 s
T_{12}	0.198 pu/rad	T_{21}	0.198 pu/rad
J_2	12	ω_2	0.05
μ_2	1	T_{g2}	0.18 s
R	$100I$	Q_f	0
Q	100100	t_f	∞
β_2	21.5	β_2	21

5.2.8 Simulation Example 5.4

The distributed power control systems is where the time delay mitigation strategies are paramount. The LFC system is where the controller task regulates the state of a power plant network.

Simulation studies were conducted to evaluate the effects of TDS attacks on the dynamics of the system. By solving the Riccati matrix equation, the close loop control can be designed in the form of state feedback. For this simulation, a discrete linear–quadratic regulator design is used from the continuous cost function called the "lqrd" function in MATLAB 2013a. For simplicity of discussion, $N = 2$, which means a two-area power system.

Table 5.1 shows the parameter values used in this process. Since simulation for a certain duration tracks a step load change, both ΔP_l^1 and ΔP_l^2 are also set to zero.

Scenario 1. The hacker injects time delays on the second and eighth states, from the time of 8 to 24 s for a delay value of 1.28 and 9 s, respectively. The LFC system equipped with the time-delay estimator performs well. The power states are being regulated to zero, and a TDS attack has been detected and time-delay tracked. Fig. 5.8A shows the detection and tracking of the time delay, and Fig. 5.8B shows all the states for the two-area interconnected power system. As clearly seen from the figures, the modified controller was able to control the LFC distributed system under TDS attack.

Scenario 2. In this scenario, a TDS attack is injected at times of 1 and 3 s for delay values of 5 and 7 s for the second and eighth states. Fig. 5.9A shows that the tracking scheme works perfectly and could track the TDS attack. Figs. 5.9B–E show the simulation results of frequency and power deviations

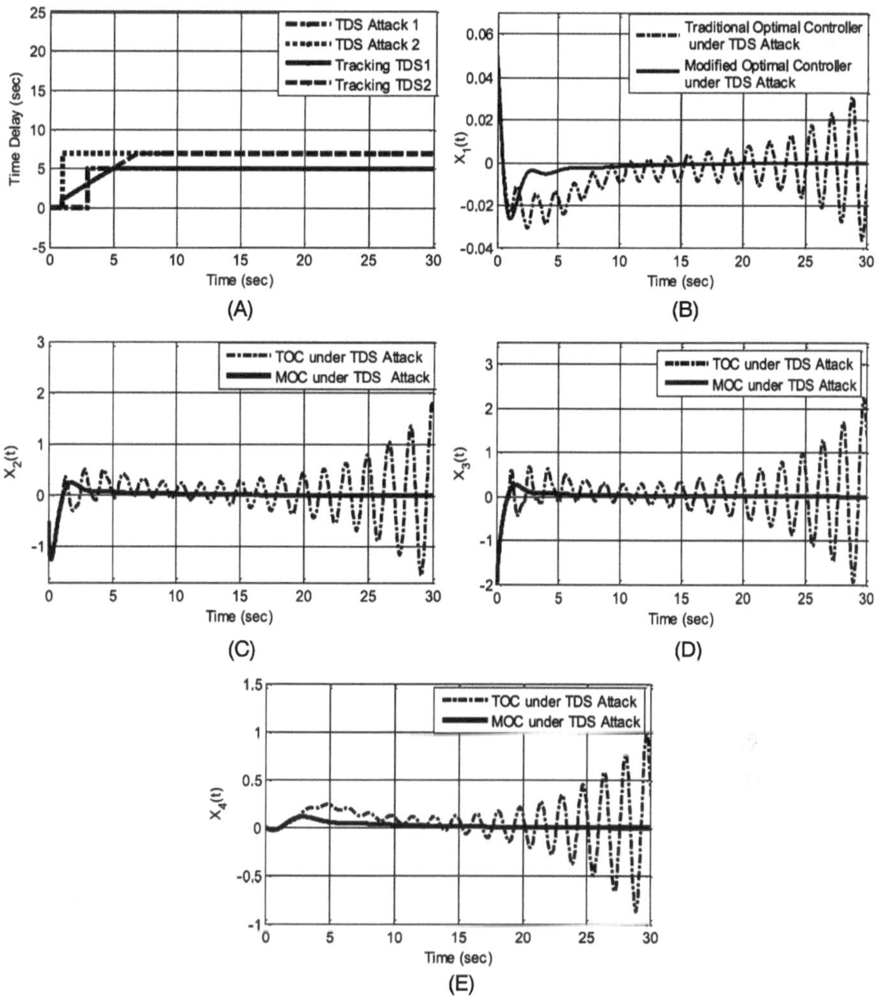

Figure 5.9 (A) Time-delay tracking under TDS attack. (B) Frequency deviation, Δf^K. (C) Power deviation of generator, ΔP_g^K. (D) Value position of the turbine, ΔP_{tu}^K. (E) Tie-line power flow, ΔP_{pf}^1

of the generator, value position of the turbine and tie-line power flow of the first power area, respectively, with and without the modified control method. It shows that the system will be unstable under a TDS attack if the modified method is not applied. In Figs. 5.9C–D, TOC and MOC denote the traditional and modified optimal controllers, respectively.

Scenario 3. This scenario is exactly the same as the second scenario, except a 15% disturbance and noise was added to the system. As shown in

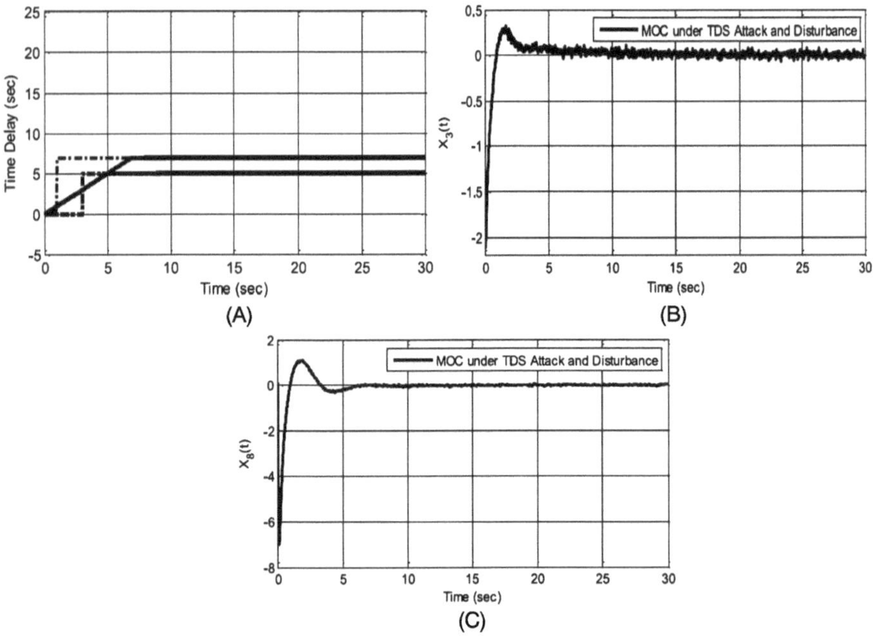

Figure 5.10 (A) TDS attack tracking. (B) Value position of the turbine system under TDS attack for the first power-area control system. (C) Value position of the turbine system under TDS attack for the second power-area control system

Fig. 5.10A, the modified control technique could detect, track and control the LFC system under TDS attack, along with some disturbances. Figs. 5.10B–C show the tracking performance of value position of the turbine system under TDS attack for the first and second power area.

5.3 SECURITY CONTROL UNDER CONSTRAINTS

Control under cyberphysical security constraints (see [44–48]) has recently generated a lot of interest in the control, networking, and cybersecurity research communities. The common focal point has been studying the effect of adversarial attacks against particular control systems. The control systems have themselves been modeled as both deterministic and probabilistic whereas different formulations to model the adversary, attacking the system, have been presented.

In the sequel, we consider the problem of designing an optimal control system subject to security constraints. A deterministic, linear, and time-invariant system is analyzed for which the terminal state can take a finite number of values. It is assumed that the controller knows the desired terminal state and applies a control sequence accordingly. An adversary makes partial noisy measurements of the state trajectory and wants to estimate the actual value of the terminal state. The adversary has knowledge of the set of possible terminal states. The task of the controller is to develop a strategy to reach the terminal state while providing minimum information to the adversary, thereby hindering its ability to estimate the terminal state.

5.3.1 Problem Formulation

Consider the following linear time-invariant system:

$$x_{k+1} = Ax_k + Bu_k, \qquad k = 0, \dots, T-1, \tag{5.46}$$

where $x_k \in \mathbb{R}^n$ is the state of the system, $u_k \in \mathbb{R}^m$ is the control input, A is an $n \times n$ matrix, and B is an $n \times m$ matrix. Without loss of generality, the initial state x_0 is assumed to be zero. The cases of both finite and continuous distributions of the terminal state are analyzed. In Section 5.3, we assume that $x_T \in \{x^0, x^1\}$ and in Section 5.4 that $x_T \in \{x^0, \dots, x^{M-1}\}$. The framework for the continuous terminal state distribution will be presented in Section 5.5. In the finite distribution case, the desired terminal state is drawn from $\{x^0, \dots, x^{M-1}\}$ with prior distribution $\pi_0, \dots, \pi_{M-1}$. The controller knows the desired terminal state and accordingly applies the appropriate control sequence, $\{u_0, \dots, u_{T-1}$, to reach it.

The adversary does not know the actual value of the terminal state, which is random, but knows its distribution. The adversary is restricted to make only the first $(k+1)$ measurements of the state with $k+1 < T$. These measurements are noisy and given as follows:

$$Y_i = Cx_i + V_i, \qquad i = 0, \dots, k, \tag{5.47}$$

where C is a $p \times n$ matrix and $V_0, \dots, V_k$ are $p \times 1$ independent and identically distributed random vectors. Using (5.46), we can write the measurement model in compact form as follows:

$$Y_{0,k} = \bar{C}U_{0,k-1} + V_{0,k}, \tag{5.48}$$

where $Y_{0,k}$, $\bar{C}$, $U_{0,k-1}$ and $V_{0,k}$ are given by

$$Y_{0,k} = \begin{bmatrix} Y_0 \\ \vdots \\ Y_k \end{bmatrix}, \quad V_{0,k} = \begin{bmatrix} V_0 \\ \vdots \\ V_k \end{bmatrix}, \quad U_{0,k-1} = \begin{bmatrix} U_0 \\ \vdots \\ U_{k-1} \end{bmatrix},$$

$$\bar{C} = \begin{bmatrix} 0_{p\times m} & 0_{p\times m} & \dots & 0_{p\times m} \\ CB & 0_{p\times m} & \dots & 0_{p\times m} \\ \vdots & \vdots & \ddots & \vdots \\ CA^{k-1}B & CA^{k-2}B & \dots & CB \end{bmatrix}. \tag{5.49}$$

We consider the case when $V_0, \dots, V_k$ have a Gaussian distribution and the case when they have a general finite mean distribution. We assume that $\pi_i > 0$, where $i = 0, \dots, M-1$. The dynamical system is assumed to be controllable and hence there are many control sequences which drive the system to the specified terminal state. We denote any control sequence which drives the system to the terminal state, x^i, by $U^i_{0,T}$. The task of the controller is to design the control sequences such that the adversary cannot estimate the actual value of the terminal state. Using the measurement model, the adversary solves a hypothesis testing problem. Under hypothesis H_i, when the terminal state is x^i, the information available to the adversary is as follows:

$$Y_{0,k} = \bar{C}U^i_{0,k} + V_{0,k}. \tag{5.50}$$

It should be noted that the security constraints are dependent on the information, $\bar{C}U^i_{0,k-1}$, which is provided by the controller to the adversary. The quadratic cost to be minimized is provided as

$$\mathbb{U}'Q\mathbb{U} = \sum_{i=0}^{M-1} \pi_i U^i_{0,T-1}{}'\tilde{Q}U^i_{0,T-1}, \tag{5.51}$$

where $\tilde{Q}$ is a $Tm \times Tm$ symmetric positive definite matrix. The symmetric positive definite matrix Q and the control vector $\mathbb{U}$ are given as follows:

$$\mathbb{U} = \begin{bmatrix} U^0_{0,T-1} \\ \vdots \\ U^{M-1}_{0,T-1} \end{bmatrix}, \quad Q = \begin{bmatrix} \pi_0\tilde{Q} & 0_{Tm\times Tm} & \dots & 0_{Tm\times Tm} \\ 0_{Tm\times Tm} & \pi_1\tilde{Q} & \dots & 0_{Tm\times Tm} \\ \vdots & \vdots & \ddots & \vdots \\ 0_{Tm\times Tm} & 0_{Tm\times Tm} & \dots & \pi_{M-1}\tilde{Q} \end{bmatrix}. \tag{5.52}$$

Since the control sequences $U^0_{0,T-1}, \dots, U^{M-1}_{T-1}$ must drive the system to the terminal states $x^0, \dots, x^{M-1}$, we need to introduce the following equality constrains:

$$B_T U^i_{0,T-1} = x^i, \qquad i = 0, \dots, M-1, \tag{5.53}$$

which can be written in compact form as follows:

$$F\mathbb{U} = b, \tag{5.54}$$

$$F = \begin{bmatrix} B_T & 0_{n\times Tm} & \dots & 0_{n\times Tm} \\ 0_{n\times Tm} & B_T & \dots & 0_{n\times Tm} \\ \vdots & \vdots & \ddots & \vdots \\ 0_{n\times Tm} & 0_{n\times Tm} & \dots & B_T \end{bmatrix}, \quad b = \begin{bmatrix} x^0 \\ \vdots \\ x^{M-1} \end{bmatrix},$$

$$B_T = \left[A^{T-1}B, A^{T-2}B, \dots, B\right]. \tag{5.55}$$

The optimization problem that we will solve involves minimization of the cost function (5.51), subject to the equality constraint (5.54), and the security constraints that we provide in the next two sections. The optimization will be performed with respect to the control variable $\mathbb{U}$. The equality constraints can be removed from this optimization problem by utilizing an appropriate re-parametrization. Let U_b be a suitable control vector which is provided to us and satisfies $FU_b = b$. Let $\tilde{F}$ be the matrix whose columns form a basis for the null space of F. Then we can write

$$U = U_b + \tilde{F}\eta, \qquad \eta \in \mathbb{R}^{\dim(\mathrm{Null}(F))}. \tag{5.56}$$

It should be noted from the definition of $\tilde{F}$ that for any η, $F(\tilde{F}\eta) = 0$. Now η becomes our optimization variable and we do not need to incorporate the equality constraint (5.54). The cost function can be written as

$$\mathbb{U}'QU = U_b'QU_b + 2U_b'Q\tilde{F}\eta + \eta'\tilde{F}'Q\tilde{F}\eta. \tag{5.57}$$

5.3.2 A Binary Framework

In this section, we consider the framework where the terminal state can take two different values x^0 and x^1 with probabilities π_0 and π_1, respectively. We consider two different cases of measurement noise distribution.

5.3.2.1 Gaussian Noise Distribution

Consider the case when $V_0, \dots, V_k$ are i.i.d. and have a Gaussian $\mathcal{N}(0, \Sigma)$ distribution. Then $V_{0,k}$ is an $\mathcal{N}(0, \Sigma)$ random vector where the covariance matrix Σ is block diagonal and has matrices $\tilde{\Sigma}$ on its diagonal. The minimum probability of error, in the estimate of the actual value of the terminal state using measurements made by the adversary, is introduced as a security constraint. Under H_0 and H_1 the measurements have the following distributions:

$$H_0 : Y_{0,k} = \bar{C}U^0_{0,k-1} + V_{0,k} \sim \mathcal{N}(\bar{C}U^0_{0,k-1}, \Sigma), \tag{5.58}$$
$$H_1 : Y_{0,k} = \bar{C}U^1_{0,k-1} + V_{0,k} \sim \mathcal{N}(\bar{C}U^1_{0,k-1}, \Sigma). \tag{5.59}$$

We use a Bayesian formulation with a uniform cost and knowledge of the priors to compute the minimum probability of error. The optimal Bayes test is a likelihood ratio test [49], which is given by

$$\begin{aligned} L(y_{0,k}) &= \frac{p(y_{0,k}|H_1)}{p(y_{0,k}|H_0)} \\ &= \exp\{(U^1_{0,k-1} - U^0_{0,k-1})'\bar{C}'\Sigma^{-1}y_{0,k} \\ &\quad - \frac{1}{2}(U^1_{0,k-1} - U^0_{0,k-1})'\bar{C}'\Sigma^{-1}\bar{C}(U^1_{0,k-1} + U^0_{0,k-1})\}. \end{aligned} \tag{5.60}$$

The optimal Bayes test γ_B is given as follows:

$$\gamma_B = \left\{ \begin{matrix} H_1 & \text{if} & L(y_{0,k}) \geq \frac{\pi_0}{\pi_1} \\ H_0 & \text{if} & L(y_{0,k}) < \frac{\pi_0}{\pi_1} \end{matrix} \right\}. \tag{5.61}$$

By taking ln of both sides, we can write the inequality $L(y_{0,k}) \geq \frac{\pi_0}{\pi_1}$ as:

$$\begin{aligned} (U^1_{0,k-1} - U^0_{0,k-1})'\bar{C}\Sigma^{-1}y_{0,k} \geq \tau, \\ \tau = \frac{1}{2}(U^1_{0,k-1} - U^0_{0,k-1})'\Sigma^{-1}\bar{C}(U^1_{0,k-1} - U^0_{0,k-1}) + \ln(\frac{\pi_0}{d\pi_1}). \end{aligned}$$

Now $y_{0,k}$ is a Gaussian random vector whose linear transformation is also Gaussian. By computing the mean and variance of $(U^1_{0,k-1} - U^0_{0,k-1})'\bar{C}'\Sigma^{-1}Y_{0,k}$ under H_0 and H_1, we get

$$\begin{aligned} H_0 : &(U^1_{0,k-1} - U^0_{0,k-1})'\bar{C}'\Sigma^{-1}Y_{0,k} \\ &\sim \mathcal{N}\left((U^1_{0,k-1} - U^0_{0,k-1})'\bar{C}'\Sigma^{-1}\bar{C}U^0_{0,k-1}, d^2\right), \end{aligned} \tag{5.62}$$
$$\begin{aligned} H_1 : &(U^1_{0,k-1} - U^0_{0,k-1})'\bar{C}'\Sigma^{-1}Y_{0,k} \\ &\sim \mathcal{N}\left((U^1_{0,k-1} - U^0_{0,k-1})'\bar{C}'\Sigma^{-1}\bar{C}U^0_{0,k-1}, d^2\right), \end{aligned} \tag{5.63}$$
$$d^2 = (U^1_{0,k-1} - U^0_{0,k-1})'\bar{C}'\Sigma^{-1}\bar{C}(U^1_{0,k-1} - U^0_{0,k-1}).$$

The probability of error in estimating the true value of the terminal state is given by

$$\begin{aligned} P_e &= \pi_0 P\left(L(Y_{0,k} \geq \frac{\pi_0}{\pi_1} | H_0\right) + \pi_1 P\left(L(Y_{0,k} \geq \frac{\pi_0}{\pi_1} | H_1\right) \\ &= \pi_0 - \pi_0 \Phi\left(\frac{1}{d}\ln\frac{\pi_0}{\pi_1} + \frac{d}{2}\right) + \pi_1 \Phi\left(\frac{1}{d}\ln\frac{\pi_0}{\pi_1} + \frac{d}{2}\right), \end{aligned} \tag{5.64}$$

where Φ is the pdf of a standard normal $\mathcal{N}(0, 1)$ distribution. For the special case of equiprobable priors, $\pi_0 = \pi_1 = \frac{1}{2}$, the above expression simplifies to

$$P_e = \rho(\frac{d}{2}). \tag{5.65}$$

Here ρ denotes the Q-function, the tail probability of the standard normal distribution. Consider the following optimization problem which introduces the probability of error as a security constraint:

Problem A. Minimize the cost function

$$U_b' Q U_b + 2U_b' Q\tilde{F}\eta + \eta' \tilde{F}' Q\tilde{F}\eta$$

subject to the constraint

$$P_e \geq \alpha,$$

where $\alpha > 0$ is a parameter that we choose. This constraint basically tells how inaccurate the estimate of the adversary is in determining the actual value of the terminal state. The value of α provides a measure on the security level of the control sequences. The following result shows that this security constraint is convex in η.

Proposition 5.1. *The probability of error constraint is convex in η, and Problem A is a convex program.*

Proof. From (5.54), we note that P_e is decreasing in d which is nonnegative. Therefore,

$$P_e \geq \alpha \Longleftrightarrow d^2 \leq \alpha_1,$$

where α_1 can be determined from the values of α, π_0, and π_1. For the special case where $\pi_0 = \pi_1$ we have $\alpha_1 = 4(\rho^{-1}(\alpha))^2$. Consider the following notation which enables us to write d^2 in terms of η:

$$\begin{aligned}
&U^i_{0,k-1} = GS_i(U_b + \tilde{F}\eta), \quad G = \begin{bmatrix} I_{km\times km} & 0_{km\times(T-k)m}\end{bmatrix}, \\
&S_0 = \begin{bmatrix} I_{Tm\times Tm} & 0_{Tm\times Tm}\end{bmatrix}, \quad S_1 = \begin{bmatrix} 0_{Tm\times Tm} & I_{Tm\times Tm}\end{bmatrix}, \\
&d^2 = (U_b + \tilde{F}\eta)'(S_1 - S_0)'G'\bar{C}'\Sigma^{-1}\bar{C}G(S_1 - S_0)(U_b + \tilde{F}\eta).
\end{aligned} \tag{5.66}$$

Clearly, d^2 is convex in η, $\{d^2 \le \alpha_1\}$ forms a convex set, and the cost is strictly convex in η. □

Using Proposition 5.1 and (5.66), we solve the following convex program:

$$\min_{\eta} U_b' Q U_b + 2U_b' Q\tilde{F}\eta + \eta'\tilde{F}' Q\tilde{F}\eta$$

subject to the constraint:

$$(U_b + \tilde{F}\eta)'(S_1 - S_0)'G'\bar{C}'\Sigma^{-1}\bar{C}G(S_1 - S_0)(U_b + \tilde{F}\eta) \le \alpha_1. \tag{5.67}$$

We can practically solve this convex program by using standard convex optimization software like cvx [50]. By making some constraint qualification assumptions, we characterize the optimal solution using Lagrangian duality.

Assumption 5.1. We assume that α is selected such that

$$\Big\{\eta \in \mathbb{R}^{\dim(\mathrm{Null}(F))} \,|\, (Ub + \tilde{F}\eta)'(S_1 - S_0)'G'\bar{C}'\Sigma^{-1}\bar{C}G(S_1 - S_0)(U_b + \tilde{F}\eta) < \alpha_1\Big\} \neq \emptyset. \quad \square$$

This assumption ensures that there exists an η such that the constraint in (5.67) is satisfied with strict inequality. This is precisely Slater's condition (see [51] for details) for this problem. Therefore, Assumption 5.1 implies that the duality gap is zero.

Proposition 5.2. *The optimal solution to Problem A, under Assumption 5.1, is nonlinear in b and is given by:*

$$\eta^* = -\left(\tilde{F}'Q\tilde{F} + \lambda^*\tilde{F}'\mathcal{L}'\mathcal{L}\tilde{F}\right)^{-1}(\tilde{F}Q + \lambda^*\tilde{F}'\mathcal{L}'\mathcal{L})U_b$$

where $\lambda^* \ge 0$ is the solution to the following equation:

$$\|\mathcal{L}U_b - \mathcal{L}\tilde{F}\left(\tilde{F}'Q\tilde{F} + \lambda\tilde{F}'\mathcal{L}'\mathcal{L}\tilde{F}\right)^{-1}\left(\tilde{F}'Q + \lambda\tilde{F}'\mathcal{L}'\mathcal{L}\right)\|_2 = \sqrt{\alpha_1}$$

where $\mathcal{L} = W'\bar{C}G(S_1 - S_0)$, $\Sigma^{-1} = WW'$, and b is the vector of terminal states.

Proof. Problem A is a Quadratically Constrained Quadratic Program (QCQP). Such problems have previously been solved in the literature (see [51, Chap. 4]). We provide a proof here for the sake of completeness.

Let $\Sigma^{-1} = WW'$ be the Cholesky decomposition of the inverse of the covariance matrix. The Lagrangian can be written as follows:

$$\begin{aligned} L(\eta, \lambda) &= U_b' Q U_b + 2U_b' Q\tilde{F}\eta + \eta'\tilde{F}\eta \\ &\quad + \lambda\left(U_b'\mathcal{L}'\mathcal{L}U_b + 2\eta'\tilde{F}'\mathcal{L}'\mathcal{L}U_b + \eta'\tilde{F}'\mathcal{L}'\mathcal{L}\tilde{F}\eta - \alpha_1\right). \end{aligned} \tag{5.68}$$

It should be noted that the Lagrangian, $L(\eta, \lambda)$, is strictly convex in η. This is due to the columns of $\tilde{F}$ being linearly independent. Strong duality holds from Slater's conditions, and we can solve the dual problem to get the optimal cost. Strict convexity, strong duality, and the fact that the optimal cost is finite ensure that we can get a solution of the primal problem through the dual problem (see [51, Chap. 5]). The Lagrange dual function is given by:

$$\begin{aligned} g(X) &= \min_{\eta} f(\eta, X), \\ f(\eta, X) &= U_b' Q U_b + 2U_b' Q\tilde{F}\eta + \eta'\tilde{F}\eta \\ &\quad + \lambda\left(U_b'\mathcal{L}'\mathcal{L}U_b + 2\eta'\tilde{F}'\mathcal{L}'\mathcal{L}U_b + \eta'\tilde{F}'\mathcal{L}'\mathcal{L}\tilde{F}\eta - \alpha_1\right). \end{aligned} \tag{5.69}$$

Computing the gradient of the Lagrangian with respect to η and setting it equal to zero we get

$$\eta = -(\tilde{F}'Q\tilde{F} + \lambda\tilde{F}'\mathcal{L}'\mathcal{L}\tilde{F})^{-1}(\tilde{F}'Q + \lambda\tilde{F}'\mathcal{L}'\mathcal{L})U_b. \tag{5.70}$$

Plugging (5.70) into (5.69) we get

$$\begin{aligned} g(X) &= U_b' Q U_b - 2U_b' Q\tilde{F}\left(\tilde{F}'Q\tilde{F} + \lambda\tilde{F}'\mathcal{L}'\mathcal{L}\tilde{F}\right)^{-1}\left(\tilde{F}'Q + \lambda\tilde{F}'\mathcal{L}'\mathcal{L}\right)U_b \\ &\quad + U_b'(\tilde{F}'Q + \lambda\tilde{F}'\mathcal{L}'\mathcal{L})'(\tilde{F}'Q\tilde{F} + \lambda\tilde{F}'\mathcal{L}'\mathcal{L}\tilde{F})^{-1}(\tilde{F}'Q + \lambda\tilde{F}'\mathcal{L}'\mathcal{L})U_b \\ &\quad + \lambda U_b'\mathcal{L}'\mathcal{L}U_b \\ &\quad - 2\lambda U_b'(\tilde{F}'Q + \lambda\tilde{F}'\mathcal{L}'\mathcal{L})'(\tilde{F}'Q\tilde{F} + \lambda\tilde{F}'\mathcal{L}'\mathcal{L}\tilde{F})^{-1}\tilde{F}'\mathcal{L}'\mathcal{L}U_b - \lambda\alpha_1. \end{aligned}$$

Now the optimal cost can be obtained by maximizing the Lagrange dual function with respect to λ which is assumed to be nonnegative. Differentiating g with respect to λ and setting the derivative equal to zero with some algebraic manipulations, we get

$$\left\| \mathcal{L}U_b - \mathcal{L}\tilde{F}\left(\tilde{F}'Q\tilde{F} + \lambda\tilde{F}'\mathcal{L}'\mathcal{L}\tilde{F}\right)^{-1}\left(\tilde{F}'Q + \lambda\tilde{F}'\mathcal{L}'\mathcal{L}\right)U_b \right\|_2 = \sqrt{\alpha_1}. \quad (5.71)$$

Plugging the value of $\lambda \geq 0$, which solves the above equation, into (5.70) provides an optimal solution to the problem, proving the claim in the proposition. It should be noted from (5.71) that λ^* is nonlinear in U_b. Therefore, the optimal solution of the problem is nonlinear in b. □

Proposition 5.2 gives us a form of the optimal control sequences which will satisfy the probability of error security constraint. The convexity of this problem makes it very easy to practically implement these results using commercial optimization solvers.

5.3.2.2 Finite Mean Noise Distribution

In this section, we consider the case where $V_0, \ldots, V_k$ are i.i.d. and have a general distribution with finite mean. Computing a closed-form expression for the probability of error is a very difficult problem even for the specific case of the exponential family of distributions [52].

We consider a new security framework based on the conditional mean which leads to a constraint very similar in structure to (5.67). Let μ be the mean of the noise vector $V_{0,k}$, defined in (5.48). Under the hypothesis H_0 and H_1, the measurement model is given as follows:

$$\begin{aligned} H_0 &: Y_{0,k} = \bar{C}GS_0(U_b + \tilde{F}\eta) + V_{0,k}, \\ H_1 &: Y_{0,k} = \bar{C}GS_1(U_b + \tilde{F}\eta) + V_{0,k}. \end{aligned} \quad (5.72)$$

Problem B. Minimize the cost function

$$U_b'QU_b + 2U_b'Q\tilde{F}\eta + \eta'\tilde{F}'Q\tilde{F}\eta$$

subject to the constraint

$$\left(E(Y_{0,k}|H_1) - E(Y_{0,k}|H_0)\right)'\left(E(Y_{0,k}|H_1) - E(Y_{0,k}|H_0)\right) \leq \alpha_2, \quad \alpha_2 \geq 0. \quad (5.73)$$

The security constraint in (5.73) has a nice operational interpretation. It basically measures how far apart the means of the observations are under the two different hypotheses. The further apart the means, the easier it will be for the adversary to determine the true value of the terminal state. It has the same intuitive interpretation as the probability of error constraint utilized in the previous section. It should be noted that $\alpha_2 = 0$ implies that the adversary does not get any useful information from the partial state observations.

Proposition 5.3. *Problem B is a convex program.*

Proof. Now the difference of the conditional means is given as follows:

$$\left(E(Y_{0,k}|H_1) - E(Y_{0,k}|H_0)\right) = \bar{C}G(S_1 - S_0)(U_b + \tilde{F}\eta). \tag{5.74}$$

Using (5.74), the constraint in (5.73) can be written as

$$U_b'\mathcal{L}'\mathcal{L}U_b + 2\eta'\tilde{F}'\mathcal{L}'\mathcal{L}U_b + \eta'\tilde{F}'\mathcal{L}'\mathcal{L}\tilde{F}\eta \leq \alpha_2, \tag{5.75}$$

where $\mathcal{L} = \bar{C}G(S_1 - S_0)$. Clearly, this security constraint is convex and Problem B is a convex program. □

The security constraint (5.75) is very similar to (5.67), as $\mathcal{L}'\mathcal{L} = \mathcal{L}'\Sigma^{-1}\mathcal{L}$, and hence this section generalizes the work developed earlier. In order to use Lagrangian duality techniques, we make the following assumption.

Assumption 5.2. We assume that α_2 is selected such that

$$\left\{\eta \in \mathbb{R}^{\dim(\mathrm{Null}(F))} | U_b'\mathcal{L}'\mathcal{L}U_b + 2\eta'\tilde{F}'\mathcal{L}'\mathcal{L}U_b + \eta'\tilde{F}'\mathcal{L}'\mathcal{L}\tilde{F}\eta < \alpha_2\right\} \neq \emptyset. \quad \square$$

Proposition 5.4. *The optimal solution to Problem B, under Assumption 5.2, is nonlinear in b and is given by*

$$\eta^* = -\left(\tilde{F}'Q\tilde{F} + \lambda^*\tilde{F}'\mathcal{L}'\mathcal{L}\tilde{F}\right)^{-1}\left(\tilde{F}'Q + \lambda^*\tilde{F}'\mathcal{L}'\mathcal{L}\right)U_b$$

where $\lambda^* > 0$ is the solution to the following equation:

$$\left\|\mathcal{L}U_b - \mathcal{L}\tilde{F}\left(\tilde{F}'Q\tilde{F} + \lambda\tilde{F}\mathcal{L}'\mathcal{L}\tilde{F}\right)^{-1}\left(\tilde{F}'Q + \lambda\tilde{F}'\mathcal{L}'\mathcal{L}\right)U_b\right\|_2 = \sqrt{\alpha_2}.$$

Proof. The proof follows exactly along the same lines as that of Proposition 5.2. □

Using these results, we can design secure controls for any system for which the adversary makes observations where the additive noise is drawn from a general finite mean distribution. The aforementioned security constraint can also be extended to the M-ary framework. We explore such an extension in the next section.

5.3.3 An M-ary Framework

In this section, we consider the framework where the terminal state can take M different values $x^0, x^1, \dots, x^{M-1}$ with probabilities $\pi_0, \pi_1, \dots, \pi_{M-1}$ respectively. We assume that the measurement noise vector $V_{0,k}$ has a distribution where the mean vector, μ, has finite components.

We consider a constraint defined using the conditional means. Now under the hypothesis H_i, $i = 0, \dots, M-1$, the measurement model is given by

$$H_i : Y_{0,k} = \bar{C}GSi(U_b + \tilde{F}\eta) + V_{0,k}, \quad i = 0, \dots, M-1. \tag{5.76}$$

Consider the following security metric:

$$\sum_{i=0}^{M-1} \pi_i \left(\mathbb{E}(Y_{0,k}|H_i) - \sum_{j=0}^{M-1} \pi_j \mathbb{E}(Y_{0,k}|H_j) \right)' \left(\mathbb{E}(Y_{0,k}|H_i) - \sum_{j=0}^{M-1} \pi_j \mathbb{E}(Y_{0,k}|H_j) \right). \tag{5.77}$$

This security metric can be considered a generalization of the difference of conditional means which was employed for the binary framework. The distribution π_i is used to assign weights to each quadratic quantity in (5.77), which provides a difference of the conditional mean given one hypothesis with the weighted sum of the conditional means given other hypotheses. It should be noted that $Y_{0,k}$ is independent of the terminal state X_T if and only if this security metric is zero. A lower level of dependence between the observations and the terminal state is indicated by a small value of this metric. A smaller value makes it more difficult for the adversary to predict the terminal state and indicates a higher level of security of the control sequence. Introducing

$$\sigma_i = \left(\mathbb{E}(Y_{0,k}|H_i) - \sum_{j=0}^{M-1} \pi_j \mathbb{E}(Y_{0,k}|H_j) \right) \tag{5.78}$$

and utilizing the foregoing security metric, we state the following optimization problem:

Problem C. Minimize the cost function

$$U_b^t Q U_b + 2U_b^t Q \tilde{F}\eta + \eta^t \tilde{F}^t Q \tilde{F}\eta$$

subject to the constraint

$$\sum_{i=0}^{M-1} \pi_i \sigma'{}^t \sigma_i \leq \alpha_3. \tag{5.79}$$

Parameter α_3 is selected to be nonnegative. A small α_3 indicates that a lower level of useful information can be transmitted to the adversary. For the case $\alpha_3 = 0$, the controller will be restricted to those control sequences which generate the same state trajectory for the first $k+1$ steps. Using the measurement model, we get

$$\mathbb{E}(Y_{0,k}|H_i) = \sum_{i=0}^{M-1} \pi_i \left(\mathbb{E}(Y_{0,k}|H_i)\right) = \bar{C}G\left(S_i - \sum_{j=0}^{M-1} \pi_j S_j\right)(U_b + \tilde{F}\eta). \tag{5.80}$$

Using (5.80), and $\mathcal{L}_i = \overline{C}G\left(S_i - \sum_{j=0}^{M-1} \pi_j S_j\right)$, we can write (5.79) as

$$\sum_{i=0}^{M-1} \pi_i \left(U_b' \mathcal{L}_i' \mathcal{L}_i U_b + 2\eta' \tilde{F}' \mathcal{L}_i' \mathcal{L}_i U_b + \eta' \tilde{F}' \mathcal{L}_i' \mathcal{L}_i \tilde{F}\eta\right) \leq \alpha_3. \tag{5.81}$$

Clearly, (5.81) is convex in η, and hence we can conclude that Problem C is a convex program. By making similar assumptions, we can extend the results to the M-ary framework.

Assumption 5.3. We assume that α_3 is selected such that $\eta \in \mathbb{R}^{\dim(\mathrm{Null}(F))}$ and

$$\left\{\sum_{i=0}^{M-1} \pi_i U_b' \mathcal{L}_i' \mathcal{L}_i U_b + 2\eta' \tilde{F}' \mathcal{L}_i' \mathcal{L}_i U_b + \eta' \tilde{F}' \mathcal{L}_i' \mathcal{L}_i \tilde{F}\eta < \alpha_3\right\} \neq \emptyset. \quad \square$$

Proposition 5.5. *The optimal solution of Problem C, under Assumption 5.3, is nonlinear in b and is given by*

$$\eta^* = -\left(\tilde{F}' Q \tilde{F} + \lambda^* \sum_{i=0}^{M-1} \pi_i \tilde{F}' \mathcal{L}_i' \mathcal{L}_i \tilde{F}\right)^{-1} \left(\tilde{F}' Q + \lambda^* \sum_{i=0}^{M-1} \pi_i \tilde{F}' \mathcal{L}_i' \mathcal{L}_i\right) U_b$$

where $\lambda^* > 0$ is the solution to the following equations:

$$\sigma_1 = \mathcal{L}_k \tilde{F}\left(\tilde{F}' Q \tilde{F} + \lambda \sum_{i=0}^{M-1} \pi_i \tilde{F}' \mathcal{L}_i' \mathcal{L}_i \tilde{F}\right)^{-1},$$

$$\sigma_2 = \left(\tilde{F}' Q + \lambda \sum_{i=0}^{M-1} \pi_i \tilde{F}' \mathcal{L}_i' \mathcal{L}_i \tilde{F} \right) U_b,$$

$$\sigma = \sum_{k=0}^{M-1} \pi_k \left\| \mathcal{L}_k U_b - \sigma_1 \sigma_2 \right\|_2 = \sqrt{\alpha_3}.$$

Proof. The proof follows exactly as that of Proposition 5.3. □

The results stated in Proposition 5.5 are very similar to the results presented in Propositions 5.1 and 5.2. These results imply that the optimal solution is nonlinear in the vector of terminal states. In order to find the optimal solution in these problems, we need to solve for the equality of a norm to a design parameter. In Proposition 5.5, we have the weighted sum of such norms equaling $\sqrt{\alpha_3}$.

5.3.4 Terminal State With a Continuous Distribution

In this section, we consider the situation where the terminal state has a continuous finite mean distribution with given density function $p(x_T)$. In addition, the components of the terminal state are assumed to have a finite variance. It is assumed that the adversary knows the distribution of the terminal state. We first recall the definition of the Gateaux differential presented in Appendix A.

Now introduce the following security metric based on the difference of conditional means:

$$\int_{\mathbb{R}^n} \int_{\mathbb{R}^n} p(x_T) p(x_T) \left(\mathbb{E}(Y_{0,k} | H_{x_T}) - \mathbb{E}(Y_{0,k} | H_{y_T}) \right)' \times \left(\mathbb{E}(Y_{0,k} | H_{x_T}) - \mathbb{E}(Y_{0,k} | H_{y_T}) \right) d_{y_T} d_{x_T},$$

where the suffix y_T is another realization of the terminal state. This security metric basically provides a measure on the difference of the conditional means. The higher the value of this metric, the easier it will be for the adversary to estimate the terminal state. Using the measurement model, this metric can be simplified as follows:

$$\int_{\mathbb{R}^n} \int_{\mathbb{R}^n} p(x_T) p(y_T) \left(U(x_T - U(y_T)) \right) \bar{G}' \bar{G} \left(U(x_T) - U(y_T) \right) d_{y_T} d_{x_T}. \tag{5.82}$$

Using the cost functional (A.10), and the security constraint (5.82), we state the following optimization problem:

Problem D. Find

$$\min_{\eta(\cdot)} \int_{\mathbb{R}^n} \left(U(x_T) + \tilde{B}\eta(x_T)\right)' \tilde{Q}\left(\tilde{U}(x_T) + \tilde{B}\eta(x_T)\right) p(x_T) d_{x_T}$$

subject to the constraint

$$\int_{\mathbb{R}^n}\int_{\mathbb{R}^n} p(x_T)p(y_T)\left(\tilde{U}(x_T) - \tilde{U}(y_T) + \tilde{B}(\eta(x_T) - \eta(y_T))\right)' \bar{G}'\bar{G} \\ \times \left(\tilde{U}(x_T) - \tilde{U}(y_T) + \tilde{B}(\eta(x_T) - \eta(y_T))\right) d_{y_T} d_{x_T} \leq \alpha_4,$$

where α_4 is assumed to be nonnegative. A small value of α_4 indicates that a lower level of useful information is provided to the adversary. If $\alpha_4 = 0$, then the first k control inputs of all admissible control sequences will be the same. Problem D is a convex infinite-dimensional optimization problem. We provide a solution to this problem by utilizing the generalized Kuhn–Tucker conditions [53].

Assumption 5.4. We consider only those values of all for which the appropriate Lagrange multiplier, λ^*, is small enough such that

$$\left[I_{q\times q} - 2\lambda^* \tilde{B}'\bar{G}'\bar{G}\tilde{B}\left(\tilde{B}'(\tilde{Q} + 2\lambda^* \bar{G}'\bar{G})\tilde{B}\right)^{-1}\right]$$

is nonsingular. □

Assumption 5.5. We assume that the constraint parameter α_4 and the system dynamics are selected such that the optimal solution $\eta^*(\cdot)$ to Problem D satisfies the security constraint and that there exists an $h(\cdot) \in \mathbb{L}^2(\mathbb{R}^n, \mathcal{B}(\mathbb{R}^n), \mu_p)$ such that

$$\int_{\mathbb{R}^n}\int_{\mathbb{R}^n} p(x_T)p(y_T)\left(\tilde{U}(x_T) - \tilde{U}(y_T) + \tilde{B}(\eta(x_T) - \eta(y_T))\right)' \bar{G}'\bar{G} \\ \times \left(\tilde{U}(x_T) - \tilde{U}(y_T) + \tilde{B}(\eta(x_T) - \eta(y_T))\right) d_{y_T} d_{x_T} \\ + 4\int_{\mathbb{R}^n} \left(\tilde{U}(x_T) + \tilde{B}\eta^*(x_T)\right)' \bar{G}'\bar{G}\tilde{B}h(x_T)p(x_T)d_{x_T} \\ - 4\int_{\mathbb{R}^n}\int_{\mathbb{R}^n} \left(\tilde{U}(x_T) + \tilde{B}\eta^*(x_T)\right)' \bar{G}'\bar{G}\tilde{B}h(y_T)p(x_T)d_{y_T}d_{x_T} < \alpha_4. \quad \square$$

Using Gateaux differentials, Assumptions 5.4 and 5.5, and the generalized Kuhn–Tucker conditions, we obtain the following result:

Proposition 5.6. *The optimal solution to Problem D, under Assumptions 5.4 and 5.5, is affine in x_i, and is given by*

$$\begin{aligned}\eta^*(x_T) = &- \left(\tilde{B}'\tilde{Q}\tilde{B} + 2\lambda^*\tilde{B}'\bar{G}'\bar{G}\tilde{B}\right)^{-1}\left(\tilde{B}'\tilde{Q} + 2\lambda^*\tilde{B}'\bar{G}'\bar{G}\right)\tilde{U}(x_T)\\ &+ 2\lambda^*\left(\tilde{B}'\tilde{Q}\tilde{B} + 2\lambda^*\tilde{B}'\bar{G}'\bar{G}\tilde{B}\right)^{-1}\\ &\times\left(\tilde{B}'\bar{G}'\bar{G} - (I_{q\times q} - 2\lambda^*\Gamma_{\lambda^*})^{-1}\Gamma_{\lambda^*}\tilde{B}'\tilde{Q}\right)\\ &\times\int p(y_T)\tilde{U}(y_T)d_{y_T},\end{aligned}$$

where $\Gamma_{\lambda^*} = \tilde{B}'\bar{G}'\bar{G}\tilde{B}\left(\tilde{B}'(\tilde{Q} + 2\lambda^*\bar{G}'\bar{G})\tilde{B}\right)^{-1}$ *is the solution to the following equation:*

$$\begin{aligned}\lambda\Bigg[&-2\int_{\mathbb{R}^n}\int_{\mathbb{R}^n} p(x_T)(y_T)\left(\tilde{U}(x_T) + \tilde{B}\eta(x_T)\right)'\bar{G}'\bar{G}\left(\tilde{U}(x_T) + \tilde{B}\eta(x_T)\right)d_{y_T}d_{x_T}\\ &+\int_{\mathbb{R}^n} p(x_T)\left(\tilde{U}(x_T) + \tilde{B}\eta(x_T)\right)'\bar{G}'\bar{G}\left(\tilde{U}(x_T) + \tilde{B}\eta(x_T)\right)d_{x_T} - \alpha_4\Bigg] = 0.\end{aligned}$$

Proof. We will first compute the Gateaux differential of the Lagrangian and then use the Kuhn–Tucker conditions to compute the optimal solution. Now the Lagrangian can be written as follows:

$$\begin{aligned}J_\lambda(\eta(\cdot)) = &\int_{\mathbb{R}^n} p(x_T)\left(\tilde{U}(x_T) + \tilde{B}\eta(x_T)\right)'\left(\tilde{Q} + 2\lambda\bar{G}'\bar{G}\right)\left(\tilde{U}(x_T) + \tilde{B}\eta(x_T)\right)'d_{x_T}\\ &-2\lambda\int_{\mathbb{R}^n}\int_{\mathbb{R}^n} p(x_T)p(y_T)\left(\tilde{U}(x_T) + \tilde{B}\eta(x_T)\right)'\bar{G}'\bar{G}\left(\tilde{U}(y_T) + \tilde{B}\eta(y_T)\right)d_{y_T}d_{x_T}\end{aligned} \tag{5.83}$$

For any admissible variation $h(\cdot) \in \mathbb{L}^2(\mathbb{R}^n, \mathcal{B}(\mathbb{R}^n), \mu_p)$, the Gateaux differential of the Lagrangian is given by

$$\begin{aligned}\delta J_\lambda(\eta(\cdot), h(\cdot)) = &\,2\int_{\mathbb{R}^n} p(x_T)\tilde{U}(x_T)'\left(\tilde{Q} + 2\lambda\bar{G}'\bar{G}\right)\tilde{B}h(x_T)d_{x_T}\\ &+2\int_{\mathbb{R}^n} p(x_T)\eta(x_T)'\tilde{B}'\left(\tilde{Q} + 2\lambda\bar{G}'\bar{G}\right)\tilde{B}h(x_T)d_{x_T}\\ &-4\lambda\int_{\mathbb{R}^n}\int_{\mathbb{R}^n} p(x_T)p(y_T)\tilde{U}(x_T)'\bar{G}'\bar{G}\tilde{B}h(y_T)d_{y_T}d_{x_T}\\ &-4\lambda\int_{\mathbb{R}^n}\int_{\mathbb{R}^n} p(x_T)p(y_T)\eta(x_T)'\tilde{B}'\bar{G}'\bar{G}h(y_T)d_{y_T}d_{x_T}.\end{aligned} \tag{5.84}$$

Setting the Gateaux differential equal to zero, we get

$$
\begin{aligned}
&\int_{\mathbb{R}^n} p(x_T) h(x_T)' \Big(\tilde{B}' \tilde{Q} \tilde{U}(x_T) \\
&\quad + 2\lambda \tilde{B}' \bar{G}' \bar{G} \tilde{U}(x_T) + \tilde{B}' \tilde{Q} \tilde{B} \eta(x_T) + 2\lambda \tilde{B}' \bar{G}' \bar{G} \tilde{B} \eta(x_T) \\
&\quad - 2\lambda \int_{\mathbb{R}^n} p(y_T) \tilde{B}' \bar{G}' \bar{G} \tilde{U}(y_T) d_{y_T} \\
&\quad - 2\lambda \int_{\mathbb{R}^n} p(y_T) \tilde{B}' \bar{G}' \bar{G} \tilde{B} \eta(y_T) d_{y_T} \Big) \, d_{x_T} = 0 \qquad \forall h(\cdot).
\end{aligned} \tag{5.85}
$$

Now (5.85) holds if and only if

$$
\begin{aligned}
&\tilde{B}' \left(\tilde{Q} + 2\lambda \bar{G}' \bar{G} \right) \tilde{U}(x_T) + \tilde{B}' \left(\tilde{Q} + 2\lambda \bar{G}' \bar{G} \right) \tilde{B} \eta(x_T) \\
&\quad - 2\lambda \int_{\mathbb{R}^n} \tilde{B}' \bar{G}' \bar{G} \tilde{U}(y_T) d_{y_T} - 2\lambda \int_{\mathbb{R}^n} p(y_T) \tilde{B}' \bar{G}' \bar{G} \tilde{B} \eta(y_T) d_{y_T} \\
&= 0, \qquad \forall_{x_T} \in \mathbb{R}^n.
\end{aligned} \tag{5.86}
$$

Now multiplying (5.86) throughout by $p(x_T)\Gamma_\lambda$ and integrating over x_T, we get

$$
\left(I_{q\times q} \quad 2\lambda \Gamma_\lambda \right) \int_{\mathbb{R}^n} p(x_T) \tilde{B}' \bar{G}' \bar{G} \tilde{B} \eta(x_T) = - \int_{\mathbb{R}^n} p(x_T) \Gamma_\lambda \tilde{B}' \tilde{Q} \tilde{U}(x_T) d_{x_T}.
$$

Using Assumption 5.4, we get

$$
\int_{\mathbb{R}^n} p(x_T) \tilde{B}' \bar{G}' \bar{G} \tilde{B} \eta(x_T) d_{x_T} = - \left(I_{q\times q} - 2\lambda \Gamma_\lambda \right)^{-1} \int_{\mathbb{R}^n} p(x_T) \Gamma_\lambda \tilde{B}' \tilde{Q} \tilde{U}(x_T) d_{x_T}. \tag{5.87}
$$

Now plugging (5.87) into (5.86), we get

$$
\begin{aligned}
F_1 &= - \left(\tilde{B}' \tilde{Q} \tilde{B} + 2\lambda^* \tilde{B}' \bar{G}' \bar{G} \tilde{B} \right)^{-1} \left(\tilde{B}' \tilde{Q} + 2\lambda^* \tilde{B}' \bar{G}' \bar{G} \right) \tilde{U}(x_T), \\
F_2 &= 2\lambda^* \left(\tilde{B}' \tilde{Q} + 2\lambda^* \tilde{B}' \bar{G}' \bar{G} \tilde{B} \right)^{-1} \left(\tilde{B}' \bar{G}' \bar{G} - (I_{q\times q} - 2\lambda^* \Gamma_{\lambda^*})^{-1} \Gamma_{\lambda^*} \tilde{B}' \tilde{Q} \right), \\
\eta^*(x_T) &= F_1 + F_2 \int_{\mathbb{R}^n} p(y_T) \tilde{U}(y_T) d_{u_T},
\end{aligned} \tag{5.88}
$$

which, due to the Kuhn–Tucker conditions, used in conjunction with Assumption 5.5, provides the optimal solution. Also from Kuhn–Tucker conditions, λ^* is given by the solution of the following equation:

$$\lambda[-2\int_{\mathbb{R}^n}\int_{\mathbb{R}^n} p(x_T)p(y_T)\left(\tilde{U}(x_T)+\tilde{B}\eta^*(x_T)\right)'\bar{G}'\bar{G}\left(\tilde{U}(y_T)+\tilde{B}\eta^*(y_T)\right)d_{y_T}d_{x_T}$$
$$+2\int_{\mathbb{R}^n} p(x_T)\left(\tilde{U}(x_T)+\tilde{B}\eta^*(x_T)\right)'\bar{G}'\bar{G}\left(\tilde{U}(x_T)+\tilde{B}\eta^*(x_T)\right)d_{x_T}-\alpha_4]=0. \tag{5.89}$$

It should be noted that λ^* depends upon $\tilde{U}(\cdot)$ and $P(\cdot)$, but not on x_T. Also $\tilde{U}(x_T)$ is selected to be linear in x_T. Therefore, we conclude from (5.82) that the optimal solution to Problem D is affine in x_T. □

Remark 5.10. *Proposition 5.6 is an important result and shows that the optimal solution to the problem is affine in the terminal state. Contrary to the results in the previous sections, we obtain an affine solution in the continuous terminal distribution case. Assumption 5.4 is somewhat restrictive as it allows us to operate only over particular regimes of the parameter α_4. Such an assumption can be removed if we consider the following simpler security constraint which also has an affine solution in the terminal state:*

$$\int_{\mathbb{R}^n} p(x_T)U(x_T)'\bar{G}'\bar{G}U(x_T)d_{x_T}\leq\alpha_5,\quad \alpha_5\geq 0. \tag{5.90}$$

The main difference between these constraints is that in (5.90) we compare the first k components of a control sequence, which drive the state to a particular terminal state, to the zero control sequence.

5.3.5 Simulation Example 5.5

In this section, we provide an analysis of the behavior of the cost function as the security constraint parameter is varied. In addition, the optimal solution is analyzed through simulations. We consider the framework of Problem A and make the additional assumption that the priors are equiprobable. Consider the system dynamics:

$$A=\begin{bmatrix}1 & 23 & 4\\ 5 & 7 & 12\\ 1 & 23 & 16\end{bmatrix},\quad B=\begin{bmatrix}1 & 4 & 4.3\\ 6 & 2.3 & 8\\ 12 & 7 & 1.8\end{bmatrix},\quad b_1=\begin{bmatrix}53\\ 75\\ 100\\ 32\\ 37\\ -8\end{bmatrix}$$

Let $\Sigma=2\times I_{12\times12}$, $Q=5\times I_{30\times30}$, and $\alpha=0.45$. We assume that $T=5$ and $k=2$. Using these values, α_1 is calculated to be $4(\rho^{-1}(0.45))^2$. We use the standard optimization software cvx to compute the optimal control sequences. The optimal solution, $\mathbb{U}$, is given in (5.91):

$$U(1:15) = \begin{bmatrix} -0.0118 \\ 0.0128 \\ 0.0076 \\ -0.0048 \\ -0.0023 \\ 0.0099 \\ ---- \\ 0.0032 \\ 0.0052 \\ -0.0221 \\ 0.5366 \\ 0.0016 \\ -0.3726 \\ 0.1369 \\ 0.0453 \\ -0.1924 \end{bmatrix}, \quad U(16:30) = \begin{bmatrix} -0.0118 \\ 0.0127 \\ 0.0076 \\ -0.0051 \\ -0.0020 \\ 0.0093 \\ ----- \\ -0.0063 \\ -0.0058 \\ 0.0251 \\ -0.5374 \\ -0.0018 \\ 0.3736 \\ -0.1371 \\ -0.0454 \\ 0.1927 \end{bmatrix} \tag{5.91}$$

It should be noted that $\mathbb{U}(1:15)$ corresponds to the first 15 elements of the control vector $\mathbb{U}$ which drive the system to the terminal state $[53\ 75\ 100]'$. Similarly $\mathbb{U}(16:30)$ corresponds to the control inputs which drive the system to the terminal state $[32\ 37\ -8]'$, respectively. It should be noted that $k = 2$, $m = 3$, and hence the adversary can only make noisy measurements of the first control inputs in (5.12), which are colored red (gray in print version), of these control vectors. Note that the control inputs which the adversary measures in both cases are very similar to one another thereby providing minimum useful information to the adversary. This is what we expected and this further signifies the importance of incorporating security constraints in control problems. The simulation in Fig. 5.11 utilizes the system described above. We plot the optimal cost against the constraint parameter α, where $P_e \geq \alpha$. As discussed earlier, if $\alpha > 0.5$, then the security constraint becomes infeasible. Clearly, as an increased problem becomes more constrained, the cost also increases. An exponential increase in the cost is observed by an increase in the value of α, for $\alpha \geq 14$. In Fig. 5.11, $[53\ 75\ 100\ 32\ 37\ \ -8]'$ was utilized as the vector of terminal states. In order to avoid any confusion, we clarify that the value of the optimal cost, for $0 \leq \alpha \leq 0.15$, is approximately 0.01 and not zero.

We now consider a different dynamical system and a different vector of terminal states to analyze the increase in the optimal cost with the constraint

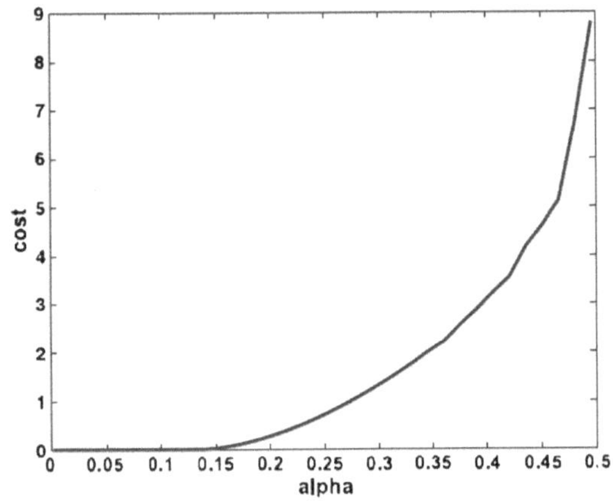

Figure 5.11 Optimal cost for the system with terminal state vector b_1

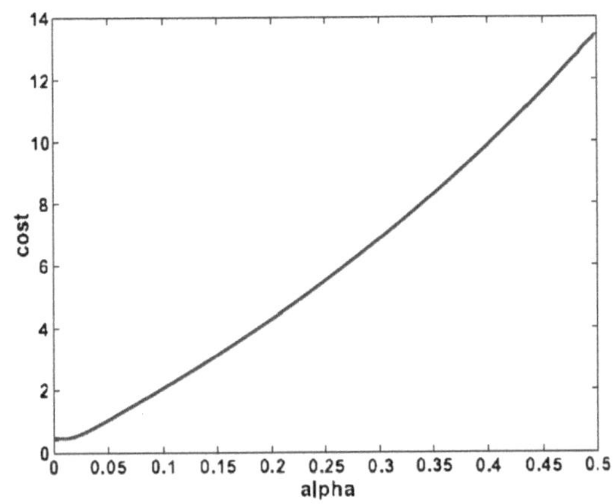

Figure 5.12 Optimal cost for the system with terminal state vector b_2

parameter. Let

$$A = \begin{bmatrix} 1 & 3 & 4 \\ 3.1 & 7 & 2 \\ 5 & 3 & 6 \end{bmatrix}, \quad B = \begin{bmatrix} 1 & 4 & 0.3 \\ 6 & 3 & 2 \\ 2 & 7 & 9 \end{bmatrix}, \quad b_2 = \begin{bmatrix} 10 \\ 12 \\ -8 \\ 4 \\ 7 \\ 5 \end{bmatrix}.$$

We take $\Sigma = 0.5 \times I_{12x12}$, $Q = 10 \times I_{30\times30}$, $T = 5$ and $k = 3$. Fig. 5.12 shows the increase in the optimal cost with the value for this system. Clearly, in this case the optimal cost increases less rapidly than in the case of Fig. 5.11. Unlike for the previous system, the optimal cost does not stay constant over a large range of values of α. Therefore, we can conclude that for this system the probability of error constraint is more tightly enforced as compared to the previous system. Also the rate of change in the optimal cost is approximately constant. These simulations further show the role of constraint parameters in the optimal solution. Similar results can be obtained when we employ constraints based on the conditional mean.

5.4 LYAPUNOV-BASED METHODS UNDER DENIAL-OF-SERVICE

In what follows, we consider networked distributed systems under DoS attacks, which has not been investigated so far under the class of DoS attacks introduced in [55,56]. Analysis of the behavior of systems in a centralized-system manner was carried out in [55–59], where the major characteristic is that all the states are assumed to be collected and sent in one transmission attempt. In this section, we analyze the problem from the distributed system point of view, where the interconnected subsystems share one communication channel and transmission attempts of the subsystems take place asynchronously.

5.4.1 Networked Distributed System

Consider a large-scale system consisting of N interacting subsystems, whose dynamics satisfy

$$\dot{x}_i(t) = A_i x_i(t) + B_i u_i(t) + \sum_{j \in N_i} H_{ij} x_j(t), \tag{5.92}$$

where A_i, B_i and H_{ij} are matrices with appropriate dimensions and $t \in \mathbb{R}_{>0}$; $x_i(t)$ and $u_i(t)$ are state and control input of subsystem i, respectively. Here we assume that all the subsystems are full state output; N_i denotes the set of neighbors of subsystem i. Subsystem i physically interacts through $\sum_{j \in N_i} H_{ij} x_j(t)$ with its neighbor subsystem(s) $j \in N_i$. Here we consider bidirectional edges, i.e., $j \in N_i$ when $i \in N_j$.

The distributed systems are controlled via a shared networked channel, through which distributed plants broadcast the measurements and controllers send control inputs. The computation of control inputs is based on the transmitted measurements. The received measurements are made in sample-and-hold fashion, so we get $x_i(t_k^i)$ where t_k^i represents the sequence of transmission instants of subsystem i. We assume that there exists a feedback matrix K_i such that $\phi_i = A_i + B_i K_i$ is Hurwitz. Therefore, the control input applied to subsystem i is given by

$$u_i(t) = K_i x_i(t_k^i) + \sum_{j \in N_i} L_{ij} x_j(t_k^j), \tag{5.93}$$

where L_{ij} is the coupling gain in the controller. Here we assume that the channel is noiseless and there is no quantization. Moreover, we assume that the network transmission delay and the computation time of control inputs are zero.

5.4.2 DoS Attacks' Frequency and Duration

We refer to Denial-of-Service as the phenomenon for which transmission attempts may fail. In this section, we do not distinguish between transmission failures due to channel unavailability and transmission failures because of DoS-induced packet corruption. Since the network is shared, DoS simultaneously affects the communication attempts of all the subsystems.

Clearly, the problem in question does not have a solution if the DoS amount is allowed to be arbitrary. Following [56], we consider a general DoS model that constrains the attacker action in time by only posing limitations on the frequency of DoS attacks and their duration. Let $\{h_n\}_{n \in \mathbb{N}_0}$, $h_0 \geq 0$, denote the sequence of DoS *off/on* transitions, i.e., the time instants at which DoS exhibits a transition from zero (transmissions are possible) to one (transmissions are not possible). Hence,

$$H_n := \{h_n\} \cup [h_n, h_n + \tau_n[\tag{5.94}$$

represents the nth DoS time-interval, of a length $\tau_n \in \mathbb{R}_{\geq 0}$, over which the network is in DoS status. If $\tau_n = 0$, then H_n takes the form of a single pulse at h_n. If $\tau_n \neq 0$, $[h_n, h_n + \tau_n[$ represents an interval from the instant h_n (including h_n) to $(h_n + \tau_n)^-$ (arbitrarily close to, but excluding $h_n + \tau_n$). Similarly, $[\tau, t[$ represents an interval from τ to t^-. Given $\tau, t \in \mathbb{R}_{\geq 0}$ with $t \geq \tau$, let $n(\tau, t)$ denote the number of DoS off/on transitions over $[\tau, t[$, and let

$$\Xi(\tau, t) := \bigcup_{n \in \mathbb{N}_0} H_n \bigcap [\tau, t] \tag{5.95}$$

denote the subset of $[\tau, t]$ where the network is in DoS status. The subset of time where DoS is absent is denoted by

$$\Theta(\tau, t) := [\tau, t] \setminus \Xi(\tau, t). \tag{5.96}$$

We make the following assumptions:

Assumption 5.6 (DoS frequency). For $\Xi(\tau, t) := \bigcup_{n \in \mathbb{N}_0} H_n \bigcap [\tau, t]$ there exist constants $\eta \in \mathbb{R}_{\geq 0}$ and $\tau_D \in \mathbb{R}_{>0}$ such that

$$n(\tau, t) \leq \eta + \frac{t - \tau}{\tau_D}, \tag{5.97}$$

for all $\tau, t \in \mathbb{R}_{\geq 0}$ with $t \geq \tau$. □

Assumption 5.7 (DoS duration). There exist constants $\kappa \in \mathbb{R}_{\geq 0}$ and $T \in \mathbb{R}_{>1}$ such that

$$|\Xi(\tau, t)| \leq \kappa + \frac{t - \tau}{T}, \tag{5.98}$$

for all $\tau, t \in \mathbb{R}_{\geq 0}$ with $t \geq \tau$. □

Remark 5.11. *Assumptions 5.6 and 5.7 do only constrain a given DoS signal in terms of its average frequency and duration. Actually, τ_D can be defined as the average dwell-time between consecutive DoS off/on transitions, while η is the chattering bound. Assumption 5.7 expresses a similar requirement with respect to the duration of DoS. It expresses the property that, on average, the total duration over which communication is interrupted does not exceed a certain fraction of time, as specified by $1/T$. Like η, constant κ plays the role of a regularization term. It is needed because during a DoS interval, one has $|\Xi(h_n, h_n + \tau_n)| = \tau_n > \tau_n/T$. Thus κ serves to make (5.98) consistent. Conditions $\tau_D > 0$ and $T > 1$ imply that a DoS cannot occur at an infinitely fast rate or be always active.*

The objective now is to find stability conditions for the networked distributed systems under DoS attacks. We first study the stabilization problem of large-scale systems under a digital communication channel in the absence of DoS.

5.4.3 A Small-Gain Approach for Networked Systems

For each subsystem i, we denote by $e_i(t)$ the error between the value of the state transmitted to its neighbors and the current state, that is,

$$e_i(t) = x_i(t_k^i) - x_i(t), \quad i = 1, 2, \ldots, N. \tag{5.99}$$

Combining (5.92), (5.93) and (5.99), the dynamics of subsystem i can be written as

$$\begin{aligned}\dot{x}_i(t) &= \Phi_i x_i(t) + B_i K_i e_i(t) + \sum j \in N_i (B_i L_{ij} + H_{ij}) x_j(t) \\ &\quad + B_i \sum j \in N_i L_{ij} e_j(t),\end{aligned} \tag{5.100}$$

from which one sees that the dynamics of subsystem i depends on the interconnected neighbors $x_j(t)$ as well as $e_i(t)$, $e_j(t)$ and the coupling parameters. Intuitively, if the couplings are weak and e remains small, then stability can be achieved. Here, the notion of "smallness" of e can be characterized by the x-dependent bound $||e_i(t)|| \leq \sigma_i ||x_i(t)||$, in which σ_i is a suitable design parameter. Notice that this is not the network update rule.

We implement a periodic sampling protocol, like round-robin, as our update law. In this respect, we make the following hypothesis:

Assumption 5.8 (Inter-sampling of round-robin)**.** In the absence of DoS attacks, there exists an inter-sampling interval Δ such that

$$||e_i(t)|| \leq \sigma_i ||x_i(t)|| \tag{5.101}$$

holds, where σ_i is a suitable design parameter. □

As mentioned in the foregoing argument, σ_i should be designed carefully. Otherwise, even if there exists a Δ under which (5.101) holds, in the event of an inappropriate σ_i, stability can be lost as well.

Given any symmetric positive definite matrix Q_i, let P_i be the unique solution of the Lyapunov equation $\Phi_i^T P_i + P_i \Phi_i + Q_i = 0$. For each i, consider the Lyapunov function $V_i = x_i^T P_i x_i$, which satisfies

$$\lambda_{min}(P_i)||x_i(t)||^2 \leq V_i(x_i(t)) \leq \lambda_{max}(P_i)||x_i(t)||^2 \tag{5.102}$$

where $\lambda_{min}(P_i)$ and $\lambda_{max}(P_i)$ represent the smallest and largest eigenvalues of P_i, respectively. The following lemma presents the design of σ_i guaranteeing stability.

Lemma 5.1. *Consider a distributed system as in* (5.92) *along with a control input as in* (5.93). *Suppose that the spectral radius $r(A^{-1}B) < 1$. The distributed system is asymptotically stable if σ_i satisfies*

$$\sigma_i < \sqrt{\frac{l_i}{j_i}}, \tag{5.103}$$

where l_i is the ith entry of row vector $L := \mu^T(A - B) = [l_1, l_2, \ldots, l_N]$ and j_i is the jth entry of row vector $J := \mu^T\Gamma[j_1, j_2, \ldots, j_N]$; $\mu \in \mathbb{R}_+^N$ is an arbitrary column vector satisfying $\mu^T(-A + B) < 0$. The matrices A, B and Γ are given by

$$A = \begin{bmatrix} \alpha_1 & & \\ & \ddots & \\ & & \alpha_N \end{bmatrix}, \quad B = \begin{bmatrix} 0 & \beta_{12} & \ldots & \beta_{1N} \\ \beta_{21} & 0 & \beta_{23} & \beta_{2N} \\ \vdots & \vdots & 0 & \vdots \\ \beta_{N1} & \beta_{2N} & \ldots & 0 \end{bmatrix},$$

$$\Gamma = \begin{bmatrix} 0 & \gamma_{12} & \ldots & \gamma_{1N} \\ \gamma_{21} & 0 & \gamma_{23} & \gamma_{2N} \\ \vdots & \vdots & 0 & \vdots \\ \gamma_{N1} & \gamma_{2N} & \ldots & 0 \end{bmatrix}, \tag{5.104}$$

with

$$\alpha_i = \lambda_{min}(Q_i) - \delta - \sum_{j \in N_i} 2\delta, \quad \beta_{ij} = \frac{||P_i||^2 ||B_i L_{ij} + H_{ij}||^2}{\delta}, \tag{5.105}$$

$$\gamma_{ii} = \frac{||P_i||^2 ||B_i K_i||^2}{\delta}, \quad \gamma_{ij} = \frac{||P_i||^2 ||B_i L_{ij}||^2}{\delta}, \tag{5.106}$$

where δ is a positive real such that $\alpha_i > 0$ and $\lambda_{min}(Q_i)$ is the smallest eigenvalue of Q_i for $i = 1, 2, \ldots, N$.

Proof. Recalling that $V_i = x_i^T P_i x_i$, the derivative of V_i along the solution to (5.100) satisfies

$$\begin{aligned} \dot{V}_i(x_i(t)) \le & -\lambda_{min}(Q_i)||x_i(t)||^2 + ||2P_i B_i K_i||\, ||x_i(t)||\, ||e_i(t)|| \\ & + \sum_{j \in N_i} ||2P_i(B_i L_{ij} + H_{ij})||\, ||x_i(t)||\, ||x_j(t)|| \\ & + \sum_{j \in N_i} ||2P_i B_i L_{ij}||\, ||x_i(t)||\, ||e_j(t)||. \end{aligned} \tag{5.107}$$

Observe that for any positive real δ, Young's inequalities yield

$$\begin{aligned} & ||2P_i B_i K_i||\, ||x_i(t)||\, ||e_i(t)|| \\ & \quad \le \delta ||x_i(t)||^2 + \frac{||P_i||^2 ||B_i K_i||^2}{\delta} ||e_i(t)||^2 \end{aligned} \tag{5.108}$$

$$\begin{aligned} & ||2P_i(B_i L_{ij} + H_{ij})||\, ||x_i(t)||\, ||x_j(t)|| \\ & \quad \le \delta ||x_i(t)||^2 + \frac{||P_i||^2 ||B_i L_{ij} + H_{ij}||^2}{\delta} ||x_j(t)||^2 \end{aligned} \tag{5.109}$$

$$\begin{aligned}&||2P_iB_iL_{ij}||\ ||x_i(t)||\ ||e_j(t)||\\&\leq \delta||x_i(t)||^2+\frac{||P_i||^2||B_iL_{ij}||^2}{\delta}||e_j(t)||^2.\end{aligned} \tag{5.110}$$

Hence, the derivative of V_i along the solution to (5.100) satisfies

$$\begin{aligned}\dot{V}_i(x(t)) \leq -\alpha_i||x_i(t)||^2 &+ \sum_{j\in N_i}\beta_{ij}||x_j(t)||^2\\&+\gamma_{ii}||e_i(t)||^2+\sum_{j\in N_i}\gamma_{ij}||e_j(t)||^2,\end{aligned} \tag{5.111}$$

where α_i, β_{ij}, γ_{ii}, and γ_{ij} are as considered in Lemma 5.1. Notice that one can always find a δ such that $\alpha_i > 0$ for $i = 1, 2, \ldots, N$.

By defining vectors

$$\begin{aligned}V_{vec}(x_i(t)) &:= [V_1(x_1(t)),\ V_2(x_2(t)), \ldots,\ V_N(x_N(t))]^T,\\||x(t)||_{vec} &:= [||x_1(t)||^2, ||x_2(t)||^2, \ldots, ||x_N(t)||^2]^T,\\||e(t)||_{vec} &:= [||e_1(t)||^2, ||e_2(t)||^2, \ldots, ||e_N(t)||^2]^T,\end{aligned}$$

inequality (5.111) can be compactly written as

$$\dot{V}_{vec}(x_i(t)) \leq (-A+B)||x(t)||_{vec}+\Gamma||e(t)||_{vec} \tag{5.112}$$

with A, B and Γ being as in Lemma 5.1.

If the spectral radius satisfies $r(A^{-1}B) < 1$, there exists a positive vector $\mu \in \mathbb{R}^n_+$ such that $\mu^T(-A+B) < 0$. We refer the readers to [62] for more details. We select a Lyapunov function $V(x(t)) := \mu^T V_{vec}(x_i(t))$. Then taking the derivative of V yields

$$\begin{aligned}\dot{V}(x(t)) &= \mu^T\dot{V}_{vec}(x_i(t))\\&\leq \mu^T(-A+B)||x(t)||_{vec}+\mu^T\Gamma||e(t)||_{vec}.\end{aligned} \tag{5.113}$$

By noticing that $\mu^T(-A+B) < 0$, we have

$$\dot{V}(x(t)) \leq -L||x(t)||_{vec}+J||e(t)||_{vec}, \tag{5.114}$$

where $L := \mu^T(A-B)$ and $J := \mu^T\Gamma$ are row vectors. Denote by l_i and j_i the entries of L and J, respectively. Then (5.114) yields

$$\dot{V}(x(t)) \leq -\sum_{i\in N} l_i||x(t)||^2+\sum_{i\in N} j_i||e(t)||^2$$

$$= -\sum_{i\in N} \left(l_i||x(t)||^2 - j_i||e(t)||^2\right), \tag{5.115}$$

which implies asymptotic stability with $\sigma_i < \sqrt{\frac{l_i}{j_i}}$. □

Remark 5.12. *Lemma 5.1 can only deal with the case where $j_i > 0$. The case $j_i = 0$ is only possible whenever every entry in the column i of Γ is zero. In fact, $j_i = 0$ implies that the error $||e_i(t)||$ never contributes to the system dynamics via (5.115), which in turn implies that $||e_i(t)||$ does not affect stability at all. Therefore, in the case $j_i = 0$, no constraint on $||e_i(t)||$ is imposed.*

5.4.4 Stabilization of Distributed Systems Under DoS

In the forgoing analysis, we have introduced the design of a suitable σ_i and hence error bound, under which the system is asymptotically stable in the absence of DoS. By hypothesis, we also assumed the existence of a round-robin transmission that satisfies such an error bound. In the presence of DoS, (5.101) is possibly violated even though the sampling strategy is still round-robin. Under such circumstances, stability can be lost. Hence, we are interested in the stabilization problem when the round-robin network is under DoS attacks.

Theorem 5.1. *Consider a distributed system as in* (5.92) *along with a control input as in* (5.93)*. The plant–controller information exchange takes place over a shared network, in which the communication protocol is round-robin with sampling interval Δ as in Assumption 5.8. The large-scale system is asymptotically stable for any DoS sequence satisfying Assumptions 5.6 and 5.7 with arbitrary η and κ, and with τ_D and T if*

$$\frac{1}{T} + \frac{\Delta_*}{\tau_D} < \frac{\omega_1}{\omega_1 + \omega_2}, \tag{5.116}$$

in which $\Delta_* = N\Delta$, $\omega_1 := \min\{\frac{l_i - \sigma_i^2 j_i}{\lambda_{max}(P_i)\mu_i}\}$ and $\omega_2 := \frac{4\max\{j_i\}}{\min\{\mu_i\lambda_{min}(P_i)\}}$; l_i, j_i, μ_i and σ_i are as in Lemma 5.1.

Proof. The proof is divided into three steps:

Step 1. Lyapunov function in DoS-free periods. In DoS-free periods, by hypothesis of Assumption 5.8, (5.101) holds true with σ_i as in

Lemma 5.1 and (5.115) negative. Therefore, the derivative of the Lyapunov function satisfies

$$\begin{aligned}\dot{V}(x(t)) &\leq -\sum_{i\in N}(l_i - j_i\sigma_i^2)||x_i(t)||^2 \\ &\leq -\sum_{i\in N}\frac{l_i - \sigma_i^2 j_i}{\lambda_{max}(P_i)\mu_i}\mu_i V_i \\ &= -\omega_1 V, \end{aligned} \tag{5.117}$$

where $\omega_1 := \min\{\frac{l_i-\sigma_i^2 j_i}{\lambda_{max}(P_i)\mu_i}\}$. Thus for $t \in [h_n + \tau_n, h_n + 1[$ (DoS-free time), the Lyapunov function yields

$$V(x(t)) \leq e^{-\omega_1(t-h_n-\tau_n)} V(x(h_n + \tau_n)). \tag{5.118}$$

Step 2. Lyapunov function in DoS-active periods. Here we let z_m^i denote the last successful sampling instant before the occurrence of DoS. Recalling the definition of $e_i(t)$, we obtain

$$e_i(t) = x_i(z_m^i) - x_i(t) = x_i(h_n) - x_i(t) \tag{5.119}$$

and

$$||e_i(t)||^2 \leq ||x_i(h_n)||^2 + 2||x_i(t)||\;||x_i(h_n)|| + ||x_i(t)||^2, \tag{5.120}$$

for $t \in H_n$. By summing up $||e_i(t)||^2$ over $i \in N$, we obtain

$$\begin{aligned}\sum_{i\in N}||e_i(t)||^2 &\leq \sum_{i\in N}||x_i(h_n)||^2 + \sum_{i\in N}||x_i(t)||^2 \\ &\quad + \sum_{i\in N}\left(||x_i(h_n)||^2 + ||x_i(t)||^2\right) \\ &= 2\sum_{i\in N}||x_i(h_n)||^2 + 2\sum_{i\in N}||x_i(t)||^2. \end{aligned} \tag{5.121}$$

If $\sum_{i\in N}||x_i(h_n)||^2 \leq \sum_{i\in N}||x_i(t)||^2$, we have

$$\sum_{i\in N}||e_i(t)||^2 \leq 4\sum_{i\in N}||x_i(t)||^2.$$

Otherwise, we have $\sum_{i\in N}||e_i(t)||^2 \leq 4\sum_{i\in N}||x_i(h_n)||^2$. Recalling (5.115), it is simple to see that

$$\dot{V}(x(t)) \leq \sum_{i\in N} j_i||e_i(t)||^2. \tag{5.122}$$

Thus, for all $t \in H_n$ (DoS-active time) in the case that $\sum_{i\in N} ||x_i(h_n)||^2 \leq \sum_{i\in N} ||x_i(t)||^2$, the derivative of the Lyapunov function yields

$$\begin{aligned}\dot{V}(x(t)) &\leq \max\{j_i\} \sum_{i\in N} ||e_i(t)||^2 \\ &\leq 4\max\{j_i\} \sum_{i\in N} ||x_i(t)||^2 \\ &\leq \frac{4\max\{j_i\}}{\min\{\mu_i \lambda_{min}(P_i)\}} \sum_{i\in N} \mu_i V(x_i(t)) \\ &= \omega_2 V(x(t)) \end{aligned} \tag{5.123}$$

with $\omega_2 := \frac{4\max\{j_i\}}{\min\{\mu_i \lambda_{min}(P_i)\}}$. On the other hand, for all $t \in H_n$ such that

$$\sum_{i\in N} ||x_i(h_n)||^2 > \sum_{i\in N} ||x_i(t)||^2,$$

one has

$$\dot{V}(x(t)) \leq \omega_2 V(x(h_n)). \tag{5.124}$$

Thus, (5.123) and (5.124) imply that the Lyapunov function during H_n satisfies

$$V(x(t)) \leq e^{\omega_2 (t-h_n)} V(x(h_n)). \tag{5.125}$$

Step 3. Switching between stable and unstable modes. Consider a DoS attack with period τ_n, at the end of which the overall system has to wait an additional period of length $N\Delta$ to have a full round of communications. Hence, the period where at least one subsystem transmission is not successful can be upper bounded by $\tau_n + N\Delta$. For all $\tau, t \in \mathbb{R}_{\geq 0}$ with $t \geq \tau$, the total length where communication is not possible over $[\tau, t[$, say $|\bar{\Xi}(\tau, t)|$, can be upper bounded by

$$\begin{aligned}|\bar{\Xi}(\tau, t)| &\leq |\Xi(\tau, t)| + (1 + n(\tau, t))\Delta_* \\ &\leq \kappa_* + \frac{t-\tau}{T_*}, \end{aligned} \tag{5.126}$$

where $\Delta_* = N\Delta$, $\kappa := \kappa + (1+\eta)\Delta_*$ and $T_* := \frac{\tau_D T}{\tau_D T + T\Delta_*}$. Considering the additional waiting time due to round-robin, the Lyapunov function in (5.118) yields

$$V(x(t)) \le e^{-\omega_1(t-h_n-\tau_n-N\Delta)} V(h_n+\tau_n+N\Delta)), \quad \forall t \in [h_n+\tau_n+N\Delta, h_{n+1}[,$$
$$V(x(t)) \le e^{\omega_2(t-h_n)} V(h_n), \quad \forall t \in [h_n, h_n+\tau_n+N\Delta[.$$

The overall behavior of the closed-loop system can therefore be regarded as a switching system with two modes. Applying simple iterations to the Lyapunov functions in and out of DoS status, one has

$$\begin{aligned} V(x(t)) &\le e^{-\omega_1|\bar{\Theta}(0,t)|} e^{2|\bar{\Xi}(0,t)|} V(x(0)) \\ &\le e^{\kappa_*(\omega_1+\omega_2)} e^{-\beta_* t} V(x(0)), \end{aligned} \tag{5.127}$$

where $\beta_* := \omega_1 - (\omega_1+\omega_2)(\frac{\Delta_*}{\tau D} + \frac{1}{T})$. By constraining $\beta_* < 0$, one obtains the desired result in (5.116). Hence, stability is implied at once. □

Remark 5.13. *The resilience of the distributed systems depends on the largeness of ω_1 and the smallness of ω_2. To achieve this, one can try to find K_i and L_{ij} such that $||B_iK_i||$ and $||B_iL_{ij}||$ are small. On the other hand, the sampling interval of round-robin also affects stability in the sense that it determines how fast the overall system can restore the communication. One can always apply smaller round-robin inter-sampling time to reduce the left-hand side of (5.116) at the expense of higher communication load.*

5.4.5 A Hybrid Transmission Strategy

In the foregoing argument (see Remark 5.13), we have shown that system resilience depends on the sampling rate of round-robin. The faster the sampling rate of round-robin, the quicker the overall system restores the communication. On the other hand, in DoS-free periods, we are interested in the possibility of reducing communication load while maintaining the comparable robustness as before. To realize this, we propose a hybrid transmission strategy: in the absence of DoS, the communications of the distributed systems are event-based; if DoS occurs, the communications switch to round-robin until the moment where every subsystem has one successful update. Observe that the advantage of event-triggered control is saving communication resources. However, the effectiveness of prolonging transmission intervals, in turn, appears to be a disadvantage in the presence of DoS.

5.4.6 Zeno-Free Event-Triggered Control in the Absence of DoS

In the sequel, we denote by $\{t_k^i\}$ the triggering time sequence of subsystem i under event-triggered control scheme. For a given initial condition $x_i(0)$, if t_k^i converges to a finite t^{i*}, we say that the event-triggered control induces Zeno behavior [54]. Hence, Zeno-freeness implies an event-triggered control scheme preventing the occurrence of Zeno behavior. The following lemma addresses the Zeno-free event-triggered control.

Lemma 5.2. *Consider a distributed system as in* (5.92) *along with a control input as in* (5.93)*. Suppose that the spectral radius* $r(A^{-1}B) < 1$*. In the absence of DoS, the distributed system is practically stable and Zeno-free if the event-triggered law satisfies*

$$||e_i(t)|| \leq \max\{\sigma_i||\xi(t)||, c_i\}, \tag{5.128}$$

in which c_i *is a positive finite real and*

$$\sigma_i < \min\{\sqrt{\frac{l_i}{j_i}}, 1\}, \tag{5.129}$$

where l_i *and* j_i *are the same as in Lemma 5.1.*

Proof. By Lemma 5.1, if spectral radius $r(A^{-1}B) < 1$, (5.115) holds true. Then one can observe that the event-triggered control law (5.128) would lead (5.115) to

$$\begin{aligned}\dot{V}(x(t)) &\leq -\sum_{i\in N}\left(l_i||x_i(t)||^2 - j_i\max\{\sigma^2||x_i(t)||^2, c_i^2\}\right)\\ &\leq \max\{-\sum_{i\in N}(l_i - j_i\sigma_i^2)||x_i(t)||^2,\\ &\quad -\sum_{i\in N} l_i||x_i(t)||^2 + \sum_{i\in N} j_i c_i^2\}\\ &\leq -\sum_{i\in N}(l_i - j_i\sigma_i^2)||x_i(t)||^2 + \sum_{i\in N} j_i c_i^2,\end{aligned} \tag{5.130}$$

which implies practical stability with $\sigma_i < \min\{\sqrt{\frac{l_i}{j_i}}, 1\}$ and finite c_i.

Then we introduce the analysis about Zeno-freeness of this distributed event-triggered control law. Since $\dot{e}_i(t) = -\dot{x}_i(t)$, the dynamics of e_i satisfies

$$\begin{aligned}\dot{e}_i(t) = A_i e_i(t) - \Phi_i x_i(t_k^i) - \sum_{i\in N}(B_i L_{ij} + H_{ij})x_j(t_k^j)\\ + \sum_{i\in N} H_{ij} e_j(t).\end{aligned} \tag{5.131}$$

From the triggering law (5.128), one can obtain

$$||x_i(t_k^i) - x_i(t)||^2 \leq \max\{\sigma_i||x_i(t)||, c_i\}.$$

Further calculations yield $||x_i(t)|| - ||x_i(t_k^i)|| \leq \sigma_i||x_i(t)|| + c_i$. Thus, it is simple to verify that $||e_i(t)|| \leq \bar{\sigma}_i||x_i(t_k^i)|| + \bar{\sigma}_i c_i$, where $\bar{\sigma}_i := \frac{\sigma_i}{1-\sigma_i}$.

For each i, at the instant t_{k+1}^i, $||e_i(t)||$ satisfies

$$\begin{aligned} ||e_i(t_{k+1}^i)|| \leq & f_i||\Phi_i||\ ||x_i(t_k^i)|| \\ & + f_i\sum_{j\in N_i}||B_iL_{ij} + H_{ij}||m \\ & + f_i\sum_{j\in N_i}||H_{ij}||\bar{\sigma}_j(m + c_j), \end{aligned} \tag{5.132}$$

where $f_i := \int_{t_k^i}^{t_{k+1}^i} e^{A(t_{k+1}^i - \tau)}d\tau$, $m = \max\{||x_j(t_p^j)||\}$ for $t_k^i \leq t_p^j < t_{k+1}^i$ and $j \in N_i$. Meanwhile, the triggering law in (5.128) implies that $||e_i(t_{k+1}^i)|| \geq c_i$. Then, one immediately sees that

$$\begin{aligned} & t_{k+1}^i - t_k^i \geq z_i, && \text{if } \mu_{A_i} \leq 0, \\ & t_{k+1}^i - t_k^i \geq \frac{1}{\mu_{A_i}}\log(z_i\mu_{A_i} + 1), && \text{if } \mu_{A_i} > 0, \end{aligned} \tag{5.133}$$

in which

$$z_i := \frac{c_i}{||\Phi_i||\ ||x_i(t_k^i)|| + m\sum_{j\in N_i}\zeta_{ij} + \sum_{j\in N_i}||H_{ij}||\bar{\sigma}_j c_j},$$

where $\zeta_{ij} := ||B_iL_{ij} + H_{ij}|| + ||H_{ij}||\bar{\sigma}_j$ and μ_{A_i} is the logarithmic norm of A_i. Notice that the system is practically stable, so that $||x_i(t_k^i)||$ and m are bounded. This implies that $z_i > 0$ and hence $t_{k+1}^i - t_k^i > 0$. □

5.4.7 Hybrid Transmission Strategy Under DoS

As a counterpart of Assumption 5.8, here we assume that there exists a round-robin sampling interval Δ satisfying (5.128). Now we are ready to present the following result.

Theorem 5.2. *Consider a distributed system as in* (5.92) *along with a control input as in* (5.93). *The plant–controller information exchange takes place over a shared network implementing the event-triggered control law* (5.128) *in the absence of DoS. Suppose that there exists a round-robin sampling interval* Δ *such* *that* (5.128) *holds. The network is subject to DoS attacks regulated by Assumptions 5.6 and 5.7, during which the communication switches to round-robin until every subsystem updates successfully. Then the distributed system is practically stable if* (5.116) *holds true.*

Proof. Similar to the proof of Theorem 5.1, considering the additional waiting time $N\Delta$ due to round-robin for the restoring of communications, in DoS-free periods the Lyapunov function satisfies

$$V(x(t)) \leq e - \omega_1(t - h_n - \tau_n - N\Delta) V(x(h_n + \tau_n + N\Delta)) + \frac{c}{\omega_1} \quad (5.134)$$

for $t \in [h_n + \tau_n + N\Delta, h_{n+1}[$, where ω_1 is as in Theorem 5.1 and $c := \sum_{i=1}^{N} j_i c_i^2$. On the other hand, (5.125) still holds for $t \in [h_n, h_n + \tau_n + N\Delta[$.

Applying similar calculation as in Step 3 in the proof of Theorem 5.1, we obtain

$$\begin{aligned} V(x(t)) &\leq e^{-\omega_1 |\bar{\Theta}(0,t)|} e^{\omega_2 |\bar{\Xi}(0,t)|} V(x(0)) \\ &\quad + \sum_{n=0}^{q} e^{-\omega_1 |\bar{\Theta}(h_n,t)|} e^{\omega_2 |\bar{\Xi}(h_n,t)|} \frac{c}{\omega_1} + \frac{c}{\omega_1} \\ &\leq e^{\kappa_*(\omega_1+\omega_2)} e^{-\beta_* t} V(x(0)) \\ &\quad + e^{\kappa_*(\omega_1+\omega_2)} \sum_{n=0}^{q} e^{-\beta_*(t-h_n)} \frac{c}{\omega_1} + \frac{c}{\omega_1}, \end{aligned} \quad (5.135)$$

where $n \in \mathbb{N}_0$, $q := \sup\{q \in \mathbb{N}_0 | h_q \leq t\}$ and β_* is as in the proof of Theorem 5.1. Notice that $t - h_n \geq \tau_D n(h_n, t) - \tau_D \eta$ by exploiting Assumption 5.6. Then, the Lyapunov function yields

$$\begin{aligned} V(x(t)) &\leq e^{\kappa_*(\omega_1+\omega_2)} e^{-\beta_* t} V(x(0)) \\ &\quad + e^{\kappa_*(\omega_1+\omega_2)+\beta_* \tau_D \eta} \sum_{n=0}^{q} e^{-\beta_* \tau_D n(h_n,t)} \frac{c}{\omega_1}. \end{aligned} \quad (5.136)$$

Recalling Assumption 5.6, one has that $n(h_n, t) - n(h_{n+1}, t) \geq 1$ for $t \geq h_{n+1}$. This implies that

$$\sum_{n=0}^{q} e^{-\beta_* \tau_D n(h_n,t)} \leq \frac{1}{1 - e^{-\beta_* \tau_D}}. \quad (5.137)$$

Finally, (5.136) can be written as

$$V(x(t)) \le e^{\kappa_*(\omega 1+\omega 2)} e^{-\beta_* t} V(x(0)) + \frac{e^{\kappa_*(\omega 1+\omega 2)+\beta_* \tau_D \eta}}{1-e^{-\beta_* \tau_D}} \frac{c}{\omega_1} + \frac{c}{\omega_1}. \tag{5.138}$$

If (5.116) holds, it is simple to verify that $\beta_* < 0$, which implies practical stability. □

5.4.8 Simulation Example 5.6

Consider an open-loop unstable system such as

$$\dot{x}_1(t) = x_1(t) + u_1(t) + x_2(t),$$
$$\dot{x}_2(t) = x_2(t) + u_2(t),$$

under distributed control inputs such that

$$u_1(t) = -4.5x_1(t_k^1) - 1.4x_2(t_k^2),$$
$$u_2(t) = -6x_2(t_k^2) - x_1(t_k^1).$$

Solutions of the Lyapunov equation $\Phi_i^T P_i + P_i \Phi_i + Q_i = 0$ with $Q_i = 1$ $(i = 1, 2)$ yields $P_1 = 0.1429$ and $P_2 = 0.1$. The matrices are $A = [0.7\ 0; 0\ 0.9]$, $B = [0\ 0.0327; 0.1\ 0]$ and $\Gamma = [4.1327\ 0.4; 0.1\ 3.6]$ according to Lemma 5.1. From these parameters, we obtain that the spectral radius $r(A^{-1}B) = 0.072$, $\sigma_1 < 0.3765$ and $\sigma_2 < 0.4657$. We let $\sigma_1 = \sigma_2 = 0.2$. Based on Assumption 5.3, we choose a round-robin sampling interval $\Delta = 0.01$ s.

With these parameters, we obtain the bound $\frac{\omega_1}{\omega_1+\omega_2} \approx 0.0175$ with $\omega_1 \approx 3.0149$ and $\omega_2 \approx 169.3061$. This implies that a maximum duty cycle of 1.75% of a sustained DoS would not destabilize our systems in the example. Actually, this bound is conservative. The systems in inspection can endure more DoS without losing stability. As shown in Fig. 5.13, lines represent states and grey stripes represent the presence of DoS. Over a simulation horizon of 20 s, the DoS corresponds to parameters of $\tau_D \approx 1.8182$ and $T \approx 2.5$, and $\sim 40\%$ of transmission failures. According to (5.116), we obtain $\frac{\Delta_*}{\tau_D} + \frac{1}{T} = 0.411$, which violates the theoretical bound, but the system is still stable.

Meanwhile, the hybrid transmission strategy is able to reduce communications effectively. As shown in Fig. 5.13, the transmissions with the hybrid

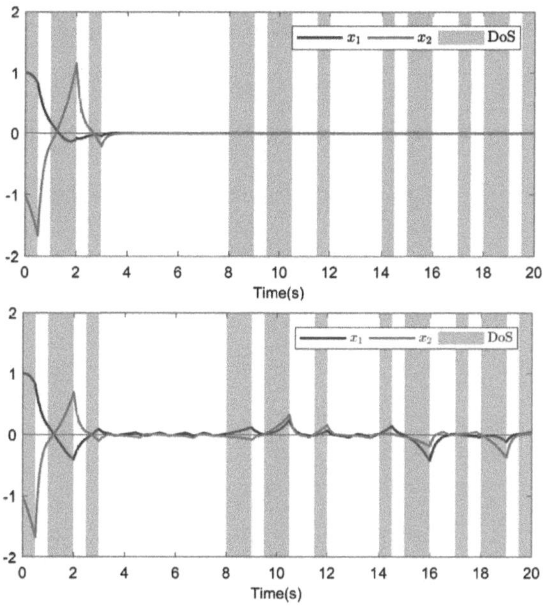

Figure 5.13 (Top) States under pure round-robin communication where there are 1200 transmissions in total; (Bottom) States under hybrid communication strategy where there are 112 transmissions

transmission strategy is only 10% of the transmissions with the pure round-robin strategy.

5.4.9 Simulation Example 5.7

In this example, we consider a physical system as in [64]. The system is composed of N inverted pendulums interconnected as a line by springs, whose states are $x_i = [\bar{x}_i, \tilde{x}_i]^T$ for $i = 1, 2, \ldots, N$. Here, we consider a simple case where $N = 3$. The parameters of the pendulums are

$$A_1 = A_3 = \begin{bmatrix} 0 & 1 \\ -3.75 & 0 \end{bmatrix}, A_2 = \begin{bmatrix} 0 & 1 \\ -2.5 & 0 \end{bmatrix},$$

$$B_1 = B_2 = B_3 = \begin{bmatrix} 0 \\ 0.25 \end{bmatrix},$$

$$H_{12} = H_{21} = H_{23} = H_{32} = \begin{bmatrix} 0 & 0 \\ 1.25 & 0 \end{bmatrix}.$$

The parameter of designed controllers are given by

$$K_1 = K_3 = [-23 \quad -12],\ K_2 = [-18 \quad -12],$$
$$L_{12} = L_{32} = [-5 \quad 0.25],\ L_{21} = L_{23} = [-4.75 \quad -0.25].$$

With the solutions of Lyapunov function $\Phi_i^T P_i + P_i \Phi_i + Q_i = 0$ where $Q_i = I$ and $i = 1, 2, 3$, we obtain

$$A = \begin{bmatrix} 0.67 & 0 & 0 \\ 0 & 0.45 & 0 \\ 0 & 0 & 0.67 \end{bmatrix},$$
$$B = \begin{bmatrix} 0 & 0.0608 & 0 \\ 0.1217 & 0 & 0.1217 \\ 0 & 0.0608 & 0 \end{bmatrix},$$
$$\Gamma = \begin{bmatrix} 47.7983 & 24.4007 & 0 \\ 22.0276 & 33.2386 & 22.0276 \\ 0 & 24.4007 & 47.7983 \end{bmatrix}.$$

With A, B and Γ, we obtain that $r(A^{-1}B) = 0.2216$, $\sigma_1 < 0.0646$, $\sigma_2 < 0.0844$ and $\sigma_3 < 0.0646$. We select $\sigma_1 = \sigma_2 = \sigma_3 = 0.01$. The round-robin sampling interval is chosen as $\Delta = 0.001$ s according to Assumption 5.8. Following the same procedures as in Simulation Example 5.1, we obtain $\frac{\omega_1}{\omega_1 + \omega_2} \approx 0.00012$, which is considerably conservative. In fact, if the systems are under the same DoS attacks as in Simulation Example 5.6, they are still stable, which can be seen from Fig. 5.14. The conservativeness is due to the unstable dynamics of the inverted pendulums, the feedback gain K_i and the coupling parameter L_{ij} in the controllers. It is worth investigating how to design suitable K_i and L_{ij} to mitigate this effect, see (Remark 5.13).

5.5 STABILIZING SECURE CONTROL

In a networked control system, attacks to the communication links can be classified as either deception attacks or denial-of-service (DoS) attacks. The former affect the trustworthiness of data by manipulating the packets transmitted over the network; see [65–69] and the references therein. DoS attacks are primarily intended to affect the timeliness of the information exchange to cause packet losses [70,71].

This section examines DoS attacks by considering a sampled-data control system in which the plant–controller communication is networked; the

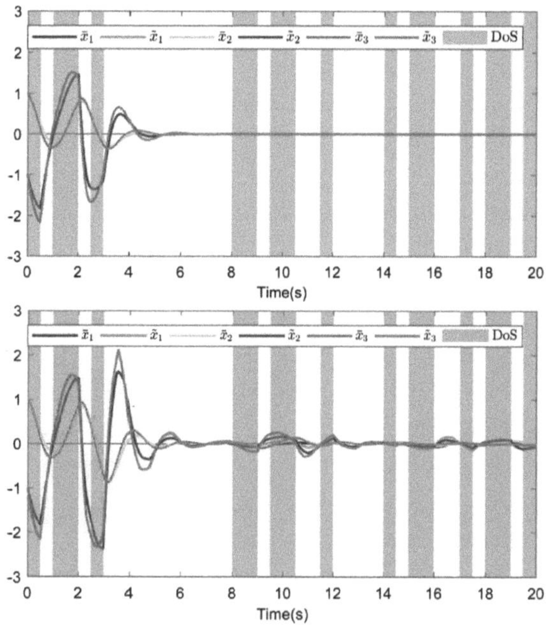

Figure 5.14 States under pure round-robin communication

attacker objective is to induce instability in the control system by denying communication on measurement (sensor-to-controller) and control (controller-to-actuator) channels. Under DoS attacks, the process evolves in open-loop according to the last transmitted control sample. The problem of interest is that of finding conditions under which closed-loop stability, in some suitably defined sense, can be preserved. We consider a general attack model that only constrains the attacker action in time by posing limitations on the frequency of DoS attacks and their duration. This makes it possible to capture many different types of DoS attacks, including trivial, periodic, random and protocol-aware jamming attacks [71–74]. It turns out that the design of the transmission times plays a key role.

To guarantee stability, the transmission times are selected in such a way that, whenever communication is possible, the closed-loop trajectories satisfy a suitable norm bound. Such a choice has two main advantages:

i. It can ensure *global exponential input-to-state stability* (ISS) with respect to process disturbances even in the presence of DoS; and

ii. It is flexible enough so as to allow the designer to choose from several implementation options that can be used for trading-off performance versus communication resources.

It is reported that the design of the network transmission times has interesting and perhaps surprising connections with the *event-based* sampling approach of [75], though substantial modifications are needed to account for the presence of DoS and disturbances. More specifically, the adoption of sampling rules that suitably constrain the closed-loop trajectories is crucial for achieving a simple Lyapunov-based analysis of the ISS property during the *on/off* periods of DoS.

Recently in [76], DoS attacks was treated in the form of pulse-width modulated signals. The goal is to identify salient features of the DoS signal such as maximum *on/off* cycle in order to suitably scheduling the transmission times. For the case of *periodic* jamming (of unknown period and duration), an identification algorithm is derived which makes it possible to desynchronize the transmission times from the on periods of DoS. This framework should be therefore looked at as complementary more than alternative to the present one, since dealing with cases where the jamming signal is "well-structured" so that de-synchronization from attacks can be achieved. Such a feature is conceptually impossible to achieve in scenarios such as the one considered in this section, where the jamming strategy is not prefixed (the attacker can modify online the attack strategy).

5.5.1 Process Dynamics and Ideal Control Action

The framework of interest is schematically represented in Fig. 5.15. The process to be controlled is described by the differential equation

$$\frac{d}{dt}x(t) = Ax(t) + Bu(t) + w(t), \tag{5.139}$$

where $t \in \mathbb{R}_{\geq 0}$; $x \in \mathbb{R}^n$ is the state and $u \in \mathbb{R}^m$ is the control input; A and B are matrices of appropriate size; $w \in \mathbb{R}^n$ is an unknown disturbance: it accounts for process input disturbances as well as noises on control (controller-to-actuator) and measurement (sensor-to-controller) channels.

The control action is implemented over a sensor/actuator network. We assume that (A, B) is stabilizable and that a state-feedback matrix K has been designed in such a way that all the eigenvalues of $A + BK$ have negative real part. The control signal is sampled using a *sample-and-hold device*. Let $\{t_k\}_{k\in\mathbb{N}_0}$, where $t_0 := 0$ by convention, represent the sequence of time instants at which it is desired to update the control action. Accordingly, whatever the logic generating the sequence $\{t_k\}_{k\in\mathbb{N}_0}$, in the ideal situation where data can be sent and received at any desired instant of time, the

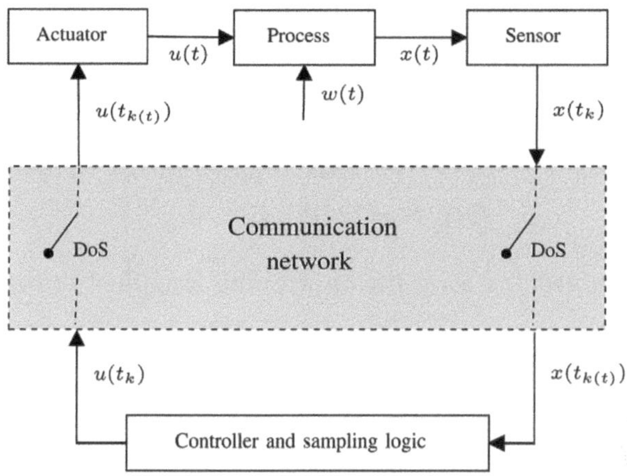

Figure 5.15 Block diagram of the closed-loop system

control input applied to the process is given by

$$u_{\text{ideal}}(t) = Kx(t_k), \tag{5.140}$$

for all $t \in \mathcal{I}_k := [t_k, t_{k+1}[$.

5.5.2 DoS and Actual Control Action

We refer to *Denial-of-Service* (DoS) as the phenomenon that may prevent (5.140) from being executed at each desired time. In principle, this phenomenon can affect measurement and control channels separately. In the sequel, we consider the case of DoS simultaneously affecting both measurement and control channels. This amounts to assuming that, in the presence of DoS, data can be *neither sent nor received*. Specifically, let $\{h_n\}_{n\in\mathbb{N}_0}$, where $h_0 \geq 0$, denote the sequence of DoS *off/on* transitions, that is, the time instants at which DoS exhibits a transition from zero (communication is possible) to one (communication is interrupted). Then

$$H_n := \{h_n\} \cup [h_n, h_n + \tau_n[\tag{5.141}$$

represents the nth DoS time-interval of length $\tau_n \in \mathbb{R}_{\geq 0}$, over which communication is not possible. If $\tau_n = 0$, the nth DoS takes the form of a single pulse at time h_n.

In the presence of DoS, the actuator generates an input that is based on the *most recently received* control signal. Given $\tau,\ t \in \mathbb{R}_{\geq 0}$ with $t \geq \tau$, let

$$\Xi(\tau, t) := \bigcup_{n \in \mathbb{N}_0} H_n \bigcap\, [\tau, t], \tag{5.142}$$

$$\Theta(\tau, t) := [\tau, t] \setminus \Xi(\tau, t). \tag{5.143}$$

Accordingly, for each $t \in \mathbb{R}_{\geq 0}$, the control input applied to the process can be expressed as

$$u(t) = Kx(t_{k(t)}), \tag{5.144}$$

where

$$k(t) := \begin{cases} -1, & \text{if } \Theta(0, t) = \emptyset, \\ \sup\{k \in \mathbb{N}_0 \mid t_k \in \Theta(0, t)\}, & \text{otherwise.} \end{cases} \tag{5.145}$$

This implies that for each $t \in \mathbb{R}_{\geq 0}$, $k(t)$ represents the last successful control update. Notice that $h_0 = 0$ implies $k(0) = -1$, which raises the question of assigning a value to the control input when communication is not possible at the process start-up. In this respect, we assume that when $h_0 = 0$ then $u(0) = 0$, and let $x(t_{-1}) := 0$ for notational consistency.

5.5.3 Control Objectives

The problem of interest is that of finding sampling logic that achieve robustness against DoS, while ensuring that the control inter-execution times are bounded away from zero. While robustness is concerned with stability and performance of the closed-loop system, positive inter-execution times are required for the control scheme to be physically implementable over a network.

The following definitions reflect the stated goals.

Definition 5.2 ([76]). Let Σ be the control system resulting from (5.139) under a control signal as in (5.144). System Σ is said to be input-to-state stable (ISS) if there exist a $\mathcal{KL}$-function β and a $\mathcal{K}_\infty$-function γ such that, for each $w \in \mathcal{L}_\infty(\mathbb{R}_{\geq 0})$ and $x(0) \in \mathbb{R}^n$,

$$\|x(t)\| \leq \beta(\|x(0)\|, t) + \gamma(\|w_t\|_\infty), \tag{5.146}$$

for all $t \in \mathbb{R}_{\geq 0}$. If (5.146) holds when $w \equiv 0$, then Σ is said to be globally asymptotically stable (GAS).

Definition 5.3. A control update sequence $\{t_k\}_{k\in\mathbb{N}_0}$ is said to have the finite sampling rate property if there exists $\underline{\Delta} \in \mathbb{R}_{>0}$ such that

$$\Delta_k := t_{k+1} - t_k \geq \underline{\Delta}, \tag{5.147}$$

for all $k \in \mathbb{N}_0$.

In the sequel, it is understood that the network can send information at the sample rate induced by $\underline{\Delta}$.

5.5.4 Stabilizing Control Update Policies

We first introduce a class of control update policies ensuring ISS in the absence of DoS. The results will serve as a basis for the subsequent developments.

Consider the closed-loop system resulting from (5.139) under a control signal as in (5.144). As a first step, we rewrite it in a form that is better suited for analysis purposes. Let

$$e(t) := x(t_{k(t)}) - x(t) \tag{5.148}$$

where $t \in \mathbb{R}_{\geq 0}$, represent the error between the value of the process state at the last successful control update and the value of the process state at the current time. The closed-loop system can be therefore rewritten as

$$\frac{d}{dt}x(t) = \Phi x(t) + BKe(t) + w(t), \tag{5.149}$$

where $\Phi := A + BK$.

The closed-loop system now depends on the control update rule through e, which enters the dynamics as an additional disturbance term. It is then intuitively clear that stability will not be destroyed if one adopts control update rules that keep e small in a suitable sense. The notion of "smallness" considered here, which characterizes the control update rules of interest, is expressed in terms of the following bound:

$$\|e(t)\| \leq \sigma \|x(t)\| + \sigma \|w_t\|_\infty, \tag{5.150}$$

where $\sigma \in \mathbb{R}_{>0}$ is a suitable design parameter. We anticipate that (5.150) is not the control update rule we are going to implement, because of its dependence on the supremum norm of the disturbance w, which in general is unknown. We will rather adopt different update rules guaranteeing that (5.150) is always satisfied. Such different update rules, motivated by Lemma 5.3 below, are discussed in detail later on.

As the next result shows, provided that σ is suitably chosen, any control update rule that restricts e to satisfy (5.150) is stabilizing. This can be proved by resorting to standard Lyapunov arguments. Given any positive definite matrix $Q = Q^\top \in \mathbb{R}^{n\times n}$, let P be the unique solution of the Lyapunov equation

$$\Phi^\top P + P\Phi + Q = 0. \tag{5.151}$$

Then, by taking $V(x) = x^\top Px$ as a Lyapunov function, and differentiating it along the solution of (5.149), it is simple to verify that

$$\alpha_1 \|x(t)\|^2 \le V(x(t)) \le \alpha_2 \|x(t)\|^2, \tag{5.152a}$$

$$\frac{d}{dt} V(x(t)) \le -\gamma_1 \|x(t)\|^2 + \gamma_2 \|x(t)\| \, \|e(t)\| + \gamma_3 \|x(t)\| \, \|w(t)\| \tag{5.152b}$$

hold for all $t \in \mathbb{R}_{\ge 0}$, with α_1 and α_2 equal to the smallest and largest eigenvalues of P, respectively, γ_1 equal to the smallest eigenvalue of Q, $\gamma_2 := \|2PBK\|$ and $\gamma_3 := \|2P\|$. It is then immediate to see that, under (5.150), the second term in (5.152b) does always satisfy a dissipation-like inequality whenever σ is chosen small enough.

Theorem 5.3. *Consider the control system* Σ *composed of* (5.139) *and control input* (5.144), *where* K *is such that all the eigenvalues of* $\Phi = A + BK$ *have negative real part. Given any positive symmetric definite matrix* $Q \in \mathbb{R}^{n\times n}$, *let* P *be the unique solution of the Lyapunov equation* $\Phi^\top P + P\Phi + Q = 0$. *Let* $V(x) = x^\top Px$. *Consider any control update sequence occurring at a finite sampling rate and satisfying* (5.150) *for all* $t \in \mathbb{R}_{\ge 0}$, *with* σ *such that*

$$\gamma_1 - \sigma\gamma_2 > 0, \tag{5.153}$$

where γ_1 *and* γ_2 *are as in* (5.152b). *Then,* Σ *is ISS.*

Proof. Substituting (5.150) into (5.152b) yields

$$\frac{d}{dt} V(x(t)) \le -\gamma_4 \|x(t)\|^2 + \gamma_5 \|x(t)\| v(t), \tag{5.154}$$

where $v(t) := \sup\{\|w(t)\|,\ \|w_t\|_\infty\}$, $\gamma_4 := (\gamma_1 - \sigma\gamma_2)$ and $\gamma_5 := (\gamma_3 + \sigma\gamma_2)$. Here, we recall that γ_1 is the minimal eigenvalue of Q, $\gamma_2 = \|2PBK\|$, $\gamma_3 =$

$\|2P\|$, and $\sigma < \gamma_1/\gamma_2$. Observe that for any positive real δ, Young's inequality (see, e.g., [88]) yields

$$2\,\|x(t)\|\,v(t) \leq \frac{1}{\delta}\,\|x(t)\|^2 + \delta v^2(t). \tag{5.155}$$

By letting $\delta := \gamma_5/\gamma_4$, we get

$$\begin{aligned}\frac{d}{dt}V(x(t)) &\leq -\frac{\gamma_4}{2}\|x(t)\|^2 + \gamma_6 v(t)^2 \\ &\leq -\omega_1 V(x(t)) + \gamma_6 v(t)^2,\end{aligned} \tag{5.156}$$

where $\omega_1 := \gamma_4/(2\alpha_2)$ and $\gamma_6 := \gamma_5^2/(2\gamma_4)$. Note now that $\|v_t\|_\infty = \|w_t\|_\infty$ for any $t \in \mathbb{R}_{>0}$. Thus, standard comparison results for differential inequalities yield

$$V(x(t)) \leq e^{-\omega_1 t}V(x(0)) + \gamma_7\|w_t\|_\infty^2, \tag{5.157}$$

where $\gamma_7 := \gamma_6/\omega_1$. Using (5.152a), we get

$$\|x(t)\|^2 \leq \frac{\alpha_2}{\alpha_1}e^{-\omega_1 t}\|x(0)\|^2 + \frac{\gamma_7}{\alpha_1}\|w_t\|_\infty^2. \tag{5.158}$$

Since $a^2 + b^2 \leq (a+b)^2$ for any pair of positive reals a and b, we finally get

$$\|x(t)\| \leq \sqrt{\frac{\alpha_2}{\alpha_1}}e^{-(\frac{\omega_1}{2})t}\|x(0)\| + \sqrt{\frac{\gamma_7}{\alpha_1}}\|w_t\|_\infty, \tag{5.159}$$

which yields the desired result. □

Inequality (5.153) can always be satisfied by selecting σ sufficiently small in that $\gamma_1 > 0$. Given any σ satisfying (5.153), the only question that arises is on the possibility of designing sampling logic that guarantees (5.150) with a finite-sampling rate. As next result shows, in the absence of DoS, this is always possible. Given a matrix $M \in \mathbb{R}^{n\times n}$, let

$$\mu_M := \max\left\{\lambda \mid \lambda \in \text{spectrum}\left\{\frac{(M+M^\top)}{2}\right\}\right\} \tag{5.160}$$

denote the logarithmic norm of M.

Lemma 5.3. *Consider the same notation as in Theorem 5.3. Then, in the absence of DoS, any control update rule with inter-sampling times smaller than or* *equal to*

$$\bar{\Delta}_\sigma := \left(\frac{\sigma}{1+\sigma}\right)\frac{1}{\max\{\|\Phi\|,1\}}, \tag{5.161}$$

when $\mu_A \le 0$, *and*

$$\bar{\Delta}_\sigma := \frac{1}{\mu_A}\log\left[\left(\frac{\sigma}{1+\sigma}\right)\frac{1}{\max\{\|\Phi\|,1\}}\mu_A + 1\right], \tag{5.162}$$

when $\mu_A > 0$, *satisfies* (5.150) *for all* $t \in \mathbb{R}_{\ge 0}$.

Proof. In the absence of DoS, any control update attempt is successful. Thus, in accordance with (5.148), the dynamics of e satisfies

$$\begin{aligned}\frac{d}{dt}e(t) &= -Ax(t) - BKx(t_k) - w(t) \\ &= Ae(t) - \Phi x(t_k) - w(t),\end{aligned} \tag{5.163}$$

for all $t \in \mathcal{I}_k$ and for all $k \in \mathbb{N}_0$, where $e(t_k) = 0$. Recall now that $\|e^{At}\| \le e^{\mu_A t}$ for all $t \in \mathbb{R}_{\ge 0}$. Using this property, we then have

$$\|e(t)\| \le \kappa_1 \int_{t_k}^{t} e^{\mu_A(t-s)}\left[\|x(t_k)\| + \|w(s)\|\right] ds, \tag{5.164}$$

for all $t \in \mathcal{I}_k$ and all $k \in \mathbb{N}_0$, where $\kappa_1 := \max\{\|\Phi\|, 1\}$.

Let $f(t-t_k) := \int_{t_k}^{t} e^{\mu_A(t-s)} ds$ and using the fact that $x(t_k) = e(t) + x(t)$, we obtain

$$\begin{aligned}\|e(t)\| &\le \kappa_1 f(t-t_k)\|e(t)\| \\ &\quad + \kappa_1 f(t-t_k)\left(\|x(t)\| + \|w_t\|_\infty\right).\end{aligned} \tag{5.165}$$

Observe now that $f(0) = 0$ and $f(t-t_k)$ is monotonically increasing with t. Accordingly, for any positive real Δ such that

$$f(\Delta) \le \frac{1}{\kappa_1}\frac{\sigma}{(1+\sigma)}, \tag{5.166}$$

any control update rule such that $\Delta_k \le \Delta$ will satisfy (5.150) for all $t \in \mathbb{R}_{\ge 0}$. To conclude the proof, we derive an explicit expression for Δ. Let first $\mu_A \le 0$. In this case, $f(\Delta) \le \Delta$, so that (5.161) yields the desired result. If instead $\mu_A > 0$, then we have

$$f(\Delta) = \frac{1}{\mu_A}(e^{\mu_A \Delta} - 1), \tag{5.167}$$

and (5.162) yields the desired result. □

5.5.5 Time-Constrained DoS

We now address the issue of determining the amount of DoS that a system can tolerate before experiencing instability. In this respect, it is simple to see that such an amount is not arbitrary, and that suitable conditions must be imposed on both DoS *frequency and duration.*

(1) *DoS Frequency.* Consider first the frequency at which DoS can occur, and let $\Lambda_n := h_{n+1} - h_n$, $n \in \mathbb{N}_0$, denote the time elapsing between any two successive DoS triggering. One immediately sees that if $\Lambda_n \leq \underline{\Delta}$ for all $n \in \mathbb{N}_0$ (DoS can occur at the same rate as the minimum possible sampling rate $\underline{\Delta}$) then stability can be lost regardless of the adopted control update policy. It is intuitively clear that, in order to get stability, the frequency at which DoS can occur must be sufficiently small compared to the minimum sampling rate. A natural way to express this requirement is via the concept of *average dwell-time.* Given $\tau, t \in \mathbb{R}_{\geq 0}$ with $t \geq \tau$, let $n(\tau, t)$ denote the number of DoS *off/on* transitions occurring on the interval $[\tau, t[$.

Assumption 5.9. There exist $\eta \in \mathbb{R}_{\geq 0}$ and $\tau_D \in \mathbb{R}_{>\underline{\Delta}}$ such that

$$n(\tau, t) \leq \eta + \frac{t - \tau}{\tau_D}, \tag{5.168}$$

for all $\tau, t \in \mathbb{R}_{\geq 0}$ with $t \geq \tau$. □

(2) *DoS Duration.* In addition to the DoS frequency, one also needs to constrain the DoS *duration,* namely the length of the intervals over which communication is interrupted. To see this, consider, for example, a DoS sequence consisting of the singleton $\{h_0\}$. Assumption 5.9 is clearly satisfied with $\eta \geq 1$. However, if $H_0 = \mathbb{R}_{\geq 0}$ (communication is never possible) then stability is lost regardless of the adopted control update policy. Recalling the definition of $\Xi(\tau, t)$ in (5.142), the assumption that follows provides a quite natural counterpart of Assumption 5.9 with respect to the DoS duration.

Assumption 5.10 (DoS Duration). There exist $\kappa \in \mathbb{R}_{\geq 0}$ and $T \in \mathbb{R}_{>1}$ such that

$$|\Xi(\tau, t)| \leq \kappa + \frac{t - \tau}{T}, \tag{5.169}$$

for all $\tau, t \in \mathbb{R}_{\geq 0}$ with $t \geq \tau$. □

Fig. 5.16 exemplifies values of $n(\tau, t)$ and $\Xi(\tau, t)$ for a given DoS pattern.

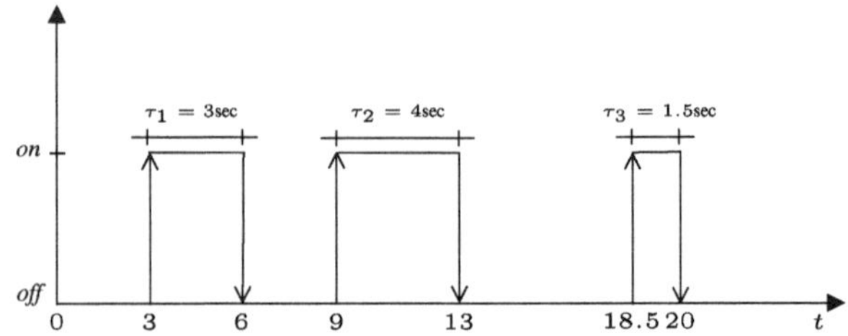

Figure 5.16 Example of DoS signal. *Off/on* transitions are represented as ↑, while *on/off* transitions are represented as ↓. *Off/on* transitions occur at 3, 9, and 18.5 s, and the corresponding DoS intervals have duration of 3, 4, and 1.5 s, respectively. This yields, for instance, $n(0,1)=0$, $n(1,10)=2$ and $n(10,20)=1$, while $\Xi(0,1)=\emptyset$, $\Xi(1,10)=[3,6[\cup[9,10[$, and $\Xi(10,20)=[10,13[\cup[18.5,20[$

5.5.6 ISS Under Denial-of-Service

We are now in position to derive the main result of this section, which can be expressed in words as follows: *any control update rule attaining the conditions of Lemma 5.3 preserves ISS for any DoS signal that satisfies Assumptions 5.9 and 5.10 with τ_D and T sufficiently large.* Although the proof of this result is rather involved, the underlying approach is very intuitive. We decompose the time axis into intervals where it possible to satisfy (5.150) and intervals where, due to the occurrence of DoS, (5.150) need not hold. We then analyze the closed-loop dynamics as a system switching between stable and unstable modes, and determine values of τ_D and T under which the stable behavior is predominant with respect to the unstable one.

Consider a sequence $\{t_k\}_{k\in\mathbb{N}_0}$ of sampling times, along with a DoS sequence $\{h_n\}_{n\in\mathbb{N}_0}$. Let

$$\mathscr{S} := \left\{ k\in\mathbb{N}_0 \mid t_k \in \bigcup_{n\in\mathbb{N}_0} H_n \right\} \tag{5.170}$$

denote the set of integers related to a control update attempt occurring under DoS. The following result holds.

Theorem 5.4. *Consider the control system Σ composed of* (5.139) *and control input* (5.144)*, where K is such that all the eigenvalues of $\Phi = A + BK$ have negative real part. Given any positive symmetric definite matrix $Q\in\mathbb{R}^{n\times n}$, let P be the unique solution of the Lyapunov equation $\Phi^\top P + P\Phi + Q = 0$. Let $V(x) = x^\top P x$. Consider any control update sequence occurring at a finite sampling rate and with inter-sampling times smaller than or equal to $\bar{\Delta}_\sigma$ as in Lemma 5.1,* *with σ satisfying* (5.153). *Consider any DoS sequence satisfying Assumptions 5.9 and 5.10 with arbitrary η and κ, and with τ_D and T such that*

$$\frac{\Delta_*}{\tau_D} + \frac{1}{T} < \frac{\omega_1}{\omega_1 + \omega_2}, \tag{5.171}$$

where Δ_* is a nonnegative constant satisfying

$$\sup_{k \in \mathcal{S}} \Delta_k \le \Delta_*, \tag{5.172}$$

with Δ_k being as in (5.147), *$\omega_1 := (\gamma_1 - \gamma_2 \sigma)/2\alpha_2$ and $\omega_2 := 2\gamma_2/\alpha_1$, where α_1, α_2, γ_1 and γ_2 are as in* (5.152). *Then, Σ is ISS.*

Proof. The idea is to decompose the time axis into intervals where it possible to satisfy (5.150) and intervals where, due to the occurrence of DoS, (5.150) need not hold. We then analyze the closed-loop dynamics as a system that switches between stable and unstable modes. For clarity of exposition, the proof is divided into three steps.

Step I. Modeling of the Intervals Related to Stable and Unstable Dynamics. In this step we characterize the intervals of time where (5.150) holds and those where it need not hold. During these intervals, the closed-loop system evolves obeying stable and possibly unstable dynamics, respectively. The characterization of these intervals is essential for the Lyapunov-based analysis we carry out in the forthcoming steps and can be formalized as follows:

Lemma 5.4. *For any $\tau, t \in \mathbb{R}_{\ge 0}$, with $0 \le \tau \le t$, the interval $[\tau, t]$ is the disjoint union of $\bar{\Theta}(\tau, t)$ and $\bar{\Xi}(\tau, t)$, where $\bar{\Theta}(\tau, t)$ (respectively, $\bar{\Xi}(\tau, t)$) is the union of subintervals of $[\tau, t]$ over which* (5.150) *holds (respectively, need not hold). Specifically, there exist two sequences of nonnegative and positive real numbers $\{\zeta_m\}_{m \in \mathbb{N}_0}$, $\{v_m\}_{m \in \mathbb{N}_0}$ such that*

$$\bar{\Xi}(\tau, t) := \bigcup_{m \in \mathbb{N}_0} Z_m \cap [\tau, t], \tag{5.173}$$

$$\bar{\Theta}(\tau, t) := \bigcup_{m \in \mathbb{N}_0} W_{m-1} \cap [\tau, t], \tag{5.174}$$

where

$$Z_m := \{\zeta_m\} \cup [\zeta_m, \zeta_m + v_m[, \tag{5.175}$$

$$W_m := \{\zeta_m + v_m\} \cup [\zeta_m + v_m, \zeta_{m+1}[, \tag{5.176}$$

and where $\zeta_{-1} = v_{-1} := 0$.

Proof. Let $\mathscr{S}_n := \{k \in \mathbb{N}_0 | t_k \in H_n\}$ denote the set of integers related to a control update attempt occurring over H_n, $n \in \mathbb{N}_0$. Define

$$\lambda_n := \begin{cases} \tau_n, & \text{if } \mathscr{S}_n = \emptyset, \\ t_{\sup\{k \in \mathbb{N}_0 : k \in \mathscr{S}_n\}} - h_n, & \text{otherwise}, \end{cases} \tag{5.177}$$

$$\Lambda_n := \begin{cases} 0, & \text{if } \mathscr{S}_n = \emptyset, \\ \Delta_{\sup\{k \in \mathbb{N}_0 : k \in \mathscr{S}_n\}}, & \text{otherwise}. \end{cases} \tag{5.178}$$

Thus,

$$\bar{H}_n := \{h_n\} \cup [h_n, h_n + \lambda_n + \Lambda_n[\tag{5.179}$$

specifies the nth time interval where (5.150) need not hold, which consists of H_n plus the corresponding DoS-induced actuation delay. Note that $\lambda_n + \Lambda_n \geq \tau_n$ for all $n \in \mathbb{N}_0$. Notice now that the intervals $\bar{H}_n$ and $\bar{H}_{n+1}$ may overlap each other in that h_{n+1} may belong to $\bar{H}_n$. For analysis purposes, it is convenient to regard these overlapping intervals as a single interval of the form (5.175). This can be done by defining an auxiliary sequence $\{\zeta_m\}_{m \in \mathbb{N}_0}$, which is recursively defined from $\{h_n\}_{n \in \mathbb{N}_0}$ as follows:

$$\zeta_0 := h_0, \tag{5.180}$$

$$\zeta_{m+1} := \inf\{\, h_n > \zeta_m \mid h_n > h_{n-1} + \lambda_{n-1} + \Lambda_{n-1}\}, \tag{5.181}$$

for all $m \in \mathbb{N}$, and letting

$$v_m := \sum_{\substack{n \in \mathbb{N}_0; \\ \zeta_m \leq h_n < \zeta_{m+1}}} |\bar{H}_n \backslash \bar{H}_{n+1}| \tag{5.182}$$

for all $m \in \mathbb{N}_0$. Given $\tau, t \in \mathbb{R}_{\geq 0}$, define $\bar{\Xi}(\tau, t)$ as in (5.173). Hence, by construction, $\bar{\Xi}(\tau, t)$ represents the union of subintervals of $[\tau, t]$ where, due to the occurrence of DoS, (5.150) need not hold.

Let W_m be as in (5.176). Given $\tau, t \in \mathbb{R}_{\geq 0}$, let $\bar{\Theta}(\tau, t)$ be as in (5.174). Note that over each $[\tau, t]$, the sets $\bar{\Xi}(\tau, t)$ and $\bar{\Theta}(\tau, t)$ are complementary since their union equals $[\tau, t]$ and their intersection is the empty set. Also notice that $\bar{\Theta}(\tau, t)$ represents the union of subintervals of $[\tau, t]$ where (5.150) holds true. In fact, by construction, for each $m \in \mathbb{N}_0$, a successful control update necessarily occurs at $\zeta_m + v_m$ and there is no DoS over W_m. □

Step II. Lyapunov Function Analysis. Consider the intervals W_m, $m \in \mathbb{N}_0$, where (5.150) holds true by construction. From (5.157) we get

$$V(x(t)) \leq e^{-\omega_1(t-\zeta_m-v_m)} V(x(\zeta_m + v_m)) + \gamma_7 \|w_t\|_\infty^2, \tag{5.183}$$

for all $t \in W_m$ and all $m \in \mathbb{N}_0$. Here and in the sequel, all the relevant constants are as in the proof of Theorem 5.3. In particular, we recall that $\gamma_7 = \alpha_2(\gamma_3 + \sigma\gamma_2)^2/(\gamma_1 - \sigma\gamma_2)^2$ with $\sigma < \gamma_1/\gamma_2$.

Consider next the intervals Z_m, $m \in \mathbb{N}_0$, over which the inequality (5.150) does not necessarily hold. To derive a bound on the growth of $V(x(t))$ in this case, some intermediate steps are needed. We first prove that for each $m \in \mathbb{N}_0$

$$\|e(t)\| \leq (1+\sigma)\|x(\zeta_m)\| + \|x(t)\| + \sigma \|w_t\|_\infty, \tag{5.184}$$

for all $t \in Z_m$. Recall that

$$e(t) = x\left(t_{k(\zeta_m)}\right) - x(t), \tag{5.185}$$

for all $t \in Z_m$, where $x(t_{k(\zeta_m)})$ represents the value of the process state at the last successful control update up to ζ_m. If $\zeta_0 = 0$, then $x(t_{k(\zeta_0)}) = 0$ by definition so that (5.184) is valid. Consider next the case $\zeta_0 > 0$. By construction, (5.150) holds true for all $t \in W_m$. Hence, by continuity of x, one has

$$\|e(\zeta_m)\| \leq \sigma \|x(\zeta_m)\| + \sigma \|w_{\zeta_m}\|_\infty, \tag{5.186}$$

for all $m \in \mathbb{N}_0$. Hence,

$$\|x(t_{k(\zeta_m)}) - x(\zeta_m)\| \leq \sigma \|x(\zeta_m)\| + \sigma \|w_{\zeta_m}\|_\infty \tag{5.187}$$

and (5.184) follows by applying the triangular inequality.

Substituting (5.184) into (5.152b) yields

$$\begin{aligned} \frac{d}{dt} V(x(t)) \leq\ & (\gamma_2 - \gamma_1)\|x(t)\|^2 \\ & + \gamma_2(1+\sigma)\|x(t)\|\,\|x(\zeta_m)\| \\ & + (\gamma_3 + \sigma\gamma_2)\|x(t)\| v(t), \end{aligned} \tag{5.188}$$

where $v(t) := \sup\{\|w(t)\|, \|w_t\|_\infty\}$. We then proceed as in the proof of Theorem 5.3. Using (5.155) with $\delta = (\gamma_3 + \gamma_2\sigma)/(\gamma_1 - \sigma\gamma_2)$, simple calculations yield

$$\frac{d}{dt}V(x(t)) \leq \gamma_2(1-\sigma)\|x(t)\|^2 \\ + \gamma_2(1+\sigma)\|x(t)\|\,\|x(\zeta_m)\| + \gamma_6 v^2(t), \tag{5.189}$$

where we recall that $\gamma_6 = (\gamma_3 + \gamma_2\sigma)^2/(2(\gamma_1 - \sigma\gamma_2))$. Note that

$$\frac{d}{dt}V(x(t)) \leq \omega_2 \max\{V(x(t)), V(x(\zeta_m))\} + \gamma_6 v^2(t), \tag{5.190}$$

where $\omega_2 := 2\gamma_2/\alpha_1$. Since $\|v_t\|_\infty = \|w_t\|_\infty$ for any $t \in \mathbb{R}_{\geq 0}$, we then have

$$V(x(t)) \leq e^{\omega_2(t-\zeta_m)} V(x(\zeta_m)) + \gamma_8 e^{\omega_2(t-\zeta_m)} \|w_t\|_\infty^2, \tag{5.191}$$

for all $t \in Z_m$, where $\gamma_8 := \gamma_6/\omega_2$.

Combining (5.183) and (5.191), we can prove the following result.

Lemma 5.5. *For all $t \in \mathbb{R}_{\geq 0}$, the Lyapunov function satisfies*

$$V(x(t)) \leq e^{-\omega_1|\bar{\Theta}(0,t)|} e^{\omega_2|\bar{\Xi}(0,t)|} V(x(0)) \\ + \gamma_* \left[1 + 2 \sum_{\substack{m\in\mathbb{N}_0; \\ \zeta_m \leq t}} e^{-\omega_1|\bar{\Theta}(\zeta_m+v_m,t)|} e^{\omega_2|\bar{\Xi}(\zeta_m,t)|} \right] \|w_t\|_\infty^2, \tag{5.192}$$

where $\gamma_* := \max\{\gamma_7, \gamma_8\}$.

Hereafter, in accordance with (5.174), it is understood that $|\bar{\Theta}(\zeta_m + v_m, t)| = 0$ whenever $t < \zeta_m + v_m$.

Proof of Lemma 5.5. We use an induction argument. First, we show that the inequality holds true over $W_{-1} = [0, \zeta_0]$. If $\zeta_0 = 0$, the claim trivially holds. Suppose $\zeta_0 > 0$. Over W_{-1} the Lyapunov function obeys (5.183); thus, (5.192) follows by noting that $|\bar{\Theta}(0,t)| = t$ and $|\bar{\Xi}(0,t)| = 0$ for all $t \in W_{-1}$ and the sum term in (5.192) is zero.

Assume next that (5.192) holds true over the interval $[0, \zeta_p]$, where $p \in \mathbb{N}_0$. By hypothesis, and since $V(x)$ is continuous, we have

$$V(x(\zeta_p)) \leq e^{-\omega_1|\bar{\Theta}(0,\zeta_p)|} e^{\omega_2|\bar{\Xi}(0,\zeta_p)|} V(x(0)) \\ + \gamma_* \left[1 + 2 \sum_{\substack{m\in\mathbb{N}_0; \\ \zeta_m < \zeta_p}} e^{-\omega_1|\bar{\Theta}(\zeta_m+v_m,\zeta_p)|} e^{\omega_2|\bar{\Xi}(\zeta_m,\zeta_p)|} \right] \|w_{\zeta_p}\|_\infty^2. \tag{5.193}$$

Then, consider first the interval Z_p, over which the Lyapunov function obeys (5.191) with

$$V(x(t)) \le e^{\omega_2(t-\zeta_p)} V\left(x(\zeta_p)\right) + \gamma_* e^{\omega_2(t-\zeta_p)} \|w_t\|_\infty^2. \tag{5.194}$$

Notice that $t-\zeta_p = |\bar{\Xi}(\zeta_p, t)|$ for all $t \in Z_p$. Hence,

$$|\bar{\Xi}(0,t)| = t - \zeta_p + |\bar{\Xi}(0,\zeta_p)|, \tag{5.195}$$

$$|\bar{\Theta}(0,t)| = |\bar{\Theta}(0,\zeta_p)|, \tag{5.196}$$

for all $t \in Z_p$. Thus, multiplying $e^{\omega_2(t-\zeta_p)}$ by the first term on the right-hand side (RHS) of (5.193) we get the first term on the RHS of (5.192). In addition, multiplying $e^{\omega_2(t-\zeta_p)}$ by the sum term in (5.193), we obtain

$$\begin{aligned} e^{\omega_2(t-\zeta_p)} &\sum_{\substack{m\in\mathbb{N}_0;\\ \zeta_m<\zeta_p}} e^{-\omega_1|\bar{\Theta}(\zeta_m+v_m,\zeta_p)|} e^{\omega_2|\bar{\Xi}(\zeta_m,\zeta_p)|} \\ &= \sum_{\substack{m\in\mathbb{N}_0;\\ \zeta_m<\zeta_p}} e^{-\omega_1|\bar{\Theta}(\zeta_m+v_m,t)|} e^{\omega_2|\bar{\Xi}(\zeta_m,t)|}, \end{aligned} \tag{5.197}$$

for all $t \in Z_p$ because $|\bar{\Theta}(\zeta_m + v_m, \zeta_p)| = |\bar{\Theta}(\zeta_m + v_m, t)|$ and $t - \zeta_p + |\bar{\Xi}(\zeta_m, \zeta_p)| = |\bar{\Xi}(\zeta_m, t)|$ for all $t \in Z_p$. Overall, the factor multiplying the disturbance term can be therefore rewritten as

$$\begin{aligned} 2\gamma_* e^{\omega_2(t-\zeta_p)} &+ 2\gamma_* \sum_{\substack{m\in\mathbb{N}_0;\\ \zeta_m<\zeta_p}} e^{-\omega_1|\bar{\Theta}(\zeta_m+v_m,t)|} e^{\omega_2|\bar{\Xi}(\zeta_m,t)|} \\ &= 2\gamma_* \sum_{\substack{m\in\mathbb{N}_0;\\ \zeta_m\le\zeta_p}} e^{-\omega_1|\bar{\Theta}(\zeta_m+v_m,t)|} e^{\omega_2|\bar{\Xi}(\zeta_m,t)|} \\ &= 2\gamma_* \sum_{\substack{m\in\mathbb{N}_0;\\ \zeta_m\le t}} e^{-\omega_1|\bar{\Theta}(\zeta_m+v_m,t)|} e^{\omega_2|\bar{\Xi}(\zeta_m,t)|}, \end{aligned} \tag{5.198}$$

where the first equality follows from $|\bar{\Theta}(\zeta_p + v_p, t)| = 0$ and $t - \zeta_p = |\bar{\Xi}(\zeta_p, t)|$ for all $t \in Z_p$, while the second equality holds because condition $\zeta_m \le \zeta_p$ is equivalent to $\zeta_m \le t$ for all $t \in Z_p$. Hence, (5.192) holds for all $t \in [0, \zeta_p + v_p]$.

Consider next the interval W_p, over which the Lyapunov function obeys (5.183). In particular,

$$\begin{aligned} V(x(t)) \le\ & e^{-\omega_1(t-\zeta_p-v_p)} e^{\omega_2 v_p} V(x(\zeta_p)) \\ &+ \gamma_* \left[1 + e^{-\omega_1(t-\zeta_p-v_p)} e^{\omega_2 v_p}\right] \|w_t\|_\infty^2, \end{aligned} \tag{5.199}$$

for all $t \in W_p$, with $V(x(\zeta_p))$ as in (5.193). To see that this implies (5.192) again, notice that $t - \zeta_p - v_p = |\bar{\Theta}(\zeta_p + v_p, t)|$ and $|\bar{\Theta}(0, \zeta_p)| = \bar{\Theta}(0, \zeta_p + v_p)|$. Hence,

$$|\bar{\Theta}(0, t)| = t - \zeta_p - v_p + |\bar{\Theta}(0, \zeta_p)|, \tag{5.200}$$

for all $t \in W_p$. Moreover, $v_p = |\bar{\Xi}(\zeta_p, \zeta_p + v_p)|$ and $|\bar{\Xi}(\zeta_p + v_p, t)| = 0$ for all $t \in W_p$. Hence,

$$|\bar{\Xi}(0, t)| = v_p + |\bar{\Xi}(0, \zeta_p)|, \tag{5.201}$$

for all $t \in W_p$. Thus, multiplying $e^{-\omega_1(t-\zeta_p-v_p)}e^{\omega_2 v_p}$ by the first term on the RHS of (5.193) we get the first term on the RHS of (5.192). Moreover, multiplying $e^{-\omega_1(t-\zeta_p-v_p)}e^{\omega_2 v_p}$ by the sum term in (5.193), we obtain

$$\begin{aligned} & e^{-\omega_1(t-\zeta_p-v_p)}e^{\omega_2 v_p} \sum_{\substack{m\in\mathbb{N}_0;\\ \zeta_m<\zeta_p}} e^{-\omega_1|\bar{\Theta}(\zeta_m+v_m,\zeta_p)|}e^{\omega_2|\bar{\Xi}(\zeta_m,\zeta_p)|} \\ & = \sum_{\substack{m\in\mathbb{N}_0;\\ \zeta_m<\zeta_p}} e^{-\omega_1|\bar{\Theta}(\zeta_m+v_m,t)|}e^{\omega_2|\bar{\Xi}(\zeta_m,t)|}. \end{aligned} \tag{5.202}$$

Overall, the factor multiplying the disturbance term can be therefore rewritten as

$$\begin{aligned} & \gamma_*\left[1 + 2e^{-\omega_1(t-\zeta_p-v_p)}e^{\omega_2 v_p}\right] \\ & \quad + 2\gamma_* \sum_{\substack{m\in\mathbb{N}_0;\\ \zeta_m<\zeta_p}} e^{-\omega_1|\bar{\Theta}(\zeta_m+v_m,t)|}e^{\omega_2|\bar{\Xi}(\zeta_m,t)|} \\ & = 2\gamma_* \sum_{\substack{m\in\mathbb{N}_0;\\ \zeta_m\le\zeta_p}} e^{-\omega_1|\bar{\Theta}(\zeta_m+v_m,t)|}e^{\omega_2|\bar{\Xi}(\zeta_m,t)|} \\ & = 2\gamma_* \sum_{\substack{m\in\mathbb{N}_0;\\ \zeta_m\le t}} e^{-\omega_1|\bar{\Theta}(\zeta_m+v_m,t)|}e^{\omega_2|\bar{\Xi}(\zeta_m,t)|}. \end{aligned} \tag{5.203}$$

As before, the second equality holds because, for all $t \in W_p$, condition $\zeta_m \le \zeta_p$ is equivalent to $\zeta_m \le t$. Thus (5.192) holds true for all $t \in [0, \zeta_{p+1}]$, which concludes the proof. □

Step III. Bounds on DoS Frequency and Duration. In order to conclude the proof, we need to bound the sum term in (5.192). The following can be proven:

Lemma 5.6. *Under Assumptions 5.9 and 5.10 and condition* (5.171), *the sum*

$$\sum_{\substack{m\in\mathbb{N}_0;\\ \zeta_m\leq t}} e^{-\omega_1|\bar{\Theta}(\zeta_m+v_m,t)|}e^{\omega_2|\bar{\Xi}(\zeta_m,t)|} \tag{5.204}$$

is bounded above by

$$e^{(\omega_1+\omega_2)\kappa_*}\frac{e^{\beta_*\tau_D\eta}}{1-e^{-\beta_*\tau_D}}, \tag{5.205}$$

where the constants ω_1, ω_2, τ_D, κ_* and β_*, are defined in Lemma 5.3, (5.165) *and* (5.166).

Proof. First notice that

$$|\bar{\Xi}(\tau,t)|\leq|\Xi(\tau,t)|+(1+n(\tau,t))\Delta_*, \tag{5.206}$$

for all $\tau, t\in\mathbb{R}_{\geq 0}$ with $t\geq\tau$. In words, $|\bar{\Xi}(\tau,t)|$ can be upper bounded by the total length of DoS over $[\tau,t]$ plus the maximum actuation delay Δ_*, which may occur once at the beginning of the interval $[\tau,t]$ (as a result of a previous DoS) plus $n(\tau,t)$ times, where $n(\tau,t)$ represents the number of *off/on* transitions of DoS occurring over $[\tau,t[$. Using Assumptions 5.9 and 5.10, we have

$$\begin{aligned}|\bar{\Xi}(\tau,t)|&\leq\kappa+\frac{t-\tau}{T}+\left(1+\eta+\frac{t-\tau}{T}\right)\Delta_*\\&=:\kappa_*+\frac{t-\tau}{T_*},\end{aligned} \tag{5.207}$$

where $\kappa_*:=\kappa+(1+\eta)\Delta_*$ and $T_*:=\tau_D T/(\tau_D+T\Delta_*)$. With this in mind, we can now analyze the sum term in (5.204). From the above inequality, we have

$$|\bar{\Xi}(\zeta_m,t)|\leq\kappa_*+\frac{t-\zeta_m}{T_*}, \tag{5.208}$$

for all $t\in\mathbb{R}_{\geq\zeta_m}$, $m\in\mathbb{N}_0$. Consider next $|\bar{\Theta}(\zeta_m+v_m,t)|$. We have

$$|\bar{\Theta}(\zeta_m+v_m,t)|=t-\zeta_m-|\bar{\Xi}(\zeta_m,t)|, \tag{5.209}$$

for all $t\in\mathbb{R}_{\geq\zeta_m}$, $m\in\mathbb{N}_0$. To see this, consider first the interval Z_m. For all $t\in Z_m$, we have $|\bar{\Xi}(\zeta_m,t)|=t-\zeta_m$ so that (5.209) holds true since, by definition, $|\bar{\Theta}(\zeta_m + \nu_m, t)| = 0$ whenever $t < \zeta_m + \nu_m$. For all $t \in \mathbb{R}_{\geq \zeta_m + \nu_m}$, we have

$$|\bar{\Theta}(\zeta_m + \nu_m, t)| = t - \zeta_m - \nu_m - |\bar{\Xi}(\zeta_m + \nu_m, t)|$$
$$= t - \zeta_m - |\bar{\Xi}(\zeta_m, t)|, \tag{5.210}$$

where the second equality follows because

$$|\bar{\Xi}(\zeta_m + \nu_m, t)| = |\bar{\Xi}(\zeta_m, t)| - |\bar{\Xi}(\zeta_m, \zeta_m + \nu_m)|$$
$$= |\bar{\Xi}(\zeta_m, t)| - \nu_m, \tag{5.211}$$

for all $t \in \mathbb{R}_{\geq \zeta_m + \nu_m}$, $m \in \mathbb{N}_0$. Hence,

$$\sum_{\substack{m \in \mathbb{N}_0; \\ \zeta_m \leq t}} e^{-\omega_1 |\bar{\Theta}(\zeta_m + \nu_m, t)|} e^{\omega_2 |\bar{\Xi}(\zeta_m, t)|}$$
$$\leq e^{(\omega_1 + \omega_2)\kappa_*} \sum_{\substack{m \in \mathbb{N}_0; \\ \zeta_m \leq t}} e^{-[\omega_1 - (\omega_1 + \omega_2)/T_*](t - \zeta_m)}$$
$$= e^{(\omega_1 + \omega_2)\kappa_*} \sum_{\substack{m \in \mathbb{N}_0; \\ \zeta_m \leq t}} e^{-\beta_*(t - \zeta_m)}, \tag{5.212}$$

where $\beta_* = \omega_1 - (\omega_1 + \omega_2)/T_*$. Under condition (5.171), we have $\beta_* > 0$. Moreover, Assumption 5.9 yields

$$t - \zeta_m \geq \tau_D n(\zeta_m, t) - \tau_D \eta. \tag{5.213}$$

This implies

$$\sum_{\substack{m \in \mathbb{N}_0; \\ \zeta_m \leq t}} e^{-\beta_*(t - \zeta_m)} \leq e^{\beta_* \tau_D \eta} \sum_{\substack{m \in \mathbb{N}_0; \\ \zeta_m \leq t}} e^{-\beta_* \tau_D n(\zeta_m, t)}. \tag{5.214}$$

Let

$$m(t) := \begin{cases} -1, & \text{if } t < \zeta_0, \\ \sup\{m \in \mathbb{N}_0 | \zeta_m \leq t\}, & \text{otherwise.} \end{cases} \tag{5.215}$$

We then have

$$\sum_{\substack{m \in \mathbb{N}_0; \\ \zeta_m \leq t}} e^{-\beta_* \tau_D n(\zeta_m, t)} = \sum_{m=0}^{m(t)} e^{-\beta_* \tau_D n(\zeta_m, t)}. \tag{5.216}$$

Since

$$n(\zeta_m, t) \geq m(t) - m, \tag{5.217}$$

we obtain

$$\sum_{m=0}^{m(t)} e^{-\beta_* \tau_D n(\zeta_m, t)} = \sum_{m=0}^{m(t)} e^{-\beta_* \tau_D (m(t)-m)} = \sum_{m=0}^{m(t)} e^{-\beta_* \tau_D m} \leq \frac{1}{1 - e^{-\beta_* \tau_D}}. \tag{5.218}$$

Relations (5.212), (5.214), (5.216), (5.218) yield the desired bound (5.205). □

We can now finalize the proof of Theorem 5.140. Observe that the first term on the RHS of (5.192) is bounded by

$$e^{\kappa_*(\omega_1+\omega_2)} e^{-(\omega_1 - \frac{\omega_1+\omega_2}{T_*})t} V(x(0)), \tag{5.219}$$

where we have exploited the equality $|\bar{\Theta}(0,t)| = t - |\bar{\Xi}(0,t)|$ and (5.207). Moreover, $\beta_* = \omega_1 - (\omega_1 + \omega_2)/T_*$. Hence, from (5.192) and (5.205), one sees that the Lyapunov function $V(x)$ computed along any trajectory $x(t)$ satisfies

$$V(x(t)) \leq e^{\kappa_*(\omega_1+\omega_2)} e^{-\beta_* t} V(x(0)) + \gamma_* \left[1 + 2e^{\kappa_*(\omega_1+\omega_2)} \frac{e^{\beta_* \tau_D \eta}}{1 - e^{-\beta * \tau_D}}\right] \|w_t\|_\infty^2. \tag{5.220}$$

Letting $\delta_* := \gamma_*/\alpha_1$, we readily get

$$\|x(t)\| \leq \sqrt{\frac{\alpha_2}{\alpha_1} \frac{e^{\kappa_*(\omega_1+\omega_2)} 2}{e}}^{-(\beta_*/2)t} \|x(0)\| + \sqrt{\delta_*} \left[1 + 2e^{\kappa_*(\omega_1+\omega_2)} \frac{e^{\beta_* \tau_D \eta}}{1 - e^{-\beta_* \tau_D}}\right]^{\frac{1}{2}} \|w_t\|_\infty. \tag{5.221}$$

Note that $\delta_* = ((\gamma_3 + \gamma_2\sigma)^2/(\gamma_1 - \gamma_2\sigma)) \max\{(\alpha_2/(\alpha_1(\gamma_1 - \gamma_2\sigma))), (1/4\gamma_2)\}$ is a positive constant, i.e., independent of the process initial condition and the disturbance w. Since Δ_* is a positive constant, $\kappa_* = \kappa + (1+\eta)\Delta_*$ and $\beta_* = \omega_1 - (\omega_1 + \omega_2)((\Delta_*/\tau_D) + (1/T))$ are also independent of the process initial condition and w. Thus ISS follows at once. □

In connection with (5.172), we note that, in accordance with the adopted notation, $\sup_{k\in\mathscr{S}} \Delta_k = 0$ when $\mathscr{S} = \emptyset$.

Remark 5.14 (Performance Bounds). *The considered class of sampling logic provides quite strong stability properties, namely* exponential *ISS with linear bounds on the disturbance-to-state map. In particular, from the proof of Theorem 5.4, one has*

$$\|x(t)\| \leq \sqrt{\frac{\alpha_2}{\alpha_1}} e^{\frac{\kappa_*(\omega_1+\omega_2)}{2}} e^{-(\frac{\beta_*}{2})t} \|x(0)\| + \sqrt{\delta_*}\left[1 + 2e^{\kappa_*(\omega_1+\omega_2)} \frac{e^{\beta_* \tau_D \eta}}{1 - e^{-\beta_* \tau_D}}\right]^{\frac{1}{2}} \|w_t\|_\infty, \tag{5.222}$$

where

$$\kappa_* := \kappa + (1+\eta)\Delta_*, \tag{5.223}$$

$$\beta_* := \omega_1 - (\omega_1 + \omega_2)\left(\frac{\Delta_*}{\tau_D} + \frac{1}{T}\right), \tag{5.224}$$

$$\delta_* := \frac{(\gamma_3 + \gamma_2\sigma)^2}{(\gamma_1 - \gamma_2\sigma)} \max\left\{\frac{\alpha_2}{\alpha_1(\gamma_1 - \gamma_2\sigma)}, \frac{1}{4\gamma_2}\right\}, \tag{5.225}$$

with α_1, α_2, γ_1, γ_2 and γ_3 as in (5.152). *One sees that the convergence rate of the closed-loop dynamics as well as the amount of DoS that one can tolerate can be increased by decreasing the sampling rate upon the DoS occurrence, i.e., by decreasing Δ_*.*

Remark 5.15 (DoS-Induced Actuation Delay). *Decreasing Δ_* can also be viewed as reducing the DoS-induced actuation delay. Assume for instance that the kth sampling instant belongs to $\mathscr{S}$. Since $t_k \in \mathscr{S}$, the process will evolve in open-loop under out-of-date control up to time t_{k+1}, even if DoS ceases before t_{k+1}. Thus decreasing Δ_* has the effect of reducing the delay in the actuation of the new control sample. In general, due to the finite sampling rate, such a delay is non-zero. In fact, $\Delta_* = 0$ only in two cases. The first one is when $\mathscr{S} = \emptyset$. Such a case is not interesting since if $\mathscr{S} = \emptyset$, then DoS is ineffective and ISS follows directly from Theorem 5.3. The second case is when $\mathscr{S} \neq \emptyset$ and, for each DoS interval H_n containing a t_k, the control update is scheduled exactly at $h_n + \tau_n$. This case is, however, unrealistic since h_n and τ_n are unknown. The case $\mathscr{S} \neq \emptyset$ and $\Delta_* = 0$ becomes of interest when considering control under DoS in a purely continuous-time framework. In fact, if the control action is continuous then $\Delta_* = \bar{\Delta}_\sigma = 0$. Under* *such circumstances, (5.171) reduces to* $1/T < \omega_1/(\omega_1 + \omega_2)$, *which is independent of* τ_D. *In fact, in a continuous-time setting, slow-on-average DoS in the form of pulses has no effect on closed-loop stability.*

5.5.7 Simulation Example 5.8

Here, we consider the situation where, on every interval which contains N communication attempts, a number $M < N$ of these attempts can be denied. A simple way to account for this situation is to regard the DoS as a train of pulses, M of which are superimposed to the sampling times. This implies that Assumption 5.10 holds true with $\kappa = 0$ and $T = \infty$. As for Assumption 5.9, assume a lower bound $\underline{\Delta}$ on the control executions. One sees that

$$n(\tau, t) \leq \left\lceil \frac{t - \tau}{\underline{\Delta} N} \right\rceil M, \tag{5.226}$$

for all $\tau, t \in \mathbb{R}_{\geq 0}$ with $t \geq \tau$. Then, for each N, (5.168) holds true with $\eta = M$ and $\tau_D = \underline{\Delta} N/M$. In connection with Theorem 5.2, this means that stability is preserved whenever $(\underline{\Delta} N)/(\Delta_* M) > (\omega_1 + \omega_2)/\omega_1$. Using a logic that samples at rate $\underline{\Delta}$ upon DoS, stability is preserved whenever $N/M > (\omega_1 + \omega_2)/\omega_1$. For instance, this means that if $(\omega_1 + \omega_2)/\omega_1 = 5$, then up to 20% of the communication attempts can be denied without destroying stability. The value of N affects both closed-loop performance and robustness against DoS: the larger the N, the larger the number of consecutive communication attempts that can be denied without destroying stability. However, this potentially results in larger overshoots since $\eta = M$.

5.5.8 Simulation Example 5.9

Another interesting scenario is when DoS is sustained. One can account for this situation by modeling the DoS signal as a rectangular wave of a (possibly) variable period and duty cycle [73]. Let P_n and $D_n = \tau_n/P_n$ denote the period and duty cycle of the nth DoS attack, respectively. Also let $P_{\min} := \inf_{n \in \mathbb{N}_0} P_n$, $D_{\max} := \sup_{n \in \mathbb{N}_0} D_n$ and $\tau_{\max} := \sup_{n \in \mathbb{N}_0} \tau_n$. Suppose that $P_{\min} > \underline{\Delta}$, $D_{\max} < 1$ and $\tau_{\max} < \infty$. Since the maximum number of *off/on* transitions of DoS during the interval $[\tau, t[$ can be upper bounded as $n(\tau, t) \leq \lceil (t - \tau)/P_{\min} \rceil$, Assumption 5.9 holds true with $\eta = 1$ and $\tau_D = P_{\min}$. Let now

$$n(t) := \begin{cases} -1, & \text{if } t < h_0, \\ \sup\{n \in \mathbb{N}_0 | h_n \leq t\}, & \text{otherwise,} \end{cases} \tag{5.227}$$

where $\tau_{-1} := 0$. For any $\tau, t \in \mathbb{R}_{\geq 0}$ with $t \geq \tau$, it is possible to write

$$|\Xi(\tau,t)| \leq \max\{0, h_{n(\tau)} + \tau_{n(\tau)} - \tau\} \\ + \min\{t - h_{n(t)}, \tau_{n(t)}\} + D_{\max} \sum_{\substack{n \in \mathbb{N}_0; \\ h_{n(\tau)} < h_n; \\ h_n < h_{n(t)}}} P_n. \tag{5.228}$$

Hence, one sees that Assumption 5.10 holds true with $\kappa = 2\tau_{\max}$ and $T = D_{\max}^{-1}$. This example includes as a special case the so-called periodic jamming where both period and duty cycle are constant.

5.5.9 Simulation Example 5.10

In connection with Simulation Example 5.9, we pointed out that the requirements $P_{\min} > \underline{\Delta}$ and $D_{\max} < 1$ are in a wide-sense necessary for closed-loop stability unless other conditions are imposed on DoS (as for $\tau_{\max} < \infty$). These conditions, however, rule out, for example, the possibility of DoS signals with *off/on* transitions occasionally faster than the maximum transmission rate. Consider first Assumption 5.9 and suppose that $P_{\min} > 0$. Let $f(\tau, t) := (t - \tau)/n(\tau, t)$ represent the average dwell-time of DoS *off/of* transitions on $[\tau, t[$. Assume that for some $\tau_D \in \mathbb{R}_{>\underline{\Delta}}$, there exists a $\gamma \in \mathbb{R}_{>0}$ such that

$$f(\tau, \tau + \gamma) \geq \tau_D, \tag{5.229}$$

for all $\tau \in \mathbb{R}_{\geq 0}$, which implies that the DoS *off/on* transitions are slower than $\underline{\Delta}$ on every sufficiently large time-interval. Then, it is easy to verify that Assumption 5.9 holds true with $\eta = \lceil \gamma / P_{\min} \rceil$. In fact, if $t - \tau \leq \gamma$ then $n(\tau, t) \leq n(\tau, \tau + \gamma)$ where $n(\tau, \tau + \gamma) \leq \lceil \gamma / P_{\min} \rceil$. If instead $t - \tau > \gamma$, let m denote the largest integer m such that $m\gamma < t - \tau$. Then

$$n(\tau, t) = \sum_{k=0}^{m-1} n(\tau + k\gamma, \tau + (k+1)\gamma) + n(\tau + m\gamma, t) \\ = \sum_{k=0}^{m-1} \gamma / f(\tau + k\gamma, \tau + (k+1)\gamma) \\ + n(\tau + m\gamma, t) \leq m\gamma / \tau_D + \lceil \gamma / P_{\min} \rceil \tag{5.230}$$

as it follows from (5.229) and the definition of m. Then the claim follows by recalling that $m\gamma < t - \tau$. As for Assumption 5.10, let $D(\tau, t) :=$

$|\Xi(\tau, t)|/(t-\tau)$, which can be thought of as the average DoS duty cycle over $[\tau, t]$. Assume that for some $T \in \mathbb{R}_{>1}$, there exists a $\delta \in \mathbb{R}_{>0}$ such that

$$D(\tau, \tau+\delta) \leq 1/T, \tag{5.231}$$

for all $\tau \in \mathbb{R}_{\geq 0}$. Reasoning as before, it is simple to verify that Assumption 5.10 holds true with $\kappa = \delta$. In connection with Example 5.9, the conditions just stated allow one to consider more general DoS classes. For instance, let

$$\tau_{ave} := \lim_{m \to \infty} \frac{1}{m+1} \sum_{n=0}^{m} (h_{n+1} - h_n), \tag{5.232}$$

$$D_{ave} := \lim_{m \to \infty} \frac{1}{m+1} \sum_{n=0}^{m} D_n \tag{5.233}$$

denote the average dwell-time of DoS *off/on* transitions and the average DoS duty cycle, respectively. Then (5.229) and (5.231) with $\tau_D = \tau_{ave}$ and $T = D_{ave}^{-1}$ are sufficient to conclude that the DoS signal is also slow-on-average in the sense of Assumptions 5.9 and 5.10 with respect to τ_{ave} and D_{ave}.

5.5.10 Simulation Example 5.11

For the sake of clarity, a numerical example illustrating the theory as well as the foregoing discussions is reported. Consider the following open-loop unstable system [80]:

$$\frac{d}{dt}x(t) = \begin{bmatrix} 1 & 1 \\ 0 & 1 \end{bmatrix} x(t) + u(t) + w(t), \tag{5.234}$$

under LQR gain

$$K = \begin{bmatrix} -2.1961 & -0.7545 \\ -0.7545 & -2.7146 \end{bmatrix}. \tag{5.235}$$

Solving the Lyapunov equation $\Phi^\top P + P\Phi + Q = 0$ with $Q = I_2$ yields $\alpha_1 = 0.2779$, $\alpha_2 = 0.4497$, $\gamma_1 = 1$, and $\gamma_2 = 2.1080$. From this we deduce that we must select σ such that $\sigma < 0.4744$. Picking, for instance, $\sigma = 0.26$, Lemma 5.3 yields $\bar{\Delta}_\sigma = 0.1005$, where $\|\Phi\| = 1.9021$ and $\mu_A = 1.5$; $\bar{\Delta}_\sigma$ specifies the intersampling time of maximal length that guarantees ISS. Further, $\omega_1/(\omega_1 + \omega_2) = 0.0321$. This latter value determines the DoS signals

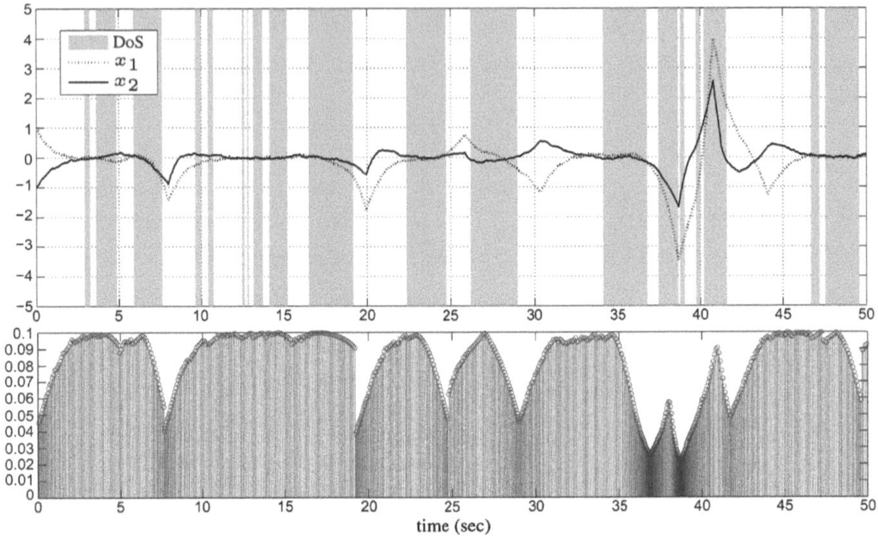

Figure 5.17 Simulation example for system (5.234) under state feedback (5.236) with $\delta_1 = 0.01, \delta_2 = 0.1$ and $\varphi(\cdot) = 2/\pi \arctan(\cdot)$

which are admissible in accordance with the present analysis. In connection with Example 5.9, this means a maximum of $\sim 3\%$ of communication denials on average. As for Example 5.10, this implies a maximum (average) duty cycle of $\sim 3\%$ in case of a sustained DoS attack. The value obtained for $\omega_1/(\omega_1 + \omega_2)$ is conservative: as shown in Fig. 5.17, the bounds can in practice be much smaller than the theoretical one. This was also confirmed by extensive simulations.

In Fig. 5.17, the top graph is the closed-loop state response under initial conditions $x = [1 \; -1]^\top$, disturbance w uniformly distributed in $[-1, 1]$, and sustained DoS attack with variable period and duty cycle, generated randomly, with $P_{\min} = 0.01$ s and $P_{\max} = 10$ s. The resulting DoS signal has an average duty cycle of $\sim 42\%$. The vertical grey stripes represent the time-intervals over which DoS is active. At the bottom we list intersampling times determined by the sampling logic.

The conservativeness of the bound comes from two main sources:

i. The bounds on the growth of the Lyapunov function under DoS (cf., (5.184)–(5.191)). In this respect, the approach in [81], which does not rely on Lyapunov functions (albeit restricted to the disturbance-free case), can provide a possible alternative to the present analysis; and

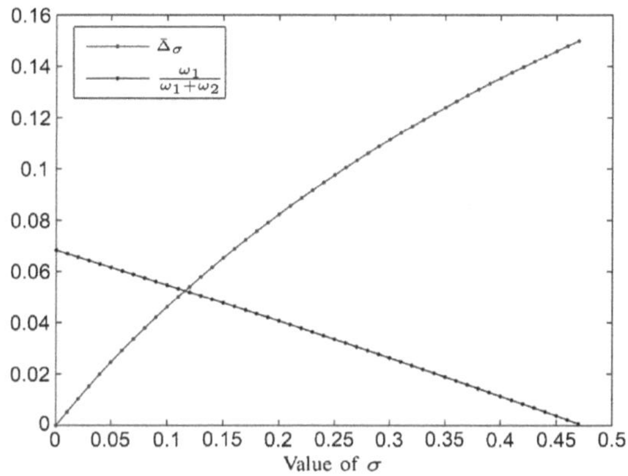

Figure 5.18 Maximum intersampling time and value of $\omega_1/(\omega_1+\omega_2)$ versus choice of σ using the LQR gain and $Q = I_2$

ii. The generality of the considered scenario. In fact, tighter bounds are likely to be obtained when more "structure" is assumed for the DoS. In this respect, interesting results in case of *periodic jamming* have been recently reported in [76].

It is interesting to observe that the value of $\omega_1/(\omega_1+\omega_2)$ also depends on a number of design parameters. In fact, it depends on the Lyapunov equation $\Phi^\top P + P\Psi + Q = 0$, and, as such, on Q and the state-feedback matrix K. For instance, the choice

$$K = \begin{bmatrix} -4.5 & -1 \\ 0 & -6 \end{bmatrix} \tag{5.236}$$

achieves a bound $\omega_1/(\omega_1+\omega_2) = 0.0971$, thus allowing for an average duty cycle $\sim 10\%$ in case of a sustained attack. This suggests investigation of analytic or numeric methods to find the Q and K that could maximize robustness against DoS. In practice, another possibility for increasing $\omega_1/(\omega_1+\omega_2)$ is to reduce the value of σ. This, however, has to be traded-off against the intersampling times. For instance, using the LQR gain and letting $Q = I_2$, the choice $\sigma = 0.1$ is sufficient to increase $\omega_1/(\omega_1+\omega_2)$ to 0.0547. As an offset, $\bar{\Delta}_\sigma$ drops to 0.0462. This phenomenon is illustrated in Fig. 5.18.

5.6 NOTES

Several security constraints have been incorporated in the control synthesis of a linear quadratic optimal control problem. The resulting optimization problems are shown to be convex. Lagrangian duality techniques have been used to compute and characterize the optimal solutions and their properties. The optimal solution is shown to be affine for the case when the terminal state has a continuous distribution. Utilizing the standard optimization software cvx, we have computed the optimal control sequences and have also validated the results via numerical simulations.

It has been shown that the LTI systems can be controlled with variable time delays, either occurring naturally or as a result of a time-delay attack by a hacker. A TDS attack can also be successfully tracked with the proposed method. One kind of delay was addressed, that is, the delay in the observed state of the controlled system. Is noted that only the LTI system in state feedback was discussed. The method is general, and in the future papers, the method will be shown that it also works for a class of nonlinear systems.

In the next section, the resilient strategy design problem has been investigated for a class of CPS under DoS attack. The system model in the delta domain under DoS attack occurring in wireless network between sensor and remote estimation has been set up. The minimax control strategy has been obtained in the delta domain under IPSM. Then, the multichannel transmission framework has been constructed, which can reduce the damage to control system from DoS attack. A two-player Markov stochastic game has been built to model the interactive decision-making of both sides under energy budget constraints. We have applied the dynamic programming and value iteration methods to derive the resilient strategies for control system and communication network. Finally, numerical simulations have been presented to illustrate the validity of the design scheme.

In the last section, we have investigated stability of networked systems in the presence of DoS attacks. One contribution of this paper is an explicit characterization of the frequency and duration of DoS attacks under which closed-loop stability can be preserved. The result is intuitive as it relates stability with the ratio between the *on/off* periods of jamming. An explicit characterization of sampling rules that achieve ISS was given. This characterization is flexible enough so as to allow the designer to choose from several implementation options that can be used for trading-off performance versus communication resources.

The results lend themselves to many possible extensions. As for the framework considered here, identifying optimal attack and counter-attack strategies with respect to some prescribed performance objective, represents an interesting research venue. Moreover, we have not investigated the effect of possible limitations on the information, such quantization and delays. As additional future research topics, we also envision the use of similar techniques to handle output feedback controllers as well as nonlinear systems. As for the latter case, preliminary results have been reported in [82]. Finally, an interesting research line is to address the case where control and measurement channels can be interrupted *asynchronously*. In this respect, the self-triggering logic described earlier within the ISS under denial-of-service, which relies on predictions of the process state, appears as a convenient tool for updating the control action in case of DoS of the measurement channel.

One of the main motivations for considering control over networks descends from problems of distributed coordination and control of large-scale systems [83–87]. Investigating our approach to control under DoS for self-triggered coordination problems such as those in [87] does also represent an interesting research venue.

REFERENCES

[1] U.S.G.A. Office, Critical Infrastructure Protection, Multiple Efforts to Secure Control Systems Are Under Way, but Challenges Remain, Technical Report GAO-07-1036, Report to Congressional Requesters, 2007.
[2] W. Xu, K. Ma, W. Trappe, Y. Zhang, Jamming sensor networks: attack and defense strategies, IEEE Netw. 20 (3) (May 2006) 41–47.
[3] M. Bishop, Computer Security, Art and Science, Addison–Wesley, 2003.
[4] N.W. Group, Internet security glossary, http://rfc.net/rfc2828.html, May 2000.
[5] V.D. Gligor, A note on denial-of-service in operating systems, IEEE Trans. Softw. Eng. SE-10 (3) (May 1984) 320–324.
[6] J. Han, A. Jain, M. Luk, A. Perrig, Don't sweat your privacy: using humidity to detect human presence, in: Proc. the 5th Int. Workshop on Privacy in UbiComp, UbiPriv'07, Sept. 2007.
[7] A. Perrig, J.A. Stankovic, D. Wagner, Security in wireless sensor networks, Commun. ACM 47 (6) (2004) 53–57.
[8] L. Eschenauer, V. Gligor, A key-management scheme for distributed sensor networks, in: Proc. the 9th ACM Conference on Computer and Communications Security, 2002, pp. 41–47.
[9] A. Perrig, R. Szewczyk, J. Tygar, V. Wen, D.E. Culler, Spins: security protocols for sensor networks, Wirel. Netw. 8 (5) (Sept. 2002) 521–534.
[10] C. Karlof, N. Sastry, D. Wagner, Tinysec: a link layer security architecture for wireless sensor networks, in: Proc. the Second ACM Conference on Embedded Networked Sensor Systems, Nov. 2004.
[11] M. Luk, G. Mezzour, A. Perrig, V. Gligor, Minisec: a secure sensor network communication architecture, in: Sixth Int. Conf. Information Processing in Sensor Networks, IPSN 2007, Apr. 2007.
[12] C. Karlof, D. Wagner, Secure routing in sensor networks: attacks and countermeasures, in: Special Issue on Sensor Network Applications and Protocols, Ad Hoc Netw. 1 (2–3) (Sept. 2003) 293–315.

[13] B. Parno, M. Luk, E. Gaustad, A. Perrig, Secure sensor network routing: a clean-slate approach, in: Proc. the 2nd Conference on Future Networking Technologies, CoNEXT 2006, Dec. 2006.
[14] J.H. Saltzer, M.D. Schroeder, The protection of information in computer systems, Proc. IEEE 63 (9) (Sept. 1975) 1278–1308.
[15] A. Avizienis, J.-C. Laprie, B. Randell, C. Landwehr, Basic concepts and taxonomy of dependable and secure computing, IEEE Trans. Dependable Secure Comput. 1 (1) (Jan.–Mar. 2004) 11–32.
[16] B. Schneier, Managed security monitoring: network security for the 21st century, Comput. Secur. 20 pp (2001) 491–503.
[17] R. Anderson, Security Engineering, Wiley, 2001.
[18] L. Adleman, An abstract theory of computer viruses, in: Advances in Cryptology–Proceedings of CRYPTO'88, Springer-Verlag New York, Inc., New York, NY, USA, 1990, pp. 354–374.
[19] H. Chan, V.D. Gligor, A. Perrig, G. Muralidharan, On the distribution and revocation of cryptographic keys in sensor networks, IEEE Trans. Dependable Secure Comput. 2 (3) (2005) 233–247.
[20] I. John, H. Marburger, E.F. Kvamme, Leadership Under Challenge: Information Technology R&D in a Competitive World. An Assessment of the Federal Networking and Information Technology R&D Program, Technical report, President's Council of Advisors on Science and Technology, Aug. 2007.
[21] M.S. Mahmoud, Control and Estimation Methods over Communication Networks, Springer-Verlag, UK, 2014.
[22] G.P. Liu, Y.Q. Xia, J. Chen, D. Rees, W.S. Hu, Networked predictive control of systems with random network delays in both forward and feedback channels, IEEE Trans. Ind. Electron. 54 (3) (2007) 1282–1297.
[23] M. Blanke, M. Kinnaert, J. Lunze, M. Staroswiecki, Diagnosis and Fault-Tolerant Control, Springer-Verlag, 2003.
[24] L. Schenato, B. Sinopoli, M. Franceschetti, K. Poolla, S. Sastry, Foundations of Control and Estimation over Lossy Networks, in: Special Issue on Networked Control Systems, Proc. IEEE 95 (1) (2007) 163–187.
[25] R. Olfat-Saber, Distributed Kalman filter with embedded consensus filter, in: Proc. CDC and ECC, Seville, Spain, 2005.
[26] A.A. Cardenas, S. Amin, S. Sastry, Research challenges for the security of control systems, in: The 3rd USENIX Workshop on Hot Topics in Security–HotSec 2008. Associated with the 17th USENIX Security Symposium, San Jose, CA, 2008.
[27] R.J. Turk, Cyber Incidents Involving Control Systems, Technical Report, Idaho National Laboratory, 2005.
[28] A. Sargolzaei, K.K. Yen, M. Abdelghani, Control of nonlinear heartbeat models under time-delay-switched feedback using emotional learning control, Int. J. Recent Trends Eng. Technol. 10 (2) (2014) 85.
[29] A. Sargolzaei, K. Yen, M.N. Abdelghani, Delayed inputs attack on load frequency control in smart grid, in: ISGT 2014, 2014.
[30] A. Sargolzaei, K.K. Yen, M. Abdelghani, Time-delay switch attack on load frequency control in smart grid, Adva. Commun. Technol. 5 (2013) 55–64.
[31] S. Amin, A.A. Cárdenas, S. Sastry, Safe and secure networked control systems under denial-of-service attacks, in: HSCC, in: Lect. Notes Comput. Sci., vol. 5469, Springer, 2009, pp. 31–45.

[32] Y. Mo, B. Sinopoli, False data injection attacks in control systems, in: 1st Workshop Secure Control Syst., Springer, Stockholm, Sweden, 2010.
[33] S. Liu, X.P. Liu, A. El Saddik, Denial-of-service (dos) attacks on load frequency control in smart grids, in: Innovative Smart Grid Technologies, ISGT, 2013 IEEE PES, IEEE, 2013, pp. 1–6.
[34] L. Jiang, W. Yao, Q. Wu, J. Wen, S. Cheng, Delay-dependent stability for load frequency control with constant and time-varying delays, IEEE Trans. Power Syst. 27 (2) (2012) 932–941.
[35] M. Ma, H. Chen, X. Liu, F. Allgöwer, Distributed model predictive load frequency control of multi-area interconnected power system, Int. J. Electr. Power Energy Syst. 62 (2014) 289–298.
[36] H. Bevrani, Robust Power System Frequency Control, Power Electron. Power Syst., vol. 85, Springer, 2009.
[37] Y. Tan, Time-varying time-delay estimation for nonlinear systems using neural networks, Int. J. Appl. Math. Comput. Sci. 14 (2004) 63–68.
[38] L. Chunmao, X. Jian, Adaptive delay estimation and control of networked control systems, in: Int. Symposium Communications and Information Technologies, ISCIT'06, 2006, pp. 707–710.
[39] N. Sadeghzadeh, A. Afshar, M.B. Menhaj, An mlp neural network for time delay prediction in networked control systems, in: Chinese Control and Decision, CCDC 2008, 2008, pp. 5314–5318.
[40] A. Sargolzaei, K.K. Yen, S. Noei, H. Ramezanpour, et al., Assessment of He's homotopy perturbation method for optimal control of linear time-delay systems, Appl. Math. Sci. 7 (8) (2013) 349–361.
[41] S.S. Haykin, Adaptive Filter Theory, Pearson Education, India, 2008.
[42] K.J. Åström, T. Hägglund, Advanced PID Control, ISA-The Instrumentation, Systems and Automation Society, 2006.
[43] K. Zhou, J.C. Doyle, K. Glover, et al., Robust and Optimal Control, vol. 40, Prentice Hall, NJ, 1996.
[44] A.A. Cardenas, S. Amin, S. Sastry, Secure control: towards survivable cyber-physical systems, in: Proc. 28th Int. Conference Distributed Computing Systems Workshops, ICDCS'08, 2008, pp. 495–500.
[45] Y. Mo, T.H.-J. Kim, K. Brancik, D. Dickinson, H. Lee, A. Perrig, B. Sinopoli, Cyber-physical security of a smart grid infrastructure, Proc. IEEE 100 (1) (2012) 195–209.
[46] H. Fawzi, P. Tabuada, S. Diggavi, Security for control systems under sensor and actuator attacks, in: Decision and Control, CDC, 2012 IEEE 51st Annual Conference on, IEEE, 2012, pp. 3412–3417.
[47] Q. Zhu, L. Bushnell, T. Başar, Game-theoretic analysis of node capture and cloning attack with multiple attackers in wireless sensor networks, in: Decision and Control, CDC, 2012 IEEE 51st Annual Conference on, 2012, pp. 3404–3411.
[48] F. Pasqualetti, F. Dörfler, F. Bullo, Cyber-physical security via geometric control: Distributed monitoring and malicious attacks, in: Proc. 51st Annual Decision and Control Conference, CDC, IEEE, 2012, pp. 3418–3425.
[49] H.V. Poor, An Introduction to Signal Detection and Estimation, second edition, Springer Science & Business Media, 1994.
[50] M. Grant, S. Boyd, Y. Ye, Cvx: Matlab software for disciplined convex programming.
[51] S. Boyd, L. Vandenberghe, Convex Optimization, Cambridge University Press, 2004.

[52] P.J. Bickel, K.A. Doksum, Mathematical Statistics: Basic Ideas and Selected Topics, vol. 1, second edition, Pearson Prentice Hall, 2007.
[53] D.G. Luenberger, Optimization by Vector Space Methods, Wiley, 1969.
[54] C. De Persis, R. Sailer, F. Wirth, Parsimonious event-triggered distributed control: a zeno free approach, Automatica 49 (7) (2013) 2116–2124.
[55] T.A. Tamba, Y.Y. Nazaruddin, B. Hu, Resilient control under denial-of-service via dynamic event triggering, in: Proc. Asian Control Conference, ASCC, Australia, July 2017, pp. 134–139.
[56] C. De Persis, P. Tesi, Input-to-state stabilizing control under denial-of-service, IEEE Trans. Autom. Control 60 (11) (2015) 134–139.
[57] V. Dolk, P. Tesi, C. De Persis, W. Heemels, Event-triggered control systems under denial-of-service attacks, in: Proc. the 54th IEEE Conference on Decision and Control, Osaka, Japan, 2015.
[58] S. Feng, P. Tesi, Resilient control under denial-of-service: robust design, Automatica 79 (2017) 42–51.
[59] S. Feng, P. Tesi, Resilient control under Denial-of-Service: robust design, in: Proc. American Control Conference, ACC, Boston, USA, July 2016.
[60] K. Ding, Y. Li, D.E. Quevedo, S. Dey, L. Shi, A multi-channel transmission schedule for remote state estimation under dos attacks, Automatica 78 (2017) 194–201.
[61] H.S. Foroush, S. Martinez, On event-triggered control of linear systems under periodic denial-of-service jamming attacks, in: Decision and Control, CDC, 2012 IEEE 51st Annual Conference on, IEEE, 2012, pp. 2551–2556.
[62] S. Dashkovskiy, H. Ito, F. Wirth, On a small gain theorem for ISS networks in dissipative Lyapunov form, Eur. J. Control 17 (4) (2011) 357–365.
[63] Z.A. Biron, S. Dey, P. Pisu, Resilient control strategy under Denial of Service in connected vehicles, in: Proc. American Control Conference, ACC, Seattle, USA, May 2017.
[64] X. Wang, M.D. Lemmon, Event-triggered broadcasting across distributed networked control systems, in: Proc. American Control Conference, 2008, pp. 3139–3144.
[65] Y. Liu, M. Reiter, P. Ning, False data injection attacks against state estimation in electric power grids, in: Proc. ACM Conf. Computer and Commun. Security, Chicago, IL, USA, 2009.
[66] G. Dan, H. Sandberg, Stealth attacks and protection schemes for state estimators in power systems, in: Proc. IEEE Int. Conf. Smart Grid Commun., Gaithersburg, MD, USA, 2010, pp. 214–219.
[67] H. Fawzi, P. Tabuada, S. Diggavi, Secure state-estimation for dynamical systems under active adversaries, in: Proc. Annu. Allerton Conf. Commun., Control, Comput., 2011.
[68] F. Pasqualetti, F. Dorfler, F. Bullo, Attack detection and identification in cyber-physical systems, IEEE Trans. Autom. Control 58 (11) (Nov. 2013) 2715–2729.
[69] Y. Fujita, T. Namerikawa, K. Uchida, Cyber attack detection and faults diagnosis in power networks by using state fault diagnosis matrix, in: Proc. Eur. Control Conf., Zurich, Switzerland, 2013.
[70] W. Xu, K. Ma, W. Trappe, Y. Zhang, Jamming sensor networks: attack and defense strategies, IEEE Netw. 20 (3) (May–Jun. 2006) 41–47.
[71] D. Thuente, M. Acharya, Intelligent jamming in wireless networks with applications to 802.11b and other networks, in: Proc. 25th IEEE Commun. Soc. Military Commun. Conf., Washington, DC, USA, 2006.

[72] W. Xu, W. Trappe, Y. Zhang, T. Wood, The feasibility of launching and detecting jamming attacks in wireless networks, in: Proc. ACM Int. Symp. Mobile Ad-Hoc Netw. Comput., 2005.
[73] B. De Bruhl, P. Tague, Digital filter design for jamming mitigation in 802.15.4 communication, in: Proc. Int. Conf. Comput. Commun. Netw., Maui, HI, USA, 2011.
[74] P. Tague, M. Li, R. Poovendran, Mitigation of control channel jamming under node capture attacks, IEEE Trans. Mob. Comput. 8 (9) (Sept. 2009) 1221–1234.
[75] P. Tabuada, Event-triggered real-time scheduling of stabilizing control tasks, IEEE Trans. Autom. Control 52 (9) (Sept. 2007) 1680–1685.
[76] H. Shisheh Foroush, S. Martínez, On multi-input controllable linear systems under unknown periodic dos jamming attacks, in: Proc. SIAM Conf. Control and Its Applications, San Diego, CA, USA, 2013.
[77] W. Heemels, K. Johansson, P. Tabuada, An introduction to event-triggered and self-triggered control, in: Proc. 51th IEEE Conf. Decision and Control, Maui, HI, USA, 2012, pp. 3270–3285.
[78] M. Abdelrahim, R. Postoyan, J. Daafouz, D. Nešić, Stabilization of nonlinear systems using event-triggered output feedback laws, in: Proc. 21th Int. Symp. Math. Theory of Netw. Syst, 2014.
[79] M. Mazo, A. Anta, P. Tabuada, An ISS self-triggered implementation of linear controllers, Automatica 46 (2010) 1310–1314.
[80] F. Forni, S. Galeani, D. Nešić, L. Zaccarian, Lazy sensors for the scheduling of measurement samples transmission in linear closed loops over networks, in: Proc. IEEE Conf. Decision Control and Eur. Control Conf., Atlanta, GA, USA, 2010, pp. 6469–6474.
[81] C. De Persis, P. Tesi, Resilient control under denial-of-service, in: Proc. IFAC World Conf., Cape Town, South Africa, 2014.
[82] C. De Persis, P. Tesi, On resilient control of nonlinear systems under denial-of-service, in: Proc. IEEE Conf. Decision and Control, Los Angeles, CA, USA, 2014, pp. 5254–5259.
[83] X. Wang, M. Lemmon, Event-triggering in distributed networked control systems, IEEE Trans. Autom. Control 56 (3) (Mar. 2011) 586–601.
[84] C. De Persis, R. Sailer, F. Wirth, Parsimonious event-triggered distributed control: a Zeno free approach, Automatica 49 (2013) 2116–2124.
[85] G. Seyboth, D. Dimarogonas, K. Johansson, Event-based broadcasting for multi-agent average consensus, Automatica 49 (2013) 245–252.
[86] C. Stöcker, J. Lunze, Distributed event-based control of physically interconnected systems, in: Proc. IEEE Conf. Decision Control, Florence, Italy, 2013, pp. 7376–7383.
[87] C. De Persis, P. Frasca, Robust self-triggered coordination with ternary controllers, IEEE Trans. Autom. Control 58 (12) (Dec. 2013) 3024–3038.
[88] G. Hardy, J. Littlewood, G. Polya, Inequalities, Cambridge Univ. Press, Cambridge, UK, 1952.

CHAPTER 6

Case Studies

6.1 HYBRID CLOUD-BASED SCADA APPROACH

It becomes increasingly apparent that interoperability of microgrid platforms allows users to exchange and process meaningful information among the energetic and automation systems, as well as to visualize and control in real time the available experimental tools in the partner platforms. In an experimental platform, interoperability among partners is imperative when facilities from a platform are needed for an application in another institution. Connecting interoperable platforms requires much less time and resources than constructing new necessary experimental modules.

Fig. 6.1 illustrates different layers in the context of interoperability between two microgrid platforms, represented on the SGAM interoperability stack [1].

- From the top layer, a harmonization of communication policies between two operational bodies should be done. Interoperability in the function

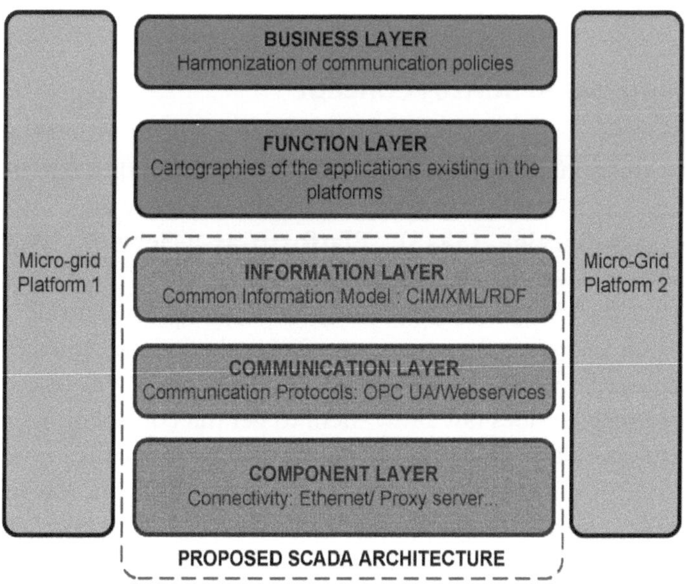

Figure 6.1 Different layers and necessary tasks

Networked Control Systems
https://doi.org/10.1016/B978-0-12-816119-7.00014-9

layers requires a communication between two parties about their cartography of experimental applications and research infrastructure. This information is used to plan for multisite projects and applications, such as cosimulation or long distance control. Hence, interoperability requires that both parties share a common information model or at least a conversion interface, so that the exchanged information is understandable to the other side.

- In the communication layer, synchronization of communication protocol is necessary to guarantee good emission and reception of information.
- The component layer concerns the physical elements of the communication. The choices in these three layers are strongly influenced by the popular communication standards and should be open for an eventual integration of more partners into the network.
- The architecture, on which these three layers are organized, should allow the seamless and reliable integration of SCADA systems while offering enough security measures to protect against cyberattacks.

The *integration of SCADA systems of partner microgrid platforms* requires a secured and distanced access to shared resources (data and control of the platforms). In this context, the cloud based SCADA concept offers great flexibility and significantly lower cost, but also provokes additional security aspects.

6.1.1 Cloud-Based SCADA Concept

The cloud is a concept of using remote network based servers to store and handle information. This information can be accessed through a network connection. Cloud computing also offers three service delivery models: **Software-as-a-Service (SaaS), Platform-as-a-Service (PaaS) and Infrastructure-as-a-Service (IaaS)** [2].

IaaS, PaaS and SaaS are classified by the level of control the users can have access to. IaaS gives the users control over the infrastructure and applications deployed on the cloud. PaaS delivery model allows users to deploy the applications, but does not allow them to get full control of the underlying infrastructure or to get access to restricted data. SaaS delivery model, on another higher level of security, allows users to use the applications running on the infrastructure, though a client interface, such as web browser. The operator, however, has complete control over the infrastructure [3].

Depending on the level of data exposure to the public, c1oudbased SCADA systems can be classified as **Public, Community or Private**.

When SCADA applications are entirely deployed on-site and run on Intranet, the system is considered as a private cloud-based SCADA. The access in this case is restricted to local level and may be granted to certain cooperators to create a community. When SCADA applications are entirely run in the cloud with remote connectivity to a control network, the architecture is considered as public and requires access authentication. The delivery models (SaaS, PaaS or laaS) are chosen according to the needs of the operators and the security restriction. Due to reliability and security issues of an electrical grid (response time, availability, etc.), critical control and protection tasks should be executed by the local PLC/RTU and the exchanged information should be selected and moderated. The applications in this scenario are directly connected to the local control network and data analysis is done on the cloud. This architecture can be considered as a *hybrid cloud-based SCADA*. Fig. 6.2 represents different elements of a simple hybrid cloud-based SCADA system.

Compared to a traditional approach, cloud-based SCADA systems offer several considerable advantages:

- Scalability;
- "Real-time" and historical data access from everywhere, with proper access authentication;
- Better collaboration, i.e., ease of information access at different levels of the system/project enabling all partners to work together more efficiently;
- **Ease of upgrading and expanding the system**;
- More **efficient disaster recovery**;
- Much lower **cost** for maintaining and expanding/upgrading system.

Cloud-based SCADA offers many attractive benefits; however, it also introduces some potential risks. This is especially true when some critical data is concerned. The most significant risks to be considered are cybersecurity and reliability:

- Cybersecurity – The cloud comes up with a secured system of data supervision and access authentication. However, a cyberattack may affect the server or eventually the data link (which is often outside the internal network).
- Reliability – Cloud connection relies on bandwidth availability and reliability.

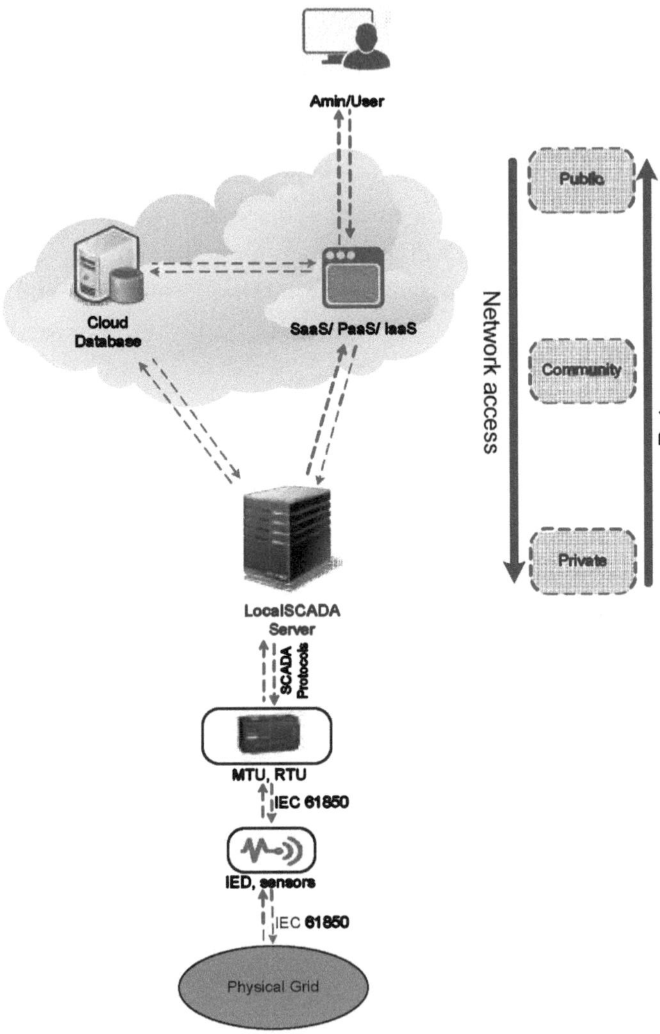

Figure 6.2 A simple hybrid cloud-based SCADA architecture

- Performance of the functions that demand high bandwidth and low latency would also be affected in cloud-based SCADA.

Due to the possible sensitive nature of the exchanged data among research institutions, the implementation of cloud-based SCADA should include a strict consideration of the aforementioned risks. In the following, we adapt the architecture to achieve cloud-based benefits while limiting the associated security and reliability drawbacks.

6.1.2 Architecture Adaptation

In order to enable the interoperability among microgrid platforms, we propose in this section an SCADA architecture, which allows a secured and meaningful communication among the platforms and provides the ability to easily integrate new partner platforms to the networks.

The architecture represents the three lower layers of Fig. 6.1. We adopt the hybrid cloud architecture and the platform-as-a-service (PaaS) delivery model. The local SCADA server supervises and controls its components via the corresponding industrial Ethernet. The physical components of both platforms are connected to their MTU and RTU via standardized protocols (IEC 61850, IEC 60870 or DNP over TCP/IP, etc.). Bulk data from the grid comes from sensors and is transmitted in different intervals (from milliseconds to several hours).

Two main issues of putting SCADA system to cloud are security and reliability. In this approach, the critical SCADA task is solely located on-site and is controlled by the local SCADA server, PLC or RTU, etc. This classical way ensures a low latency (as the communication is done via LAN) and strong protection (the information exchange is local) for these critical tasks. On the other hand, applications like AMI, DMS, VAR optimization and outage management are suitable candidates for putting them on the cloud [2]. These applications are delivered via the PaaS model (or SaaS as an alternative).

The interoperability of partner microgrid platforms is actualized through a common private cloud SCADA server. Data of the shared applications is transferred from the local service to a common private platform. It can be a physical SCADA server or a virtual PaaS/SaaS based server. This server communicates with local SCADA server in the platforms with WAN network. Ethernet with TCP/IP and Web-service protocols is the simplest and probably cheapest choice for implementation. However, according to the requirement of speed and latency, other options can be considered.

In this architecture with PaaS delivery model, the SCADA application is running on-site and is directly connected to the platform control network. The critical functions of the SCADA server are isolated at a local platform and only selected non-critical information is transferred to the common cloud SCADA server that provides visualization, reporting and limited access to a range of applications, to remote authorized users and partner institutions. This (physical or virtual) common cloud SCADA server should be private (for the partner network), located at a partner platform or a site,

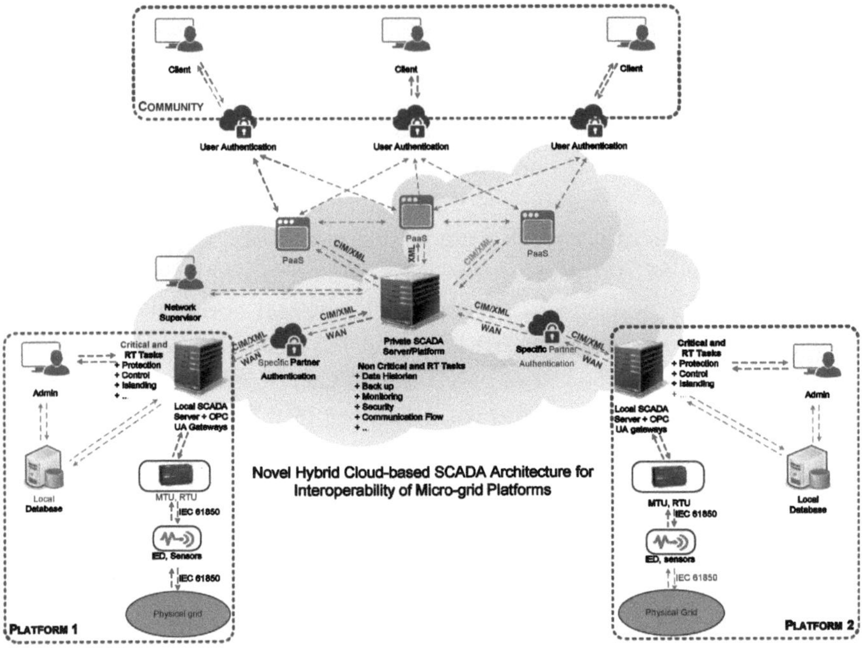

Figure 6.3 Proposed hybrid-cloud based SCADA architecture

agreed by the partners, to provide optimized latency for the possible applications and is strictly moderated by a common council and technical staff, in charge of the interoperability within the network. Using PaaS delivery model, it is also possible to remotely demand to launch a certain function (setting values, starting simulating, etc.) and visualize the result, provided that the demander is allowed by the platform owner. This property is very important in the context of interoperability among microgrid platforms of research institutions, because it enables the possibility to make experiments on the shared resources, without having to come to the platform in person.

CIMIXML/RDF [4] is chosen as the common information model in the architecture for the partners, because of its independence of platform and transport. OPC UA gateways are used as a protocol conversion for the legacy DCOM to meet the XML and SOAP. The proposed architecture considers also the mapping of CIM semantic to OPC UA data model and the implementation of OPC UA into the PaaS delivery model. They are, however, out of the scope of this chapter and will be presented in a following publication. See Fig. 6.3.

During the implementation of the proposed architecture, it is important to determine which applications are suitable to access from the cloud. In general, Markovich [2] discussed some applications that the cloud SCADA server is expected to address. In our specific context of interoperability among microgrid platforms, these applications are decided according to the confidential policies and agreement of the partners. The SCADA system is dependent on the bandwidth and latency of the network connection. Losing functionality to real-time monitoring and control for a few minutes or even seconds may wreak havoc on the platform. Therefore, the critical tasks should only be accessed from the cloud after a **strict risk evaluation**. The following criteria should be considered:

- Performance fluctuation,
- Latency and latency variability, and
- Effect of network inaccessibility to the platform and data.

For each application, the response time cannot deviate from the requested value by more than a defined difference:

$$T < T_{Req} + \triangle T.$$

Also, the total traffic of active applications and the dedicated gateway for protocol conversion cannot exceed the overall bandwidth the connection.

$$\sum P_{App} + P_{gateway} < P_{Req}.$$

These requirements need to be met in both the WAN connection to the cloud SCADA server and LAN connection in local platforms. Therefore, a test of communication and risk evaluation are necessary before implementing the architecture. This will also be used as an additional criterion to decide which services will be available to the partners, besides the security, confidentiality, communication and sharing policies of the institutions. If the involved risk is too high, only non-real-time applications should be available from the cloud. Each situation has to be evaluated on its own terms. The data should also be alternatively stored in the local database.

In term of reliability, **critical functions are processed on site by the local SCADA server**. Latency and bandwidth are no longer an issue. The system, on the other hand, can benefit from the aforementioned advantages of the hybrid-cloud architecture and PaaS delivery model. For the security issue, the common cloud server is actually **private for the partners in the network and is strictly moderated**. It is not open for public and can only be accessed after a successful user authentication. A partner in the network can choose to grant access to a certain part of its platform to a specific partner and not the others via the partner authorization procedure. These two authentications can be separated (user is not associated to the institution) or be associated to the institution (the system will auto-detect the institution and check for access right).

6.1.3 Risk Evaluation

Interoperability of the microgrid platforms and research institutions should only be done within strict security consideration. The proposed architecture demands two kinds of authentication. User authentication is required to grant access to the common cloud SCADA server, which provides selected PaaS or SaaS applications. When access to a specific platform is necessary, the second authentication is required to check if the corresponding partner institution is authorized to access the local SCADA services.

The security risks in the local front-end SCADA server are mainly caused by the lack of cryptographic capacity. Until recently, secure DNP3 and IEC 61850 introduced the ability to validate the authenticity of the messages [2]. Sharing the known security risks with the classical SCADA system, the proposed SCADA architecture is potentially vulnerable to attacks on the common cloud server. Some risks can be addressed:

- Denial-of-Service (DoS) and Distributed-Denial-of-Service (DDoS), the main goal of which is to flood the system with demands of service and make the system unable to function as intended. A DoS attack in the server can cause unavailability to shared resources and disruption in communication over collaboration activities.
- Data security – even though the common server is secured, the communication link and intermediate servers are not, especially when the network uses the public Ethernet.

To properly address these risks and enforce the security of the SCADA architecture, several solutions can be implemented:

- DoS Detection – several methods can be implemented to detect a DoS attack: using low entropy [5,6], signal strength [7], sensing time measurement [8], transmission failure count [9] or signatures [7].
- DoS mitigation – it is often done over two layers, network and physical layers, to protect the nodes and minimize the outage time. The action may vary from push-back, limiting the traffic from the attackers [7,9], topology reconfiguration [10] or frequency hopping technologies [11].
- Authentication – preliminary access identification.

- Authorization – a mechanism to implement control access by user profile.
- Notification – capacity to inform the operator when the resource is accessed through the collaboration network.
- Data Encryption – both symmetric [10] and public key encryption [12] can be used.

6.2 SMART GRID UNDER TIME DELAY SWITCH ATTACKS

The control of power systems with time delays has been previously explored [13,14]. Researchers, however, considered either the construction of controllers that are robust to time delays or controllers that use offline estimation. At this time, apparently, there are no control methods and adaptive communication protocols that implement an online estimation of dynamic time delays and real-time control of power systems to overcome TDS attacks.

The stability of power control systems with time delays was studied in [15–20], which proposed a controller for power systems with delayed states. Related work on time-delay estimation includes [21,22].

6.2.1 System Model

The following section introduces the system and adversarial model used to validate the CF-TDSR protocol. Background information of this research is also discussed. Fig. 6.4 illustrates a two-area power plant with automatic generation control (AGC). The LFC component sends control signals to the plant and receives state feedback through the communication channel. Different attack types can be launched against an LFC system, including DoS, False Date Injection (FDI) and Time Delay Switch (TDS) attacks. The LFC is a large scale NCS that regulates the power flow between different areas while holding the frequency constant. Power systems are usually large-scale systems with complex nonlinear dynamics. Modern power grids are divided into various areas. Each area is connected to its neighboring areas by transmission lines called tie-lines. Tie-lines facilitate power sharing between neighboring areas. An LFC is used to make sure the power grid is stable and efficient. More information about the technical part of an LFC can be found in [23,24].

LFCs are usually designed as optimal feedback controllers. In order to operate on an optimal level, the LFC requires power state estimates to be telemetered in real time. In the case when an adversary injects a TDS attack

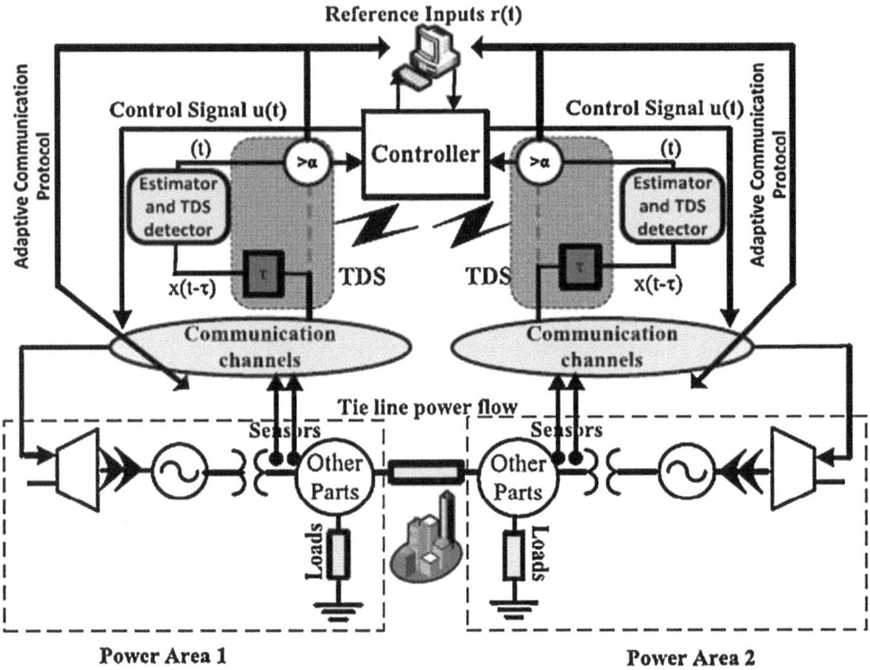

Figure 6.4 Two-area power system controlled under TDS attack

to the telemetered control signals or communication channel of feedback loop, the LFC will diverge from its optimality, and in most cases, depending on the amount and duration of the attack, the system will even get unstable if there is no prevention and stabilizer in place, such as in our proposed protocol.

Consider the multiarea LFC power system with the attack model as follows:

$$\begin{cases} \dot{X}(t) = AX(t) + BU(t) + D\Delta P_l, \\ X(0) = X_0, \end{cases} \tag{6.1}$$

where ΔP_l is the power deviation of the load. The optimal feedback controller can be found as

$$U = -K\hat{X} \tag{6.2}$$

and the new state after the time delay attack is given by

$$\hat{X}(t) = X(t - \tau), \tag{6.3}$$

where $\tau = [t_{d1}, t_{d2}, \ldots, t_{dN}]^T$ are different/random positive value time delays and t is the time vector which is the same for all states.

While the system is in its normal operation, $t_{d1}, t_{d2}, \ldots, t_{dN}$ are all zero. An adversary can get access to the communication channel or sensors and add delays to the channel to make the system unstable. In (6.1), $X = [x_1, x_2, \ldots, x_N]^T$ denotes the states in each power area. The state vector in the ith power area is described as

$$x_i(t) = [\Delta f^i(t) \quad \Delta P_g^i(t) \quad \Delta P_{tu}^i(t) \quad \Delta P_{pf}^i(t) \quad \Lambda^i(t)]^T, \tag{6.4}$$

where $\Delta f^i(t)$, $\Delta P_g^i(t)$, $\Delta P_{tu}^i(t)$, $\Delta P_{pf}^i(t)$, and $\Lambda^i(t)$ are the frequency deviation and deviation of the generator, position value of the turbine, tie-line power flow, and the control error on the ith power area, respectively [17]. The control error of the ith power area is expressed as

$$\Lambda^i(t) = \int_0^t \beta_i \Delta f^i(s) dt, \tag{6.5}$$

where β_i denotes the frequency bias factor. The dynamic model of the multiarea LFC of (6.1) can be expanded using

$$A = \begin{bmatrix} A_{11} & A_{12} & A_{13} & \ldots & A_{1N} \\ A_{21} & A_{22} & A_{23} & \ldots & A_{2N} \\ A_{31} & A_{32} & A_{33} & \ldots & A_{3N} \\ \vdots & \vdots & \vdots & \ddots & \vdots \\ A_{N1} & A_{N2} & A_{N3} & \ldots & A_{NN} \end{bmatrix}, \tag{6.6}$$

$$B = \text{diag}\{\begin{bmatrix} B_1^T & B_2^T & B_3^T & \ldots & B_N^T \end{bmatrix}^T\}, \tag{6.7}$$

$$B = \text{diag}\{\begin{bmatrix} D_1^T & D_2^T & D_3^T & \ldots & D_N^T \end{bmatrix}^T\}, \tag{6.8}$$

where A_{ii}, $A_{ij} B_i$ and D_i are represented by

$$B_i = \begin{bmatrix} 0 & 0 & \frac{1}{T_{gi}} & 0 & 0 \end{bmatrix}^T, \tag{6.9}$$

$$A_{ij} = \begin{bmatrix} 0 & 0 & 0 & 0 & 0 \\ 0 & 0 & 0 & 0 & 0 \\ 0 & 0 & 0 & 0 & 0 \\ -2\pi T_{ij} & 0 & 0 & 0 & 0 \\ 0 & 0 & 0 & 0 & 0 \end{bmatrix}, \tag{6.10}$$

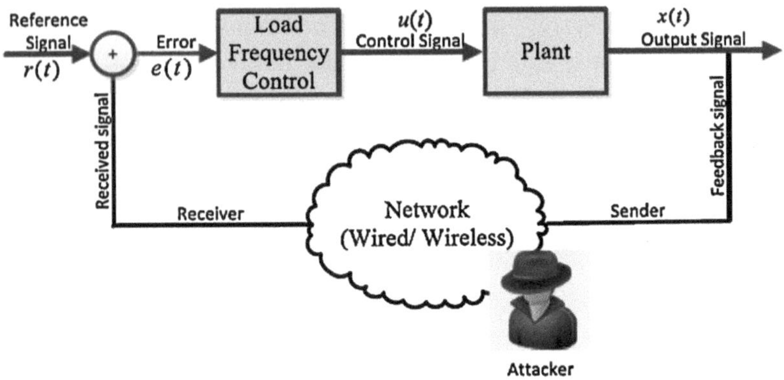

Figure 6.5 A simplified LFC system under TDS attack

$$A_{ij} = \begin{bmatrix} -\frac{\mu_i}{J_i} & \frac{1}{J_i} & 0 & \frac{1}{J_i} & 0 \\ 0 & -\frac{1}{T_{tui}} & \frac{1}{T_{tui}} & 0 & 0 \\ -\frac{1}{\omega_i T_g i} & 0 - \frac{1}{T_{gi}} & 0 & 0 & \\ \sum_{i \neq j, j=1}^{N} 2\pi T_{ij} & 0 & 0 & 0 & 0 \\ \beta_i & 0 & 0 & 0 & 1 \end{bmatrix}, \tag{6.11}$$

$$D_j = \begin{bmatrix} -\frac{1}{J_i} & 0 & 0 & 0 & 0 \end{bmatrix}^T. \tag{6.12}$$

Here N denotes the number of power areas, J_i, ω_i, μ_{1i}, T_{gi}, and T_{tu} are the generator moment of inertia, coefficient for droop speed, coefficient for generator damping, time constant of governor, time constant of turbine in the ith power area, and T_{ij} is the tie-line synchronizing coefficient between the ith and the jth power areas, respectively.

6.2.2 Time-Delay Attack

To describe the TDS attack, a simplified version of Fig. 6.4 is shown in Fig. 6.5. The plant (P) sends state variables $x(t)$ as information packets to the controller (C). The state variables $x(t)$ are compared with reference inputs $r(t)$, to produce the error value $e(t)Dr(t) - x(t)$. The error values are injected into the controller to calculate the control signals $u(t)$. This section assumes that all packets sent by controller include data and time-stamps (either encrypted/authenticated, or not). Also, for simplicity, it is assumed that the attacker can only target the communication channel between the

Table 6.1 Sequence of events during a replay TDS attack.

TS	$\{TS, P(t)\}$	$\{TS, C(t)\}$	**Controller Input** ($e(t)$)
0.1	$\{0.1, x(0.1)\}$	$\{0.1, x(0.1)\}$	$e(t) = r(0.1) - x(0.1)$ or $e(t) = r(0.1) - 0$
0.2	$\{0.2, x(0.2)\}$	$\{0.2, x(0.2)\}$	$e(t) = r(0.2) - x(0.2 - 0.1)$ or $e(t) = r(0.2) - 0$
0.3	$\{0.3, x(0.3)\}$	$\{0.3, x(0.3)\}$	$e(t) = r(0.3) - x(0.3 - 0.1)$ or $e(t) = r(0.3) - 0$

Table 6.2 Sequence of events during time-stamps alter TDS attack.

TS	$\{TS, P(t)\}$	$\{TS, C(t)\}$	**Controller Input** ($e(t)$)
0.1	$\{0.1, x_1(0.1)\}$	$\{0.1, x_1(0.1)\}$	$e(t) = r_1(0.1) - x_1(0.1)$
0.2	$\{0.2, x_2(0.2)\}$	$\{0.2, x_2(0.2)\}$	$e(t) = r_2(0.2) - x_1(0.2) = r_2(0.2) - x_2(0.2 - 0.1)$
0.3	$\{0.3, x_3(0.3)\}$	$\{0.3, x_3(0.3)\}$	$e(t) = r_3(0.3) - x_2(0.3) = r_3(0.3) - x_3(0.3 - 0.1)$

plant and the controller. The attacker can launch the different TDS attack, as follows:

(1) *Replay-based TDS attack*

The attacker leaves the first message $x(0.1)$ intact. It then records but drops the second message, $x(0.2)$, and resend the first massage again. Subsequently, it sends $x(0.2)$ instead of the third message, etc. The attacker can generalize this attack by introducing different time delays. The control input (error signal) under 0.1 seconds of delay will be $r(0:2) - x(0:1)$ or $r(0:2)$ in case the receiver detects an attack or fault. Table 6.1 illustrates the steps of this attack where the attacker has added a delay of 0.1 seconds. In the table, "TS" denotes time-stamp.

(2) *Time-stamp-based TDS attack*

An attacker reconstructs the packet and fixes the time-stamp. Subsequently, the time-stamping detector will not be able to find the time delay attack. The attacker receives the first message from the plant and copies the value in the buffer, then substitutes the state value of the first packet inside the second message and reconstructs the packet. Then, the adversary sends the packet to the controller. Table 6.2 illustrates this attack scenario. In this table, x_1 denotes the first state value, x_2 the second, and so forth.

Let's say, the plant sends x_2 at time 0.2, and the attacker copies the state value (packet). Now consider that the controller gets x_3, at time 0.3. During this time attacker sends x_2 instead of x_3 with a corrected 0.3 time-stamp. This can occur even in encrypted scenarios if the attacker can decrypt the packet and reconstruct it with a new wrong state and correct time-stamps as shown in Table 6.2.

(3) *Noise-based TDS attack*
An attacker injects fake packages into the system, making the system delay the transmission of system packages. Also temporally communication channel jamming can be classified as this type of attack. This way, the packets of the plant are delivered to the controller with a delay.

In all the above variants, the attacker injects time delays into the control system, making the system unstable or inefficient.

6.2.3 TDS Attack as Denial-of-Service (DoS) Attack

In this section, a unified approach to model a special case of TDS attacks as DoS attacks is proposed. Then, techniques will be investigated that address TDS attacks. Consider a linear time-invariant (LTI) system described by

$$\begin{aligned}\dot{x}(t) &= Ax(t) + Bu(t) + w(t),\\ u(t) &= -K\,x(t),\end{aligned} \tag{6.13}$$

where $x \in \mathbb{R}^n$ and $u \in \mathbb{R}^m$ are state and control signals, respectively. Matrices A, B and K are constant with appropriate dimensions. The vector $w \in \mathbb{R}^n$ is an n-dimensional zero-mean Gaussian white noise process. Suppose that a TDS attack occurs with probability p. Then

$$\dot{x}(t) = \begin{cases}(A - BK)x(t) + w(t) & \text{with probability } 1 - p,\\ Ax(t) - BKx(t - \tau) + w(t) & \text{with probability } p.\end{cases} \tag{6.14}$$

To simplify this explanation, in (6.14) we assume that the same probability of attack occurs on different channels and states. If a TDS attack occurs and packets are dropped, then (6.14) is formulated in the form of a DoS attack as follows:

$$\dot{x}(t) = \begin{cases}(A - BK)x(t) + w(t) & \text{with probability } 1 - p,\\ Ax(t) + w(t) & \text{with probability } p.\end{cases} \tag{6.15}$$

Let us calculate the expected value of Px in (6.15) as:

$$\begin{aligned}\mathbb{E}\{\dot{x}(t)\} &= [(A - BK)\mathbb{E}\{x(t)\} + \mathbb{E}\{w(t)\}](1 - p)\\ &\quad + [A\mathbb{E}\{x(t)\} + \mathbb{E}\{w(t)\}]p.\end{aligned} \tag{6.16}$$

Let $\mu(t) = \mathbb{E}x(t)$, then one can write (6.16) as

$$\begin{aligned}\dot{\mu}(t) &= [(A - BK)\mu(t)](1 - p) + A\mu(t)p\\ &= [A - (1 - p)BK]\mu(t),\end{aligned} \tag{6.17}$$

since $\mathbb{E}w = 0$. Next, the stability of (6.17) is investigated. In order for the system described in (6.17) to be stable, the mean should be bounded. Therefore, this equation must satisfy

$$A - BK + pBK < 0. \tag{6.18}$$

If there is no attack on the system, then the following condition appears:

$$A - BK < 0. \tag{6.19}$$

To satisfy the stability requirement, condition (6.18) must be made as close as possible to (6.19). To achieve this, two possibilities are available. First, the probability of a TDS attack on the communication channel is decreased. Second, the controller gain is changed. In the following probability equations, both cases are described.

Case A. *Decreasing probability of TDS attack*

This implies that the outcome of an attack is investigated when the protocol resends each packet ℓ times over independent channels. Then, the probability a TDS attack decreases by a power of ℓ, that is,

$$\dot{x}(t) = \begin{cases} (A - BK)x(t) + w(t) & \text{with probability } 1 - p^{\ell}, \\ Ax(t) + w(t) & \text{with probability } p^{\ell} \end{cases} \tag{6.20}$$

Thus (6.18) will take the form

$$(A - BK)^{-1} - I + (pI)^{\ell} < 0. \tag{6.21}$$

Therefore,

$$\begin{aligned} (A - BK)^{-1} - I &< (A - BK)^{-1} - I + (pI)^{\ell} \\ &< (A - BK)^{-1} - I + pI < 0. \end{aligned} \tag{6.22}$$

The condition in (6.22) shows that if $\ell > 1$, the channel is allocated to increase the redundancy of transmitted plant data, the total probability of faults will decrease as a result of a TDS attack. Also, the control system will get closer to its original state, that is, $A - BK < 0$. Therefore, by adaptively adding another communication channel(s), the LFC system can be stabilized. The cost of channel redundancy limits the number of communication channels that can be added to address TDS attacks. The alternative is to change the controller gain.

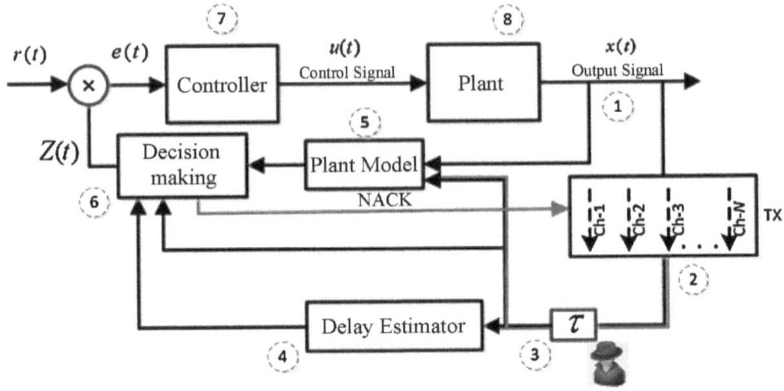

Figure 6.6 Block diagram of CF-TDSR

Case B. *Manipulating controller gain*

Changing the controller gain K to stabilize the NCS is proposed along with the first method. The new controller gain parameter is set to be $K_p = K/(1-p)$. In this case, a limited number of channels only need to be added. However, adjusting the controller gain K is subject to the probability of attack p, which is difficult to estimate. This control methodology is described before.

6.2.4 A Crypto-Free TDS Recovery Protocol

In this section, the CF-TDSR, a communication and control protocol that thwarts TDS attacks on NCS, is introduced. The CF-TDSR leverages methods to detect different types of TDS attacks introduced by a hacker. It also compensates negative effects of attacks on system. The CF-TDSR consists of the following components (see Fig. 6.6 for the system diagram). First, the smart data transmitter (Tx) adaptively allocates transmission channels on demand. Second, the plant model estimates the current plant states and helps to stabilize the system under attack. Third, the time-delay estimator continuously estimates time delays on the channels. Fourth, the time-delay detector and decision making unit (DMU) that determines if delays are detrimental to the system. The DMU also detects faults and other types of attack such as FDI accordingly. In this case, it will issue commands to inform the transmitter and controller. The last component is the controller to control the system (controller block in Fig. 6.6). The τ block in the diagram represents a hacker which injects a TDS attack to the communication channels. The adversary can inject other attacks as well.

The CF-TDSR protocol works as follows: State variables ($x(t)$) are sensed using sensors (point 1 in the diagram) and will be sent to the transmitter (Tx) unit (point 2). The transmitter unit constructs the packet and allocates the communication channel, and then transmits the packet to the delay estimator and plant model unit. The Tx unit can transmit the constructed packet through one channel or more. The attacker can inject the TDS attack to the communication channel after packets are sent from transmitter unit (point 3). The plant model calculates the estimated states (point 5) and sends them to DMU. The amount of delay will be estimated using the delay estimator unit at the same time (point 6). The DMU (point 6) receives the estimated states along with a delay estimate and, based on predefined rules, makes a decision to whether request a new redundant channel or not. It also informs the controller (point 7) to adapt itself under attack until the next healthy packet is received. The controller generates a control signal and sends it to the plant (point 8). The CF-TDSR will detect and track time delays introduced by a hacker and guide the plant to track the reference signal to improve the system performance.

The CF-TDSR is flexible and able to support communication between the plant and the controller with and without time-stamps. In the case where time-stamps are used, the controller compares the controller clock and the packet time-stamp and the state of the plant with the state predicted by the plant model. If there are any differences, the packet is dropped. If this is the case, the controller sends a negative acknowledgment (NACK) signal to the communication transmitter to use an adaptive channel allocation. Finally, the controller uses the state predicted by the plant model instead of the state received in the packet to control the system while waiting for the corrected future packets.

In the case where time-stamps are not used, the time delay estimator continuously estimates time delays, while the plant model determines the appropriate plant state values. If the estimated time delays are larger than the tolerable time delay, or if the plant state estimates are different from the received plant states, the communicated packet is dropped. Similar to the previous case, in this case, the DMU unit signals the communication transmitter to use adaptive channels allocation and its internal state estimates for control purposes.

6.2.5 Decision Making Unit (DMU)

The functionality of the delay detector which is part of DMU unit can be captured with the following mathematical formula:

$$D(t) = \begin{cases} 1 & (|t_c - t_s| > \tau_{stable}) \quad \text{or} \quad (|e_m(t)| > \epsilon) \quad \text{or} \quad (\hat{\tau} \geq \tau_{stable}), \\ 0 & \text{otherwise}, \end{cases} \tag{6.23}$$

where $D(t)$ is the detection function, t_c is the time value of the clock maintained by the controller, t_s is the time-stamp of the packet generated at the transmitter, and τ_{stable} is the tolerable time delay, or the maximum time delay for which the system remains in the stable region. This can be calculated from the eigenvalues of the system. Here, $e_m(t) = x(t) - \hat{x}(t)$ is the difference between the transmitted state of the plant $x(t)$ and the plant estimator record of the system state, $\hat{x}(t)$, and ϵ is the maximum tolerable error value. Since TDS attacks occur with probability p, $D = 1$ with probability p and $D = 0$ with probability $1 - p$ can occur. Observe that (6.23) enables the detection of TDS attacks and other types of attack such as FDI, irrespective of the use of time-stamps, even if the time-stamps are modified by the hacker.

The DMU generates the $Z(t)$ signal based on predefined rules, which can be an estimated or received state value regardless if its faulty or not. This would address other types of attack in the system such as FDI attack. Due to this feature of DMU, two types of CF-TDSR protocol will be presented. The first only detects the TDS and other types of attack and only sends an NACK signal to the transmitter to request an additional redundant communication channel; we call it CF-TDSR-Type1. The second protocol type benefits from both adaptive communication channel and controller by sending an NACK signal to the transmitter along with estimated state to the controller to adapt it; we call it CF-TDSR-Type2. While CF-TDSR-Type2 is more accurate and cost efficient, CF-TDSR-Type1 is more applicable to highly nonlinear and complex systems.

6.2.6 Delay Estimator Unit (DEU)

Consider a system that can be approximated in a region of interest by an LTI system. Note that (6.3), without the noise term, can be described by

$$\dot{x}(t) = Ax(t) + Bu(t). \tag{6.24}$$

The solution of this equation is

$$x(t) = e^{At}x_0 + \int_0^t e^{A(t-s)} Bu(s)ds. \tag{6.25}$$

Given a time delay τ, introduced by a TDS attack or by natural causes, the solution becomes

$$x(t-\tau) = e^{A(t-\tau)}x_0 + \int_0^{t-\tau} e^{A(t-\tau-s)}Bu(s)ds. \tag{6.26}$$

The modeling error signals in states can be described by $e_m(t) = x(t) - \hat{x}(t)$ and

$$e_m(t; \tau, \hat{\tau}) = x(t-\tau) - \hat{x}(t-\tau). \tag{6.27}$$

The idea is to estimate $\hat{\tau}$ in time as fast as possible to minimize the modeling error $e_m(t; \tau, \hat{\tau})$. To do so, let us assume $v = e_m^2/2$. The equation that minimizes the error is given by

$$\frac{d\hat{\tau}}{dt} = \eta \frac{\delta v}{\delta \hat{\tau}}, \tag{6.28}$$

where η is the learning parameter to be determined in conjunction with the PID or the optimal controller coefficients. Then

$$\begin{aligned} \frac{d\hat{\tau}}{dt} &= -\eta \frac{\delta v}{\delta \hat{\tau}} = -\eta e_m \frac{\delta e_m}{\delta \hat{\tau}} \\ &= \eta e_m \left[Bu(t-\hat{\tau}) - e^{A(t-\hat{\tau})}Bu(0) - Ae^{A(t-\hat{\tau})}x_0 \right]. \end{aligned} \tag{6.29}$$

In this section, we assume $u(0) = 0$ at the initial time. Then,

$$\frac{d\hat{\tau}}{dt} = -\eta e_m Bu(t-\hat{\tau}) - Ae^{A(t-\hat{\tau})}x_0, \quad 0 \le \hat{\tau} \le t. \tag{6.30}$$

Note that (6.30) is used to estimate the time delay τ. This process has some practical issues that should be considered. Computing machines have temporal resolution and finite memory. Therefore, (6.30) cannot be implemented without appropriate discrete approximation and boundedness assumptions. To assure the calculations stability and limit the memory usage, the following condition should be added, $\tau < \tau_{max}$. This condition will create a finite buffer to store the history of $u(t)$ from t to $t - \tau_{max}$ and also to prevent a runaway condition on $\hat{\tau}$.

In the following section, a novel method to prevent TDS attacks on LFC systems is illustrated.

6.2.7 Stability of the LFC Under TDS Attack

Consider a power-area LFC system of the form

$$\dot{x}(t) = Ax(t) + Bu(t) + w(t), \tag{6.31}$$

with the optimal controller given by

$$u(t) = -K\hat{x}(t) = \begin{cases} -Kx(t) & 1 - D(t), \\ -K\hat{x}(t) & D(t), \end{cases} \tag{6.32}$$

where $D(t)$ is a digital random process: $D(t)$ is 1 when a TDS attack is detected (see (6.23)) and is zero otherwise. TDS attacks are detected by comparing the received time-stamp from the plant against the controller time, or by using the time-delay estimator (see (6.30)). The new state estimate, $\hat{x}(t)$, is given by

$$\dot{x}(t) = \begin{cases} Ax(t) + Bu(t) & 1 - D(t), \\ A\hat{x}(t) + Bu(t) & D(t). \end{cases} \tag{6.33}$$

Let the estimation error be $e_m(t) = \hat{x}(t) - x(t)$. The dynamics of the closed loop can be described as

$$\dot{x}(t) = \begin{cases} (A - BK)x(t) + w(t) & 1 - D(t) \text{ and } e_m(t) = 0, \\ (A - BK)x(t) - BKe_m(t) + w(t) & D(t). \end{cases} \tag{6.34}$$

Now, let us investigate the stability of (6.34). For a reasonable stability criterion and for the covariance of $e_m(t)$ to remain bounded, the mean of the estimation error $e_m(t)$ should converge to zero. If the covariance of $e_m(t)$ is bounded, then it has convergence for state $x(t)$. The total expectation value is computed over both $x(t)$ and $D(t)$, knowing that $D(t) = 1$ with probability p when there is an attack on one channel. Thus, the equation of the system under attack can be expressed as

$$\dot{x}(t) = \begin{cases} (A - BK)x(t) + w(t) & \text{with probability } 1 - p, \\ (A - BK)x(t) - BKe_m(t) + w(t) & \text{with probability } p. \end{cases} \tag{6.35}$$

Therefore, computing the total expectation yields

$$\begin{aligned} \hat{\mu}(t) &= (A - BK)\mu(t)(1 - p) + (A - BK)\mu(t)p - BK\mu_m(t)p \\ &= (A - BK)\mu(t) - BK\mu_m(t)p. \end{aligned} \tag{6.36}$$

Table 6.3 Parameter values for a two-area power system with optimal controller [23].

Parameter	Value	Parameter	Value
$J_1(s)$	10 s	ω_1	0.05
μ_1	1.5	T_{g1}	0.12 s
T_{tu1}	0.2 s	T_{tu2}	0.45 s
T_{12}	0.198 pu/rad	T_{21}	0.198 pu/rad
$J_2(s)$	12 s	ω_2	0.05
μ_2	1	T_{g2}	0.18 s
R	$100I$	Q_f	0
Q	$100I$	t_f	∞
β_1	21.5	β_2	21

Let us now assume that ℓ channels are added to the communication channel:

$$\dot{\mu}(t) = (A - BK)\mu(t) - BK\mu_m(t)p^l. \tag{6.37}$$

If the term $BK\mu_m(t)p^l$ approaches zero, (6.37) will converge to zero and the system will be stable. Therefore, the idea is to make that term as small as possible by choosing a large l or by using a good estimator for the system that can make this term get closer to zero or stay bounded.

If the delay injected by the attacker exceeds τ_{max}, a trap condition signal is sent to the supervisory control and data acquisition (SCADA) center. The controller switches to open loop control until problem gets solved.

6.2.8 Simulation Example 6.1

Simulations are conducted to evaluate the performance of the CF-TDSR protocol under TDS attacks. The discrete linear quadratic regulator design from the *lqrd* continuous cost function (MATLAB 2013a) is used to generate the optimal control law for the system in normal operation. Two-area power systems are modeled as described earlier. Table 6.3 shows the parameter values used in the simulation. Also, ΔP_l^1 and ΔP_l^2 are set to zero.

The goal of the simulation is to demonstrate the ability of CF-TDSR to quickly respond to the TDS attacks. The total simulation time is set to 50 seconds.

The example assumes that a hacker has access to the communication channel. The attacker starts a TDS attack with values of $\tau = [t_{d1}\ t_{d2}\ \dots\ t_{dn}]^T$. Each power area has five states. Since a two-area power system is considered, the total number of states in the interconnected model is 10. Consider that the attack starts at time t_a.

The simulation is performed in three main scenarios: composite TDS attack, single power plant attack, and simultaneous composite TDS attack on a noisy system and limited available channel.

A. Composite TDS Attack

In the first investigation, a case is simulated where a hacker attacks the third state of both power areas. Since the third state of each area provides the feedback, it is thus an ideal place for the TDS attack. We assumed that there is only one extra channel available to allocate in this scenario. The attack starts at $t_a = 1$ s for the first power area and at $t_a = 3$ s for the second power area, with time-delay injected values of $t_{d3} = 1.5$ s (time delay associated to third state of first plant) and $t_{d8} = 3$ s, respectively. Fig. 6.7A illustrates the TDS attack and its tracking using our proposed delay estimator unit. In Fig. 6.7A, the dashed–dotted line shows the attack on the second power, and the solid line indicates the TDS attack injected to the first power area which are called "TDS attack 1" and "TDS attack 2", respectively. The dashed–dotted lines illustrate the delay estimation of TDS attacks 2 and 1, respectively. This figure demonstrates that CF-TDSR is capable of detecting and tracking TDS attack accurately in real-time.

The behavior of the LFC distributed power system under attack in three scenarios was evaluated. The first scenario, called "Baseline", runs without any modification to the communication protocol and to the controller (dotted line). The second scenario, called "CF-TDSR-Type1", evaluates the LFC under the attack using an adaptive communication protocol while the controller is not adaptive (dashed line). The third scenario evaluates the LFC system using "CF-TDSR-Type2", i.e., both the adaptive communication protocol and adaptive controller design defenses (solid line).

Fig. 6.7 shows that CF-TDSR-Type2 is capable of quickly detecting the TDS attack and adapting the communication protocol and controller. Note that when CF-TDSR detects a delay larger than 0.4 s, it sends an NACK to the sender. Figs. 6.7–6.9B–F show the frequency and power deviation of the generator, value position of the turbine, and the tie-line power flow of the first power area, respectively. They show that the system becomes stable when using CFTDSR-Type1 or CFTDSR-Type2, and it gets unstable in the Baseline scenario.

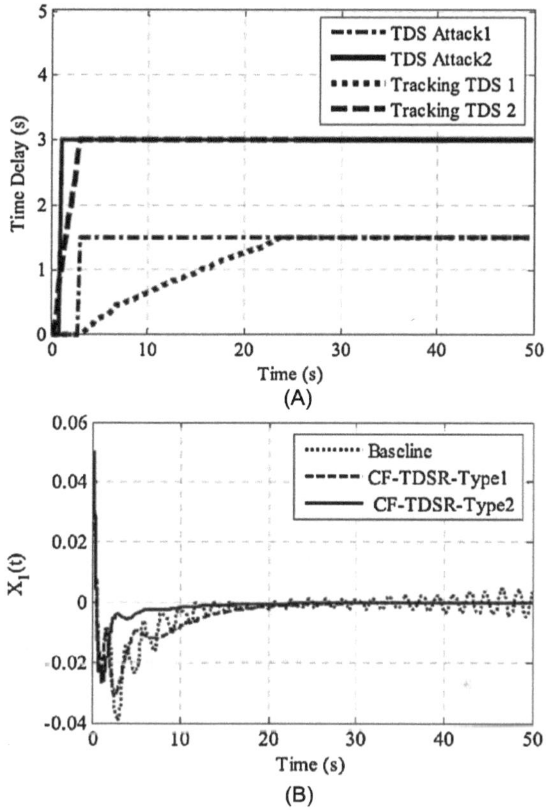

Figure 6.7 (A) Time-delay attack on system (the solid line is an attack on the third state of the first power area, and the dashed line shows the TDS attack on the third state of the second power area). (B) Frequency deviation during the attacks

Based on the results, we conclude that if CF-TDSR detects a TDS attack on the second channel and there is no more communication channel available, the estimator turns on and stays alive for the entire time and guarantees the stability of the system. The results show that CF-TDSR works very well, even with strict limitations on the number of available channels, which is evidenced by all states converging to zero as expected. The results also indicate that the system becomes unstable under a TDS attack without any defense mechanism, in the "Baseline" scenario. Furthermore, Figs. 6.7B–F indicate that CF-TDSR-Type2 has better performance than CF-TDSR-Type1, while both of them stabilize the system fairly.

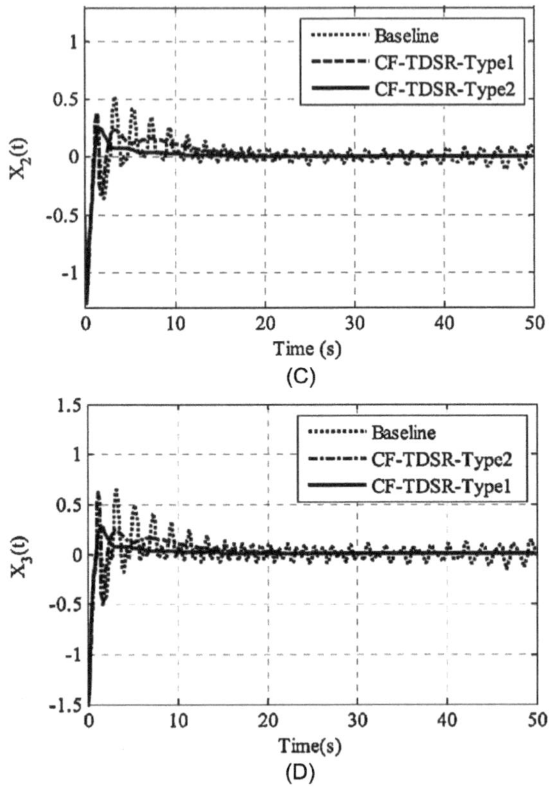

Figure 6.8 (C) Power deviation of the generator during the attack. (D) Value position of the turbine during the attack

B. Single-Plant TDS Attack

In the second case, we assume that a hacker only has access to the first power area, and can only launch TDS attacks on the specific state variable. The existence of multiple channels for allocation is assumed, but the CF-TDSR-Type2 method only needs a limited number of channels due to its adaptive controller feature of the CF-TDSR protocol. It's assumed that a powerful hacker can launch multiple and sequential TDS attacks.

Figs. 6.10–6.12 show the results of the simulation. Fig. 6.10A describes the attack that was launched: the dashed–dotted line denotes a TDS attack on the first communication channel that occurs in the time interval from 1 to 3 seconds with a delay of 4 seconds, followed by a TDS attack on channel 2 between 3 and 4 seconds, with a de-

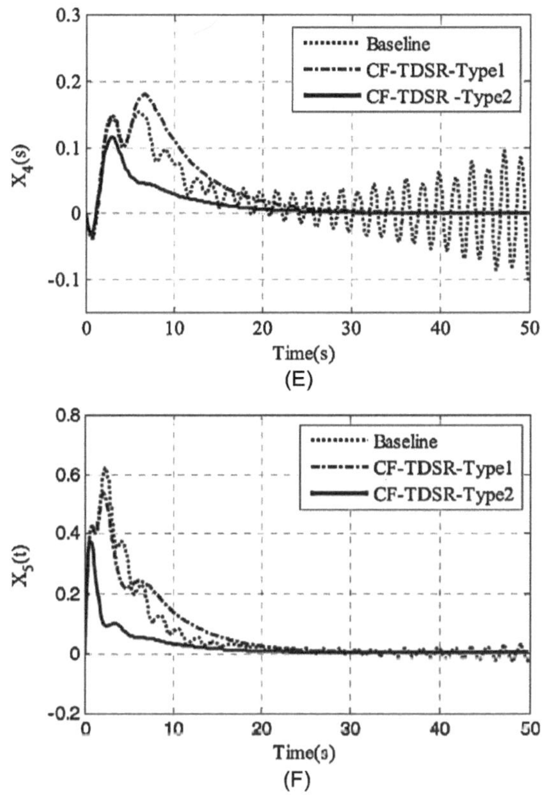

Figure 6.9 Composite TDS attack on one state of a two-area power system: (E) tie-line power flow, (F) control error

lay of 2.5 seconds (dotted line), and then a TDS attack of channel 3 between 9 and 20 seconds (dashed line) and a TDS attack on channel 4, with a delay of 9 seconds, in the time interval from 25 to 50 seconds (solid line). Fig. 6.10B shows that the CF-TDSR protocol detects each attack and requests a change of channel accurately. The attack on channel 4 is not effective because the system has already approached the stable region. The conclusion is that when the system is at optimal value (close to zero), it is more difficult for the TDS or other types of attack to destabilize the system.

Fig. 6.11C shows the frequency deviation, Δf^K, of the power system. Fig. 6.11D shows the power deviation of the generator, Δf_g^K. Fig. 6.12E shows the value position of the turbine, $\Delta f_{t_u}^K$. Fig. 6.12F shows tie-line power flow, Δf_{pf}^l. These figures assure that

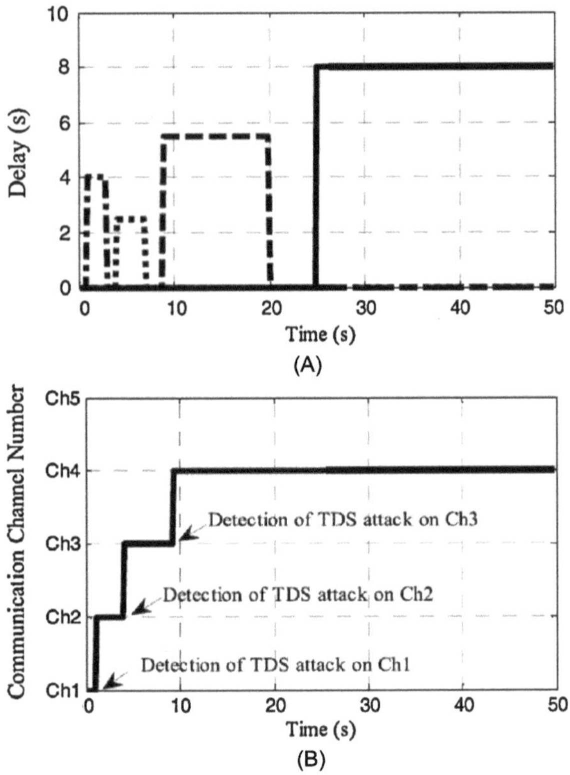

Figure 6.10 (A) Illustration of a sophisticated, sequential, multichannel attack. (B) Evolution in time of CF-TDSR

the states remain stable and converge to zero under a TDS attack. Figs. 6.11C–6.12F compare the results of a scenario where the state estimator is on and the controller is adaptive (CF-TDSR-Type2, shown in solid line) to the results of a scenario where the state estimator is off (dashed line). In both cases, the time-delay detector and channel adaptation are on. The figures show that the CF-TDSR protocol is clearly superior. The results indicate that the cost function value is improved, ($\Delta J = J_{No\ Estimation} - J_{With\ Estimation} = 5.21$), when the present state estimator is running, which take care of a TDS attack while the NACK signal is received by the transmitter and a new channel is added to the system.

A comparison of single plant TDS attack and composite TDS attack indicated that CF-TDSR is capable of detecting and compen-

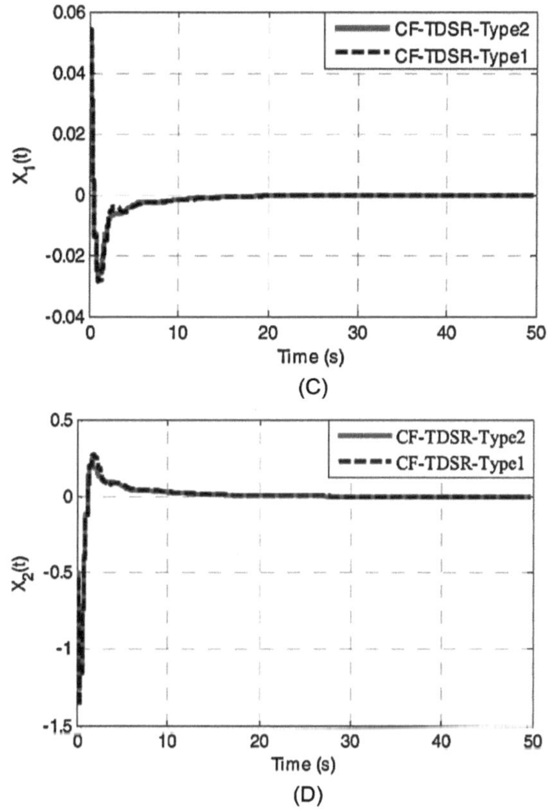

Figure 6.11 (C) The frequency deviation of the power system during the attacks. (D) The power deviation of the generator during the attacks

sating for the effect of TDS attack, in general. It also shows that CF-TDSR-Type2 is better than CF-TDSR-Type1 in both cases but, it's much more powerful than CF-TDSR-Type1 in the case when there is a limitation on the existence of redundant communication channels.

C. Simultaneous TDS Attack for Noisy Systems and Limited Available Channels

In the last experiment, the system behavior under the noise is studied. In order to do this, first, 20% of white Gaussian noise is added to the communication channel. Then, a TDS attack is launched on both power areas: The attacker simultaneously launches the attack on the third state of both the first and second power areas at a time of 1 and

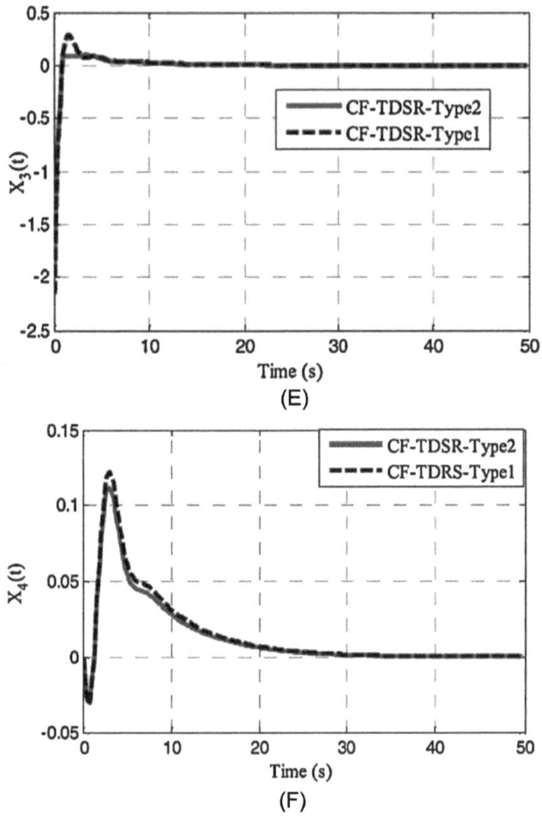

Figure 6.12 TDS attack on one power area. (E) The evolution in time of the value position of the turbine during the attacks. (F) The tie-line power flow during the attack

4 s, with a 2-second delay. Then, the delay value is increased at the time of 7 s to the value of 5 and 6.5 s, respectively. In this experiment, the availability of a single communication channel is assumed. This assumption severely restricts the CF-TDSR's options, so we can only use CF-TDSR-Type2. Fig. 6.13 shows that even under such restrictions, the CF-TDSR is able to accurately detect and reduce the effects of the noise based TDS attack. Specifically, Fig. 6.13A shows how CF-TDSR detects and tracks the TDS attack in real time. Figs. 6.13B–C show the third states of the first and second power areas under the noise based TDS attack. They show that CF-TDSR performs very well even in the absence of additional communication channels and under existence of noise.

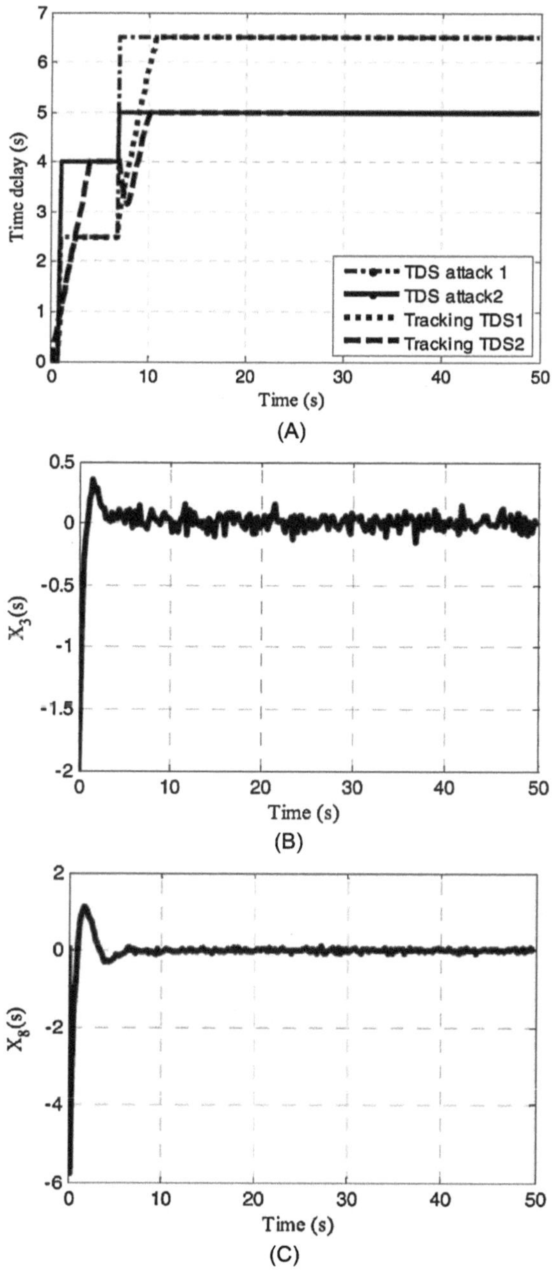

Figure 6.13 TDS attack injected at the same time on both power areas

6.3 MULTISENSOR TRACK FUSION-BASED MODEL PREDICTION

In this section, the focus is toward cyberattacks in the form of data injections [25–30]. Abnormal data superimposed into collected synchrophasor measurements can cause false system information to be interpreted by installed monitoring algorithms. This can then lead to delays in mitigation actions. Among monitoring schemes using PMU measurements, state estimation and oscillation detection are more popular applications. Despite several methods that have been proposed for bad data detection in state estimation [26], none explored in the field of oscillation detection.

Power oscillations are electromechanical dynamics between synchronous generators in an interconnected grid. The frequency of local oscillation ranges from 0.8 to 2 Hz, while the frequency of intraarea mode is from 0.1 to 0.8 Hz [31]. Interarea oscillations are difficult to monitor and are prone in systems that are operating near their technical transfer capacity. As a result, monitoring algorithms to detect interarea oscillation using synchrophasor measurements are proposed in recent studies [31–36]. The objective is to detect lightly damped oscillations at an early stage before they trigger angular and voltage instabilities. An interarea oscillation was responsible for the Northwestern blackout in North America [31]. The present research trend is moving toward recursively monitoring oscillations under ambient situations. Recursive techniques can be categorized into: (i) curve-fitting and (ii) *a priori* knowledge-based. The first refers to publications that extract oscillatory parameters directly from measurements [32–34]. The latter techniques are associated with methods that approximate parameters using previous knowledge of the system as well as collected measurements [35]. An *a priori* knowledge-based approach provides higher estimation accuracy under ambient or noisy conditions when an accurate model is provided [36]. In this case, approximating electromechanical oscillations as a sum of exponentially damped sinusoidal signals is considered an accurate model representation in oscillation monitoring research.

Despite published methods in oscillation detection that can operate under noisy conditions, they are not proven to be resilient against data-injection attacks. Such an attack is an emerging threat due to the increasing dependency of digital measurements for monitoring and control applications in recent years [27]. A majority of published monitoring methods are formulated based on the assumption of the measurements are not contaminated by human interventions. According to [25] and [28], cyberattacks that introduce periodic or continuous bias to system measurements are possible. There are no guarantees that all cyberattacks can be prevented. Any successful attack will cause existing monitoring schemes to generate inaccurate system information, which may then lead to cascading failures [27, 29,30]. In recent literature, several methods are proposed to identify abnormal data segments and isolate attacked sensors [27–30]. However, they usually require a large data batch and are computationally intensive. Although an attacked sensor can be eventually identified, the time between the start of the attack until successful isolation can be in minutes or hours. This is a significant time window to trigger wide-area blackouts as operators are still being fed with false information. Referring to [31] and [37], it only takes minutes to make interarea oscillation become lightly damped and generate wide-area angular and voltage instabilities. Coming from the system operational perspective, the key objective is to minimize the potential damage of a data-injection attack through novel processing of information collected from distributed sensors. To the best of authors' knowledge, such an enhancement in oscillation monitoring algorithms has not been proposed.

Therefore, this section contributes toward proposing a signal processing solution to enhance the resilience of existing oscillation monitoring methods against contaminated measurements. Since data-injection attacks in electrical grids can be considered as a regional event, the use of distributed architecture such as in [36] is an adequate option against data contamination. However, given the uncertainties of a data injection attack in the prescribed error statistics, it can be inappropriate to spend a huge amount of computational power to filter erroneous information as used by the algorithmic structure. Monitoring algorithms should:

1. be robust against random fluctuations and bias; and
2. have low computational cost of the propagation of estimation of each electromechanical oscillation.

To achieve the robustness, while optimizing the computational complexity, constraints on perturbation and random fluctuations shall be considered. The aim is to maintain the accuracy of extracting oscillatory parameters as well as detect potential monitoring nodes that are being attacked.

To understand the integration of data-injection attacks into the oscillation monitoring application, an overview of the proposed multisensor TFMP is illustrated in Fig. 6.14. The considered scenario assumes that the attacker is smart enough to inject data that can imitate regular variations of small-signal system dynamics. TFMP can resolve this concern by ma-

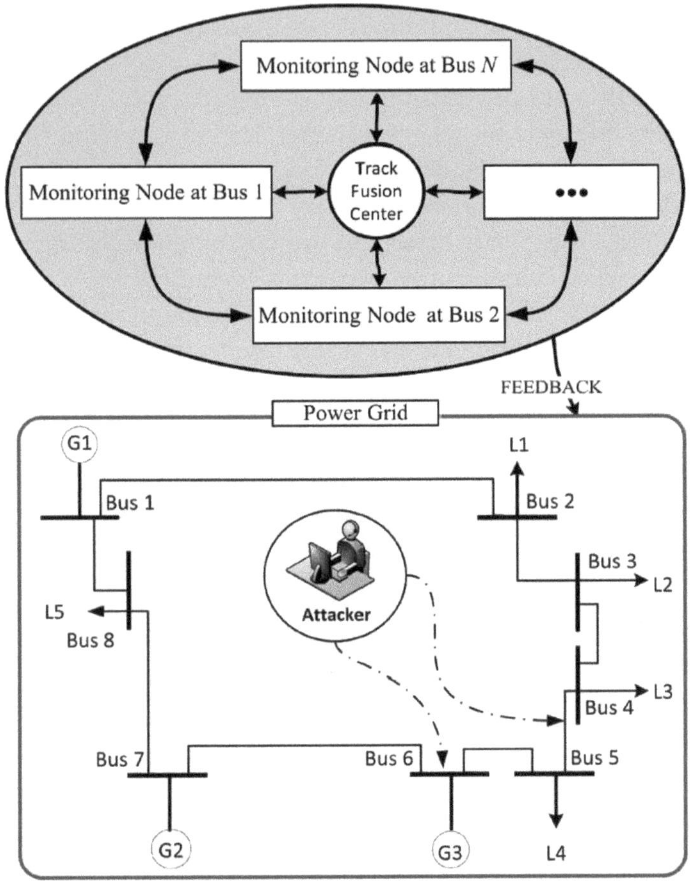

Figure 6.14 Proposed TFMP scheme to estimate and detect data-injection attacks during power oscillations monitoring

nipulating estimated oscillation parameters from all local sensor monitoring nodes. In this paper, a local sensor monitoring node refers to a site where KLPF-based smoother will be applied to extract oscillation parameters from PMU measurements collected at a substation. Furthermore, each monitoring node is assumed to be able to interact with its neighbors through substation communication channels. The estimated parameters are then communicated to the TFC and followed by track association and fusion at the global level. Note the TFC is developed to compute and minimize the errors of filtering, prediction, and smoothing within each local sensor monitoring node.

6.3.1 State Representation of Observation Model

A power grid prone to data-injection attacks can be expressed as a nonlinear dynamical system model. Perturbations and random fluctuations are part of noise induced transitions in a nonlinear system with dynamics. It is expressed as

$$\alpha x_{t+1} = f(x_t, w_t), \quad t = 0, 1, \ldots, T, \tag{6.38}$$

where α is the constant matrix with compatible dimensions to the model dynamics, $f(\cdot)$ is the nonlinear function representing the state transition model, $x_0 \in \mathbb{R}^r$ is the initial condition of the oscillation state, and r is the size of the oscillation state vector in a subspace of $\mathbb{R}$. In addition, $w_t \in \mathbb{R}^r$ is the random process noise, t is the time instant, and T is the number of time instants. Note that (6.38) represents the equation of a system which has nonlinear dynamics. Perturbations and random fluctuations are part of noise-induced transitions in a nonlinear system. These can be from load variations or switching transients of installed devices. Eq. (6.38) can also be represented by any other dynamical system model. It is not only limited to power systems.

It is assumed that the power grid described in (6.38) will be monitored by N number of synchronized sensors in a track-level measurement fusion environment. Computation is conducted at a central station, i.e., TFC, which involves control signals at each local node and predictive estimation sequences are generated in the presence of random noise fluctuations. These local sensors will basically be PMUs installed in high voltage substations, and all will operate at the same sampling rate. The observations vector for extracting electromechanical oscillations at the ith node possibly affected by the attack can be defined as

$$z_t^i = h_t^i(x_t) + v_t^i, \quad i = 1, \ldots, N, \tag{6.39}$$

where $z_t^i \in \mathbb{R}^{p^i}$, p^i is the number of synchrophasor observations made by the ith sensor, $h^i(\cdot)$ is a nonlinear function representing the local observation matrix of the ith sensor, x_t is the state matrix for oscillations, and $v_t^i \in \mathbb{R}^{p^i}$ is the observation noise of the ith sensor. A dynamical power grid will be governed by the following constraints:

$$x_t \in \mathbf{X}_t, \; w_t \in \mathbf{W}_t, \; v_t \in \mathbf{V}_t, \tag{6.40}$$

where $\mathbf{X}_t$, $\mathbf{V}_t$, and $\mathbf{W}_t$ are assumed to have Gaussian probability distribution function.

Assumption 6.1. The noises w_t and v_t are all initially assumed to be uncorrelated zero-mean white Gaussian such that $\mathbb{E}[w_t] = \mathbb{E}[v_t] = \mathbb{E}[w g v_h^T] = 0$, $\forall t$. Note $\mathbb{E}$ denotes the expectation operator, and superscript T denotes the transpose operator. Also, $\mathbb{E}[w_g w_h^T] = R_t \delta_{gh}$, $\mathbb{E}[v_g v_h^T] = Q_t \delta_{gh}$, $\forall t$, where R_t represents the residual covariance and δ_{gh} is a Kronecker delta which is 1 when variables g and h are the same; Q_t is the process noise correlation factor.

Once the observation model is constructed from synchrophasor measurements collected from the affected location, the corresponding state representation of electromechanical oscillations can be formulated in the frequency domain.

6.3.2 Electromechanical Oscillation Model Formulation

Suppose a measured noise-induced signal contained K electromechanical oscillations. Referring to (6.39), the observation output signal z_t^i from the ith sensor at time t can be modeled in the frequency domain as

$$z_t^i = \sum_{k=1}^{k} a_k e^{(-\sigma_k + j2\pi f_k)tT_s} + v_t^i, \quad t = 1, 2, \ldots, T, \tag{6.41}$$

where a_k is the complex amplitude of kth mode, σ_k is the damping factor, f_k is the oscillatory frequency, and T_s is the sampling time [35]. Eq. (6.39) has been transformed to (6.41), i.e., time domain to the frequency domain, using the Laplace transform. The system's poles and zeros are then analyzed in the complex plane. Moreover, it is especially important to transform the system into the frequency domain to ensure that the poles and zeros are in the left or right half-planes, i.e., have real part greater than or less than zero. For convenience, the term $-\sigma_k + j2\pi f_k$ is represented in the rectangular form as λ_k. In this section, the kth oscillation or eigenvalue within a mentioned signal is described by two states denoted as $x_{k,t}$ and $x_{k+1,t}$, respectively. They can also be expressed for the ith sensor as

$$x_{k,t}^i = e^{(-\sigma_k + j2\pi f_k)tT_s}, \quad x_{k+1,t}^i = b_{k+1} e^{(-\sigma_{k+1} + j2\pi f_{k+1})tT_s}. \tag{6.42}$$

The term b_k represents the complex amplitude of the kth mode. Based on (6.42), a signal consisting of K exponentially damped sinusoids will be modeled by $2K$ states. Note that the kth eigenvalue of a particular signal is described by two states denoted as $x_{k,t}$ and $x_{k+1,t}$, i.e., for the kth and $(k+1)$th mode, respectively. The eigenvalue represents the electromechanical oscillations between synchronous generators in the physical world. Details can be found in [31]. In addition, the damping factor σ_k and the corresponding frequency f_k of each oscillation will be computed from the state x_t. Estimating oscillatory parameters in the presence of a random data-injection attack will require the complete observability of the oscillation observation matrix. For a nominal case without data injections, this was previously achieved by using an expectation maximization (EM) algorithm that utilized the initial correlation information extracted from KLPF [36]. Initial correlation information can be defined as the information collected from the initial estimates of the observation model $\hat{H}_t^0$. The superscript 0 represents the initial estimates. However, considering a data-injection attack situation, taking an averaged form of the log-likelihood function to improve estimate as in [36] is not sufficient. Instead, the initial correlation shall be iteratively calculated by: (i) using the first and second moments of the input model for a node i; (ii) getting *a priori* information from the constraints; and (iii) getting observation estimates through time and frequency correlation for each sensor. Note that in this section, monitoring power oscillation is used as an application. The proposed scheme can also be utilized by any other application.

6.3.3 Initial Correlation Information

Estimating x_t in (6.38) and (6.39) from z_t^i at node i may be a difficult problem. This is both due to the nonlinear grid dynamics and the noise constraints outlined in (6.40). Referring to (6.38) and (6.39), a reasonable estimate will be to linearize the system to smooth-out nonlinearities. Thus, the linearized model of the power grid will be

$$f_t(x_t, w_t) \approx f_t(\tilde{x}_{t|t}, 0) + \kappa_t(x_t - \tilde{x}_{t|t}) + \Psi_t w_t, \tag{6.43}$$

$$h_t^i(x_t) \approx h_t^i(\tilde{x}_{t|t-1}) + H_t^i(x_t - \tilde{x}_{t|t-1}), \tag{6.44}$$

where κ, Ψ, and H^i are the Jacobean matrices with compatible dimensions used to linearize the nonlinear dynamics

$$\kappa_t = \frac{\partial}{\partial x} f_t(x, 0)|_{x=\hat{x}_{t|t}}, \quad \Psi_t = \frac{\partial}{\partial w} f_t(\hat{t|t}, w)|_{w=0}, \tag{6.45}$$
$$H_t = \frac{\partial}{\partial x} h_t(x, 0)|_{x=\hat{x}_{t|t-1}},$$

and $\tilde{x}_t$ is the linearized approximation of the system state x_t. This transforms (6.38) and (6.39) into

$$\alpha x_{t+1} = \kappa_t x_t + \Psi w_t, \tag{6.46}$$

$$z_t^i = H^i x_t + v_t^i, \quad i = 1, 2, \ldots, N, \tag{6.47}$$

where the oscillation system state $x_t \in \mathbb{R}^n$, the synchrophasor measurements $z_t^i \in \mathbb{R}^{mi}$, $w_t \in \mathbb{R}^r$, and $v_t^i \in \mathbb{R}^{mi}$. Variables α, κ, Ψ, and H^i are the constant matrices with compatible dimensions. The system model described by (6.46) and (6.47) is derived based on the following assumptions.

1. Rank $\alpha = n_1 < n$ and rank $\kappa \geq n_2$, where $n_1 + n_2 = n$.
2. System (6.46) is regular, i.e., $\det(s\alpha - \kappa) \not\equiv 0$, where s is an arbitrary complex number which can be expressed as a sum of real and imaginary components.
3. The initial state x_0, with mean μ_0 and variance P_0, is independent of w_t^i and v_t^i.

1. *Diagonalization of System Model at Node i Into Subsystems.*

Accurate monitoring of power oscillations in the presence of data-injection attacks can prove to be computationally expensive. However, to tackle the additional computational cost due to the calculation of initial estimates and the error covariance matrix $P_{t|t-1}$ may be demanding. This is due to the size of error covariance matrix which is equal to the size of the state vector, and therefore directly proportional to the size of the modeled power grid. To reduce the computational cost of the initial estimates and the error covariance matrix, diagonalizing the main system model into subsystems is proposed. This is derived on the structure for the KLPF-based smoother which can be referred to [36, (6.56)–(6.63)]. In [36], the KLPF-based smoother $\hat{x}_{t|t}^S$ of the state x_t is calculated based on measurements $(z_t^i, \ldots, z_T^i)$. Note that the attacked system at node i can be diagonalized up to N subsystems. To simplify the formulation, diagonalization of $N = 2$ subsystems is considered in this section. This reflects that each node consists of two subsystems. Using the theory of robust eigenvalue assignment from [40], the system described by (6.46) and (6.47) can be decomposed into L and R nonsingular matrices

$$L\alpha R = \begin{bmatrix} \alpha_1 & 0 \\ \alpha_2 & 0 \end{bmatrix}, \; L\kappa R = \begin{bmatrix} \kappa_1 & 0 \\ \kappa_2 & \kappa \end{bmatrix},$$

$$L\Psi R = \begin{bmatrix} \Psi_1 \\ \Psi_2 \end{bmatrix}, \; H^i R = \begin{bmatrix} H_1^t \\ H_2^t \end{bmatrix}, \tag{6.48}$$

where $\alpha_1 \in \mathbb{R}^{n_1 \times n_2}$ is nonsingular lower-triangular, $\kappa_1 \in \mathbb{R}^{n_1 \times n_1}$ is quasi-lower-triangular, and $\kappa_3 \in \mathbb{R}^{n_2 \times n_2}$ is nonsingular lower-triangular. Transforming $x_t = R[x_{1,t}^*, x_{2,t}^*]^*$, where $x_{1,t} \in \mathbb{R}^{n_1}$ and $x_{2,t} \in \mathbb{R}^{n_2}$. The system can be transformed into the following two diagonalizable subsystems by taking the inverse of high-dimensional matrices of (6.38) and (6.39) using a linear minimum variance [41]:

$$x_{1,t+1} = \kappa_0 x_{1,t} + \Psi_0 w_t, \tag{6.49}$$

$$x_{2,t} = \bar{\kappa} x_{1,t} + \bar{\Psi} w_t, \tag{6.50}$$

$$z_t^i = \bar{H}_t^i x_{1,t} + \bar{v}_t^i, \tag{6.51}$$

where $x_{1,t}$ and $x_{2,t}$ are the states of subsystems 1 and 2, respectively; κ_0, Ψ_0, $\bar{\kappa}$, $\bar{\Psi}$, $\bar{H}$, and $\bar{v}$ are diagonalized variables, which are computed from the inverse of weighted matrices α_1 and κ_3 as shown in Appendix A. In the subsystem transformation, only the first subsystem will have the prediction and filtering stage, whereas the other $N-1$ subsystems will only have a filtering stage. Referring to (6.49)–(6.51), the resultant noises $\bar{w}_t$ and $\bar{v}_t$ will have the diagonalizable expected value

$$\mathbb{E}\left\{ \lfloor \begin{matrix} \bar{w}_t \\ \bar{v}_t^{-1} \end{matrix} \rceil, \lceil \begin{matrix} \bar{w}_t^* & \bar{v}_t^{2*} \end{matrix} \rceil \right\} = Q_t^{1,2} \delta_t^{1,2}, \tag{6.52}$$

where $Q_t^{1,2}$ is the process noise correlation factor between subsystem 1 and 2, $\delta_t^{1,2}$ is the Kronecker delta function used for shifting the integer variable in the presence or absence of noise; $Q_t^{1,2}$ can be expressed as

$$Q_t^{1,2} = [\begin{matrix} Q_{\bar{w}} & \alpha^2 \\ \alpha^{1*} & Q_{\bar{v}1,2} \end{matrix}], \tag{6.53}$$

where $\alpha^1 = Q\bar{w}\Psi_3^{i*}$, $Q\{\bar{v}_i\} = \Psi_3^i Q\bar{w}\Psi_3^{i*} + Q_{v^i}$, $Q_{\bar{v}1,2} = \Psi_3^i Q\bar{w}\Gamma_3^{i*}$, and Ψ_3^i is defined in Appendix A. Once the subsystems are constructed from the system affected by the data-injection attacks, the interactions between them shall be evaluated. This will require extracting the signature of random variations, which can be obtained by comparing measurements with known system dynamics. This interaction is evaluated here by using crosscovariance analysis. It is proposed to improve the goodness of fit of random variations, while enhancing the predictive accuracy and covariance estimates of the KLPF.

6.3.4 Computation of Crosscovariance

From (6.49)–(6.53) and considering [36, (6.56)–(6.63)], the state of subsystem 1, $x_{1,t}$, and subsystem 2, $x_{2,t}$, can be derived. This is to have complete observability on the dynamics of power oscillations in the presence of random data-injection attacks. First, suppose the a *priori* equation of state $x_{1,t}$ at node i is computed as

$$\tilde{x}^i_{1,t+1|t} = \bar{\kappa}^i_0(I_{n_1} - \beta^i_t \bar{H}^i_t)\tilde{x}^i_{1,t|t-1} + \Psi_0 w_t - (\kappa^i_0\beta^i_t + J^i)\bar{v}^i_t \tag{6.54}$$

where $\tilde{x}^i_{1,t|t-1}$ is the difference between $x_{1,t}$ and $\hat{x}^i_{1,t|t-1}$, $\bar{\kappa}^i_0 = \kappa_0 - J^i\bar{H}^i_t$, $J^i = \Psi_0\alpha^i Q^{-1}_{\bar{v}^i}$ and $\kappa^i_t = \bar{\kappa}(I_{n_1-\beta^i_t\bar{H}^i_t}) - \bar{\Psi}\alpha^i Q^{-1}_{\varepsilon^i}$; I_{n_1} is and $n_1 \times n_1$ identity matrix. For notation convenience, $\beta^i_t = (P^i_t H^{i*}_t/(H^i_t P^i_t H^{i*}_t + \sigma^2_v))$ The corresponding updated a posteriori equation of state $x_{1,t}$ at node i will be

$$\tilde{x}^i_{1,t|t} = ((I_{n_1} - \beta^i_t\bar{H}^i_t)\tilde{x}^i_{1,t|t-1} - \beta^i_t)v^i_t, \tag{6.55}$$

where $\tilde{x}^i_{1,t|t}$ is the difference between $x_{1,t}$ and $\hat{x}^i_{1,t|t}$. Thus, the updated and predicted error equations of state $x_{1,t}$ are achieved. The state $x_{2,t}$ of second diagonalized subsystem at node i can also be expressed as

$$\tilde{x}^i_{2,t|t} = F^i_t\tilde{x}^i_{1,t|t-1} + D^i_t[\bar{w}^*_t, \bar{v}^{i*}_t]^*, \tag{6.56}$$

where $\tilde{x}^i_{2,t|t} = x_{2,t} - \hat{x}^i_{2,t|t}$, $F^i_t = \bar{\kappa}(I_{n_1} - \beta^i_t\bar{H}^i_t) - \bar{\Psi}\alpha^i Q^{-1}_{\varepsilon^i,t}\bar{H}^i_t$ and $D^i = [\bar{\Psi} - \bar{H}_t\beta^i_t - \bar{\Psi}\alpha^i Q^{-1}_{\varepsilon^i,t}]$ Once the subsystem states are derived, the crosscovariance between them can be formulated to filter any random variations caused due to the attack within collected synchrophasor measurements.

1. *Crosscovariance of State* $x_{1,t}$ (6.49) *for Subsystem 1*

Using the projection theory proposed in [42], the crosscovariance equation of the prediction and filtering errors of state $x_{1,t}$ between the subsystems 1 and 2 of the ith node can be computed as

$$\begin{aligned} P^{1,2}_{1,t+1|t} &= \bar{\kappa}^1_0[I_{n_1} - \beta^1_t\bar{H}^1_t]P^{1,2}_{1,t|t-1}[I_{n_1} - \beta^2_t\bar{H}^2_t]^* \\ &+ [\Psi_0 - \bar{\kappa}^1_0\beta^1_t - J^1]Q^{1,2}[\Psi_0 - \bar{\kappa}^2_0\beta^2_t - J^2]^*, \end{aligned} \tag{6.57}$$

where the subscripts 1 and 2 represent the subsystems 1 and 2 for ith node, respectively. The initial value of $P^{1,2}_{1,t+1|t}$ is $P^{1,2}_{1,0|t}$, which is the first $n_1 \times n_1$ block of $\mathbb{R}^{-1}P^{1,2}_{1,0|0}\mathbb{R}^{-1*}$. Subsequently, the error equation of the updated a posteriori estimates will be

$$P^{1,2}_{1,t|t} = [I_{n_1} - \beta^1_t\bar{H}^1_t]P^{1,2}_{1,t|t-1}[I_{n_1} - \beta^2_t\bar{H}^2]^* + \beta^1_t Q_{\bar{v}^{1,2}}\beta^2_t. \tag{6.58}$$

2. *Crosscovariance of State* $x_{2,t}$ (6.50) *for Subsystem 2*

The covariance matrix of the filtering errors for state $x_{2,t}$ between the subsystems 1 and 2 for the ith node can be expressed as

$$P_{2,t|t}^{1,2} = F_t^1 P_{1,t|t-1}^{1,2} F_t^{2^*} + D_t^1 Q^{1,2} D_t^{2^*} \tag{6.59}$$

where $P_{2,t|t}^{1,2}$ is the filtering error covariance of $x_{2,t}$ based on the subsystem 1 of the ith node.

Considering the cross-covariance computation of subsystems 1 and 2, the smoother of [36] is rederived here to further improve the estimation of suboptimal correlation information provided by the diagonalized subsystems. Moreover, the estimated output of the smoother will be more superior in providing insight to the power oscillation dynamics to those obtained from the subsystems 1 and 2 as it extrapolates backwards in time.

3. *Crosscovariance of Smoothing*

The crosscovariance of the smoothed *a posteriori* estimate between the subsystems 1 and 2 of the ith nodes is extended in (6.55), (6.62) and (6.63). The state estimate $\hat{x}_{t|T}$, given the whole time sequence, can be represented as

$$\hat{x}_{t|T}^{1,2} = \hat{x}_{t|t-1}^{1,2} + P_{t|t-1}^{1,2} \mathbb{R}_{t,T}^{1,2}, \tag{6.60}$$

where $t = N-1, N-2, \ldots, 1$. Here r is an $n \times n$ vector that satisfies the backward recursive equation where $r_{T+1|T} = 0$. $P_{t|T}^{\alpha 1,2}$ is the covariance matrix of $r_{t|T}$ with a size of $n \times n$ and satisfies the backward recursive equation. The resultant crosscovariance of $r_{t|T}$ will be

$$\begin{aligned}\Re_{t|T}^{1,2} &= \bar{\kappa}_p^{1^*} [I_{n_1} - \beta_t^{1,2} \bar{H}_t^2] r_{t|t-1} \\ &\quad + H^{2^*} [H^2 P_{t|t-1}^{1,2} H_t^{2^*} + R_t]^{-1} (\tilde{z}_{t+1}^2 \tilde{x}_{t+1}^2),\end{aligned} \tag{6.61}$$

where $\kappa_{t+1|t} = \kappa_{t+1|t}[I - K_t H_t]$. According to the smoothing property, the covariance matrix $P_{t|T}$ shall depend on the time sequence T such that

$$P_{t|T}^{1,2} = P_{t|t-1}^{1,2} - P_{t|t-1}^{1,2} P_{t|T}^{S_{1,2}} P_{t|t-1}^{1,2}. \tag{6.62}$$

This is followed by the smoothed-run updated a posteriori estimate, which will update the error covariance matrix in the smoothed run:

$$P_{t|T}^{S_{1,2}} = \bar{\kappa}_p^{1^*} [I_{n_1} - \beta_t^{1,2} \bar{H}^2]^* P_{t|t-1}^{\alpha_{1,2}} \bar{\kappa}_p^{2^*} [I_{n_1} - \beta_t^{1,2} \bar{H}^1]$$

$$+H^{1^*}[H^2 P_{t|t-1} H^{2^*} + R_t]^{-1} H^2. \tag{6.63}$$

The derived crosscovariance matrices for smoothing the state x_t between subsystems 1 and 2 are

$$P_{1,t}^{S_{1,2}} = I_{n_1} P_{1,t}^{1,2} I_{n_1}^* + (H^1 P_{t|t-1} H^{2^*})^{-1}, \tag{6.64}$$

$$P_{2,t}^{S_{1,2}} = F_t^1 P_{1,t}^{1,2} I_t^{t^*} + D_t^1 (H^1 P_{t|t-1} H^{2^*})^{-1} D_t^{1^*}, \tag{6.65}$$

where $P_{1,t|t}^{S_{1,2}}$ and $P_{2,t|t}^{S_{1,2}}$ are the smoothing error covariances of state $x_{1,t}$ and $x_{2,t}$, respectively. Up to now, the formulations of the crosscovariance for prediction, filtering, and smoothing error for the subsystems are derived. The next step will be to combine them into an interaction filter so that the variance of interaction errors among the state $\hat{x}_{2,t|t}^{\bar{\omega}}$ can be determined.

4. *Interaction Filter Structure Based on Crosscovariance Computation*

Based on (6.49)–(6.51) for subsystems 1 and 2, the interacted filter can be stated for state $x_{1,t}$ of subsystem 1 as

$$\hat{x}_{1,t|t}^{\bar{\omega}} = (e_{1,t}^* \Upsilon_{1,t}^{-1} e_{1,t})^{-1} e_{1,t}^* \Upsilon_{1,t}^{-1} [\hat{x}_{1,t}^{t^*}, \hat{x}_{2,t}^{t^*}, \ldots, \hat{x}_{t,t}^{N^*}] \tag{6.66}$$

where the superscript denotes the interaction between subsystems 1 and 2; $e_{1,t} = [I_{n,1}, \ldots, I_{n,1}]$ is an $n_1 N \times n_1$ matrix, $\Upsilon_{1,t} = P_{1,t|t}^{1,2}$ is an $n_1 N \times n_1 N$ positive definite matrix. Similarly, the resultant diagonalized interacted filter for state $x_{2,t}$ of subsystem 2 becomes

$$\hat{x}_{2,t|t}^{\bar{\omega}} = (e_{3,t}^* \Upsilon_{2,t}^{-1} e_{2,t})^{-1} e_{2,t}^* \Upsilon_{2,t}^{-1} [\hat{x}_{1,t}^{t^*}, \hat{x}_{2,t}^{t^*}, \ldots, \hat{x}_{t,t}^{N^*}], \tag{6.67}$$

where $e_{2,t} = [I_{n,2}, \ldots, I_{n,2}]$ is an $n_2 N \times n_2$ matrix; $\Upsilon_{2,t} = P_{2,t|t}^{1,2}$ is an $n_2 N \times n_2 N$ positive definite matrix. Variances of $\hat{x}_{1,t|t}^{\bar{\omega}}$ and $\hat{x}^{\bar{\omega}}2, t|t$ are given by

$$P_{1,t|t}^{\bar{\omega}} = (e_{1,t}^* \Upsilon_{1,t}^{-1} e_{1,t})^{-1}, \quad P_{1,t|t}^{\bar{\omega}} = (e_{2,t}^* \Upsilon_2^{-1} e_2)^{-1}, \tag{6.68}$$

where $P_{1,t|t}^{\bar{\omega}} \leq P_{1,t|t}^{i}$ and $P_{2,t|t}^{\bar{\omega}} \leq P_{2,t|t}^{i}$. Restoring the variances of (6.49)–(6.51) to the main singular system described by (6.46) and (6.47) makes the filter into

$$\hat{x}_{t|t}^{\bar{\omega}} = \mathbb{R}[\hat{x}_{1,t|t}^{\bar{\omega}^*}, \hat{x}_{2,t|t}^{\bar{\omega}^*}]^*. \tag{6.69}$$

The variance of the filtering error of $\hat{x}_{t|t}^{\bar{\omega}}$ in (6.69) can be computed by

$$P_{t|t}^{\bar{\omega}} = \mathbb{R}[\begin{matrix} P_{1,t|t}^{1,2} & P_{12,t|t}^{1,2} \\ P_{21,t|t}^{1,2} & P_{2,t|t}^{1,2} \end{matrix}]\mathbb{R}^* \tag{6.70}$$

where covariance matrices $P^{1,2}_{1,t|t}$ and $P^{1,2}_{2,t|t}$ are computed by (6.58) and (6.59), respectively. The covariance matrix $P^{1,2}_{1,t|t}$ between filtering errors $\tilde{x}^{\bar{\omega}}_{1,t|t}$ and $\tilde{x}^{\bar{\omega}}_{2,t|t}$ can then be defined as

$$P^{\bar{\omega}1,2}_{t|t} = P^{\bar{\omega}}_{t|t} e^{*}_{1} \Upsilon^{-1}_{1,t} \Upsilon^{1,2}_{t} \Upsilon^{-1}_{2,t} e_2 P^{\bar{\omega}}_{2,t|t}, \tag{6.71}$$

where $P^{\bar{\omega}1,2}_{t|t} = P^{\bar{\omega}^*_{1,2}}_{t|t}$ and $\Upsilon^{1,2}_{t} = P^{1,2}_{t|t}$, which can be computed as follows:

$$P^{\bar{\omega}1,2}_{t|t} = (I_{n_1} - \beta^1_t \bar{H}^1_t) P^{1,2}_{1,t|t-1} F^{2^*}_t + [0, -\beta^1 1_t] Q^{1,2} D^{2^*}, \tag{6.72}$$

where $P^{1,2}_{t|t} = P^{2,1^*}_{t|t}$. Likewise, the variance of smoothing error of $\hat{x}^S_{t|t}$ is computed by (6.63).

The developed diagonalized interacted filters will be used to determine the initial correlation information. However, up to now, the initial correlation information has not considered the constraints outlined in (6.40). This means that the initial correlation will only be good enough to give the first estimates of the oscillation monitoring procedure, where its performance is enhanced by computing the interaction parameter between the subsystems. To take the constraints into account, the maximum a posteriori (MAP) estimate shall be calculated. Note that by handling the noise and state constraints of (6.40), the immunity of the estimation results during data injection can be increased. To achieve this, an MHE is proposed. It involved a state prediction stage to mitigate data-injection attacks.

6.3.5 Moving Horizon Estimate

Given the observation measurement sequence $(z_1, \ldots, z_T)$ at time t, the MAP criteria for calculating the oscillation estimate with constraints can be expressed as

$$\hat{x}^{MAP}_t = \arg \max_{x_0,\ldots,x_T} P(x_0, \ldots, x_T | z_0, \ldots, z_{T-1}). \tag{6.73}$$

Considering the constraints (6.40) and the observation vector (6.39), the log-likelihood is implemented with state variables as

$$= \arg \min_{x_0,\ldots,x_T} \sum_{t=0}^{T-1} ||v||^2_{R_t^{-1}} + ||w_t||^2_{Q_k^{-1}} + ||x_0 + \bar{x}_0||^2_{P_0^{-1}}. \tag{6.74}$$

Considering (6.74), the minimization problem can be formulated as

$$\min_{x_0, w-t, v_t} \sum_{t=0}^{T-1} L_t(w_t, v_t) + \Theta(x_0), \tag{6.75}$$

where L_t and are positive functions, $L_t(w_t, v_t) = ||v||^2_{R_k^{-1}} + ||w_t||^2_{Q_t^{-1}}$ and $\Theta(x_0) = ||x_0 - \hat{x}_0||^2_{P_0^{-1}}$. The MAP estimate from each local ith sensor can then be gathered. Using previously derived expressions and computed information, the track fusion architecture can now be established.

6.3.6 Track Fusion Center

The TFC functions to estimate oscillatory parameters from all the local monitoring nodes in the presence of data-injection attacks. Its purpose is to improve the accuracy of the covariance and estimated states in each node. Subsequently, all local sensor observations from N sensors are integrated into the track observation vector $z_t^{TF} \in \mathbb{R}^{pTF}$. The superscript TF denotes the track fusion, and pTF is the number of track fusion-based observation measurements collected from N sensors. Thus, the track fusion-based observation model at time instant t can be represented as

$$z_t^{TF} = H_t^{TF} x_t + w_t^{TF}. \tag{6.76}$$

Similar to (6.39), the corresponding observation model is H_t^{TF}, and the noise vector is w_t^{TF}. They can also be expressed as an array of information collected from all substations as $z_t^{TF} = [z_t^1, \ldots, z_t^N]^*$, $H_t^{TF} = [H_t^1, \ldots, H_t^N]^*$, and $w_t^{TF} = [w_t^1, \ldots, w_t^N]^*$, where N is the number of sensors. Considering the track estimation-based variables z_t^{TF}, H_t^{TF}, and w_t^{TF}, the oscillation state estimate at TFC can be presented as

$$\hat{x}_{t|t}^{TF} = P_{t|t}^{TF} \sum_{i=1}^{N} P^{i-1_{t|t}} \hat{x}_{t|t}^{i}, \tag{6.77}$$

where $P_{t|t}^{TF}[\sum_{i=1}^{N} P_{t|t}^{i-1}]^{-1}$. Apart from calculating the crosscovariance of subsystems at each node, TFC also calculates the interactions of neighboring sensors. Considering the interactions between local sensors, the covariance matrix for the ith and jth sensors can be expressed as

$$P_{t|t}^{ij} = \mathbb{E}[\tilde{x}_{t|t}^{i} \tilde{x}_{t|t}^{j*}] = [1 - w_t^i H_t^i] P_{t|t-1}^{ij} [1 - w_t^j H_t^j]^*, \tag{6.78}$$

where $\tilde{x}_{t|t} = x_{t|t} - \hat{x}_{t|t}$. This is derived based on the same principles as the covariance of the subsystems within one monitoring node. Hence, $P^{ij}_{t|t-1}$ can be calculated based on the diagonalized subsystem variance by (6.70), while its smoothed variance by (6.63). The TFC will provide estimation of the oscillation parameters in the presence of data injections. To detect the occurrence of injected data, residuals can be continuously generated and evaluated from each sensor. Note that the TFC receives tracked measurements from each sensor node. This may cause processing and communication delays between the local sensors and fusion center. This delay has been tackled by the model-prediction property of the proposed scheme. It can be observed from (6.73)–(6.75) that the MHE considers the whole time sequence for the calculation of model prediction. This idea covers any time delays which are actually measurement delay less than, equal to, or more than one sampling period. Moreover, to tackle this problem at a large scale, the delayed measurements can be determined by deriving the crosscovariance for each delayed measurement and at each time interval.

6.3.7 Evaluation of Residuals

The residual of the estimated parameters is generated to detect any variations due to system bias and injected faults. To detect variations from a residual generation for each measurement, there exists an L_0 such that for any norm bounded $x_{1,t}, x_{2,t} \in \mathbb{R}^n$, the inequality $||(u_t, z_t, x_{1,t}) - (u_t, z_t, x_{2,t})|| \leq L_0||x_{1,t} - x_{2,t}||$ holds. Considering the simplified form of system as (6.46), the transfer function matrix $H_t[sI - (\kappa_t - K_tH_t)]^{-1}\Psi_t$ is strictly positive real, where $K_t \in \mathbb{R}^{n\times r}$ is chosen such that $A_t - K_tH_t$ is stable. Thus, the following expression is constructed:

$$\hat{x}_t = \kappa\hat{x}_t + (u_t, z_t)\xi_{f,t}(u_t, z_t, \hat{x}_t) + K_t(z_t - \hat{z}_t) \tag{6.79}$$

where $\xi_t \in \mathbb{R}$ is a parameter that changes unexpectedly when a fault occurs, K_t is the gain matrix; $\hat{z}_t = H_t\hat{x}_t$ and $r_t = V(z_t - \hat{z}_t)$, where the variable V is the residual weighting matrix. Since the pair (κ_t, H_t) is assumed to be observable, K_t can be selected to ensure $\kappa_t - K_tH_t$ is a stable matrix. This can be defined as

$$e_{x,t} = x_t - \hat{x}_t, \quad e_{z,t} = z_t - \hat{z}_t. \tag{6.80}$$

Error equations will then become

$$e_{x,t+1} = (\kappa_t - K_tH_t)e_{x,t} + [\xi_t(u_t, z_t, x_t) - \xi_{f,t}(u_t, z_t, \hat{x}_t)], \tag{6.81}$$

$$e_{z,t} = H_t e_{x,t}. \tag{6.82}$$

Once the residual is found, evaluations are required to determine the threshold selection for identifying a fault.

The residual evaluation is performed by a coherence function [43,44]. A function based on magnitude of squared coherence spectrum is employed to determine the fault injection status of a power grid at its outputs. Let $\hat{G}(\omega)$ and $\hat{G}_f(\omega)$ be the estimates of the frequency response of the power grid under normal fault-free and faulty operating output regimes, respectively. Here ω is the frequency in rad/s. The magnitude-squared coherence spectrum of the two signals can be defined as

$$c(\hat{G}(\omega), \hat{G}_f(\omega)) = \frac{|\hat{G}(\omega)\hat{G}_f(\omega)|^2}{|\hat{G}(\omega)|^2 |\hat{G}_f(\omega)|^2}, \tag{6.83}$$

where $c(\hat{G}(\omega), \hat{G}_f(\omega))$ is the magnitude-squared coherence spectrum, and $\hat{G}^*(\omega)$ is the complex conjugate of $\hat{G}(\omega)$. In the presence of noise, a threshold value is estimated to give a high probability of detection and a low probability of false alarms. The test statistic *teststat* is chosen to be the mean value of the coherence spectrum $teststat = \mu_0 1/2(c(\hat{G}(\omega), \hat{G}_f(\omega)))$ as

$$test_{stat} = \begin{cases} \leq th & \forall \omega \in \Omega, \quad \text{fault}, \\ > th & \forall \omega \in \Omega, \quad \text{no fault}, \end{cases} \tag{6.84}$$

where $0 \leq th \leq 1$ is a threshold value, Ω is the relevant spectral region, e.g., bandwidth. This gives the coherence function-based thresholds for detection of fault injections.

6.3.8 Simulation Studies

Validation of the proposed TFMP estimation scheme is conducted using simulated synchrophasor measurements collected from the IEEE 39-Bus New England system as shown in Fig. 6.15. Modeling details are based on [45] and [46]. In these papers, synchrophasor measurements are collected from buses 15–17, 29, 30, 35, and 37–39. From these data, three dominant electromechanical modes are detected using Welch power spectral density. Their predisturbance values are: (1) 0.69 Hz with a damping ratio of 3.90%; (2) 1.12 Hz with a damping ratio of 5.71%; and (3) 1.17 Hz with a damping ratio of 5.62%. The 0.69 Hz mode will be considered as an interarea oscillation. All loads are continuously being subjected to random small magnitude fluctuations of up to 10 MW/s. Furthermore, the

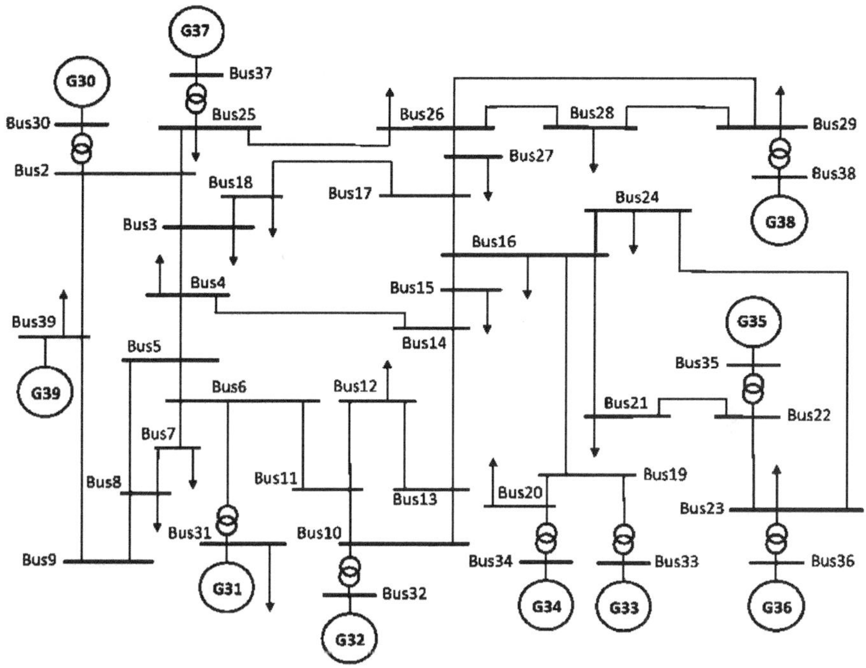

Figure 6.15 Single line diagram of the IEEE 39-Bus New England System

system is excited by four large-signal disturbances over a period of 60 s. First, a three-phase-to-ground fault occurred at bus 24 at 5 s and is cleared after 0.1 s. Second, the active and reactive power demands of the load connected at bus 21 is ramped up by 30% and 10% over 10 s, respectively. Third, the line connecting buses 16 and 17 is disconnected at 25 s and reconnected after 5 s. Lastly, the active and reactive load demands at bus 4 are increased by 20% and 10%, respectively. This occurred over a 5 s ramp. All simulations are performed using DIgSILENT PowerFactory Ver. 15.1 [47]. From the collected measurements, monitoring schemes updated the averaged oscillatory parameters every 5 s. In this example, the proposed method is evaluated against the distributed technique of [36]. See Fig. 6.15.

To simulate deliberate attack scenarios, data injections are carried out in the collected synchrophasor measurements. Since all three electromechanical modes are observable at buses 16 and 17, these two locals are selected as attack nodes. Their neighboring nature as shown in Fig. 6.16 helped to create a situation of regional attacks on measured data. Simulated attack

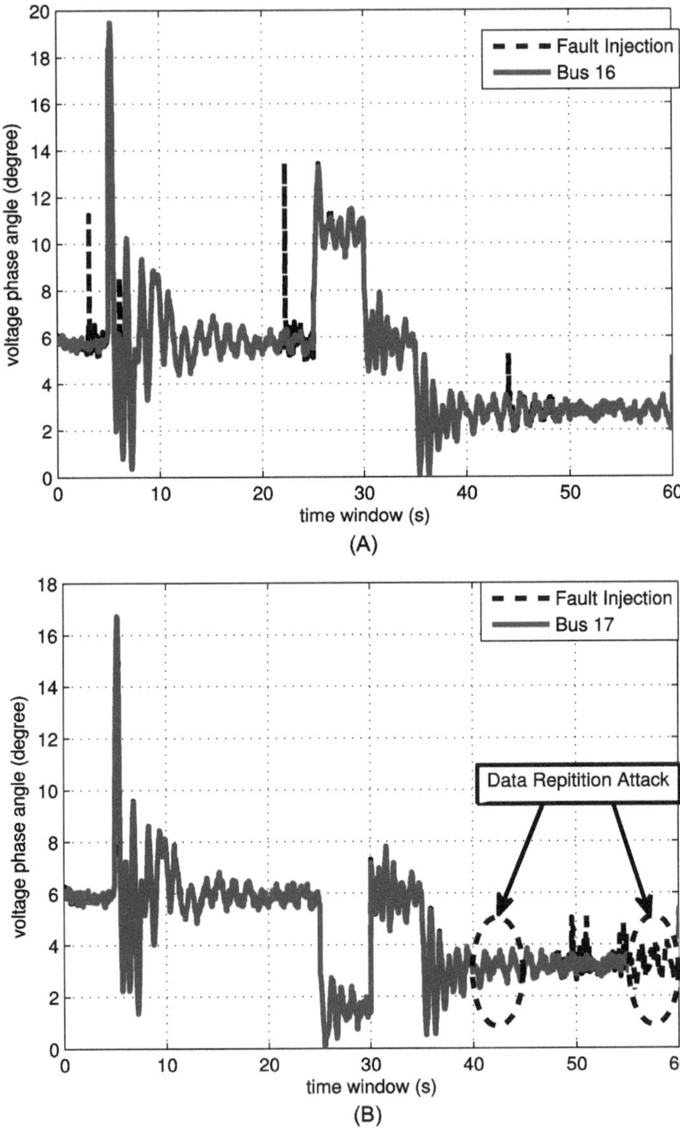

Figure 6.16 Random fault injections profile: (A) bus 16 and (B) bus 17

scenarios at buses 16 and 17 are as follows:

1. First Injection. Random data injections are introduced at bus 16 from 7 to 12 s.
2. Second Injection. Signals with relatively high energy potency are injected at bus 16 from 22 to 27 s.

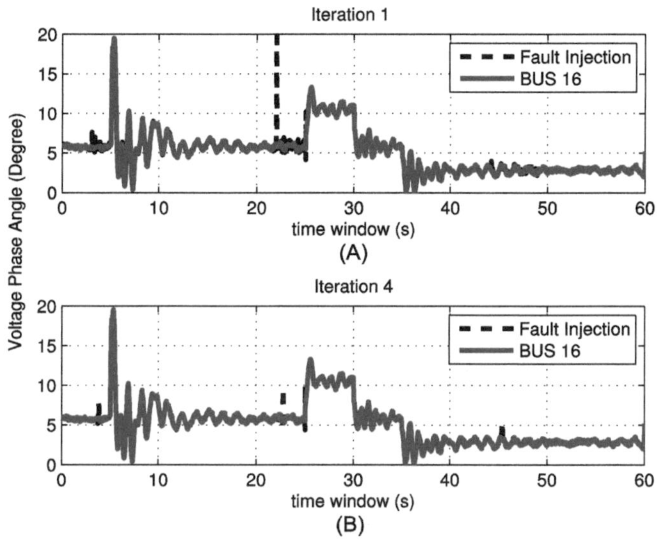

Figure 6.17 Performance of the proposed method for bus 16 at a particular iteration: (A) 1; (B) 4

3. Third Injection. Small signatures of random sinusoidal waveforms are introduced at bus 16 from 44 to 49 s. Also, ambient disturbance-like injections are introduced at bus 17 from 48 to 55 s.
4. Fourth Injection. Small signatures of random sinusoidal waveforms are introduced at bus 16 from 44 to 49 s. Also, a data-repetition attack was introduced at bus 17 from 55 to 60 s. This attack replaces the normal oscillation behavior with that recorded at bus 17 from 40 to 45 s.

The first injection illustrates a pure random attack with no bias toward any signal characteristics. The second injection imitates an attack attempting to bring down a local/regional network. The third injection represents an ambient attack with the aim to generate a cascading failure in the longer form that can often led to wide-area blackouts. Ambient disturbance-like injections are also introduced at bus 17 to create a multisensor injection attack situation as part of the third injection scenario. The fourth injection represents a data-repetition attack at bus 17. The purpose of injecting different nature of signals and at multiple locations are to assess the robustness of the proposed scheme. These data segments, outlined in dashed black line, are added to the original synchrophasor measurements, depicted in solid red line, as shown in Fig. 6.16A and B. See Fig. 6.17.

Table 6.4 Estimated case I–New England System: Detecting multiple oscillations in the presence of random data-injection attacks.

Measurements	ζ_D	f_D	ζ_{TFMP}	f_{TFMP}	ζ_D	f_D	ζ_{TFMP}	f_{TFMP}	ζ_D	f_D	ζ_{TFMP}	f_{TFMP}	ζ_D	f_D	ζ_{TFMP}	f_{TFMP}
Time	**0 s–5 s**				**5 s–10 s**				**10 s–15 s**				**15 s–20 s**			
	4.1	0.69	3.7	0.69	3.9	0.73	3.8	0.70	4.0	0.69	3.8	0.69	4.2	0.68	3.9	0.69
	6.0	1.07	5.9	1.07	–	–	5.9	1.05	5.5	1.12	5.7	1.12	5.9	1.07	5.9	1.07
	5.8	1.12	5.7	1.13	5.4	1.17	5.5	1.17	5.4	1.17	5.6	1.17	5.7	1.11	5.8	1.12
MSE	2.1×10^{-2}		1.8×10^{-3}		5.5×10^{-1}		4.8×10^{-2}		2.5×10^{-2}		3.1×10^{-3}		2.2×10^{-2}		2.1×10^{-3}	
Time	**20 s–25 s**				**25 s–30 s**				**30 s–35 s**				**35 s–40 s**			
	3.9	0.71	3.9	0.69	4.0	0.72	3.9	0.68	4.3	0.69	3.9	0.69	4.0	0.69	3.9	0.69
	5.8	1.11	5.6	1.11	5.9	1.07	5.9	1.07	5.7	1.08	5.8	1.08	5.6	1.11	5.6	1.11
	–	–	5.6	1.17	5.7	1.12	5.7	1.14	5.2	1.12	5.7	1.13	5.5	1.15	5.5	1.16
MSE	2.1×10^{-1}		2.4×10^{-3}		4.2×10^{-1}		3.6×10^{-3}		5.5×10^{-2}		2.5×10^{-3}		4.5×10^{-2}		3.9×10^{-3}	
Time	**40 s–45 s**				**45 s–50 s**				**50 s–55 s**				**55 s–60 s**			
	4.3	0.70	3.9	0.69	4.3	0.69	3.9	0.70	4.5	0.66	3.9	0.69	4.4	0.69	3.9	0.69
	–	–	5.6	1.11	–	–	5.8	1.10	5.7	1.13	5.9	1.09	–	–	5.6	1.11
	5.2	1.14	5.6	1.15	5.8	1.14	5.7	1.14	5.9	1.00	5.7	1.14	5.1	1.11	5.6	1.17
MSE	8.3×10^{-1}		6.8×10^{-3}		6.5×10^{-1}		5.7×10^{-3}		2.2×10^{-1}		1.9×10^{-3}		9.8×10^{-1}		2.7×10^{-3}	

In this table, ζ is the damping ratio i.e. $\zeta = \frac{-\sigma}{\sqrt{\sigma^2+(2\pi f)^2}} \times 100$. f is the frequency in hertz, MSE is the mean-square error, subscript D and TFMP are the acronyms of Distributed approach [18] and the proposed TFMP, respectively.

The monitoring performance is summarized in Table 6.4. First, the tracking performance in windows without the presence of data injections is discussed. Overall, both methods are able to accurately estimate all three electromechanical parameters. A slight increase in mean squared error (MSE) values for the distributed method is observed between 30 and 40 s period. They are in parameters with the case of the line outage event in the previous time window. The reason can be due to the dominance of nonlinear dynamics in the measurements, which caused the linear-based distributed monitoring scheme to struggle. In contrast, the proposed scheme is less influenced due to its crosscovariance computation at each local sensor. Overall, both methods generated low MSE while the proposed TFMP estimation scheme achieved higher accuracy. Next, the performance under deliberate data injections is analyzed as follows. The first injection scenario consisted of a few large spikes spread across two monitoring windows. As a result, the accuracy of 5–10 and 10–15 s windows is impacted. Since the larger spike and the three-phase-to-ground fault occurred in the 5–10 s period, both methods incurred their highest MSE values. However, the proposed scheme is still able to provide oscillatory parameters with adequate precision whereas the distributed method failed to track one electromechanical mode. This is due to initial estimates collected from the interaction of neighboring sensors. In the second injection scenario, the system contained less nonlinear dynamics and oscillations were more dominant in measurements. Here, the largest spike was introduced during the 20–25 s window, which caused the distributed scheme to fail to track one oscillation. Although high energy signals were injected, they did not flood the entire monitoring window. Hence, the distributed scheme managed to track all three oscillations in the following time window. Nevertheless, the lowest frequency mode (0.69 Hz) incurred noticeable estimation errors. For the proposed TFMP, the removal of abnormal data segments through subsystem diagonalization helps to avoid them from incurring errors into the estimation stage. As a result, the proposed scheme maintained similar estimation accuracy as in the case of without data injections. Next, the third injection scenario is examined. Since the injected measurements from buses 16 and 17 contained amplitudes and characteristics similar to collected synchrophasor measurements, extracting oscillatory parameters is more challenging than previous scenarios. This can be reflected by the consecutive high MSE values generated by both methods during the time of 40–50 s. The filtering stage of the distributed method was not able to remove injected data, which caused it to lose track of one oscillatory mode due to slow convergence of EM. In contrast, the proposed TFMP estimation scheme still computed accurate oscillation parameters. An interesting observation is made during the sole injection of ambient data at bus 17 throughout the entire 50–55 s window. Despite the fact that the distributed method from [36] has detected all three oscillations, the frequency and associated damping factor of the interarea oscillation (0.69 Hz mode) incurred most errors compared with all other windows. The estimated incorrect higher damping ratio can delay subsequent damping strategies and reduce the effectiveness of the system damping capability. As a result, a cascading failure leading to wide-area blackouts can potentially occur at a later stage. In comparison, the proposed TFMP estimation was able to mitigate such abnormalities and maintained reasonable estimation accuracy for all oscillatory modes. The removals of ambient grid-like dynamics are illustrated in Fig. 6.17 for bus 16. Referring to these plots, the proposed scheme iteratively minimized data abnormalities by removing them as outliers using the derived crosscovariance relationships. Finally, the fourth injection scenario is presented. There is a more realistic and challenging scenario to mitigate the previous events. During the attack at bus 16 from 44 to 49 s, both schemes performed well due to their property of retrieving missing measurements to make an accurate estimation of oscillations. However, during the data-repetition attack at bus 17 from 55 to 60 s window, the distributed scheme of [36] was unable to predict the model and tackle the noise variances independently. This resulted in a high MSE value generated by the distributed algorithm. Moreover, the algorithm was also not able to track one of the nearby oscillation modes. In contrast, the proposed TFMP scheme estimated all oscillations accurately. The incorporation of the model prediction demonstrates that using MHE and calculating the covariances of each local sensor helps to achieve better MSE values. To observe the impact of bus 16 during all these data injections, an MSE-based comparison has been made between the proposed TFMP scheme, the distributed scheme of [36], and the distributed scheme proposed of [38]. Results are shown in Fig. 6.18, where all three schemes performed reasonably well. However, if the level of precision is considered, the proposed TFMP superseded the other two schemes especially when estimating the oscillatory modes between 40 and 60 s. Once the estimation accuracy is achieved, the residuals are generated to quantify any system variations. Referring to Fig. 6.19, all attacks have been detected using the threshold selection by a coherence function. The threshold selected for buses 16 and 17 residuals were ± 5 and ± 10, respectively, using the coherence spectrum function. As observed

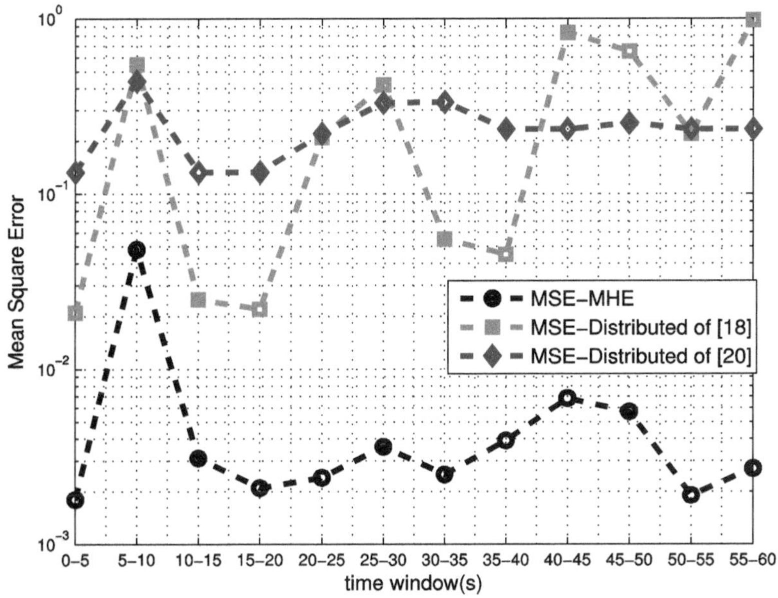

Figure 6.18 Estimation comparison analysis of different methods at bus 16

in Fig. 6.19A, some wiggles within the threshold limits correspond to the dynamics of real-time system. However, they can also be mistaken as system faults if inappropriate thresholds are selected. In this case, the threshold selection algorithm is adequate to detect the system faults while avoiding the false alarms. Overall, residuals exceeding the thresholds coincide with the time of data-injection events. Meanwhile, the residual profile of bus 17 shows less variations as observed in Fig. 6.19B. This is because the location has been subjected to less data-injection attacks than bus 16. Nevertheless, the more challenging data repetition attack has been well-detected by the coherence function-based threshold. See Fig. 6.19.

6.4 NOTES

In the context of strong development and investment in smart grid research, interoperability of microgrid platforms, particularly in research institutions, is necessary to enable the collaboration and information exchange among research and industrial institutions. Interoperability of microgrid platforms provides a common support for multisite research and development projects.

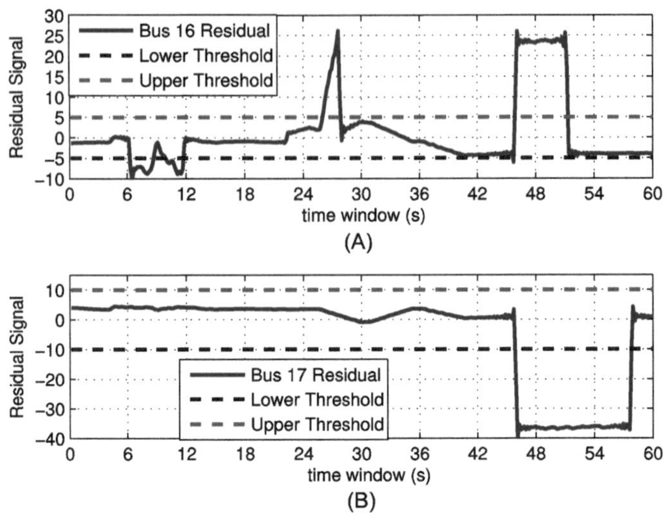

Figure 6.19 Fault residual evaluation in (A) bus 16 and (B) bus 17

In the first section, a novel hybrid cloud-based SCADA architecture is proposed as a support for the interoperability of microgrid platforms in research and industrial institutions. Providing the benefits of hybrid-cloud SCADA and PaaS delivery model, this architecture can be common research infrastructure for the partners of the working network. Efficient security control measures should be considered to ensure a secured data exchange and interoperability of the platforms.

CIM, over OPC UA protocol, is chosen as the common information model in the architecture for the partners, to bring semantics to the exchanged data and to assure a mutual understanding among partner platforms.

In the second section, we discussed networked control systems used in power systems that share information via a variety of communication protocols, making them vulnerable to attacks by hackers at any infrastructure point. In this section, different types of time delay switch attacks were the main focus. The CF-TDSR, a communication protocol that uses adaptive channel redundancy techniques, as well as a novel state estimator, was developed to detect and obviate unstable effects of a TDS attack. It was demonstrated that the CF-TDSR enabled the linear time-invariant control systems to remain stable. The simulation experiments show that CF-TDSR enables the multiarea load frequency control component to quickly stabilize the system under a suite of TDS attacks.

Next, a proposed TFMP-based monitoring scheme is proposed and demonstrated to estimate power oscillation modes during data-injection attacks. The model prediction property of the algorithm has helped to remove bias and noise while accurately extracting the system parameters. It is further facilitated by the derived diagonalized interaction filter, which tackles the error covariance in the form of subsystems, and thus improves the initial oscillatory state estimates. As a result, the incorporation of the proposed algorithm into oscillation detection has provided more accurate results than existing oscillation monitoring schemes in the presence of data-injection attacks. The immunity of monitoring applications against intentional data injections has been enhanced. In the future, studies to quantitatively verify the effectiveness and robustness of the proposed method to more adverse nonregional threats will be conducted.

Finally, a Bayesian-based approximation filter has been proposed and demonstrated to improve the immunity of the monitoring applications against data-injection attacks. The predictive distribution property of the algorithm has helped to monitor power oscillation even in the presence of information loss. Mathematical derivations demonstrated the ability to identify attacks through interactions with neighboring monitoring nodes. In this paper, the proposed scheme has been applied to a mature wide-area monitoring application known as oscillation detection. Manipulating recorded and simulated measurements collected from Phasor Measurement Unit, the proposed method was able to extract accurate oscillatory parameters in the presence of data-injection attacks.

REFERENCES

[1] F.E. Catalin, A. Miicea, V. Julija, M. Anna, F. Gianluca, A. Eleterios, Smart Grid Projects Outlook 2014, JRC Science and Policy Reports, European Commission – Joint Research Center, 2014.

[2] D.S. Markovic, D. Zivkovic, I. Branovic, R. Popovic, D. Cvetkovic, Smart power grid and cloud computing, Renew. Sustain. Energy Rev. 24 (2013) 566–577.

[3] M. Peter, G. Timothy, The NTST Definition of Cloud Computing, Recommendations of the National Institute of Standards and Technology Special Publication 800-145, National Institute of Standards and Technology, Sept. 2011.

[4] E. Electric Power Research Institute, Common Information Model Primer: Third Edition, Technical Results 3002006001, Electric Power Research Institute, Jan. 2015.

[5] M. Gao, N. Wang, A network intrusion detection method based on improved K-means algorithm, Adv. Sci. Technol. Lett. 53 (2013) 429–433.

[6] S. Shin, S. Lee, H. Kim, S. Kim, Advanced probabilistic approach for network intrusion forecasting and detection, Expert Syst. Appl. 40 (1) (2013) 315–322.

[7] D. Lin, Network Intrusion Detection and Mitigation Against Denial of Service Attack, MS-CIS-13-04, University of Pennsylvania, Department of Computer & Information Science, Jan. 2013.
[8] W. Xu, W. Trappe, Y. Zhang, T. Wood, The feasibility of launching and detecting jamming attacks in wireless networks, in: 6th ACM Int. Symposium on Mobile Adhoc Networking and Computing, 2005.
[9] S. Shapsough, F. Quaten, R. Aburukba, F. Aloul, Smart grid cyber security: challenges and solutions, in: International Conference on Smart Grid and Clean Energy Technologies, Offenburg, 2015.
[10] W. Wang, Z. Lu, Cyber security in the smart grid: survey and challenges, Comput. Netw. 57 (5) (2013) 1344–1371.
[11] E. Pooper, M. Strasser, S. Capkun, Anti jamming broadcast communication using uncoordinated spread spectrum techniques, IEEE J. Sel. Areas Commun. 28 (5) (2010) 703–715.
[12] M. Line, J. Tondel, M. Jaatun, Cyber security challenges in smart grids, in: ISGT Europe, Manchester, 2011.
[13] L. Schenato, Optimal estimation in networked control systems subject to random delay and packet drop, IEEE Trans. Autom. Control 53 (5) (2008) 1311–1317.
[14] M.S. Mahmoud, Robust Control and Filtering for Time-Delay Systems, CRC Press, 2000.
[15] I. Kamwa, R. Grondin, Y. Hébert, Wide-area measurement based stabilizing control of large power systems – a decentralized/hierarchical approach, IEEE Trans. Power Syst. 16 (1) (2001) 136–153.
[16] H. Wu, K.S. Tsakalis, G.T. Heydt, Evaluation of time delay effects to wide-area power system stabilizer design, IEEE Trans. Power Syst. 19 (4) (2004) 1935–1941.
[17] B. Chaudhuri, R. Majumder, B.C. Pal, Wide-area measurement-based stabilizing control of power system considering signal transmission delay, IEEE Trans. Power Syst. 19 (4) (2004) 1971–1979.
[18] F. Milano, M. Anghel, Impact of time delays on power system stability, IEEE Trans. Circuits Syst. I, Regul. Pap. 59 (4) (2012) 889–900.
[19] S. Ray, G.K. Venayagamoorthy, Real-time implementation of a measurement-based adaptive wide-area control system considering communication delays, IET Gener. Transm. Distrib. 2 (1) (2008) 62–70.
[20] M.T. Alrifai, M. Zribi, M. Rayan, M.S. Mahmoud, On the control of time delay power systems, Int. J. Innov. Comput. Inf. Control 9 (2) (2013) 769–792.
[21] L. Chunmao, X. Jian, Adaptive delay estimation and control of networked control systems, in: Int. IEEE Symposium Communications and Information Technologies, ISCIT'06, 2006, pp. 707–710.
[22] A. Sargolzaei, K.K. Yen, M.N. Abdelghani, Preventing time-delay switch attack on load frequency control in distributed power systems, IEEE Trans. Smart Grid 7 (2) (2016) 1176–1185.
[23] L. Jiang, W. Yao, Q. Wu, J. Wen, S. Cheng, Delay-dependent stability for load frequency control with constant and time-varying delays, IEEE Trans. Power Syst. 27 (2) (2012) 932–941.
[24] H. Bevrani, Robust Power System Frequency Control, Power Electron. Power Syst., vol. 85, Springer, 2009.
[25] T.T. Kim, H.V. Poor, Strategic protection against data-injection attacks on power grids, IEEE Trans. Smart Grid 2 (2) (June 2011) 326–333.

[26] M. Ozay, I. Esnaola, F.T.Y. Vural, S.R. Kulkarni, H.V. Poor, Sparse attack construction and state estimation in the smart grid: centralized and distributed models, IEEE J. Sel. Areas Commun. 31 (7) (July 2013) 1306–1318.
[27] S. Sridhar, A. Hahn, M. Govindarasu, Cyber-physical system security for the electric power grid, Proc. IEEE 100 (1) (Jan. 2012) 210–224.
[28] O. Kosut, L. Jia, R.J. Thomas, L. Tong, Malicious data attacks on the smart grid, IEEE Trans. Smart Grid 2 (4) (Dec. 2011) 645–658.
[29] S. Cui, et al., Coordinated data-injection attack and detection in the smart grid: a detailed look at enriching detection solutions, IEEE Signal Process. Mag. 29 (5) (Sept. 2012) 106–115.
[30] L. Xie, Y. Mo, B. Sinopoli, Integrity data attacks in power market operations, IEEE Trans. Smart Grid 2 (4) (Dec. 2011) 659–666.
[31] G. Rogers, Power System Oscillations, Kluwer, Boston, MA, USA, 2000.
[32] J.J. Sanchez-Gasca, J.H. Chow, Performance comparison of three identification methods for the analysis of electromechanical oscillations, IEEE Trans. Power Syst. 14 (3) (Aug. 1999) 995–1002.
[33] S.A.N. Sarmadi, V. Venkatasubramanian, Electromechanical mode estimation using recursive adaptive stochastic subspace identification, IEEE Trans. Power Syst. 29 (1) (Jan. 2014) 349–358.
[34] R.W. Wies, J.W. Pierre, Use of least-mean square (LMS) adaptive filtering technique for estimating low-frequency electromechanical modes of power systems, in: Proc. Amer. Control Conf., vol. 6, Anchorage, AK, USA, May 2002, pp. 4867–4873.
[35] J.C.-H. Peng, N.C. Nair, Enhancing Kalman filter for tracking ring down electromechanical oscillations, IEEE Trans. Power Syst. 27 (2) (May 2012) 1042–1050.
[36] H.M. Khalid, J.C.-H. Peng, Improved recursive electromechanical oscillations monitoring scheme: a novel distributed approach, IEEE Trans. Power Syst. 30 (2) (Mar. 2015) 680–688.
[37] D.N. Kosterev, C.W. Taylor, W.A. Mittelstadt, Model validation for the August 10, 1996 WSCC system outage, IEEE Trans. Power Syst. 14 (3) (Aug. 1999) 967–979.
[38] L. Xie, D.-H. Choi, S. Kar, H.V. Poor, Fully distributed state estimation for wide-area monitoring systems, IEEE Trans. Smart Grid 3 (3) (Sep. 2012) 1154–1169.
[39] H.B. Mitchell, Multi-Sensor Data Fusion: An Introduction, Springer, New York, NY, USA, 2007.
[40] V.L. Syrmos, F.L. Lewis, Robust eigenvalue assignment for generalized systems, Automatica 28 (6) (1992) 1223–1228.
[41] U. Shaked, C.E. de Souza, Robust minimum variance filtering, IEEE Trans. Signal Process. 43 (11) (Nov. 1995) 2474–2483.
[42] Z.L. Deng, Kalman Filtering and Wiener Filtering – Modern Time Series Analysis Method, Harbin Inst. Technol., Harbin, China, 2001.
[43] T. Yanagisawa, H. Takayama, Coherence coefficient measuring system and its application to some acoustic measurements, Appl. Acoust. 16 (2) (Mar. 1983) 105–119.
[44] C. Zheng, H. Yang, X. Li, On generalized auto-spectral coherence function and its applications to signal detection, IEEE Signal Process. Lett. 21 (5) (May 2014) 559–563.
[45] B. Pal, B. Chaudhuri, Robust Control in Power Systems, Springer, New York, NY, USA, 2005.
[46] J.C.-H. Peng, J.L. Kirtley, An improved empirical mode decomposition method for monitoring electromechanical oscillations, in: Proc. IEEE PES Innov. Smart Grid Technol. Conf., ISGT, Washington, DC, USA, 2014, pp. 1–5.
[47] DIgSILENT, Power Factory 15 User Manual, Gomaringen, Germany, 2013.

CHAPTER 7

Smart Grid Infrastructures

7.1 CYBERPHYSICAL SECURITY

The electric grid is arguably the world's largest engineered system. Vital to human life, its reliability is a major and often understated accomplishment of the humankind. It is the motor of the economy and a major driver of progress. In its current state, the grid consists of four major components:

1. Generation produces electric energy in different manners, e.g., by burning fossil fuels, inducing nuclear reaction, harnessing water (hydroelectric dams), employing wind, solar, and tidal forces;
2. Transmission moves electricity via a very high voltage infrastructure;
3. Distribution steps down current and spreads out for consumption; and
4. Consumption, which can be industrial, commercial, and residential, uses electric energy in a multitude of ways.

7.1.1 Introduction

Given the wide variety of systems, their numerous owners, and a diverse range of regulators, a number of weaknesses have emerged.

Outages are often recognized only after consumers report. Matching generation to demand is challenging because utilities do not have clear cut methods to predict demand and to request demand reduction (load shedding). As a consequence, they need to overgenerate power for peak demand, which is expensive and contributes to greenhouse gas (GhG) emissions. For similar reasons it is difficult to incorporate variable generation, such as wind and solar power, into the grid. Last, there is a dearth of information available for consumers to determine how and when to use energy.

To address these challenges, the smart grid concept has evolved. The smart grid uses communications and information technologies to provide better situational awareness to utilities regarding the state of the grid. Smart grid provides numerous benefits [1–4]. Using intelligent communications, load shedding can be implemented so that peak demand can be flattened, which reduces the need to bring additional (expensive) generation plants online. Using information systems to perform predictive analysis, including when wind and solar resources will produce less power, the utilities can

Networked Control Systems
https://doi.org/10.1016/B978-0-12-816119-7.00015-0

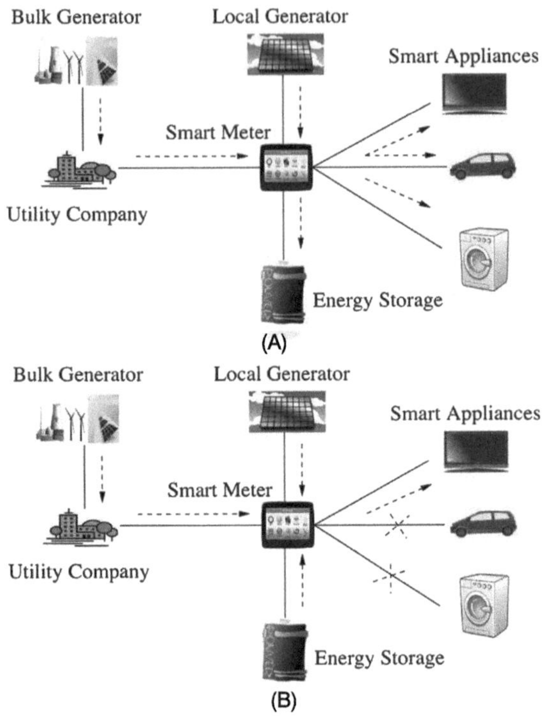

Figure 7.1 (A) Power usage during off-peak time period; (B) Power usage during peak time period

keep power appropriately balanced. As new storage technologies emerge at the utility scale, incorporation of these devices will likewise benefit from intelligent demand prediction. Last, the ability for consumers to receive and respond to price signals will help them manage their energy costs, while helping utilities avoid building additional generation plants.

With all these approaches, the smart grid enables a drastic cost reduction for both power generation and consumption.

7.1.2 Pricing and Generation

Dynamic pricing and distributed generation with local generators can significantly reduce the electricity bill. In particular, Fig. 7.1A shows how to use electricity during off-peak periods when the price is low. Conversely, Fig. 7.1B shows load shedding during peak times and utilization of energy storage to meet customer demand. The effect of peak demand reduction by "demand management" is shown in Fig. 7.2. Pilot projects in the states of

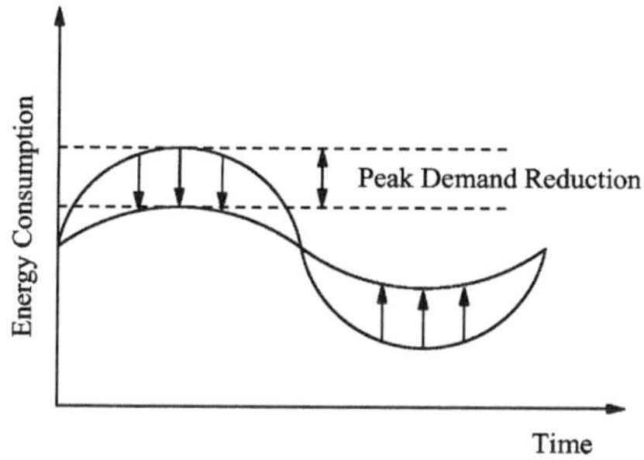

Figure 7.2 The peak demand profile for electricity

California and Washington [1] indicate that scheduling appliances based on price information can reduce electricity costs by 10% for consumers. More advanced smart grid technologies promise to provide even larger savings.

To establish the smart grid vision, widespread sensing and communications between all grid components (generation, transmission, distribution, storage) and consumers must be created and managed by information technology systems. Furthermore, sophisticated estimation, control, and pricing algorithms need to be implemented to support the increasing functionality of the grid while maintaining reliable operations. It is the greatly increased incorporation of IT systems that supports the vision, but unfortunately also creates exploitable vulnerabilities for the grid and its users.

7.1.3 Cyberphysical Approach to Smart Grid Security

A wide variety of motivations exist for launching an attack on the power grid, ranging from economic reasons (e.g., reducing electricity bills), to pranks, and all the way to terrorism (e.g., threatening people by controlling electricity and other life-critical resources). The emerging smart grid, while benefiting the benign participants (consumers, utility companies), also provides powerful tools for adversaries.

The smart grid will reach every house and building, giving potential attackers easy access to some of the grid components. While incorporating information technology (IT) systems and networks, the smart grid will be exposed to a wide range of security threats [5]. Its large scale also makes it nearly impossible to guarantee security for every single subsystem. Furthermore, the smart grid will be not only large but also very complex. It needs to connect different systems and networks, from generation facilities and distribution equipment to intelligent end points and communication networks, which are possibly deregulated and owned by several entities. It can be expected that the heterogeneity, diversity, and complexity of smart grid components may introduce new vulnerabilities, in addition to the common ones in interconnected networks and stand-alone microgrids [3]. To make the situation even worse, the sophisticated control, estimation, and pricing algorithms incorporated in the grid may also create additional vulnerabilities.

The first-ever control system malware called *Stuxnet* was found in July 2010. This malware, targeting vulnerable SCADA systems, raises new questions about power grid security [6]. SCADA systems are currently isolated, preventing external access. Malware, however, can spread using USB drives and can be specifically crafted to sabotage SCADA systems that control electric grids. Furthermore, increasingly interconnected smart grids will unfortunately provide external access which in turn can lead to compromise and infection of components.

Many warnings concerning the security of smart grids are appearing [7–12] and some guidelines have been published, such as NISTIR 7628 [3] and NIST SP 1108 [13]. We argue that a new approach to security, bringing together cybersecurity and system theory under the name of cyberphysical security (CPS), is needed to address the requirements of complex, large-scale infrastructures like the smart grid. In such systems, cyberattacks can cause disruptions that transcend the cyberrealm and affect the physical world. Stuxnet is a clear example of a cyberattack used to induce physical consequences. Conversely physical attacks can affect the cybersystem. For example, the integrity of a meter can be compromised by using a shunt to bypass it. Secrecy can be broken by placing a compromised sensor beside a legitimate one. As physical protection of all assets of large-scale physical systems, such as the smart grid, is economically infeasible, there arises the need to develop methods and algorithms that can detect and counter hybrid attacks. Based on the discussions at the Army Research Office workshop on CPS security in 2009, we classify current attacks on cyberphysical systems into four categories and provide examples to illustrate our classification in Table 7.1. Although cybersecurity and system theory have achieved remarkable success in defending against pure cyber or pure physical attacks, neither of them alone is sufficient to ensure smart grid security, due to hybrid attacks. Cybersecurity is not equipped to provide an analysis of the possible consequences of attacks on physical systems. System theory is usually concerned with properties such as performance, stability, and safety of physical systems. Its theoretical framework, while well consolidated, does not provide a complete modeling of the IT infrastructure.

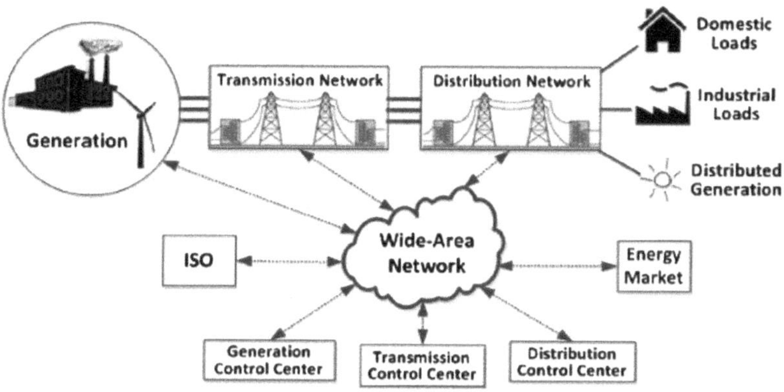

Figure 7.3 Power grid cyberphysical infrastructure

In this section, we explore combining system theory and cybersecurity to ultimately build a science of cyberphysical security. Toward this goal, it is important to develop cyberphysical security models capable of integrating dynamic systems and threat models within a unified framework. We believe that cyberphysical security can not only address problems that cannot be currently solved but also provide new improved solutions for detection, response, reconfiguration, and restoration of system functionalities while keeping the system operating. We also believe that some existing modeling formalisms can be used as a starting point toward a systematic treatment of cyberphysical security. Game theory [14] can capture the adversarial nature of the interaction between an attacker and a defender. Networked control systems [15] aim at integrating computing and communication technologies with system theory, providing a common modeling framework for cyberphysical systems. Finally, hybrid dynamic systems [16] can capture the discrete nature of events such as attacks on control systems.

In the sequel, ingredients of cybersecurity to smart grid security will be delineated.

7.1.4 System Model

As illustrated in Fig. 7.3, power grids consist of four components: generation, transmission, distribution, and consumption. In the consumption component, customers use electric devices (that is, smart appliances, electric vehicles), and their usage of electricity will be measured by an enhanced metering device, called a smart meter. The smart meter is one of the core components of the advanced metering infrastructure (AMI) [17]. The me-

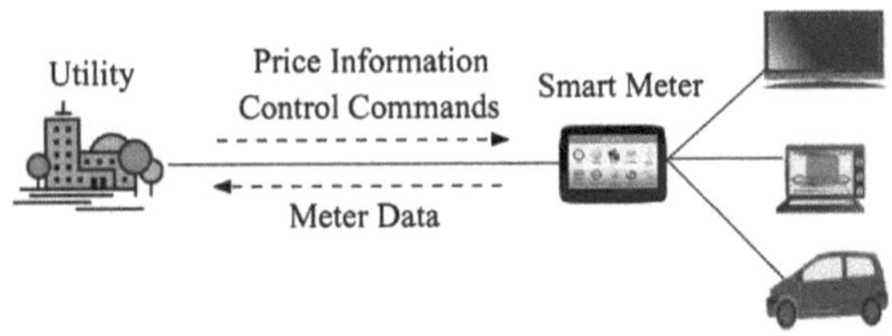

Figure 7.4 Information flows to/from a smart meter including price information, control commands, and meter data

ter can be collocated and interact with a gateway of a home-area network (HAN) or a business-area network (BAN). For a simple illustration, we denote a smart meter in the figure as a gateway of an HAN. A neighbor-area network (NAN) is formed under one substation, where multiple HANs are hosted. Finally, a utility company may leverage a wide-area network (WAN) to connect distributed NANs.

7.1.5 Cybersecurity Requirements

In the sequel, we analyze the information security requirements for smart grids. In general, information security requirements for a system include three main security properties: confidentiality, integrity, and availability. Confidentiality prevents an unauthorized user from obtaining secret or private information. Integrity prevents an unauthorized user from modifying the information. Availability ensures that the resource can be used when requested. As shown in Fig. 7.4, price information, meter data, and control commands are the core information exchanged in smart grids which we consider in this paper.

While more types of information are exchanged in reality, these core information types provide a comprehensive sample of security issues.

We now examine the importance of protecting the core information types with respect to the main security properties. The degree of importance for price information, control commands, and meter data is equivalent to the use cases of NISTIR 7628 [3], to which we added the degree of importance for software. The most important requirement for protecting smart grids are outlined below.

- *Confidentiality of power usage.* Confidentiality of meter data is important, because power usage data provides information about the usage patterns for individual appliances, which can reveal personal activities through non-intrusive appliance monitoring [18]. Confidentiality of price information and control commands are not important in cases where it is public knowledge. Confidentiality of software should not be critical, because the security of the system should not rely on the secrecy of the software, but only on the secrecy of the keys, according to Kirchhoff principle [19].

- *Integrity of data, commands, and software.* Integrity of price information is critical. For instance, negative prices injected by an attacker can cause an electricity utilization spike as numerous devices would simultaneously turn on to take advantage of the low price. Although integrity of meter data and commands is important, their impact is mostly limited to revenue loss. On the other hand, integrity of software is critical since compromised software or malware can control any device and grid component.
- *Availability against DoS/DDoS attacks.* Denial-of-service (DoS) attacks are resource consumption attacks that send fake requests to a server or a network, and distributed DoS (DDoS) attacks are accomplished by utilizing distributed attacking sources such as compromised smart meters and appliances. In smart grids, availability of information and power is a key aspect [20]. More specifically, availability of price information is critical due to serious financial and possibly legal implications. Moreover, outdated price information can adversely affect demand. Availability of commands is also important, especially when turning a meter back on after completing the payment of an electric bill. On the other hand, availability of meter data (e.g., power usage) may not be as critical because the data can usually be read at a later point.

From the above discussion, we can summarize the importance of data, commands, and software, which are shown in Table 7.2. "High" risk implies that a property of certain information is very important/critical, and "medium" and "low" risks classify properties that are important and noncritical, respectively. This classification enables prioritization of risks, to focus effort on the most critical aspects first. For example, integrity of price information is more important than its confidentiality; consequently, we need to focus on efficient cryptographic authentication mechanisms before encryption.

7.1.6 Attack Model

To launch an attack, an adversary must first exploit entry points, and upon successful entry, an adversary can deliver specific cyberattacks on the smart grid infrastructure. In the following sections, we describe this attacker model in detail.

7.1.6.1 Attack Entry Points

In general, strong perimeter defense is used to prevent external adversaries from accessing information or devices within the trusted grid zone.

Unfortunately, the size and complexity of grid networks provide numerous potential entry points as follows.

- *Inadvertent infiltration through infected devices.* Malicious media or devices may be inadvertently infiltrated inside the trusted perimeter by personnel. For example, USB memory sticks have become a popular tool to circumvent perimeter defenses: a few stray USB sticks left in public spaces are picked up by employees and plugged into previously secure devices inside the trusted perimeter, enabling malware on the USB sticks to immediately infect the devices. Similarly, devices used both inside and outside the trusted perimeter can get infected with malware when outside, and infiltrate that malware when used inside. Common examples are corporate laptops that are privately used at home over the weekend.
- *Network-based intrusion.* Perhaps the most common mechanism to penetrate a trusted perimeter is through a network-based attack vector. Exploiting poorly configured firewalls for both misconfigured inbound and faulty outbound rules is a common entry point, enabling an adversary to insert a malicious payload onto the control system.

Back-doors and holes in the network perimeter may be caused by components of the IT infrastructure with vulnerabilities or misconfigurations. Networking devices at the perimeter (e.g., fax machines, forgotten but still connected modems) can be manipulated for bypassing proper access control mechanisms. In particular, dial-up access to remote terminal units (RTUs) is used for remote management, and an adversary can directly dial into modems attached to field equipment, where many units do not require a password for authentication or have unchanged default passwords. Further, adversaries can exploit vulnerabilities of the devices and install back-doors for future access to the prohibited area. Exploiting trusted peer utility links is another potential network-based entry point.

An attacker could wait for a legitimate user to connect to the trusted control system network via VPN and then hijack that VPN connection. The network-based intrusions described above are particularly dangerous because they enable a remote adversary to enter the trusted control system network.

- *Compromised supply chain.* An attacker can preinstall malicious codes or back-doors into a device prior to shipment to a target location, called supply chain attacks. Consequently, the need for security assurance in the development and manufacturing process for sourced software, firmware, and equipment is critical for safeguarding the cybersupply chain involving technology vendors and developers.

- *Malicious insider.* An employee or legitimate user who is authorized to access system resources can perform actions that are difficult to detect and prevent. Privileged insiders also have intimate knowledge of the deployed defense mechanisms, which they can often easily circumvent. Trivial accessibility to smart grid components will increase the possibility of escalating an authorized access to a powerful attack.

7.1.6.2 Adversary Actions

Once an adversary gains access to the power control network, he/she can perform a wide range of attacks. In what follows, we list actions that an adversary can perform to violate the main security properties (confidentiality, integrity, availability) for the core types of information. We classify more specific cyberattacks that lead to either cyber or physical consequences.

A. Cyberconsequences

- *Malware spreading and controlling devices.* An adversary can develop malware and spread it to infect smart meters [21] or company servers. Malware can be used to replace or add any function to a device or a system such as sending sensitive information or controlling devices.
- *Vulnerabilities in common protocols.* Smart grid components will use existing protocols, inheriting the vulnerabilities on the protocols. Common protocols may include TCP/IP, and remote procedure call (RPC).
- *Access through database links.* Control systems record their activities onto a database on the control system network then mirror logs into the business network. A skilled attacker can gain access to the database on the business network, and the business network gives a path to the control system network. Modern database architectures allow this type of attack if they are improperly configured.
- *Compromising communication equipments.* An attacker can potentially reconfigure or compromise some of the communication equipment, such as multiplexers.
- *Injecting false information on price and meter data.* An adversary can send packets to inject false information on current or future prices, or send wrong meter data to a utility company. Results of injecting false prices, such as negative pricing, will be power shortage or other significant damages on the target region. Results of sending wrong data include

reduced electric bills for economic damages due to the loss of revenue of a utility company. Also, fake information can give huge financial impacts on electricity markets [12].

- *Eavesdropping attacks.* An adversary can obtain sensitive information by monitoring network traffic, which results in privacy breaches by stealing power usage, disclosure of the controlling structure of smart grids and future price information. Such eavesdropping can be used for gathering information to perpetrate further crimes. For example, an attacker can gather and examine network traffic to deduce information from communication patterns, and even encrypted communication can be susceptible to traffic analysis attacks.
- *Modbus security issues.* A SCADA protocol of noteworthy concern is the Modbus protocol [22], which is widely used in industrial control applications such as in water, oil, and gas infrastructures. The Modbus protocol defines the message structure and communication rules used by process control systems to exchange SCADA information for operating and controlling industrial processes. Modbus is a simple client–server protocol that was originally designed for low-speed serial communication in process control networks. Given that the Modbus protocol was not designed for highly security-critical environments, several attacks are possible.
  1. *Broadcast message spoofing.* This attack involves sending fake broadcast messages to slave devices.
  2. *Baseline response replay.* This attack involves recording genuine traffic between a master and a field device, and replaying some of the recorded messages back to the master.
  3. *Direct slave control.* This attack involves locking out a master and controlling one or more field devices.
  4. *Modbus network scanning.* This attack involves sending benign messages to all possible addresses on a Modbus network to obtain information about field devices.
  5. *Passive reconnaissance.* This attack involves passively reading Modbus messages or network traffic.
  6. *Response delay.* This attack involves delaying response messages so that the master receives out-of-date information from slave devices.
  7. *Rogue interloper.* This attack involves attacking a computer with the appropriate (serial or Ethernet) adapters to an unprotected communication link.

B. Physical consequences

- *Interception of SCADA frames.* An attacker can use a protocol analysis tool for sniffing network traffic to intercept SCADA Distributed Network Protocol 3.0 (DNP3) frames and collect unencrypted plain text frames that would provide valuable information, such as source and destination addresses. This intercepted data, which include control and setting information, could then be used at a later date on another SCADA system or intelligent equipment device (IED), thereby shutting services down at worst or at the minimum causing service disruptions.
- *Malware targeting industrial control systems.* An attacker can successfully inject worms into vulnerable control systems and reprogram industrial control systems. A well-known example is Stuxnet as discussed in Section 7.1.3.
- *DoS/DDoS attacks on networks and servers.* An adversary can launch a DoS/DDoS attack against various grid components including smart meters, networking devices, communication links, and utility business servers. If the attack is successful, then electricity cannot be controlled in the target region. Furthermore, power supply can be stopped from the result of the attack.
- *Sending fake commands to smart meters in a region.* An adversary can send fake commands to a device or a group of devices in a target region. For example, sending disconnect messages to smart meters in a region will stop power delivery to that region. Furthermore, invalid switching of electric devices can result in unsafe connections which may lead to setting the target place on fire. Thus, insecure communication in smart grids may be able to threaten human life.

The attacks mentioned above are not exhaustive, yet they serve well to illustrate risks and help develop secure grid systems. Additional examples of SCADA threats are available at the web site of US-CERT. See Table 7.1.

7.1.7 Countermeasures

1. *Key Management.* Key management is a fundamental approach for information security. Shared secret keys or authentic public keys can be used to achieve secrecy and authenticity for communication. Authenticity is especially important to verify the origin which in turn is key for access control.

   The key setup in a system defines the root of trust. For example, a system based on public/private keys may define the public key of a trust

Table 7.1 Threat type classification as caused by attacking security properties.

	Price information	**Control command**
Confidentiality	Leakage of price info	Exposure of control structure
Integrity	Incorrect price info	Changes of control command
Availability	Unavailability of price info	Inability to control grid
	Meter data	**Software**
Confidentiality	Unauthorized access to meter data	Theft of proprietary software
Integrity	Incorrect meter data	Malicious software
Availability	Unavailability of billing info	N/A

center as the root of trust, and the trust center private key is used to sign certificates and delegate trust to other public keys. In a symmetric-key system, each entity and the trust center would set up shared secret keys and establish additional trust relationships among other nodes by leveraging the trust center, as in Kerberos.

The challenge in this space is key management across a very broad and diverse infrastructure. As a recent NIST report documents [3], several dozens of secure communication scenarios are required, ranging from communication between the power distributor and the smart meter to communication between equipment and field crews. For all these communication scenarios, keys need to be set up to ensure secrecy and authenticity. Besides the tremendous diversity of equipment, there is also a wide variety of stakeholders: government, corporations, and consumers. Even secure e-mail communication among different corporations is a challenge today; the secure communication between equipment from one corporation and a field crew of another one poses numerous additional challenges. By adding a variety of key management operations to the mix (e.g., key refresh, revocation, backup, and recovery), the complexity of key management becomes truly formidable. Moreover, business, policy, and legal aspects also need to be considered, as a message signed by a private key can hold the key owner liable for the contents. A recent publication from NIST provides a good guideline for designing cryptographic key management systems to support an organization [23], but the diverse requirements of smart grid infrastructures are not considered.

2. *Secure Communication Architecture.* Designing a highly resilient communication architecture for a smart grid is critical to mitigate attacks while achieving high-level availability. Here are the required components.

- *Network topology design.* A network topology represents the connectivity structure among nodes, which can have an impact on the robustness against attacks [24]. Thus, connecting networking nodes to be highly resilient under attack can be the basis to build a secure communication architecture.
- *Secure routing protocol.* A routing protocol on a network is to build logical connectivity among nodes, and one simplest way to prevent communication is by attacking the routing protocol. By compromising a single router and by injecting bogus routes, all communication in the entire network can come to a standstill. Thus, we need to consider the security of a routing protocol running on top of a network topology.
- *Secure forwarding.* An adversary who controls a router can alter, drop, and delay existing data packets or inject new packets. Thus, securing individual routers and detecting malicious behaviors will be required to achieve secure forwarding.
- *End-to-end communication.* From the end-to-end perspective, secrecy and authenticity of data are the most crucial properties. Secrecy prevents an eavesdropper from learning the data content, while authenticity (sometimes referred to as integrity) enables the receiver to verify that the data indeed originated from the sender, thus preventing an attacker from altering the data.

While numerous protocols exist (e.g., SSL/TLS, IPsec, SSH), some low-power devices may need lightweight protocols to perform the associated cryptography.

- *Secure broadcasting.* Many smart grid environments rely on broadcast communication. Especially for price dissemination, authenticity of the information is important, because an adversary could inject a negative cost and cause an electricity utilization to spike when numerous devices simultaneously turn on to take advantage of the low price.
- *DoS defense.* Given all the above mechanisms, an adversary can still prevent communication by mounting a DoS attack. For example, if an adversary controls many end points after compromising them, he can use these end points to send data to flood the network. Hence, enabling communication under these circumstances is crucial, for example, to perform network management operations to

defend against the attack. Moreover, electricity itself, rather than communication networks, can be a target of DoS attacks [25].

- *Jamming defense.* To prevent an external adversary from jamming the wireless network, jamming detection mechanisms can be used to detect attacks and raise alarms. A multitude of methods to counter jamming attacks has been developed [26], enabling operation during jamming.

3. *System and Device Security.* An important area is to address vulnerabilities that enable exploitation through software-based attacks, where an adversary either exploits a software vulnerability to inject malicious code into a system, or where a malicious insider uses administrative privileges to install and execute malicious code. The challenge in such an environment is to obtain "ground truth" when communicating with a potentially compromised system: Is the response sent by legitimate code or by malware? An illustration of this problem is when we attempt to run a virus scanner on a potentially compromised system – If the virus scanner returns the result that no virus is present, is that really because no virus could be identified or is it because the virus has disabled the virus scanner? A related problem is that current virus scanners contain an incomplete list of virus signatures, and the absence of a virus detection could be because the virus scanner does not yet recognize the new virus.

   In the context of smart grids, researchers have proposed several techniques to provide prevention and detection mechanisms against malware. McLaughlin et al. have proposed diversity for embedded firmware [27] to avoid an apocalyptic scenario where malware pervasively compromises equipment, because each device executes different software, thus avoiding common vulnerabilities.

   A promising new approach to provide remote code verification is a technology called attestation. Code attestation enables an external entity to inquire the software that is executing on a system in a way that prevents malware from hiding. Since attestation reveals a signature of executing code, even unknown malware will alter that signature and can thus be detected. In this direction, hardware-based approaches for attestation have been studied in [28,29]. Software-based attestation is an approach that does not rely on specialized hardware, but makes some assumptions that the verifier can uniquely communicate with the device under verification [30]. Shah et al. have demonstrated the feasibility of this concept on SCADA devices [31].

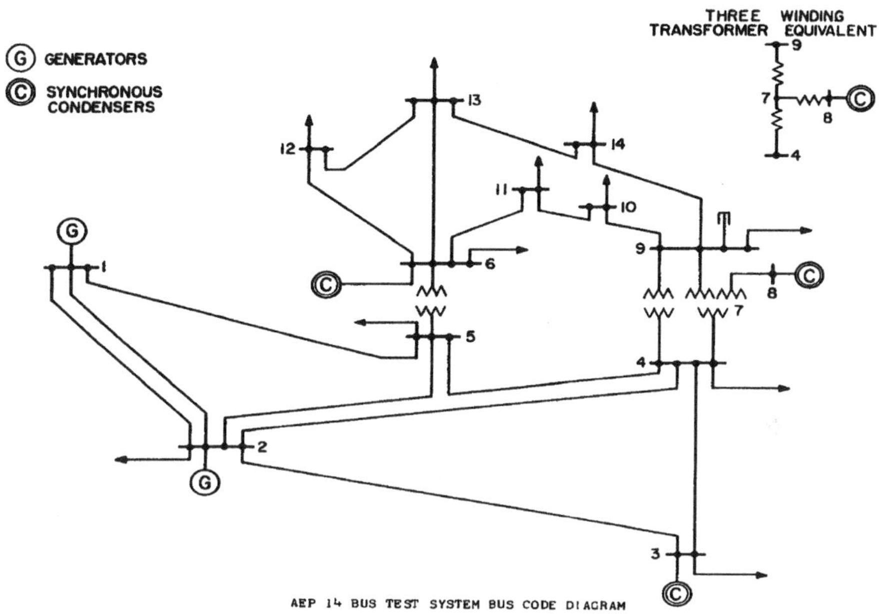

Figure 7.5 System diagram

7.2 SYSTEM-THEORETIC APPROACHES

In this section, we want to focus on system-theoretic approaches to the real-time security of smart grids, which encompasses two main parts: contingency analysis (CA) and system monitoring [32].

7.2.1 System Model

Fig. 7.5 shows a typical system-theoretic view of an IEEE 14-bus system. The focus of such a view is the physical interactions between each component in the grid, while the cyberview focuses on the modeling of IT infrastructures.

Suppose the grid consists of N buses. Let us define the active power flow, reactive power flow, the voltage magnitude, and phase angle for each bus as P_i, Q_i, V_i, and θ_i, respectively. Let us define vectors P, Q, V, and θ as the collections of P_i, Q_i, V_i, and θ_i, respectively.

The relationship between node current I_k and voltage $V_k e_{j\theta_k}$ is given by the following linear equations [33]:

$$I_k = \sum_{i=1}^{N} Y_{ki} V_i e^{i\theta_i},$$

where Y_{ki} is the admittance between bus k and i. As a result, the active and reactive power at node k are given by

$$P_k + jQ_k = V_k e^{j\theta_i} \times \bar{I}_k = V_k e^{j\theta_i} \sum_{i=1}^{N} Y_{ki} \bar{V}_i e^{j\theta_i}, \tag{7.1}$$

where $\bar{I}_k$ means complex conjugate. It can be seen that V and θ are the states of the system since they completely determine power flow P and Q. Let us define the state x as $x = [V', \theta_1; \ldots; \theta_{N-1}]' \in \mathbb{R}^{2N-1}$. The remote terminal units (RTUs) provide the system measurements. Let us denote as $z \in \mathbb{R}^m$ the collection of all measurements, assumed to satisfy the following equation:

$$z = h(x) + \nu, \tag{7.2}$$

where $h : \mathbb{R}^{2N-1} \rightarrow \mathbb{R}^m$ represents the sensor model and $\nu \in \mathbb{R}^m$ denotes the measurement noise, which is further assumed to be Gaussian with mean 0 and covariance R. Here we briefly introduce the weighted least squares (WLS) estimator [34], as it is widely used in practice. Define the estimated state as $\hat{x}$, and the residue vector as $r = z - h(\hat{x})$, which measures the inconsistency between state estimation $\hat{x}$ and measurements z. A WLS estimator tries to find the best estimate $\hat{x}$ with minimum inconsistency. In particular, the WLS estimator computes $\hat{x}$ based on the following minimization problem:

$$\hat{x} = \arg\min_{\hat{x}} r^T R^{-1} r. \tag{7.3}$$

7.2.2 Security Requirements

The US Department of Energy (DoE) Smart Grid System Report [35] summarizes six characteristics of the smart grid, which were further developed from the seven characteristics of the Modern Grid [36] published by the National Energy Technology Laboratory (NETL). With respect to security, the most important characteristic identified by DoE is to operate resiliently even during disturbances, attacks, and natural disasters.

In real-time security settings, the following properties are essential for the resilience of smart grids:

1. The power system should withstand a prespecified list of contingencies;
2. The accuracy of state estimation should degrade gracefully with respect to sensor failures or attacks.

Table 7.2 Comparison between cyber- and system-theoretic security.

	Cybersecurity
System Model	WAN/NAN/HAN model
Requirements	Confidentiality
	Integrity
	Availability
Attack Model	DoS attack
	Network-Based Intrusion
Countermeasures	Key Management
	Secure Communication
	System and Device Security
	System-Theoretic Security
System Model	Power Flow Model Sensor Model
Requirements	Robust to Prespecified Contingency
	Accurate State Estimation
Attack Model	Contingencies
	Sensor Failures
	False Data Injection
Countermeasures	Contingency Analysis
	Bad Data Detection

The first property is passive and prevention based. The second property enables the detection of attacks or abnormalities and helps the system operator actively mitigate the damage.

7.2.3 Attack Model

A contingency can usually be modeled as a change in vectors P; Q; V; θ (such as a loss of a generator) or as a change in the admittance Y_{ki} (such as an opening transmission line). For system monitoring, corrupted measurements can be modeled as an additional term in (7.2), that is,

$$z^a = z + u = h(x) + v + u, \tag{7.4}$$

where $u = [u_1, \ldots, u_m]' \in \mathbb{R}^m$ and $u_i \neq 0$ only if the sensor i is corrupted. See Table 7.2.

7.2.4 Countermeasures

1. *Contingency Analysis.* Contingency analysis checks if the steady-state system is outside operating region for each contingency [32]. However, the number of potential contingencies is high for large power grids. Due to realtime constraints, it is impossible to evaluate each contingency. As a result, in practice, usually only "$N-1$" contingencies are evaluated, via considering single failure cases instead of multiple ones. Moreover, the list of possible contingencies is usually screened and ranked. After that, a selected number of contingencies is evaluated. If a violation occurs, the system needs to determine the control actions that can mitigate or completely eliminate the violation.

2. *Bad Data Detection.* Bad data detector such as χ^2 or largest normalized residue detector [34] detects the corruption in measurement z by checking the residue vector r. For uncorrupted measurements, it is expected that the residue vector r will be small since z should be consistent with (7.2). However, such a detection scheme has an inherent vulnerability as different z vectors can generate the same residue r. By exploiting this vulnerability, it is shown [10] that an adversary can inject a stealthy input u into the measurements to change the state estimate $\hat{x}$ and fool the bad data detector at the same time. In [37], it is reported how to find a sparse stealthy u, which enables the adversary to launch an attack with a minimum number of compromised sensors. To counter such a vulnerability, it is proposed in [38] to use the prior knowledge of the state x to help detect malicious sensors.

7.2.5 Needs for Cyberphysical Security

The cybersecurity approaches focus on the IT infrastructures of the smart grid while system-theoretic approaches focus more on the physical aspects. We argue that pure cyber or system-theoretic approaches are insufficient to guarantee security of the smart grid, for the following reasons:

1. *The system and attack models of both approaches are incomplete.* Cybersecurity does not model the physical system. Therefore, cybersecurity can hardly defend against physical attacks. For example, cybersecurity protects the integrity of measurements data by using secure devices and communication protocols. However, integrity of sensors can be broken by modifying the physical state of the system locally, e.g., shunt connectors can be placed in parallel with a meter to bypass it and cause energy theft. In that case, no purely cybersecurity method can be employed to effectively detect and counter such attacks, since the cyberportion of the system is not compromised. Thus, even the goals of cybersecurity cannot be achieved by pure cyberapproaches in cyberphysical systems. Moreover, cybersecurity is not well equipped to predict the

effect of cyberattacks and countermeasures on the physical system. For example, the DoS attacks can cause drops of measurements data and control command, which can lead to instability of the grid. A countermeasure to DoS attacks is to isolate some of the compromised nodes from the network, which may result in even more severe stability issues. Thus, an understanding of the physical system is crucial even for defending against cyberattacks. On the other hand, the system-theoretic model does not model the whole IT infrastructures, but usually just a high level abstraction. As a result of this oversimplification of the cyberworld, it difficult to analyze the effect of cyberattacks on physical systems. For example, in DoS attacks, some control commands may be dropped due to limited bandwidth. However, the effect of the lossy communication cannot be evaluated in a pure power flow model.

2. *The security requirements of both approaches are incomplete and the security of the smart grid requires both of them.* System level concerns, such as stability, safety, and performance, have to be guaranteed in the event of cyberattacks. Cybersecurity metrics do not currently include the aforementioned metrics. On the other hand, system theory is not concerned with secrecy of information. Furthermore, it usually treats integrity and availability of information as intermediate steps to achieve stability, safety, or better performance. In the design of secure smart grid it is important to identify a set of metrics that combines and addresses the concerns of the two communities.
3. *The countermeasures of both approaches have drawbacks.* System-theoretic methods will not be able to detect any attack until it acts on the physical system. Furthermore, since system theory is based on approximate models and is subject to unknown disturbances, there will always be a discrepancy between the observed and the expected behavior. Most of the attack can bypass system theory-based intrusion detection algorithms with a small probability, which could be detrimental. Last, contingency analysis generally focuses on "$N-1$" contingencies, which is usually enough for independent equipment failures. However, as we integrate the IT infrastructures into the smart grid, it is possible that several contingencies will happen simultaneously during an attack. On the other hand, cybercountermeasures alone are not sufficient to guarantee security of the smart grid. History has so far taught that cybersecurity is not always bulletproof. As operational continuity is essential, the system must be built to withstand and operate even in the event of zero day vulnerabilities or insider threats, resorting to

rapid reconfiguration to provide graceful degradation of performance in the face of an attack. As a large blackout can happen in a few minutes [39], it is questionable that pure cybersecurity approaches can react fast enough to withstand zero-day vulnerability exploits or insider attacks.

7.2.6 Typical Cases

As shown earlier, both cyber and system-theoretic approaches are essential for the security of smart grids. In this section, we want to use two examples to show how the combination of cyber and system-theoretic approaches together can provide better security level than traditional methods. In the first example, we show how system-theoretic countermeasures can be used to defend against a replay attack, which is a cyber attack on the integrity of the measurement data. In the second example, we show how system theory can guide cybersecurity investment strategies.

7.2.7 Defense Against Replay Attacks

In this example, we consider defense against replay attack, where an adversary records a sequence of sensor measurements and replays the sequence afterwards. Replay attacks are cyberattacks which break the integrity or more precisely the freshness of measurements data. It is worth mentioning that Stuxnet [40] employed a replay attack of this type to cover its goal of damaging the centrifuges in a nuclear facility by inducing excessive vibrations or distortions. While acting on the physical system, the malware was reporting old measurements indicating normal operations. This integrity attack, clearly conceived and operated in the cyberrealm, exploited four zero-day vulnerabilities to break the cyberinfrastructures and it remained undiscovered for several months after its release. Therefore, a pure cyber-approach to replay attacks may not be able to react fast enough before the system is damaged.

Next we develop the concept of physical authentication, a methodology that can detect such attacks independently of the type of attack used to gain access to the control system. This algorithm [41] was developed long before Stuxnet appeared and preceded it. We are reporting a summary below. To achieve greater generality, the method is presented for a generic control system. We assume the sensors are monitoring a system with the following state dynamics:

$$x_k + 1 = Fx_k + Bu_k + w_k \tag{7.5}$$

where $x_k \in \mathbb{R}^n$ is the vector of state variables at time k, $w_k \in \mathbb{R}^n$ is the process noise at time k, and x_0 is the initial state. We assume w_k, x_0 are independent Gaussian random variables, $x_0 \sim \mathcal{N}(\bar{x_0}, \sum)$, $w_k \sim \mathcal{N}(0, Q)$.

For each sampling period k, the true measurement equation of the sensors can be written as

$$z_k = Hx_k + v_k, \tag{7.6}$$

where $z_k \in \mathbb{R}^m$ is a collection of all the measurements from sensors at time k and $v_k \sim \mathcal{N}(0, R)$ is the measurement noise independent of x_0 and w_k.

We assume that an attacker records a sequence of measurements from time T_0 to time $T_0 + T - 1$ and replays it from time $T_0 + T$ to time $T_0 + 2T - 1$, where $T_0 \geq 0$, $T \geq 1$. As a result, the corrupted measurements z_k^a received by the system operator are

$$z_k^a = \begin{cases} z_k, & 0 \leq k \leq T_0 + T - 1, \\ z_{k-T}, & T_0 + T \leq k \leq T_0 + 2T - 1. \end{cases} \tag{7.7}$$

Our goal is to design an estimator, a controller and a detector such that:

1. The system is stable when there is no replay attack;

2. The detector can detect the replay attack with a high probability.

We propose the following design of a fixed gain estimator, a fixed gain controller with random disturbance and a χ^2 detector. In particular, our estimator takes the following form:

$$\hat{x}_{k+1} = F\hat{x}_k + Bu_k + Kr_{k+1}, \tag{7.8}$$

where K is the observation gain matrix and the residue r_k is computed as

$$r_{k+1} = z_{k+1}^a - C(F\hat{x}_k + Bu_k). \tag{7.9}$$

Our controller takes the following form:

$$u_k = I\hat{x}_k + \Delta u_k, \tag{7.10}$$

where L is the control gain matrix and Δu_k are independent identically distributed (i.i.d.) Gaussian noises generated by the controller, with zero mean and covariance $\mathcal{Q}$. It can be easily shown that the residue r_k is a Gaussian random variable with zero mean when there is no replay. As a

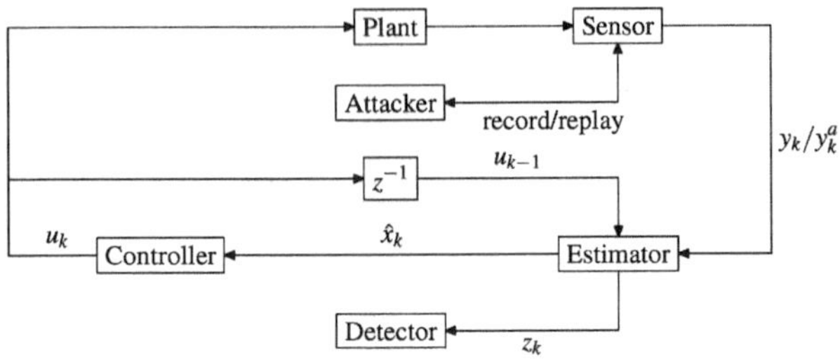

Figure 7.6 System diagram

result, with large probability it cannot be far away from 0. Therefore, we design our filter to trigger an alarm at time k based on the following event:

$$\{g_k = r_k' \mathcal{P} r_k \geq \text{threshold}\}, \tag{7.11}$$

where $\mathcal{P}$ is a predefined weight matrix. Fig. 7.6 shows the diagram of the proposed system.

We first consider the stability of the proposed system. It is well known that without Δu_k, the closed-loop system without replay is stable if and only if both $F - KCF$ and $F + BL$ are stable. Moreover, one can easily prove that adding Δu_k does not affect the stability of the system since Δu_k is i.i.d. Gaussian distributed. Hence, to ensure that the system is closed-loop stable without replay, we only need to make $F - KCF$ and $F + BL$ stable, which can be easily done as long as the system is both detectable and stabilizable. Now we want to show our system design can successfully detect replay attacks. Consider the residue r_k, where $T_0 + T \leq k \leq T_0 + 2T - 1$, then one can prove that

$$\begin{aligned} r_k = {} & r_{k-T} + C\mathcal{A}^{k-T_0-T}(I - KC)(\hat{x}_{T_0} - \hat{x}_{T_0+T}) \\ & + \sum_{i=0}^{k-T-T_0-1} C\mathcal{A}^i B(\Delta u_{k-T-1-i} - \Delta u_{k-1-i}), \end{aligned}$$

where $\mathcal{A} = (F + BL)(I - KC)$. The second term above converges to 0 exponentially fast if $\mathcal{A}$ is stable. As a result, if we do not introduce any random control disturbance, i.e., $\Delta u_k = 0$, then the third term vanishes and the residue r_k under replay attack converges to the residue r_{k-T} when no replay attack is present. Therefore, the detection rate of the replay attack will be the same as the false alarm rate. In other words, the detector cannot distinguish between healthy and corrupted measurements. However, if $\Delta u_k \neq 0$,

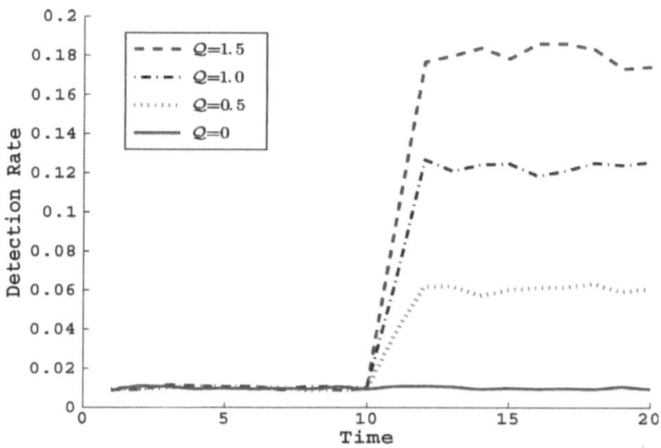

Figure 7.7 Detection rate over time

then the third term will always be present and therefore the detector can detect replay attacks with a probability larger than the false alarm rate.

It is worth mentioning that the role of Δu_k is similar to an authentication signal on the measurements. When the system is under normal operation, it is expected that the measurements z_k will reflect the random disturbances Δu_k. On the other hand, when the replay begins, z_k and Δu_k become independent of each other. Therefore, the integrity and freshness of the measurements can be protected by checking the correlation between z_k and Δu_k. This technique is cyberphysical as it uses the physics of the system to authenticate data coming from the cyberportion. We now wish to provide a numerical example to illustrate the performance of our detection algorithm. We impose the following parameters: $F = B = Q = R = \mathcal{P} = 1$, $K = 0.9161$, $L = 0.618$. One can verify that $\mathcal{A} = 0.0321 < 1$. The threshold of the filter is chosen such that the false alarm rate is 1%. We assume that the recording starts at time 1 and replay starts at time 11. Fig. 7.7 shows different detection rate over time as $\mathcal{Q}$ increases. It can be seen that the detection fails when there is no disturbance. Moreover, a larger disturbance can increase the performance of the detector.

7.2.8 Cybersecurity Investment

In this example, we show how system theory can be used to expose the critical assets to protect and thus provide important insights toward the allocation of security investments. In particular, we consider how to deploy secure sensors to help detect corrupted measurements. We assume the true measurements of sensors follow a linearized model of (7.2):

$$z = Hx + v \tag{7.12}$$

where $z \in \mathbb{R}^m$ and $x \in \mathbb{R}^{2N-1}$ and $H \in \mathbb{R}^{m\times(2N-1)}$ is assumed to be of full column rank. For linearized models, (7.3) can be solved analytically as

$$\hat{x}(x) = (H'R^{-1}H)^{-1}H'R^{-1}z = Kz. \tag{7.13}$$

Therefore, the residue can be calculated explicitly as

$$r(z) = z - H\hat{x}(z) = (I - HK)z = Sz, \tag{7.14}$$

where $S = I - HK$. Suppose that an attacker is able to modify the readings of a subset of sensors. As a result, the corrupted measurements take the following form:

$$z^a = z + u = Hx + v + u, \tag{7.15}$$

where $u = [u_1, \ldots, u_m]' \in \mathbb{R}^m$ indicates the error introduced by the attacker and $u_i \neq 0$ only if sensor i is compromised. An attack is called stealthy if the residue r does not change during the attack. In mathematical terms, a stealthy attack u satisfies $r(z) = r(z + u)$. Since $r(z)$ is linear with respect to z, we can simplify the above equation to

$$r(u) = Su = 0 \tag{7.16}$$

without loss of generality.

As shown by Liu et al. [10], the χ^2 detectors fail to detect a stealthy input u. In fact, any detector based on r is ineffective against stealthy attacks as they do not change the residue r. On the other hand, the stealthy attack can introduce estimation error to $\hat{x}$.

To defend against such attacks, we deploy secure devices, such as tamper resistant devices, to protect the sensors. To this end, we define a sensor i to be secure if it cannot be compromised, i.e., the corresponding u_i is guaranteed to be 0. Let us also define the set of secure sensors to be $S_e \subseteq 1, \ldots, m$. An attack u is feasible if and only if $u_i = 0$ for all $i \in S_e$.

Our security goal is to deploy the minimum number of secure sensors such that the system can detect the compromised nodes. In other words, we want to find the smallest set S_e such that there is no nonzero feasible and stealthy u.

This problem is of practical importance in the smart grid as the current insecure sensors can only be replaced gradually by secure sensors due to the scale of the grids. As a result, it is crucial to know which set of sensors to replace first to achieve better security.

Let us define $\Gamma(S_e) = \text{diag}(\gamma_1, \ldots, \gamma_m)$ where $\gamma_i = 1$ if and only if $i \in S_e$. A set S_e is called observable if and only if $\Gamma(S_e)H$ is of full column rank. In other words, if a vector $p \in \mathbb{R}^{2N-1} \neq 0$, then $\Gamma(S_e)Hp \neq 0$. The following theorem relates the observability of secure sensor set S_e with the existence of a feasible and stealthy attack u.

Theorem 7.1. *The only feasible and stealthy attack is $u = 0$ if and only if S_e is observable.*

Proof. First suppose that S_e is observable and u is stealthy and feasible. As a result, $\Gamma(S_e)u = 0$. On the other hand, since u is stealthy, $S_u = 0$, which implies that

$$HKu = (I - S)u = u.$$

Therefore

$$\Gamma(S_e)HKu = \Gamma(S_e)u = 0.$$

Since $\Gamma(S_e)H$ is full column rank, we know that $K_u = 0$, which implies that $HKu = 0$. Thus

$$u = (I - HK + HK)u = Su + HKu = 0.$$

On the other hand, suppose that S_e is not observable. Find $x \neq 0$ such that $\Gamma(S_e)Hx = 0$. Choose $u = Hx$. Since H is full column rank, $u \neq 0$. Moreover, $\Gamma(S_e)u = \Gamma(S_e)Hx = 0$. Hence, u is feasible. Finally,

$$\begin{aligned} Su = (I - HK)u &= u - H(H'R^{-1}H)^{-1}H'R^{-1}u \\ &= Hx - H(H'R^{-1}H)^{-1}H'R^{-1}Hx = 0, \end{aligned}$$

which implies that u is stealthy. □

Therefore, finding the smallest S_e such that there is no nonzero feasible and stealthy u is equivalent to finding the smallest observable S_e, which can be achieved using the following theorem:

Theorem 7.2. *If S_e is observable and* $\text{rank}(\Gamma(S_e)) > 2N - 1$, *then there exists an observable S'_e, which is a proper subset of S_e.*

Proof. Let $H' = [H_1, \ldots, H_m]$, where $H_i \in \mathbb{R}^{2N-1}$. Since S_e is observable,

$$\text{rank}(\gamma_1 H, \ldots, \gamma_m H_m) = 2N - 1.$$

Without lost of generality, let us assume that $S_e = \{1, \ldots, 1\}$ thus $\gamma_1 = \cdots = \gamma_l = 1$ and $\gamma_{l+1} = \cdots = \gamma_m = 0$, where $l > 2N - 1$. Since $H_i \in \mathbb{R}^{2N-1}$, $H_1, \ldots, H_l$ are not linearly independent. Hence, there exist $\alpha_1, \ldots, \alpha_l \in \mathbb{R}$ that are not all zero such that $\alpha_1 H_1 + \cdots + \alpha_l H_l = 0$. Without loss of generality, let us assume that $\alpha_l \neq 0$. Therefore

$$\text{span}(H_1, \ldots, Hl - 1) = \text{span}(H_1, \ldots, Hl) = \mathbb{R}^{2N-1},$$

which implies that $S_e' = \{1, \ldots, l - 1\}$ is observable. □

It is easy to see that $\text{rank}(\Gamma(S_e))$ must be no less than $2N - 1$ to make S_e observable. As a result, one can use the procedure described in the proof of Theorem 7.2 to find the smallest observable set. Analyses of this kind are essential to prioritize security investments.

Remark 7.1. *It is worth noticing that the attacks we discussed in this section are cyberattacks which have physical consequences. The replay attack itself can render the system unstable if the original system is open-loop unstable or it can enable future attacks on the physical system, as in the case of Stuxnet. A stealthy integrity attack can cause large estimation error and potentially damage the system. Furthermore, our approaches to security are hybrid in nature. In the first example, we use system-theoretic models and countermeasures to detect replay attacks, which is a cyberattack. Our detection algorithm complements the pure cybersecurity approaches and provides an additional layer of protection. In the second example, we use a system-theoretic model of the grid to develop an optimal cybersecurity countermeasure to integrity attacks. The results illustrate that combining cybersecurity and system theory can provide a better level of security for the smart grid.*

7.3 WIDE-AREA MONITORING, PROTECTION AND CONTROL

Smart grid technologies utilize recent cyberadvancements to increase control and monitoring functions throughout the electric power grid. The smart grid incorporates various individual technical initiatives such as:

- Advanced Metering Infrastructure (AMI),
- Demand Response (DR),

- Wide-Area Monitoring, Protection and Control systems (WAMPAC) based on Phasor Measurement Units (PMU),
- Large scale renewable integration in the form of wind and solar generation, and
- Plug-in Hybrid Electric Vehicles (PHEV).

Of these initiatives, AMI and WAMPAC depend heavily on the cyberinfrastructure and its data transported through several communication protocols to utility control centers and the consumers.

7.3.1 Introduction

As we learned in the previous chapters, cybersecurity concerns within the communication and computation infrastructure may allow attackers to manipulate either the power applications or physical system. Cyberattacks can take many forms depending on their objective. Attackers can perform various intrusions by exploiting software vulnerabilities or misconfigurations. System resources can also be rendered unavailable through denial of service (DoS) attacks by congesting the network or system with unnecessary data. In principle, secure cybersystems can be attacked due to insider threats, where a trusted individual can leverage system privileges to steal data or impact system operations. Also, weaknesses in communication protocols allow attackers to steal or manipulate data in transit.

AMI is based on the deployment of smart meters at consumer locations to provide two-way communication between the meter and utility. This provides the utility with the ability to push real-time pricing data to consumers, collect information on current usage, and perform more advanced analysis of faults within the distribution system. Since AMI is associated with the distribution system, typically many consumer meters need to be compromised to create a substantial impact in the bulk power system reliability. This is in strong contrast to the impact a coordinated cyberattack on WAMPAC would have on bulk power system reliability.

In what follows, we study pertinent issues in cyberphysical security of WAMPAC. It is to be noted, however, that important several cybersecurity and privacy issues do exist with respect to AMI.

7.3.2 Wide-Area Monitoring, Protection and Control

Wide Area Monitoring, Protection and Control systems (WAMPAC), leverages the Phasor Measurements Units (PMUs) to gain real-time awareness of current grid operations and also provides real-time protection and control functions such as Special Protection Schemes (SPS) and Automatic Generation Control (AGC), besides other emerging applications such as oscillation detection, and transient stability predictions.

While communication is the key to a smarter grid, developing and securing the appropriate cyberinfrastructures and their communication protocols is crucial. WAMPAC can be subdivided further into its constituent components namely, Wide-Area Monitoring Systems (WAMS), Wide-Area Protection Systems (WAP), and Wide-Area Control (WAC). PMUs utilize high sampling rates and accurate GPS-based timing to provide very accurate, synchronized grid readings. While PMUs provide increasingly accurate situational awareness capabilities, their full potential will not be realized unless these measurement data can be shared among other utilities and regulators. Additionally, power system applications need to be reexamined to determine the extent to which these enhancements can improve the grid efficiency and reliability. The development of advanced control applications will depend on WAMS that can effectively distribute information in a secure and reliable manner. An example of WAMS deployment is *NASPInet*, which is the development of a separate network for PMU data transmission and data sharing including realtime control, Quality-of-Service and cybersecurity requirements [42].

Wide-Area Protection (WAP) involves the use of system wide information collected over a wide geographic area to perform fast decision-making and switching actions in order to counteract the propagation of large disturbances. The advent of Phasor Measurement Units (PMUs) has transformed protection from a local concept into a system level wide-area concept to handle disturbances. Several protection applications fall under the umbrella of Wide-Area Protection (WAP), but the most common one among them is Special Protection Schemes (SPSs). The North American Electric Reliability Council (NERC) defines SPS as an automatic protection system designed to detect abnormal or predetermined system conditions, and takes corrective actions other than and/or in addition to the isolation of faulted components to maintain system reliability [43]. Such action may include changes in demand, generation (MW and MVAR), or system configuration to maintain system stability, acceptable voltage, or power flows. Some of the most common SPS applications are as follows: generator rejection, load rejection, under frequency load shedding, under voltage load shedding, out-of step relaying, VAR compensation, discrete excitation control, HVDC controls.

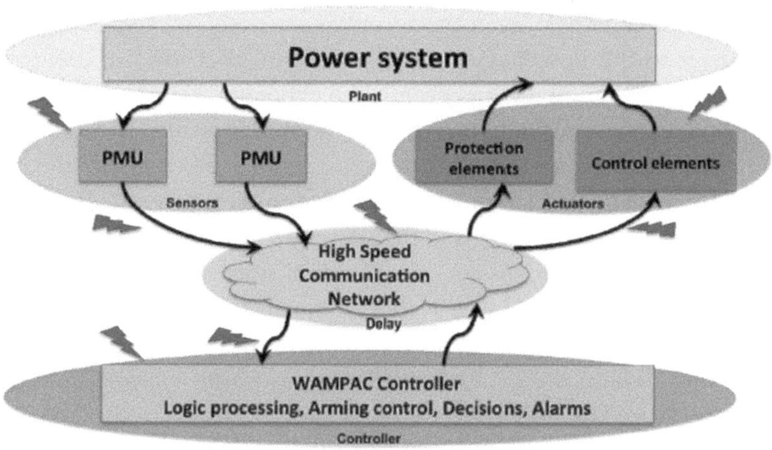

Figure 7.8 Generic architecture of wide-area monitoring, protection and control

Until the advent of PMUs, the only major Wide-Area Control mechanism in the power grid was Automatic Generation Control (AGC). The AGC functions with the help of tie line flow measurements, frequency and generation data obtained from Supervisory Control and Data Acquisition (SCADA) infrastructure. The purpose of the AGC in a power system is to correct system generation in accordance with load changes in order to maintain grid frequency at 60 Hz. Currently, the concept of real-time WAC using PMU data is still in its infancy and there are no standardized applications that are widely deployed on a system wide scale, though there are several pilot projects in that area [44]. Some of the potential WAC applications are secondary voltage control using PMU data, Static VAR Compensator (SVC) control using PMUs, and inter-area oscillation damping.

7.3.3 Cyberattack Taxonomy

A generic Wide-Area Monitoring, Protection and Control (WAMPAC) architecture is shown in Fig. 7.8 along with the various components involved. The system conditions are measured using measurement devices (mostly PMUs), these measurements are communicated to a logic processor to determine corrective actions for each contingency, and then appropriate actions are initiated, usually through high speed communication links. The inherent wide-area nature of these schemes presents several vulnerabilities in terms of possible cyberintrusions to hinder or alter the normal functioning of these schemes. Even though SPSs are designed to cause minimal or no impact to the power system under failures, they are not designed to handle failures due to malicious events like cyberattacks. Also, as more and more SPSs are added in the power system, it introduces unexpected dependencies in the operation of the various schemes and this increases the risk of increased impacts like system wide collapse, due to a cyberattack. It therefore becomes critical to reexamine the design of the Wide-Area Protection schemes with a specific focus on cyberphysical system security.

Also presented in Fig. 7.8 is a control systems view of the power system and the wide-area protection scheme. The power system is the plant under control, where the parameters like currents and voltages at different places are measured using sensors (PMUs) and sent through the high-speed communication network to the Wide-Area Protection controller for appropriate decision making. The controller decides based on the system conditions and sends corresponding commands to the actuators which are the protection elements and VAR control elements like SVC and FACTS devices for voltage control related applications. There are different places where a cyberattack can take place in this control system model. The cyberattack could affect the delays experienced in the forward or the feedback path or it could directly affect the data corresponding to sensors, the actuators or the controller.

7.3.4 Cyberattack Classification

Conceptually, we identify two three classes of attacks on this control system model for WAMPAC. They are timing based attacks, integrity attacks and replay attacks.

Timing attacks. Timing is a crucial component in any dynamic system (here a protection scheme) and in our case the control actions should be executed on the order of 50–100 ms after the disturbance. This system therefore cannot tolerate any type of delay in communications and therefore are vulnerable to timing based attacks. Timing attacks tend to flood the communication network with packets and this slows the network down in several cases and also shuts them down in some cases, both of which are not acceptable. These types of attacks are commonly known as denial of service (DoS) attacks.

Data integrity attacks. Data integrity attacks are attacks where the data is corrupted in the forward or the reverse path in the control flow. This means that there could be an attack which directly corrupts the sensor data, which in this case is the PMU data, or the actuator data, which is the command given to the protection elements or the VAR control elements. This translates to actions like blocking of the trip signals in scenarios where the controller actually sent a trip command to the protection elements or the controller commanded to increase VAR injection while the attack caused the injection to decrease or vice versa.

Replay attacks. Replay attacks are similar to data integrity attacks, where the attacker manipulates the PMU measurements or the control messages by hijacking the packets in transit between the PMU and the Phasor Data Concentrator (PDC) or the control center. In several cases, a replay attack is possible even under encrypted communication as the attack packets are valid packets with the message data integrity being intact except for the time stamp information.

7.3.5 Coordinated Attacks on WAMPAC

Intelligent coordinated attacks can significantly affect a power system security and adequacy by negating the effect of system redundancy and other existing defense mechanisms. North American Electric Reliability Council (NERC) has instituted the Cyberattack Task Force (CATF) to gauge system risk from such attacks and develop feasible, and cost-effective mitigation techniques. NERC CATF identifies intelligent coordinated cyber attacks as a category of events that are classified as High Impact Low Frequency (HILF), which cause significant impacts to power system reliability beyond acceptable margins.

The failure of any single element in the power system, such as a transformer or a transmission line, is a credible contingency ($N - 1$). The possibility of simultaneous failures of more than one element in the system is also taken into account when they are either electrically or physically linked. However, the definition of a "credible" contingency changes when potential failures from coordinated cyberattacks are considered. Also, an intelligent coordinated attack has two dimensions, where attacks can be coordinated in space and/or time. For example, elements that do not share electrical or physical relationships can be forced to fail simultaneously, or in a staggered manner at appropriate time intervals depending on the system response, which could result in unanticipated consequences. See Fig. 7.9.

7.4 NOTES

With the proliferation of remote management and control of cyberphysical systems, security plays a critically important role, because the convenience

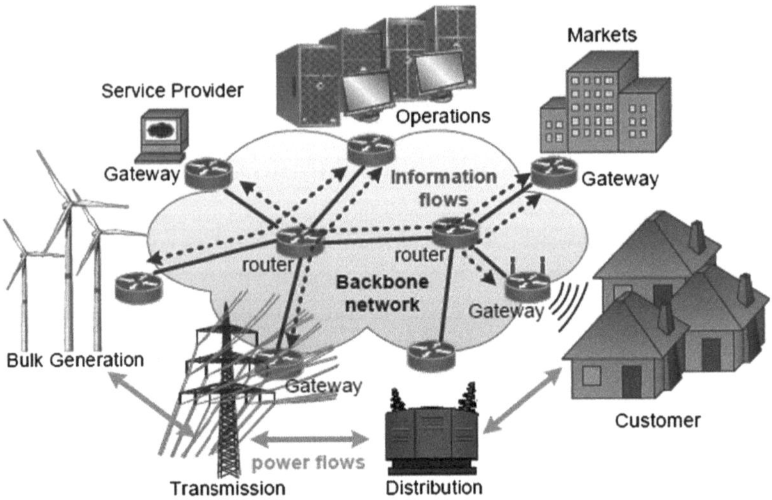

Figure 7.9 The network architecture in the smart grid: backbone and local-area networks

of remote management can be exploited by adversaries for nefarious purposes from the comfort of their homes.

Compared to current cyberinfrastructures, the physical component of cyberphysical infrastructures adds significant complexity that greatly complicates security. On the one hand, the increased complexity will require more effort from the adversary to understand the system, but on the other hand, this increased complexity also introduces numerous opportunities for exploitation. From the perspective of the defender, more complex systems require dramatically more effort to analyze and defend, because of the state-space explosion when considering combinations of events.

Current approaches to secure cyberinfrastructures are certainly applicable to securing cyberphysical systems: techniques for key management, secure communication (offering secrecy, authenticity, and availability), secure code execution, intrusion detection systems, etc. Unfortunately, these approaches are largely unaware of the physical aspects of cyberphysical systems.

System-theoretic approaches already consider physical aspects in more detail than the traditional security and cryptographic approaches. These approaches model the malicious behaviors as either components' failures, external inputs, or noises, analyze their effects on the system, and design detection algorithms or counter measures to the attacks. The strength of model-based approaches lies in a unified framework to model, analyze, detect, and counter various kinds of cyber and physical attacks. However, the physical world is modeled with approximations and is subject to noise, which can result in a deviation of any model to the reality. Therefore, system-theoretic approaches are nondeterministic as compared to information security.

As discussed in this chapter, cyberphysical system security demands additional security requirements, such as continuity of power delivery and accuracy of dynamic pricing, introduced by the physical system. Such requirements are usually closely related to the models and states of the system, which are difficult to address by information security alone. Therefore, both information security and system-theory-based security are essential to securing cyberphysical systems, offering exciting research challenges for many years to come.

REFERENCES

[1] E. Marris, Upgrading the grid, Nature 454 (2008) 570–573.

[2] S.M. Amin, For the good of the grid, IEEE Power Energy Mag. 6 (6) (Nov./Dec. 2008) 48–59.

[3] NIST, Guidelines for Smart Grid Cyber Security, Draft NISTIR 7628, July 2010.

[4] NETL, Understanding the Benefits of the Smart Grid, June 2010.

[5] US-DOE, NERCH, High-Impact, Low-Frequency Event Risk to the North American Bulk Power System, June 2010.

[6] J. Vijayan, Stuxnet Renews Power Grid Security Concerns, Computerworld, July 26, 2010 [Online]. Available: http://www.computerworld.com/s/article/9179689/Stuxnet_renews_power_grid_security_concerns.

[7] F. Cleveland, Cyber security issues for advanced metering infrastructure (AMI), in: Proc. Power Energy Soc. Gen. Meeting – Conv. Delivery Electr. Energy 21st Century, Apr. 2008.

[8] S.M. Amin, Securing the electricity grid, The Bridge 40 (Spring 2010) 13–20.

[9] P. McDaniel, S. McLaughlin, Security and privacy challenges in the smart grid, IEEE Secur. Priv. 7 (3) (May/June 2009) 75–77.

[10] Y. Liu, M. Reiter, P. Ning, False data injection attacks against state estimation in electric power grids, in: Proc. 16th ACM Conf. Comput. Commun. Security, Nov. 2009.

[11] H. Khurana, M. Hadley, N. Lu, D.A. Frincke, Smart-grid security issues, IEEE Secur. Priv. 8 (1) (Jan./Feb. 2010) 81–85.

[12] L. Xie, Y. Mo, B. Sinopoli, False data injection attacks in electricity markets, in: Proc. IEEE Int. Conf. Smart Grid Commun, Oct. 2010, pp. 226–231.

[13] NIST, NIST Framework and Roadmap for Smart Grid Interoperability Standards, Release 1.0, NIST Special Publication 1108, Jan. 2010.

[14] T. Alpcan, T. Başar, Network Security: A Decision and Game Theoretic Approach, Cambridge Univ. Press, Cambridge, UK, 2011.

[15] J.P. Hespanha, P. Naghshtabrizi, Y. Xu, BA survey of recent results in networked control systems, Proc. IEEE 95 (1) (Jan. 2007) 138–162.

[16] R. Goebel, R. Sanfelice, A. Teel, Hybrid dynamical systems, IEEE Control Syst. 29 (2) (2009) 28–93.

[17] EPRI, Advanced Metering Infrastructure (AMI), Feb. 2007.

[18] E.L. Quinn, Smart Metering & Privacy: Existing Law and Competing Policies, May 2009.

[19] A. Kerckhoffs, La cryptographie militairie, J. Sci. Militaires IX (1883) 5–38.

[20] D. Kundur, X. Feng, S. Liu, T. Zourntos, K.L. Utler-Purry, Towards a framework for cyber attack impact analysis of the electric smart grid, in: Proc. IEEE Int. Conf. Smart Grid Commun, Oct. 2010, pp. 244–249.

[21] S.S.S.R. Depuru, L. Wang, V. Devabhaktuni, N. Gudi, Smart meters for power grid: challenges, issues, advantages and status, in: Proc. IEEE/PES Power Syst. Conf. Expo., 2011.

[22] P. Huitsing, R. Chandia, M. Papa, S. Shenoi, Attack taxonomies for the Modbus protocols, Int. J. Crit. Infrastructure Prot. 1 (Dec. 2008) 37–44.

[23] E. Barker, D. Branstad, S. Chokhani, M. Smid, A Framework for Designing Cryptographic Key Management Systems, NIST DRAFT Special Publication 800-130, June 2010.

[24] H. Lee, J. Kim, W. Lee, Resiliency of network topologies under path-based attacks, IEICE Trans. Commun. E89-B (Oct. 2006) 2878–2884.

[25] D. Seo, H. Lee, A. Perrig, Secure and efficient capability-based power management in the smart grid, in: Proc. Int. Workshop Smart Grid Security Commun, May 2011, pp. 119–126.

[26] R.L. Pickholtz, D.L. Schilling, L. Milstein, Theory of spread spectrum communications – a tutorial, IEEE Trans. Commun. 30 (5) (May 1982) 855–884.

[27] S. McLaughlin, D. Podkuiko, A. Delozier, S. Mizdzvezhanka, P. McDaniel, Embedded firmware diversity for smart electric meters, in: Proc. USENIX Workshop Hot Topics in Security, 2010.

[28] M. LeMay, G. Gross, C. Gunter, S. Garg, Unified architecture for large-scale attested metering, in: Proc. Annu. Hawaii Int. Conf. Syst. Sci, Jan. 2007.

[29] M. LeMay, C.A. Gunter, Cumulative attestation kernels for embedded systems, in: Proc. Eur. Symp. Res. Comput. Security, Sept. 2009, pp. 655–670.

[30] A. Seshadri, A. Perrig, L. van Doorn, P. Khosla, SWATT: software-based attestation for embedded devices, in: Proc. IEEE Symp. Security Privacy, May 2004, pp. 272–282.

[31] A. Shah, A. Perrig, B. Sinopoli, Mechanisms to provide integrity in SCADA and PCS devices, in: Proc. Int. Workshop Cyber-Physical Syst. Challenges Appl., June 2008.

[32] M. Shahidehpour, F. Tinney, Y. Fu, Impact of security on power systems operation, Proc. IEEE 93 (11) (Nov. 2005) 2013–2025.

[33] W.F. Tinney, C.E. Hart, Power flow solution by newton method, IEEE Trans. Power Appar. Syst. PAS-86 (11) (Nov. 1967) 1449–1460.

[34] A. Abur, A.G. Exposito, Power System State Estimation: Theory and Implementation, CRC Press, Boca Raton, FL, 2004.

[35] US-DOE, Smart Grid System, Report Characteristics of the Smart Grid, July 2009.

[36] NETL, Characteristics of the Modern Grid, July 2008.

[37] H. Sandberg, A. Teixeira, K.H. Johansson, On security indices for state estimators in power networks, in: Proc. 1st Workshop Secure Control Syst, 2010.

[38] O. Kosut, L. Jia, R.J. Thomas, L. Tong, Limiting false data attacks on power system state estimation, in: Proc. 44th Annu. Conf. Inf. Sci. Syst, 2010.

[39] NERC, Technical Analysis of the August 14, 2003, Blackout: What Happened, Why, and What Did We Learn?, Tech. Rep., 2004.
[40] N. Falliere, L.O. Murchu, E. Chien, BW32.stuxnet, Tech. Rep., Dossier Symantec Corporation, 2011.
[41] Y. Mo, B. Sinopoli, Secure control against replay attacks, in: Proc. 47th Ann. Allerton Conf. Commun. Control Comput., 2009, pp. 911–918.
[42] V. Terzija, G. Valverde, D. Cai, P. Regulski, V. Madani, J. Fitch, Wide-area monitoring, protection, and control of future electric power networks, Proc. IEEE 99 (1) (2011) 80–93.
[43] V. Madani, D. Novosel, S. Horowitz, M. Adamiak, J. Amantegui, D. Karlsson, IEEE PSRC report on global industry experiences with system integrity protection schemes (SIPS), IEEE Trans. Power Deliv. 25 (4) (2010).
[44] North American Synchrophasor Initiative (NASPI), Phasor Data Applications Table [Internet]. Available from: https://www.naspi.org/File.aspx?fileID=537, 2009.

CHAPTER 8

Secure Resilient Control Strategies

8.1 BASIS FOR RESILIENT CYBERPHYSICAL SYSTEM

Admittedly, the ultimate purpose of using cyberinfrastructure (including sensing, computing and communication hardware/software) is to intelligently monitor (from physical to cyber) and control (from cyber to physical) our physical world. A system with a tight coupling of cyber and physical objects is called a cyberphysical system (CPS) [1], which has become one of the most important and popular computer applications today.

In this section, we set forth a formal definition of a resilient cyberphysical system (RCPS). Then follows a detailed discussion of the concept and strategies for building an RCPS. Initially, we adopt the statement *resilience refers to a 3S-oriented design*, that is, *stability, security*, and *systematicness*. In this regard, *stability* means the CPS can achieve a stable sensing-actuation close-loop control even though the inputs (sensing data) have noise or attacks; *security* means that the system can overcome the cyberphysical interaction attacks; and *systematicness* means that the system has a seamless integration of sensors and actuators.

A CPS can be thought of as the utilization of the logical and discrete properties of the computers to control and oversee the continuous and dynamic properties of physical systems. Using precise computations to control a seemingly unpredictable physical environment is a great challenge. A CPS often relies on sensors and actuators (or called actors, in some cases even called controllers) to implement tight interactions between cyber and physical objects. Essentially, the sensors (cyberobjects) can be used to monitor the physical environments, and the actuators/controllers can be used to change the physical parameters. Regarding the interactions between sensors and controllers, consider Fig. 8.1 in which a wind power system is used as a demo. There exist three types of communication among sensors and controllers:

1. Sensor-to-sensor (S–S) coordination, in which the sensors in a power cluster (with hundreds of wind turbines) need to communicate with each other to find an electromagnetic distribution map for power flow analysis.

Networked Control Systems
https://doi.org/10.1016/B978-0-12-816119-7.00016-2

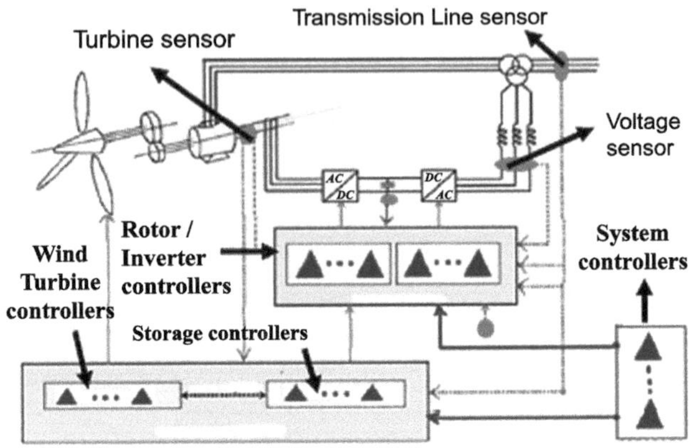

Figure 8.1 Sensors/controllers locations in a CPS

2. Sensor-to-controller (S–C) coordination, in which a controller makes decision based on the collected sensor data. A controller may need both local and remote sensors' data.
3. Controller-to-controller (C–C) coordination, in which a controller may need to coordinate with other controllers to make a coherent decision.

Typically in Fig. 8.1, a storage controller needs to work with other controllers (that control loads and renewable sources) to decide whether the storage unit should be charged or discharged and how much electricity load it should handle. A controller for a renewable energy source (such as a wind turbine or solar PV) needs direction from a system controller to control its power production so as to assure a stable and reliable system operation.

8.1.1 Networked Control System Models

The sensor and controller relationship can be represented as a networked control system with inputs (sensors' data) and outputs (control commands) [2]. As shown in Fig. 8.2, a wireless sensor and controller (WSCN) with delay and packet loss can be used to describe a CPS. It has state transitions based on the control results from $T \rightarrow T+1$.

To illustrate the CPS architecture, an intelligent water distribution network is shown in Fig. 8.3 [3]. Among the physical components, there are pipes, values, and reservoirs. Using this system, researchers are able to track water use. They are also able to predict where the majority of water will

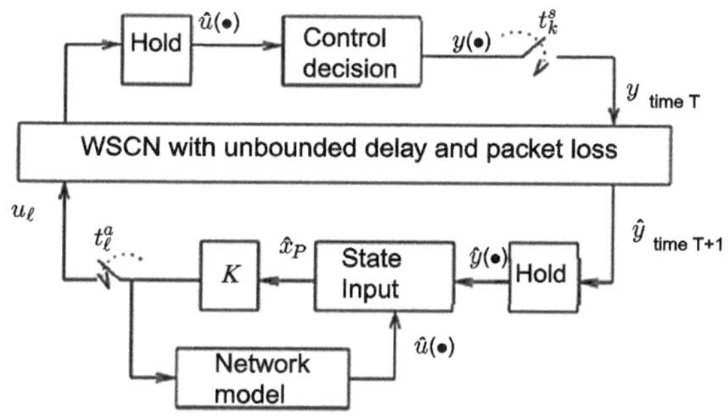

Figure 8.2 CPS state transition

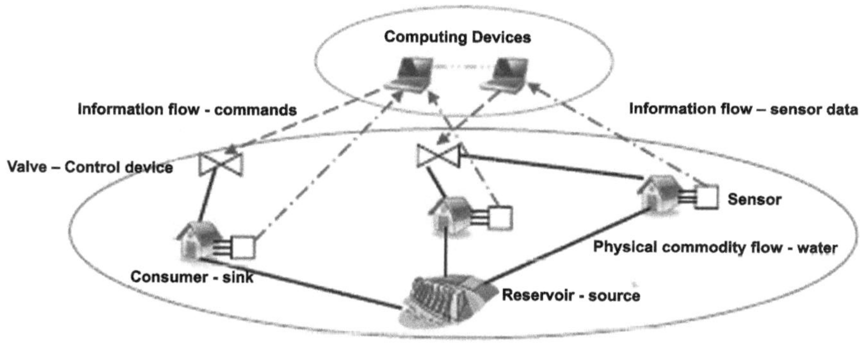

Figure 8.3 Water distribution system as CPS

be consumed. It has a multilayer architecture. One layer is the actual water flow, such as a reservoir of a sink. This layer has cyberobjects (sensors) that communicate to the higher level cyberobjects such as computer devices on how much and when the water will be used. This allows the computers to allocate water to where it will be needed at the correct times. It also allows monitoring the maintenance side of the water flow. It achieves this by monitoring what amount is being used at a house and how much water is being sent to that section. If there is a leak, then more will be sent to that section than what is being used, and so the system will know there is leak or malfunction. See Fig. 8.4

8.2 RESILIENT DESIGN APPROACH

In cyberphysical systems, the strong dependence on the cyberinfrastructure in the overall system increases the risk of cyberattacks, such as injection

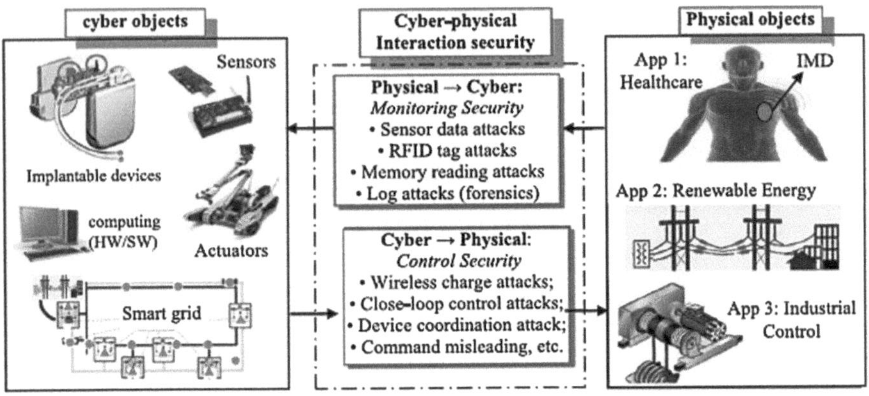

Figure 8.4 Consistency checking design flow

attacks. Security of cyberphysical systems needs to be considered and designed carefully [4]. Cyberphysical security has now developed into a comprehensive discipline that extends classic fault detection, and complements cybersecurity [5]. The aim in this area is to build a resilient mechanism to make the system aware of any attacks in progress and to adjust its dynamics if needed to guarantee desired performance even with attacks or failures.

Much interesting research has been done in this field, and we can provide only a brief summary. Numerous papers have emphasized modeling and detecting adversarial attacks [6–9]. In particular, Pasqualetti et al. [10] characterize detectability of an attack using the system output. The authors of [11] give the threshold value of the number of channels that when attacked can be successfully corrected. Zhang et al. [12] provide a stability condition for a system under Bernoulli DoS attack. Li et al. [13] formulate a game-theoretical framework and provide optimal strategies for the attacker and the plant using Markov chain theory for the case when the channel between sensor and estimator is jammed. Mo and Sinopoli [14] analyze the effect of replay attacks and the trade-off between linear quadratic Gaussian (LQG) performance and the accuracy of a failure detector.

8.2.1 Introduction

Guaranteeing system performance even in the presence of an attack still remains an open challenge. Specifically, most of the existing work focuses on detecting an ongoing attack with the assumption that, once detected, the attacker will be removed from the system. However, often the plant must be run even in the presence of an attack by altering the controller suitably, if needed. This would require a joint design of an attack detector and a controller that switches to maintain suitable performance both in the absence and presence of an attack. Such an architecture has not yet received sufficient attention in the literature. While some recent work [15–17] has considered this architecture in the context of fault detection, such works typically do not consider security threats and impose restrictions on the form and evolution of the disturbance signals. Thus such architectures are more useful for situations when assumptions on attack signals having linear or stochastic dynamics stand, see [10,18,6,19]. A design that simultaneously guarantees system security and control performance has not been explored much.

8.2.2 Background

Dissipativity theory, in general, and passivity, in particular, provide a fundamental perspective for design and analysis of dynamical systems based on a generalized energy concept. Of note, passivity implies stability under weak assumptions.

Consider a continuous system with dynamics given by

$$\begin{aligned} \dot{x} &= f(x, u), \\ y &= h(x, u), \end{aligned} \tag{8.1}$$

where $x \in \mathcal{X} \subseteq \mathbb{R}^n$, $u \in \mathcal{U} \subseteq \mathbb{R}^m$, and $y \in \mathcal{Y} \subseteq \mathbb{R}^m$ are the system state, input and output spaces, respectively; f and h are smooth mappings of appropriate dimensions.

Definition 8.1 ([20]). A state-space system (8.1) is said to be dissipative with respect to supply rate $\omega(u(t), y(t))$, if there exists a nonnegative storage function $V(x) : \mathcal{X} \to \mathbb{R}_{\geq 0}$, satisfying $V(0) = 0$, such that for all $x_0 \in \mathcal{X}$, all $t_1 \geq t_0$, and $u \in \mathcal{U}$,

$$V(x(t_1)) \leq V(x(t_0)) + \int_{t_0}^{t_1} \omega(u(t), y(t))dt, \tag{8.2}$$

where $x(t_0) = x_0$ and $x(t_1) = \phi(t_1, t_0, x_0, u)$.

Definition 8.2 ([20,21]). Suppose system (8.1) is dissipative. It is called:

1. *Passive,* if (8.2) holds for

$$\omega(u, y) = u^T y. \tag{8.3}$$

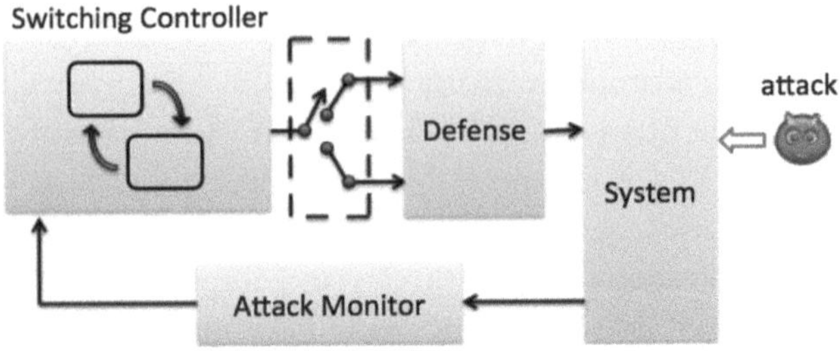

Figure 8.5 System framework considered

2. *QSR-dissipative*, if (8.2) holds and there exist matrices $Q = Q^T$, $S, R = R^T$ such that

$$\omega(u, y) = u^T Ru + 2y^T Su + y^T Qy. \tag{8.4}$$

3. $\mathcal{L}_2$ stable with finite gain $\gamma > 0$, if the system is dissipative with supply rate given by (8.4) where $R = \gamma^2 I$, $S = 0$, $Q = -I$ such that with a $\beta \leq 0$,

$$y^T y \geq -\beta + \gamma^2 u^T u. \tag{8.5}$$

8.2.3 Problem Statement

In this paper, we consider the detection of a data injection attack and operation of the plant even after an attack has been detected by switching the controller. Through an appropriate passivation approach, local passivity and exponential stability are guaranteed even under attack. Because passivity is compositional, this provides a preliminary setup for possible passivity-based control of large scale interconnected networked systems. The overall system framework is shown in Fig. 8.5.

We consider a linear model (8.6) for the plant and assume that an attacker knows the parameters $\{A, B, C, D\}$ for the system under attack, see (8.6). The intelligent attacker intends to corrupt the system state and the measurements based on this knowledge. To this end, he/she injects data through the external control inputs. Conversely, the controller seeks to monitor the measured output to identify if an attacker is present. If an attack is detected, the controller switches to another configuration to maintain performance in spite of the attack.

One possible strategy for designing the controller that guarantees a desired level of performance in spite of the presence of an attack is to design

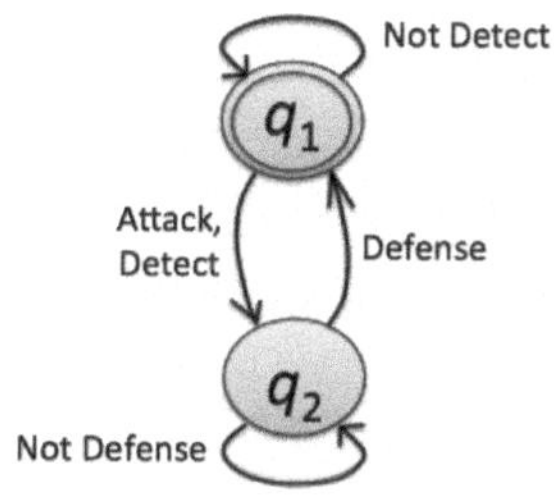

Figure 8.6 A hybrid automaton framework. Event *attack* is triggered if the attacker injects attack signal into the system. Event *detect* is triggered if the attack monitor successfully detects the existence of attack signal. Event *defense* is triggered if the system becomes passive by switching the controller

using a non-switching H^∞ controller. However, this procedure may lead to a design that is too conservative when an attack is only rarely present. Instead, we design a controller using a passivity framework that ensures that the passivity levels of the closed-loop system are guaranteed even when an attack is present. For this, we use the input–output transformation matrix M as introduced in [22] that does not require knowledge of the passivity levels of either the plant or the controller.

8.2.4 System Model

Consider a system with dynamics given as follows:

$$\begin{aligned} \dot{x}(t) &= Ax(t) + Bu(t) + w(t), \\ y(t) &= Cx(t) + Du(t), \end{aligned} \tag{8.6}$$

where $x(t) \in \mathbb{R}^n$, $u(t) \in \mathbb{R}^m$, $y(t) \in \mathbb{R}^p$ are state, system input and output, respectively; w is the unknown external control input that the attacker may possibly inject into the system. This term is set to 0 when the system is in normal operation. We refer to the signal $w(t)$ as the attack signal. We assume that the system input signal is smooth. With a switching controller, the evolution of the system is described in Fig. 8.6.

8.2.5 Attack Monitor

Lemma 8.1 ([10]). *An attack is undetectable by any attack monitor if there exist initial conditions x_1, x_2 and an attack signal w such that, for all $t \geq 0$, the input injected by this attack generates zero-dynamics on the plant as*

$$y(x_1, w, t) = y(x_2, 0, t). \tag{8.7}$$

Remark 8.1. *In view of the fundamental limitation of an attack monitor as illustrated in [10], we limit our consideration to detectable attacks. We use a modified Luenberger observer for attack monitoring.*

Assume that the unknown external control input that the attacker injects into the system is expressed as

$$w(t) = \hat{w}(t) + \Delta w, \tag{8.8}$$

where $\hat{w}(t)$ is the estimate of $w(t)$ and Δw is the corresponding estimation error. Let L denote the observer gain matrix. The classic Luenberger type disturbance observer can be written as

$$\dot{\hat{w}}(t) = -L\hat{w}(t) + L(\dot{x} - Ax - Bu), \tag{8.9}$$

from which we can get the equalities

$$\dot{\hat{w}}(t) = L\Delta w, \tag{8.10}$$

and

$$\Delta\dot{w} = \dot{w}(t) - L\Delta w. \tag{8.11}$$

Since a nonlinear dependency on x is not realizable in practice, see [23], we need to modify the design for the monitor as follows.

Theorem 8.1. *Consider system* (8.6) *and assume that the attacks are detectable. The attack detection filter can be obtained as*

$$\begin{aligned}
\dot{\epsilon} &= -l(x)\epsilon + l(x)(-Ax - Bu - \rho(x)),\\
\hat{w} &= \epsilon + \rho(x),\\
\frac{d}{dt}\rho(x) &= l(x)\dot{x},
\end{aligned} \tag{8.12}$$

where ϵ is the internal state variable of the monitor, $\rho(x)$ is a nonlinear function to be designed and $l(x)$ is the gain of the modified detection filter that can possibly depend on x. The output of the detection filter is the residual signal

$$v(t) = \hat{w}(t). \tag{8.13}$$

Proof. We let the internal state variable

$$\epsilon = \hat{w} - \rho(x), \tag{8.14}$$

where $\rho(x)$ can be determined by the modified detection filter gain $l(x)$ as

$$\frac{d}{dt}\rho(x) = l(x)\dot{x}.$$

Then according to (8.6), (8.8), (8.10) and (8.14), we have

$$\begin{aligned}
\dot{\epsilon} &= \dot{\hat{w}} - \dot{\rho}(x) \\
&= l(x)\Delta w - l(x)\dot{x} \\
&= -l(x)[\epsilon + \rho(x)] + l(x)[Ax + Bu + \rho(x)] - l(x)\dot{x} \\
&= -l(x)\epsilon + l(x)(-Ax - Bu - \rho(x)).
\end{aligned}$$

Thus, the modified attack detection filter (8.12) can be achieved. □

The error dynamics is given by

$$\begin{aligned}
\Delta\dot{w} &= \dot{w}(t) - \dot{\hat{w}}(t) = \dot{w} - \dot{\epsilon} - \frac{d\rho(x)}{dt} \\
&= l(x)[\hat{w} - \rho(x)] + l(x)[Ax + Bu + \rho(x)] + \dot{w} - l(x)\dot{x} \\
&= \dot{w}(t) - l(x)[w(t) - \hat{w}(t)].
\end{aligned}$$

Thus, we obtain that

$$\Delta\dot{w} = \dot{w}(t) - l(x)\Delta w. \tag{8.15}$$

Remark 8.2. *Note that the modified attack monitor does not need the derivative term $\dot{x}$; yet, it has similar error dynamics as the basic disturbance observer error dynamics. This attack monitor can not only detect the existence of an attack, but also track its trajectory. We can also use the detected signal as an estimated value of the unexpected input.*

The following theorem gives us a criterion to design the filter gain $l(x)$.

Theorem 8.2. *If there exists an invertible matrix X and a positive definite matrix Γ such that with the gain $l(x)$,*

$$l(x)^T X^T X + X^T X l(x) \geq \Gamma, \tag{8.16}$$

and further the derivative of $w(t)$ is negligible compared with Δw as in estimation error dynamics (8.8), i.e., $\dot{w}(t) \approx 0$, then the designed disturbance observer is exponentially stable.

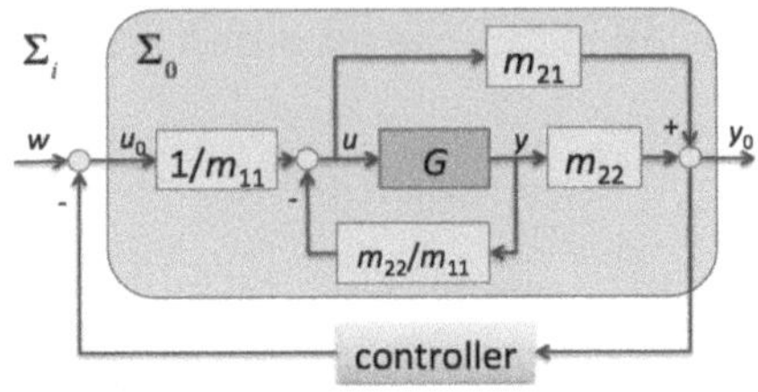

Figure 8.7 System framework with transformation matrix M

Proof. Consider the candidate Lyapunov function as

$$W(\Delta w, x) = (X\Delta w)^T(X\Delta w) = \Delta w^T X^T X \Delta w. \tag{8.17}$$

We can see that the scalar function W is positive definite. With the equations (8.15), (8.16) and (8.17), when $\dot{w}(t) \approx 0$, we get

$$\begin{aligned}\dot{W}(\Delta w, x) &= \Delta\dot{w}^T X^T X\Delta w + \Delta w^T X^T X \Delta\dot{w}\\ &= (\dot{w} - l(x)\Delta w)^T X^T X\Delta w\\ &\quad + \Delta w^T X^T X(\dot{w}) - l(x)\Delta w)\\ &= -\Delta w^T (l(x)^T X^T X + X^T X l(x))\Delta w.\end{aligned}$$

Since the condition in Theorem 8.2 is satisfied, $\dot{W}(\Delta w, x)$ is negative definite for all Δw and $\lim_{t\to\infty} \Delta w = 0$ for all $\forall \Delta w \in \mathbb{R}^n$. Moreover, from (8.15), the disturbance tracking error will converge exponentially to zero for all $\forall\Delta w \in \mathbb{R}^n$. It implies $\hat{w}$ will exponentially approach w if the detection gain $l(x)$ is chosen as in (8.16) regardless of x. □

8.2.6 Switching the Controller

Once an attack is detected, the controller switches to a new structure that ensures that the system continues to be stable and perform well. For this purpose, we use the framework of transformation M-matrix that is pictorially depicted in Fig. 8.7. The parameters m_{11}, m_{12}, m_{21}, and m_{22} are chosen such that closed-loop system guarantees desired passivity level even when no a priori knowledge of passivity indices of the system and controller is available.

We proceed as follows. Consider the unforced system [24]

$$\begin{aligned}\dot{x}(t) &= \Sigma_{i=1}^{2}\sigma_i A_i x(t) + \Sigma_{i=1}^{2}\sigma_x w(t),\\ y(t) &= \Sigma_{i=1}^{2}\sigma_i C_i x(t),\end{aligned} \tag{8.18}$$

where $\sigma_i(t) = \begin{cases} 1, & \text{when } \Sigma_i \text{ is active,} \\ 0, & \text{otherwise.} \end{cases}$ We define the indicator function $\sigma(t) = [\sigma_1(t)\ \sigma_2(t)]^T$ with $\sigma_1(t) + \sigma_2(t) = 1$.

The optimal control goal is to guarantee the passivity of the closed-loop switching system. If the passivation transformation M is chosen appropriately such that the hybrid automaton is passive, we can say that event *defense* is triggered.

A quadratic cost function is defined as

$$L_r(u, w) = \int |y(t)|^2 dt - \gamma^2 \int |w(t)|^2 dt. \tag{8.19}$$

An optimal control policy [25] guarantees the desired system performance

$$\frac{\|y\|}{\|w\|} \leq \gamma, \tag{8.20}$$

where $\|w(t)\|^2 = \int |w(t)|^2 dt$, $\|y(t)\|^2 = \int |y(t)|^2 dt$. The following theorem illustrates the condition that automaton mode Σ_i is stable when that mode is active.

Theorem 8.3. *For a fixed $\gamma \geq 0$, the system has stable performance* (8.20) *under attack, if there exist $\beta_{ij} \geq 0$ and symmetric positive definite matrix P_i, $i, j \in \overline{M} = \{1, 2\}$ such that*

$$\begin{bmatrix} A^T P_i + P_i A + \gamma^{-1} C^T C + \Sigma_{i=1}^2 \beta_{ij}(P_j - P_i) & P_i \\ P_i & -\gamma I \end{bmatrix} < 0. \tag{8.21}$$

Proof. Choose a transition law between Σ_1 and Σ_2 mode as

$$\sigma(t) = \arg\min_{i \in \overline{\mathcal{M}}} \theta(t) P_i \theta(t), \tag{8.22}$$

where $\theta(t)^T = [x(t)^T \quad w(t)^T]$. Choose a candidate Lyapunov function for the hybrid system as

$$V(t, x(t)) = x(t)^T (\Sigma_{i=1}^2 \sigma_i(t) P_i) x(t). \tag{8.23}$$

Based on whether the transition between modes happens or not, we consider two cases:

1. When $\sigma(t+\Delta t)=\sigma(t)=i$, i.e., there is no transition between two modes, we can write, based on (8.18) and (8.23),

$$
\begin{aligned}
\dot{V} &= \dot{x}^T(\Sigma_{i=1}^2\sigma_i P_i)x(t) + x(t)^T(\Sigma_{i=1}^2\sigma_i P_i)\dot{x} \\
&= [\Sigma A x + \Sigma w]^T(\Sigma\sigma P_i)x + x^T(\Sigma\sigma P_i)[\Sigma A x + \Sigma w] \\
&= \theta(t)^T\begin{bmatrix} A^T P_i + P_i A & P_i \\ P_i & 0 \end{bmatrix}\theta(t).
\end{aligned}
$$

2. When $\sigma(t+\Delta t)=\sigma(\tilde{t})=j$, $\sigma(t)=i\neq j$, i.e., the transition between modes happens, using (8.22), we get

$$
\begin{aligned}
\dot{V}(t,x) &= \lim_{\Delta t\to 0}\frac{x(\tilde{t})^T P_j x(\tilde{t}) - x(t)^T P_i x(t)}{\Delta t} \\
&\leq \lim_{\Delta t\to 0}\frac{x(\tilde{t})^T P_i x(\tilde{t}) - x(t)^T P_i x(t)}{\Delta t} \\
&= \begin{bmatrix} A^T P_i + P_i A & P_i \\ P_i & 0 \end{bmatrix}\theta.
\end{aligned}
$$

Define $L(y(t),w(t)) = \gamma w(t)^T w(t) - \gamma^{-1}y(t)^T y(t)$. Then, based on (8.18),

$$
\begin{aligned}
&\dot{V}(t,x) - L(y,w) \\
&\leq \theta^T\begin{bmatrix} A^T P_i + P_i A & P_i \\ P_i & 0 \end{bmatrix}\theta - \gamma w^T w + \gamma^{-1}y^T y \\
&= \theta^T\begin{bmatrix} A^T P_i + P_i A + \gamma^{-1}C^T C + \Sigma\beta(P_j - P_i) & P_i \\ P_i & -\gamma I \end{bmatrix}\theta.
\end{aligned}
$$

If (8.21) is satisfied, $\dot{V}(x) - L(y,w) < \theta^T(\Sigma\sigma_i(P_i - P_j))\theta$. So, based on (8.22), we get

$$
\dot{V}(t,x) < L(y,w), \quad \forall t. \tag{8.24}
$$

1. When $w=0$, $\dot{V}(t,x) < L(y,w) = -\gamma^{-1}y(t)^T y(t)$, and the system is stable.
2. Under zero initial condition, we have $\int_0^\infty \dot{V}(t,x)dt < \int_0^\infty w(t)dt$. Since $\int_0^\infty \dot{V}dt = V(\infty)$, this implies that

$$
\int_0^\infty |y(t)|^2 dt < \gamma^2\int_0^\infty |w(t)|^2 dt, \tag{8.25}
$$

that is, optimal performance for the given γ is assured. □

Under condition (8.21), Σ_i is finite gain stable. The objective is to obtain the criteria of how to choose passivation parameters in Fig. 8.7 such that the closed-loop hybrid system is passive without any prior knowledge of the passivity levels for plant G and the preset controller H. The following theorem discusses how to select the M-matrix in order to render active automaton mode Σ_i passive.

Theorem 8.4. *Consider system Σ_i with* (8.21) *being satisfied. Assume the pre-designed controller H is passive. If we select passivation transformation M-matrix with*

$$\begin{aligned}
&\frac{m_{11}m_{12}}{\gamma_1^2} + m_{12}m_{22} \geq 0,\\
&m_{12}m_{21} + m_{11}m_{22} = 0,
\end{aligned} \tag{8.26}$$

then the interconnected feedback system Σ_i under attack is passive.

Proof. Now $u_0 = w - y_2$, $u_1 = \frac{1}{m_{11}}u_0 - \frac{m_{12}}{m_{11}}y_1$, $y_0 = u_2 = m_{21}u_1 + m_{22}y_1$. We have

$$w = u_0 + y_2 = m_{11}u_1 + m_{12}y_1 + y_2. \tag{8.27}$$

Since system G is finite gain stable under (8.21), let its finite gain be $\gamma_1 > 0$. We have supply rate

$$\omega(u_1, y_1) = \gamma_1^2 u_1^T u_1 - y_1^T y_1 \geq 0. \tag{8.28}$$

When interconnected system is passive, we have

$$\begin{aligned}
\omega(w, y_0) &= w^T y_0\\
&= m_{11}m_{21}u_1^T u_1 + m_{12}m_{22}y_1^T y_1\\
&\quad + (m_{12}m_{21} + m_{11}m_{22})u_1^T y_1\\
&\quad + m_{21}y_2^T u_1 + m_{22}y_2^T y_1\\
&= \begin{bmatrix} u_1^T & y_1^T \end{bmatrix} \begin{bmatrix} m_{11}m_{21} & m_{11}m_{22}\\ m_{12}m_{21} & m_{12}m_{22} \end{bmatrix} \begin{bmatrix} u_1\\ y_1 \end{bmatrix}\\
&\quad + y_2^T u_2 \geq 0.
\end{aligned}$$

The interconnected system is passive, if (8.26) is satisfied with assumption that the predesigned controller H is passive. □

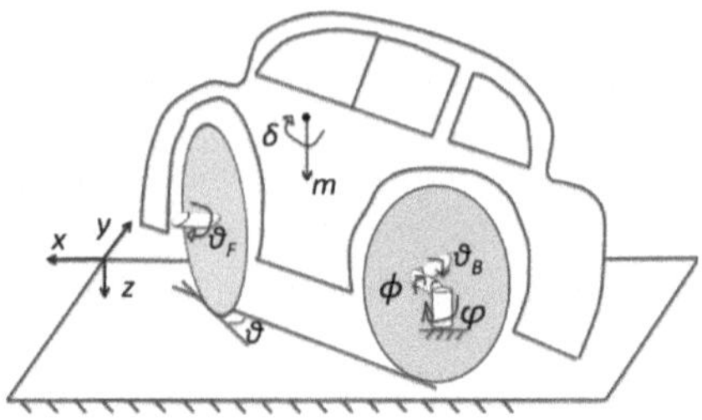

Figure 8.8 A vehicle dynamical system

Theorem 8.5. *Consider the automaton framework in Fig. 8.6. If we select passivation transformation M-matrix with criteria* (8.26)*, passivity of the overall hybrid automaton is guaranteed under condition* (8.21).

Proof. If we select transformation M-matrix with (8.26), each active mode Σ_i is passive under condition (8.21). Similarly, considering inactive mode $\Sigma_j, j \neq i$, we have $u_0 = m_{11}u_1 + m_{12}y_1$, $y_0 = m_{21}u_1 + m_{22}y_1$, and $w = u_0 + y_2$.

Arbitrary choosing the M-matrix with $m_{11} \neq 0$, there exists a supply rate

$$\omega(w, y_0) = y_0^T y_0 = u_2^T u_2 \geq 0. \tag{8.29}$$

Each mode Σ_i is dissipative when it is inactive. Moreover, since the accumulated energy flows from an active subsystem i to an inactive subsystem j at each switching instant t_{i_k}, $\forall i, k$, it is finite without external supply, i.e.,

$$V_0(x(t_0)) + \sum_{i_k=1}^{\infty}[V_{i_k}(x(t_{i_k})) - V_{i_{k-1}}(x(t_{i_k}))] \leq \infty. \tag{8.30}$$

According to Definition 3.3 in [26], the hybrid automaton is passive. □

8.2.7 Simulation Example 8.1

Consider a cyberphysical vehicle as in Fig. 8.8.

Let the dynamics of the vehicle dynamical system be [27]:

$$\begin{bmatrix} \ddot{\phi} \\ \ddot{\delta} \\ \dot{\phi} \\ \dot{\delta} \end{bmatrix} = \begin{bmatrix} -2.11 & -6.61 & 9.48 & -357.05 \\ 73.54 & -61.70 & 11.71 & -757.81 \\ 1 & 0 & 0 & 0 \\ 0 & 1 & 0 & 0 \end{bmatrix} \begin{bmatrix} \dot{\phi} \\ \dot{\delta} \\ \phi \\ \delta \end{bmatrix}$$

$$+\begin{bmatrix} T_\phi & T_\delta & 0 & 0 \end{bmatrix}^T u + \begin{bmatrix} 0 & 8 & 0 & 0 \end{bmatrix}^T w,$$

$$y(t) = \begin{bmatrix} 20 & 8 & 0 & 1 \end{bmatrix}\begin{bmatrix} \dot{\phi} & \dot{\delta} & \phi & \delta \end{bmatrix}^T, \tag{8.31}$$

Lean rotation ϕ is the angular rotation about x-axis. Steering angle δ is the rotation of the front tires with respect to the rear tires about the steering axis. T_ϕ represents the right lean torque and T_δ is an action–reaction steering torque. Here we set $T_\phi = 1.2$ and $T_\delta = 10$; w is the unknown external input that is injected by an attacker. Assume the forward speed $V = 20\,\mathrm{ms}^{-1}$.

In this example, first we choose the detector gain independently of x based on Theorem 8.1 as $l(x) = X^{-1} = 2$. The attack detector can be designed as

$$\dot{\epsilon} = -2\epsilon + \begin{bmatrix} 0.22 & 13.22 & -18.96 & 714.1 \\ -147.07 & 119.39 & -23.42 & 1515.62 \\ -2 & 0 & -4 & 0 \\ 0 & -2 & 0 & -4 \end{bmatrix}\begin{bmatrix} \dot{\phi} \\ \dot{\delta} \\ \phi \\ \delta \end{bmatrix}$$

$$\begin{bmatrix} -2.4 \\ -20 \\ 0 \\ 0 \end{bmatrix} u,$$

$$\hat{w} = \epsilon + \rho(x),$$

$$\frac{d}{dt}\rho(x) = l(x)\dot{x} = 2\dot{x}. \tag{8.32}$$

We implement the designed attack monitor. The attacker generates an aperiodic rectangular signal with sample time 5 seconds. The simulation result for the designed attack monitor is shown in Fig. 8.9.

The first part in Fig. 8.9 represents the dynamic response of the system under irregular pulse injected by the attacker. The blue dashed line depicts the dynamic of angular lean velocity $\dot{\phi}$, and the red solid line represents the response of angular steering velocity $\dot{\phi}$ for the initial condition $(\dot{\phi}, \dot{\delta}) = (8\,\mathrm{rad/s}, 10\,\mathrm{rad/s})$.

In the second part of Fig. 8.9, the blue dashed line represents the real signal the attacker injected to the system, and the red solid line represents the output of the attack monitor. We can see that the designed monitor tracks the unknown input well.

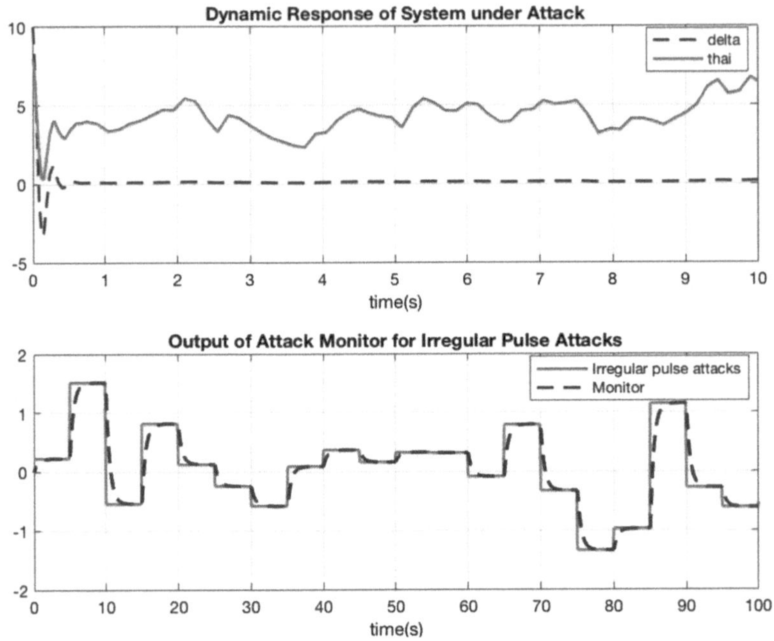

Figure 8.9 Dynamic response of a system under attack

Furthermore, choosing $\beta_{21} = 2$, condition (8.21) is satisfied. We get $\gamma = 1.008$. Select transformation M-matrix as $m_{11} = 1$, $m_{12} = 1.2$, $m_{21} = -1.2$, and $m_{22} = 6$ such that condition (8.26) is satisfied. Set the stable state feedback controller with feedback gain as

$$K = \begin{bmatrix} 0.1728 & 0.0074 & -0.0008 & 2.1023 \end{bmatrix}.$$

The defense is realized by M-matrix passivation and the dynamic response of system after correction is shown in the first part of Fig. 8.10. The second part of Fig. 8.10 represents the supply rate of the system after the defense mechanism has been triggered.

The first part of Fig. 8.10 depicts the dynamic response of $\dot{\phi}$ after correction with blue dashed line and the response of $\dot{\delta}$ after defense mechanism being triggered with red solid line. As we can see, the velocities become smoother after correction. At the mean time, the angular lean velocity becomes more gentle so that the driving process is robust in spite of the irregular force applied to the vehicle. The discomfort of the driver and passengers in the vehicle after being attacked has been reduced. The second

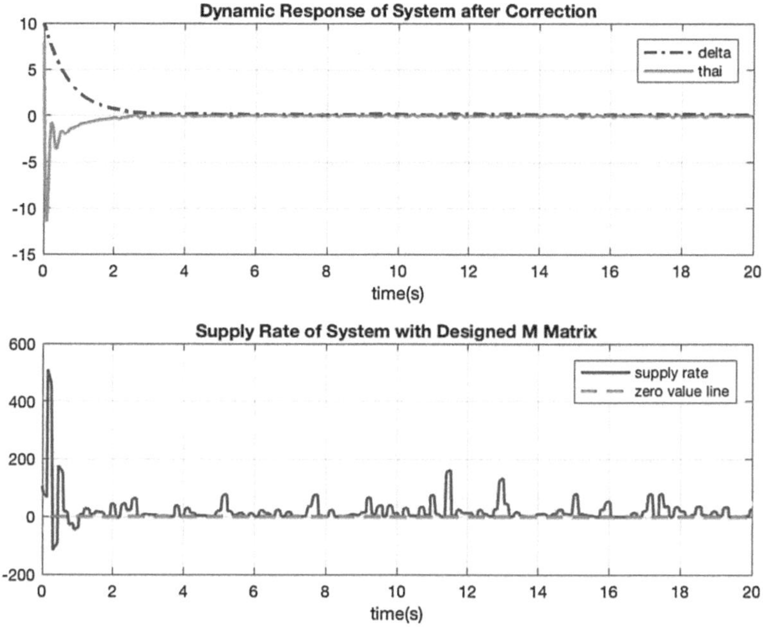

Figure 8.10 Dynamic response and supply rate after correction

part of Fig. 8.10 depicts the supply rate of the system after correction. Define the supply rate as the inner product of injected input and output of the system. The output supply rate of the system shown here is always positive after the settling time. This demonstrates that the system under attack is passive with designed passivation M-matrix.

8.3 REMOTE STATE ESTIMATION UNDER DOS ATTACKS

In order to enrich human–machine interactions in the physical/virtual world, various areas, such as aerospace, healthcare, transportation, civil infrastructure and smart grids, have increasingly adopted cyberphysical systems (CPSs) during recent years. A CPS is a system connecting the cyberworld (e.g., information, communication) with the physical world by integrating computation, communication and control. In a CPS, multiple static/mobile sensors and actuators, which communicate with each other, interact with physical processes under the control of an intelligent decision system [31]. Traditionally, the sensing data is transmitted to the control centers through a single channel. However, this technique cannot provide reliable and timely communication in the competitive communication environment brought by distributed wireless sensor networks (WSNs) in CPSs as a result of limited bandwidth. A multichannel network is an efficient technology to alleviate bursty communication traffic by transmitting the signals over several channels [34]. Besides this, multichannel networks can increase the resilience ability of a system under some rare events, e.g., severe weather or natural disasters [42]. Consequently, current MAC protocols, WSN hardware and commercially available radios for sensors support multichannel communication. Hence, it is natural and important to consider multichannel networks in the study of CPSs.

8.3.1 Introduction

Although multichannel techniques improve the throughput and capacity of the network, they cannot alleviate severe security challenges brought by the cybercomponents of the CPSs, especially vulnerabilities of wireless connection among sensors, estimators and actuators. Since CPSs are connected to many safety-critical infrastructure systems, attacks can lead to severe damage, as in the reported incident of the advanced computer worm Stuxnet infecting industrial control systems [47]. Therefore, careful defense-strategy design is essential for assuring the safe operation of CPSs in the face of adversarial attacks. In the present work, we mainly deal with DoS attacks [40], which block the information flow between the sender (sensor) and the receiver to increase the packet-drop rate, and propose a defense-strategy design under DoS attacks for the multi-channel CPSs.

Energy limitation is inevitable [44] as sensing, computation and transmission power are restricted for sensor nodes and, moreover, replacing or recharging batteries may not always be possible in some WSNs. Consequently, a complicated security issue in CPSs (see Fig. 8.11) arises. To be more specific, facing DoS attacks launched by an intelligent attacker, the receiver (estimator) should choose a channel which is not just energy-efficient, but also dodges the attacks. Simultaneously, the smart attacker may notice the evasive actions taken by the sensor, and then modify its attack mode. Thus, an interaction between the sensor and the attacker occurs incessantly. The interaction can be captured by a game-theoretical approach, under which several recent studies about CPS security have been carried out, see [28,38,39,41,45,47].

Based on cooperative game theory, the survey conducted by [28] has proposed a new method for clustering sensors to provide a higher level of security. Instead of adopting model-based attacks to networks, in the

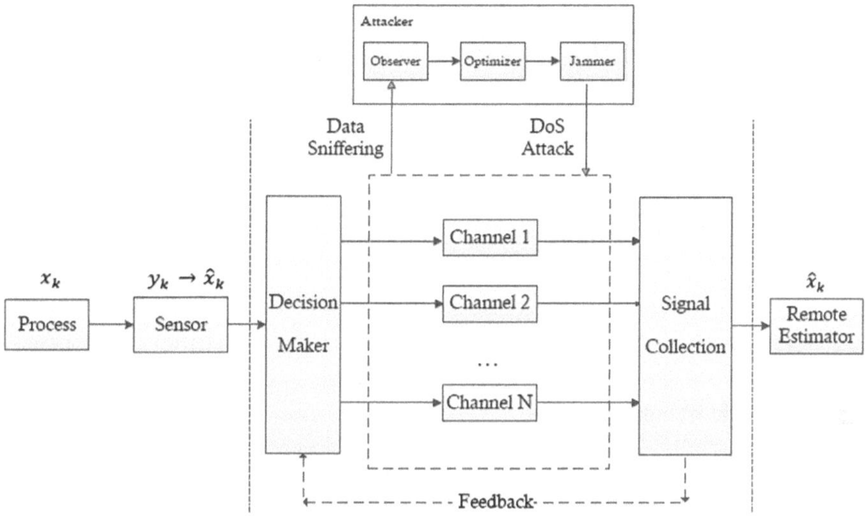

Figure 8.11 Multichannel remote estimation with an attacker

formulation of [41], an intelligent attacker with a dynamic and random attack strategy was considered in developing a two-player zero-sum stochastic game. Our recent work [40] studied an interactive power scheduling between a sensor and a jammer, and used Markov chain theory to obtain the equilibrium solution. The recent work by [47] introduced a crosslayer system design problem which results in solving a zero-sum differential game for robust control, coupled with a zero-sum stochastic game for a security policy in the absent of power constraints. In [37], Li et al. used an ideal multichannel system between a sensor and a remote estimator to avoid DoS attacks in a smart grid.

Compared with previous works, the main contributions of our current work are summarized as follows:

1. With few studies undertaking an analysis of multichannel networks for the secure state estimation problem in CPSs, this work sheds new light on the practical multichannel-based estimation issue under malicious DoS attacks. In communication theory, the transmitted signal is typically assumed to be independent and identically distributed or at least stationary and ergodic; however, in this paper we consider a dynamical process and focus on improving estimation quality in an effective and defensive way.

2. We investigate the iterative process between the sensor and the attacker by developing a two-player zero-sum stochastic game. By involving an elastic data arrival rate matrix to mathematically represent the relationship between the mixed strategies and the average packet loss rate, we obtain the reward function of the game. Moreover, by developing a Nash Q-learning algorithm, we acquire the optimal rational strategies for the sensor and the attacker.
3. The stochastic game presented can be extended to include the channel choices and the power level selections, providing an important opportunity for the analysis of power allocation in multichannel networks.

8.3.2 Problem Setup

In this section, we will introduce the mathematical models of the multichannel structure depicted in Fig. 8.11, which comprises sensing, communication and estimation parts.

8.3.3 Process and Sensor Model

Consider the following linear time-invariant system:

$$\begin{aligned} x_{k+1} &= Ax_k + w_k, \\ y_k &= Cx_k + v_k, \end{aligned} \tag{8.33}$$

where $x_k \in \mathbb{R}^n$ is the system state vector, $y_k \in \mathbb{R}^m$ is the measurement taken by the sensor at time k, $w_k \in \mathbb{R}^n$ and $v_k \in \mathbb{R}^m$ are zero-mean i.i.d. Gaussian random noises with $\mathbb{E}[w_k w_j'] = \delta_{kj} Q$ $(Q \geq 0)$, $\mathbb{E}[v_k v_j'] = \delta_{kj} R$ $(R \geq 0)$, $\forall j, k$. The initial state x_0 is a zero-mean Gaussian random vector with covariance $\Pi_0 \geq 0$, and which is uncorrelated with w_k and v_k. The pair (A, C) is assumed to be detectable and $(A, \sqrt{Q})$ is stabilizable.

In current CPSs, sensors are usually designed to be "smart" [32] to improve the estimation/control performance of the system. Advanced embedded systems-on-chip render the improvement possible as they equip sensors with storage and computing capabilities, and they also allow sensors to process the collected information by executing some simple recursive algorithms. Thus, the sensor in Fig. 8.11, after taking measurements at time step k, runs a Kalman filter to estimate the state x_k of the process locally based on the overall collected measurements, instead of transmitting them to the remote estimator directly. The local minimum mean-squared error (MMSE) estimate of the process x_k is denoted by $\hat{x}_k^s = \mathbb{E}[x_k | y_0, \ldots, y_k]$. The corresponding estimation error e_k^s and the error covariance matrix P_k^s are defined as $e_k^s \triangleq x_k - \hat{x}_k^s$ and $P_k^s \triangleq \mathbb{E}[(e_k^s)(e_k^s)'|y_0, \ldots, y_k]$. These terms are computed via the Kalman filter and the iteration starts from $\hat{x}_0^s = 0$ and $P_0^s = \Pi_0$. For convenience, we define the Lyapunov and Riccati operators h and $\tilde{g} : \mathbb{S}_+^n \to \mathbb{S}_+^n$ as $h(X) \triangleq AXA' + Q$ and $\tilde{g}(X) \triangleq X - XC'[CXC' + R]^{-1}CX$. The error covariance P_k^s converges exponentially to a unique fixed point $\bar{P}$ of $h \circ \tilde{g}$ [29]. For simplicity, we ignore the transient periods and assume that the Kalman filter at the sensor has entered steady state, i.e., we assume that

$$P_k^s = \bar{P}, \quad k \geq 1. \tag{8.34}$$

8.3.4 Multichannel Communication and Attack Model

After obtaining the state estimate $\hat{x}_k^s$, the sensor transmits this value as a data packet to the remote estimator through an N-channel communication network. We assume that all N channels have independent additive white Gaussian noise. In practice, the data packets may arrive unsuccessfully to the remote estimator because of noise, signal fading, etc. The packet arrival reliability measured in packet-error-rate (PER) is closely related to the signal-to-noise ratio (SNR) and the intrinsic channel characteristics (such as the channel coding method, the design of the receiver, etc.). Considering a DoS attack for a channel (in which an adversary can congest the communication network), the corresponding $SNR_i = \frac{p_s}{\sigma_i}$ is revised to be SINR (signal-to-interference-and-noise-ratio):

$$SNR_i = \frac{p_s}{p_a + \sigma_i}, \quad \text{if interfered,} \tag{8.35}$$

where p_s and p_a correspond to the transmission power of the sensor and the interference power of the attacker, respectively, and σ_i is the additive white noise power of the ith channel. We assume that the channel gains are taken to be unity; therefore, the definition of received SINR can be defined based on transmission powers instead of the actual received power.

In practice, different channel coding methods (such as cyclic redundancy check) will be used to detect the symbol error of the data before the decoder uploads the data to the receiver. As a result, some packets will be dropped if they contain errors. Furthermore, packet dropouts can also be caused by network congestion, timeout and other channel characteristics. This leads to the general form:

$$PER_i \triangleq \begin{cases} F(SINR_i), & \text{if the } i\text{th channel is attacked,} \\ F(SNR_i), & \text{otherwise,} \end{cases}$$

$$\beta_i = 1 - PER_i \tag{8.36}$$

where β_i represents the packet arrival rate for the ith channel, and $F(\cdot)$ is a nonincreasing function.

8.3.5 Remote State Estimation

Let $\hat{x}_k$ be the MMSE estimate of the process x_k at the remote estimator, with corresponding error covariance P_k. The remote estimator obtains the state estimate as follows [46]: after receiving x_k^s successfully, the estimator synchronizes its estimate x_k with it; otherwise, the estimator simply predicts the estimate based on its previous estimate using the system model (8.33). To be precise,

$$\hat{x}_k \triangleq \begin{cases} \hat{x}_k^s, & \text{if data is received successfully,} \\ A\hat{x}_{k-1}, & \text{otherwise.} \end{cases} \tag{8.37}$$

As a result, the error covariance P_k at time k obeys

$$\begin{aligned} P_k &\triangleq \mathbb{E}[(x_k - \hat{x}_k)(x_k - \hat{x}_k)'] \\ &\triangleq \begin{cases} \bar{P}, & \text{if data is received successfully,} \\ h(P_{k-1}), & \text{otherwise.} \end{cases} \end{aligned} \tag{8.38}$$

Assume that the initial value of the error covariance at the remote estimator also starts from $\bar{P}$, that is, $P_0 = \bar{P}$. This corresponds to receiving a local state estimate (with estimation error covariance $\bar{P}$) at time $k = 0$. According to (8.38), P_k can only take values in the finite set $\{\bar{P}, h(\bar{P}), h^2(\bar{P}), \ldots, h^k(\bar{P})\}$ at a given time k.

8.3.6 Problem of Interest

Given a tight budget of transmission and congestion power, the strategy design of the sensor or the attacker is important for efficient operation. In general, the task of the sensor is to make sure that its end user (i.e., the remote estimator) is sufficiently informed of the process, without wasting energy; and as for the attacker, it tends to disrupt the reliable communication between the sensor and the user, also without expending more effort than required. With opposite goals, the decision-making procedures of the sensor and the attacker are interactively linked. In the sequel, we will investigate the decision-making procedures for the defending sensor and the malicious attacker in a multichannel network with power constraints and develop a game played over time to model this interactive process.

8.3.7 Stochastic Game Framework

Depending on the targets of the attacker, the information available to the defending sensor and the security mechanism, the adopted game may take different forms, such as a static game, Bayesian game, etc., [36]. In order to improve their performance, it is more meaningful for both players to adopt dynamic cost-effective strategies based on the real-time information of the process rather than adopting static "offline" schemes. Motivated by this, we introduce a two-player stochastic game in this section to design the defense-attack strategies.

Define a stochastic game as $\mathfrak{g} = \langle \mathfrak{L}, \mathfrak{S}, \mathfrak{M}, \mathfrak{P}, \mathfrak{F} \rangle$, where $\mathfrak{L}$ is the set of two players; $\mathfrak{S} = (\mathfrak{S}1, \mathfrak{S}2)$ represents the action set for each player in $\mathfrak{L}$; the term $\mathfrak{M}$ denotes the state space of the game; the transition probability $p(m_{k+1}|m_k, s) \in \mathfrak{P} : \mathfrak{M}_k \times \mathfrak{S} \to \mathfrak{M}_{k+1}$ represents the probability of the next state being m_{k+1} given the current state m_k and the current action profile s; and the function $\mathfrak{J} : \mathfrak{M} \times \mathfrak{S} \to \mathbb{R}^{\mathfrak{I}}$ denotes the payoff of each player $\mathrm{i} \in \mathfrak{I}$. At time k, an action pair $s = (s_1, s_2)$ is chosen by the attacker and the sensor according to a mixed strategy pair [36]. The joint actions affect the transition probability $p(m_{k+1}|m_k, s)$ for the next states and also incur a cost. In our problem setup, we have the following specifications.

Player The sensor $\mathfrak{L}_s$ and the attacker $\mathfrak{L}_a$ are assumed to be rational players; i.e., each of them makes the best choice in terms of their own preference among all available actions for them.

Strategy The game is played over time; namely, at every time k, the sensor needs to choose through which channel to send the data packets; the attacker faces the same situation, except with the hostile motivation to block a channel. Hence, the strategies of the sensor and attacker, denoted by $(\vec{\gamma_k}, \vec{\lambda_k}) \in \mathfrak{S}$, are composed of different probabilities of choosing corresponding channels. That is, $\vec{\lambda_k} = [Pr_0, Pr_1, \ldots, Pr_N]'$ and $\vec{\gamma_k} = [\widetilde{Pr}_0, \widetilde{Pr}_1, \ldots, \widetilde{Pr}_N]'$. Note that the extra term Pr_0 describes the likelihood of choosing the inactive state (no data is sent) by the sensor for energy saving; similarly for the attacker, $\widetilde{Pr}_0$ stands for launching no attack with some probability. The two collections of strategies are denoted by $\theta_s \triangleq \{\vec{\gamma}_1, \vec{\gamma}_2, \ldots\}$ and $\theta_a \triangleq \{\vec{\lambda}_1, \vec{\lambda}_2, \ldots\}$ for the sensor and the attacker, respectively.

Remark 8.3. *The representation of the probability for channel selection consists of pure/deterministic strategies (a pure strategy specifies a choice of action) and mixed/randomized strategies (a probability distribution over pure strategies). We are interested in the design of the strategies $\vec{\gamma}_k$, $\vec{\lambda}_2$ (i.e.,* *the probability distribution) for the two players following from the action history.*

State Define the state as the error covariance, i.e., $m_k \triangleq P_k$. Generally, the network uses acknowledgment (ACK) mechanisms for reliable data transfer, i.e., a feedback loop containing the control message from the receiver to the sender (see Fig. 8.11). The estimator explicitly informs the sensor whether the data packet is received successfully or not in an ACK. Thus, the sensor has full knowledge of the state. In the current work, the attacker is assumed to infer the state s_k as well by eavesdropping the ACKs.

Transition probability Note that the transition probability depends on the packet arrival rate in (8.36). To analyze this relationship under the multichannel network, we introduce a matrix describing the packet arrival rate under different actions for each player and then link the game concept with the communication mode. Consider a time-varying matrix

$$M_k = \begin{pmatrix} 0 & 0 & \cdots & o \\ \beta'_{10} & \beta'_{11} & \cdots & \beta'_{1N} \\ \vdots & \vdots & \ddots & \cdots \\ \beta'_{N0} & \beta'_{N1} & \cdots & \beta'_{NN} \end{pmatrix} \in \mathbb{R}^{(N+1)\times(N+1)} \tag{8.39}$$

where the element β'_{nm}, $n, m \in \{0, 1, \ldots, N\}$ denotes the successful packet arrival rate when the sensor transmits data through the n-channel, and the attacker interferes the m-channel. Note that if the sensor decides not to send data, then the arrival rate $\beta'_{0m} = 0$. Moreover, based on (8.36), only the diagonal elements of the matrix M_k are functions of the $SINR$, that is,

$$\beta'_{nm} = \begin{cases} \beta_n(SINR_n), & \text{if } m = n, \\ \beta_n(SNR_n), & \text{if } m \neq n, \end{cases} \tag{8.40}$$

where $n, m \in \{1, \ldots, N\}$. We can obtain the expected packet arrival rate at time k via

$$R_k(M_k, \vec{\lambda}_k, \vec{\gamma}_k) = \vec{\gamma}'_k M_k \vec{\gamma}_k. \tag{8.41}$$

Last, we can derive the transition probability as follows:

$$p(P_{k+1}|P_k, \vec{\lambda}_k, \vec{\gamma}_k, M_k) = \begin{cases} R_k & \text{if } P_{k+1} = \bar{P}, \\ 1 - R_k & \text{if } P_{k+1} = h(P_k), \\ 0 & \text{otherwise.} \end{cases} \tag{8.42}$$

Energy constraints One of the most important goals of network design is to have effective operations with limited energy. For simplicity, we fix the transmission power level (i.e., $\overrightarrow{E_s}$) and attack power level (i.e., $\overrightarrow{E_a}$) with respect to different channels: $\overrightarrow{E_s} = [e_0, e_1, \ldots, e_N]$ and $\overrightarrow{E_a} = [\widetilde{e_0}, \widetilde{e_1}, \ldots, \widetilde{e_N}]$ (an extended form can be found in Section 2.2.3). Then, the expected transmission energy is denoted by $\bar{E}_s = \overrightarrow{E_s} \cdot \overrightarrow{\gamma_k}$ for the sensor, and $\bar{E}_a = \overrightarrow{E_a} \cdot \overrightarrow{\lambda_k}$ applied by the attacker.

Payoff According to the foregoing results, the stochastic game starts at the initial state $P_0 = \bar{P}$. Thus, the game flow is defined as $g = \{P_0, s_0, P_1, s_1, \ldots\}$. A one-stage reward can be defined for each player based on the trade-off between the average performance of the system at the next state and the consumed energy. The average performance is quantified by the expectation $\mathbb{E}(P_{k+1})$. This formulation depends on historical states and needs additional mathematical tools, which will be provided later.

With the sensor aiming to maximize the system performance without too much energy expenditure, the one-stage reward function for the sensor is given by

$$\begin{aligned} r_k(P_{k+1}, \vec{\gamma}_k, \vec{\lambda}_k) \triangleq & -\text{Tr}[\mathbb{E}(P_{k+1})] \\ & - \delta_s \frac{\bar{E}_s}{R_k} + \delta_a \frac{\bar{E}_a}{1 - R_k}, \end{aligned} \tag{8.43}$$

where the function $r_k : \mathfrak{M}_{k+1} \times \mathfrak{S} \to \mathbb{R}$ is in an average sense and the parameters δ_a, δ_s represent the proportions of the energy term in the reward. Similarly, the reward function for the attacker here is taken as $-r_k$. Note that the reward depends on the energy efficiency (i.e., the terms $\delta_s \frac{\bar{E}_s}{R_k}$ and $\delta_a \frac{\bar{E}_a}{1-R_k}$ for the sensor and the attacker) rather than the actual energy, as often used in communication systems [34].

After obtaining a stream of reward functions $\{r_0, r_1, \ldots\}$, the payoff function for the sensor is obtained through a generalization from a Markov decision processes (MDP) to the infinite horizon stochastic game

$$\mathfrak{F}(\theta_s, \theta_a, P_0) = \sum_{k=0}^{+\infty} \rho^k r_k(P_k, \theta_s, \theta_a), \tag{8.44}$$

where the parameter $\rho \in [0, 1)$ stands for the discount factor. The payoff function of the attacker is opposite to that of the sensor. Thus, we have derived a two-player zero-sum stochastic game to interpret the interactive process between the sensor and the attacker.

Assumption 8.1. The sensor and the attacker adopt stationary strategies. Moreover, the transition probability and the reward function are stationary, that is, independent of time k.

Remark 8.4. *Observe that the probabilities over actions for the stationary strategies only depend on the state, that is,* $\vec{\gamma}_k \triangleq \vec{\gamma}_k(s_k)$ *and* $\vec{\delta}_k \triangleq \vec{\delta}_k(s_k)$. *With the evolution of the system state, the strategies are different in every stage and the rational actions determined by the strategies may likewise be time-varying.*

8.3.8 Markov Chain Model

To explain the intrinsic properties of the dynamics of the receiver state P_k, we introduce a Markov decision process, which enables us to characterize $\mathbb{E}(P_{k+1})$. To be precise, we define the chain state of the MDP to be exactly the error covariance P_k. Hence, at time k, the possible values of the error covariance P_k denoted by the state set $\mathbf{Z}_k$ are represented as $\mathbf{Z}_k = \{P_k : \bar{P}, h(\bar{P}), \ldots, h^k(\bar{P})\}$. From (8.42), we describe the transitions from elements in the state set $\mathbf{Z}_k$ to values in $\mathbf{Z}_{k+1}$ via the matrix $\mathbf{T}_k$ with elements

$$\mathbf{T}_k \triangleq \mathbb{P}[\mathbf{Z}_{k+1}|\mathbf{Z}_k] = \begin{cases} R_k & \text{if } j = 1, \\ 1 - R_k & \text{if } j = i + 1, \\ 0 & \text{otherwise,} \end{cases}$$

where $i \in \{1, 2, \ldots, k\}$ and $j \in \{1, 2, \ldots, k+1\}$ stand for the index of the possible values of the states in $\mathbf{Z}_k$ and $\mathbf{Z}_{k+1}$. Define a matrix Π, in which $\Pi(i, j)$ measures the probability of state P_j at time j be equal to one of possible values $h^i(\bar{P})$, and based on $\Pi(\cdot, j) = \Pi(\cdot, j-1)'\mathbf{T}_j$, we have

$$\Pi = \begin{pmatrix} R_1 & \cdots & R_k \\ 1-R_1 & \cdots & (1-R_k)R_{k-1} \\ 0 & \cdots & (1-R_k)(1-R_{k-1})R_{k-2} \\ \vdots & \ddots & \vdots \\ 0 & \cdots & (1-R_k)(1-R_{k-1})\cdots(1-R_1) \end{pmatrix}, \tag{8.45}$$

where $i \in \{0, 1, \ldots, k\}$ and $j \in \{1, 2, \ldots, k\}$. Considering the probability distribution of P_k, i.e., the kth column of the matrix Π, we can evaluate the expected error covariance at time k as follows:

$$\mathrm{Tr}[\mathbb{E}(P_k)] = \mathrm{Tr}\left[\sum_{i=1}^{k+1} \Pi(i,k) h^{i-1}(\bar{P})\right], \quad \forall k \in \mathbb{Z}^+. \tag{8.46}$$

By now, we have an explicit expression for all the game elements. The solution of the optimal strategies for each player will be given later on.

8.3.9 Extension to Power-Level Selection

In the original formulation, only one power level was utilized for each channel. The above strategies for the sensor and attacker can be easily extended to the case of multiple power levels. For example, if there are l_i, $i \in \{1, \ldots, N\}$ transmission power levels e_i^j for each channel, then the whole power level set is denoted by $\overrightarrow{E_s} = 0 \cup \{e_i^j : j \in \{1, 2, \ldots, l_i\}, i \in \{1, 2, \ldots, N\}\}$, covering the no-data-transmission mode. Respectively, we have $\overrightarrow{\gamma_k} = [Pr_0, Pr_1^1, \ldots, Pr_i^j, \ldots, Pr_N^{l_N}]'$ and $\overrightarrow{\lambda_k} = [\widetilde{Pr_0}, \widetilde{Pr_1^1}, \ldots, \widetilde{Pr_i^j}, \ldots, \widetilde{Pr_N^{l_N}}]'$, where Pr_i^j represents the probability for the sensor to choose channel i and transmit data packets with power e_i^j. Hence, the pure strategy s for the sensor considers the $1 + \sum_{i=1}^{N} l_i$ power-level choices, which covers the $1 + N$ channel selections.

8.3.10 Equilibrium Analysis

To solve the strategy-design problem, we first introduce the equilibria of the stochastic game. Then, we prove the existence of the optimal strategies in Theorem 8.6. Eventually, the strategy-design problem is equivalent to finding the NE of the stochastic game.

Definition 8.3 (Nash Equilibrium [36]). In the two-player zero-sum stochastic game with the initial state P_0 between the sensor and attacker, a strategy profile π_s^* (or π_a^*) for the sensor (or attacker) is a Nash equilibrium if no player can benefit from changing strategies while the other keeps its own unchanged, i.e.,

$$\mathfrak{F}_S^*(P_0) \doteq \mathfrak{F}_S(P_0, \pi_s^*, \pi_a^*) \geq \mathfrak{F}_S(P_0, \pi_s, \pi_a^*), \qquad \forall \pi_s$$

and

$$\mathfrak{F}_A^*(P_0) \doteq \mathfrak{F}_A(P_0, \pi_s^*, \pi_a^*) \geq \mathfrak{F}_A(P_0, \pi_s^*, \pi_a), \qquad \forall \pi_a.$$

Theorem 8.6. *The described stochastic game between the sensor and attacker has at least one NE.*

Proof. The considered game with discounted rewards has two players and $N+1$ pure strategies for each player. Moreover, from Assumption 8.1, players all deploy stationary strategies. The result follows from [36, Theorem 6.3.5]. □

8.3.11 Learning Methodology

We now present a method to obtain the NE of the two-player zero-sum stochastic game. Denote the expected future payoffs (in which the players adopt the NE strategies from time $k+1$ onwards) for the sensor and the attacker as $\mathfrak{J}_S^*(m')$ and $\mathfrak{J}_A^*(m')$, where the notation $*$ represents the payoff derivation using the NE strategies and the term m' stands for the next state. Then, based on (8.44), each player will play the following zero-sum single-decision game at every time k:

Problem 8.1. For the sensor,

$$\begin{aligned}\mathfrak{J}_S^*(m) = \max_{\overrightarrow{\gamma_k} \in \mathfrak{S}} (r_k(m, \overrightarrow{\gamma_k}, \overrightarrow{\lambda_k^*}) \\ + \rho \sum_{m \in \mathfrak{M}} p(m'|m, \overrightarrow{\gamma_k}, \overrightarrow{\lambda_k^*}) \times \mathfrak{J}_S^*(m'))\end{aligned}$$

$$\text{s.t.} \sum_{j=1}^{N+1} \gamma_{k,j} = 1, \quad \sum_{j=1}^{N+1} \lambda_{k,j} = 1, \quad \gamma_{k,j} \geq 0, \quad \lambda_{k,j} \geq 0. \quad \square$$

Problem 8.2. For the attacker,

$$\begin{aligned}\mathfrak{J}_A^*(m) = \max_{\overrightarrow{\gamma_k} \in \mathfrak{S}} (r_k(m, \overrightarrow{\gamma_k^*}, \overrightarrow{\lambda_k}) \\ + \rho \sum_{m \in \mathfrak{M}} p(m'|m, \overrightarrow{\gamma_k^*}, \overrightarrow{\lambda_k}) \times \mathfrak{J}_S^*(m'))\end{aligned}$$

$$\text{s.t.} \sum_{j=1}^{N+1} \gamma_{k,j} = 1, \quad \sum_{j=1}^{N+1} \lambda_{k,j} = 1, \quad \gamma_{k,j} \geq 0, \quad \lambda_{k,j} \geq 0. \quad \square$$

In the above, notations $\overrightarrow{\gamma_k^*}$, $\overrightarrow{\lambda_k^*}$ represent the actions determined by the NE action rules π_s^*, π_a^* and can be obtained by solving the joint optimization problems. Unfortunately, there is no straightforward method for solving these problems in general because of the tight coupling between both problems. Classic algorithmic techniques for solving a stochastic game include value iteration and strategy improvement [30], quadratic programming, etc. However, these algorithms require information of the reward for all states, which may be inaccessible for the sensor and attacker. In order to overcome these disadvantages, we use the following Nash Q-learning algorithm [33] to obtain the NE solution. Executing this algorithm just requires knowledge of the system parameters.

First, based on the objective function in Problems 8.1 and 8.2, we define Q-values for each player and denote the Q-value for the sensor as follows (the Q-value of the attacker is negative to that of the sensor):

$$\mathfrak{Q}^*(m, \overrightarrow{\gamma_k}, \overrightarrow{\lambda_k}) = r(m, \overrightarrow{\gamma_k}, \overrightarrow{\lambda_k}) + \rho[R(m, \pi_s^*, \pi_a^*) \times \mathfrak{F}_S^*(\bar{P}) \\ + (1 - R(m, \pi_s^*, \pi_a^*)) \times \mathfrak{F}_S^*(h(m))], \tag{8.47}$$

where r is the one-stage reward defined in (8.43) with the initial state m and the joint actions $(\overrightarrow{\gamma}, \overrightarrow{\lambda})$. Note that from Assumption 8.1, the subscript k of the reward and strategies can be omitted. The value $\mathfrak{Q}^*$ combines the current reward and future reward when the two players play specified rational actions $\overrightarrow{\gamma}$ and $\overrightarrow{\lambda}$ according to the NE strategies from the next period onwards. Comparing (8.47) with Problem 8.1, we have

$$\mathfrak{F}_S^* = \underset{\overrightarrow{\gamma_k}, \overrightarrow{\lambda_k}}{Nash}\, \mathfrak{Q}^*(m, \overrightarrow{\gamma_k}, \overrightarrow{\lambda_k}), \tag{8.48}$$

where the notation *Nash* represents the process of finding the NE of the one-stage game with reward $\mathfrak{Q}^*$. Therefore, searching the NE of the stochastic game is translated from solving the joint optimization problem defined in Problems 8.1 and 8.2 to finding the NE of a single stage game with reward $\mathfrak{Q}^*$.

Next, we develop a learning process to calculate $\mathfrak{Q}^*$ through repeated plays, in which each player updates its Q-value by using the Q-value information of itself and its opponent from an arbitrary guess at the beginning. The updated equation of the Q-value for the sensor is thus defined as follows:

$$\mathfrak{Q}_{k+1}(m, \overrightarrow{\gamma_k}, \overrightarrow{\lambda_k}) = (1 - \alpha_k)\mathfrak{Q}_k(m, \overrightarrow{\gamma_k}, \overrightarrow{\lambda_k}) \\ + \alpha_k(r_k + \rho\, Nash\, \mathfrak{Q}(m')) \tag{8.49}$$

where $Nash\ \mathfrak{Q}(m')$ is as in (8.48), α_k (or $\alpha_k(m, \overrightarrow{\gamma_k}, \overrightarrow{\lambda_k})$) is the learning rate to be designed and it is related to the state m and the actions. From the standard formulation of a two-player zero-sum single stage game, $Nash\ \mathfrak{Q}(m')$ can be represented in the following min–max form and can be easily solved using linear programming:

$$
\begin{aligned}
Nash\ \mathfrak{Q}(m') &\triangleq \mathfrak{F}(m') \\
&= \max_{\pi_s} \min_{\pi_a} \sum_{\overrightarrow{\gamma_k}, \overrightarrow{\lambda_k}} \mathfrak{Q}_k(m', \overrightarrow{\gamma_k}, \overrightarrow{\lambda_k}) \\
&\quad \pi_s \mathfrak{Q}_k(m', \overrightarrow{\gamma_k}) \pi_a \mathfrak{Q}_k(m', \overrightarrow{\lambda_k}).
\end{aligned} \tag{8.50}
$$

Finally, with a large number of repeated plays and given the randomness of adopted actions in every step, the Q-value $\mathfrak{Q}_k(m, \overrightarrow{\gamma_k}, \overrightarrow{\lambda_k})$ may converge to $\mathfrak{Q}^*$ (see Theorem 8.7). The optimal strategies for the stochastic game can be obtained by (8.48). The generalized version of the Nash Q-learning algorithm is given in Algorithm 8.1. Note that $||\cdot||$ represents the matrix norm and δ is the accuracy requirement.

Algorithm 8.1 Nash Q-learning algorithm

1: **Initialization:**
2: $k = 0$ and set the initial state $m \in \mathcal{M}$
3: Initialize the Q-value $\mathcal{Q}^i_k(m, \overrightarrow{\lambda_k}, \overrightarrow{\gamma_k})$ for all states m and arbitrary $\overrightarrow{\lambda_k}$, $\overrightarrow{\gamma_k}$, where $i = 1, 2$ represents the two players respectively
4: **While** $||\mathcal{Q}_{k+1} - \mathcal{Q}_k|| < \delta$
5: Find the NE Q-value $Nash\ \mathcal{Q}(m')$ based on (8.50) and the corresponding optimal mixed strategies
6: Randomly select actions for the two players based on the optimal mixed strategy profiles
7: Observe the next state m' and update the Q-value in (8.49)
8: $k := k + 1$
9: **End**

Theorem 8.7 (Convergence Conditions). *The Nash Q-learning process (i.e., the iteration defined in* (8.49)*) will converge to* $\mathfrak{Q}^*$ *with probability 1, if*

1. *Every state* $m \in \mathfrak{M}$ *with different actions* $(\overrightarrow{\gamma_k}, \overrightarrow{\lambda_k}) \in \mathfrak{S}$ *is visited infinitely often during the learning process.*
2. *The learning rate* α_k *is such that* $\alpha_k \in [0, 1)$, $\sum_{k=0}^{\infty} \alpha_k = \infty$ *and* $\sum_{k=0}^{\infty} \alpha_k^2 < \infty$; $\alpha_k(m, \overrightarrow{\gamma_k}, \overrightarrow{\lambda_k}) \neq 0$ *if* $(m, \overrightarrow{\gamma_k}, \overrightarrow{\lambda_k})$.
3. *A saddle point exists in every stage of the game for all times k and states m, and each player uses the NE payoffs to update his Q-values.*

The proof of Theorem 8.7 is similar to those in [33]. The two former conditions can be satisfied according to Remark 8.5. Due to the existence of a saddle point in any two-player zero-sum stage game, condition 3 is always satisfied.

Remark 8.5. *Condition 1 indicates that as long as every state and action is successfully traversed, the convergence result is independent of the random actions of each player during the learning process. Through a large number of iterations and random actions in the learning process, condition 1 can be satisfied. In condition 2, the first term restricts the convergence of the learning rate, whereas the second states that the players only update the Q-values corresponding to the current time and state. To satisfy these conditions, we design the learning rate as a non-zero decreasing function of time k and the current state and actions, which is illustrated in the simulation part. It is reasonable to decrease the learning rate because the learning process starts with little information, while after many steps, sufficient information is available for the sensor. This consequently reduces the proportion of the learning term in the updated equation* (8.49).

We now outline stability issues of the estimation process (when the game is in equilibrium) in terms of the asymptotic state estimation error covariance. Based on (8.38) and the stationarity assumption of the error covariance (which was proved in [35]), we have $\lim_{k\to\infty}\sup\frac{1}{T}\sum_{k=1}^{T}\mathrm{Tr}[\mathbb{E}(P_k)]=\lim_{k\to\infty}\sup\mathrm{Tr}[\mathbb{E}(P_k)]=\sum_{i=1}^{+\infty}\pi_i\mathrm{Tr}[h^{i-1}(\bar{P})]$, $\sum_{i=1}^{+\infty}\pi_i=1$ where $\pi_i\triangleq\mathbb{P}(P_{+\infty}=h^i(\bar{P}))=\Pi_{j=0}^{i-1}(1-\epsilon_j)\epsilon_i$ and $\epsilon_i=1-\overrightarrow{\gamma_k^{*}}'(m=h^i(\bar{P}))M_k\overrightarrow{\lambda_k^{*}}(m=h^i(\bar{P}))$ represents the arrival rate when the NE strategies are adopted. Based on properties of the Lyapunov operator $h(X)=AXA'+Q$ [43], a sufficient condition for the estimation stability is $\min_{i=0,\dots,K}\epsilon_i>\rho(A)^{-2}$, where $\rho(A)$ stands for the spectral radius of A. Note that we only consider the first $K+1$ rates ϵ_i since there is a final state $h^K(\bar{P})^2$ which represents $h^i(\bar{P})$ with $i\geq K$ during the learning process and the strategies used for these states are the same as those of the state $h^K(\bar{P})$.

8.3.12 Simulations Example 8.2

In this section, we will illustrate the results using some examples. We consider the following vector system with parameters $A=\begin{pmatrix}1 & 0.5\\ 0 & 1.05\end{pmatrix}$, $C=\begin{pmatrix}1 & 0\end{pmatrix}$, $Q=\begin{pmatrix}0.5 & 0\\ 0 & 0.5\end{pmatrix}$ and $R=0.5$, where the steady-state error

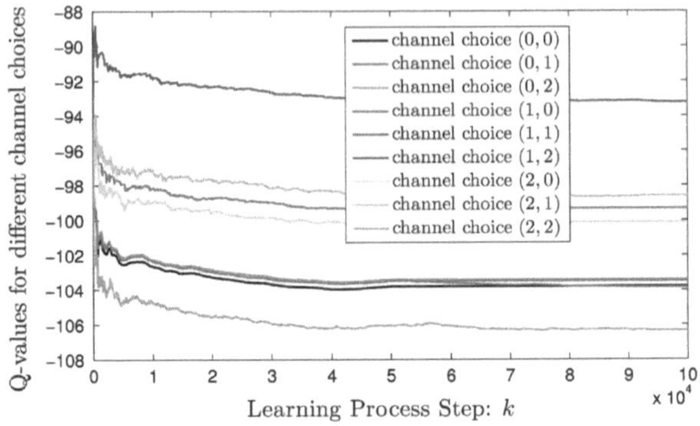

Figure 8.12 Q-value in state $\bar{P}$ for all joint channel choices

covariance is $\bar{P} = \begin{pmatrix} 0.38 & 0.28 \\ 0.28 & 1.69 \end{pmatrix}$. Assuming that the channels in the examples are wireless fast-fading channels, we adopt the general form of $F(\cdot)$ in (8.36), i.e., $F(x) = cx^{-L_i}$, where c, L_i are constants related to the channel characteristics. In the examples, we have two channels with different characteristics, $L_1 = 2$, $L_2 = 1$, but with the same noise parameters, $\sigma_i = 0.1$, $i = 1, 2$.

8.3.12.1 Multichannel Strategies

For these two channels, we design the transmission energy levels $p_s = (0.5\ \ 0.6)'$ for the sensor and the attack energy levels $p_a = (0.3\ \ 0.3)'$ for the attacker, respectively. The other parameter settings for the reward function and the learning process are $\delta_s = 1$, $\delta_a = 1$, $\rho = 0.96$ and $\alpha_t = 10/[15 + \text{count}(m, \vec{\gamma_t}, \vec{\lambda_t})]$, where count is a function to calculate the occurrence of the pair $(m, \vec{\gamma_t}, \vec{\lambda_t})$. Before the learning process, to reduce the computational burden, we impose the restriction that the states set $\mathfrak{M}$ is finite and equals $\{\bar{P}, \ldots, h^4(\bar{P})\}$. We apply the Nash Q-learning algorithm 100 000 times (about 10 min using MATLAB) and obtain the following results (see Fig. 8.12):

- **Result 1.** After 100 000 learning steps, we can see that the Q-values in different states all converge. Moreover, state $\bar{P}$ is more likely to be achieved. We take state $\bar{P}$ as an example, in which the converged Q-values are concluded in Fig. 8.12.
- **Result 2.** We observe that not only do the Q-values converge, but the channel-choice probabilities also tend to be stable for each player under different states (for example, see Fig. 8.13). Hence, we can obtain the

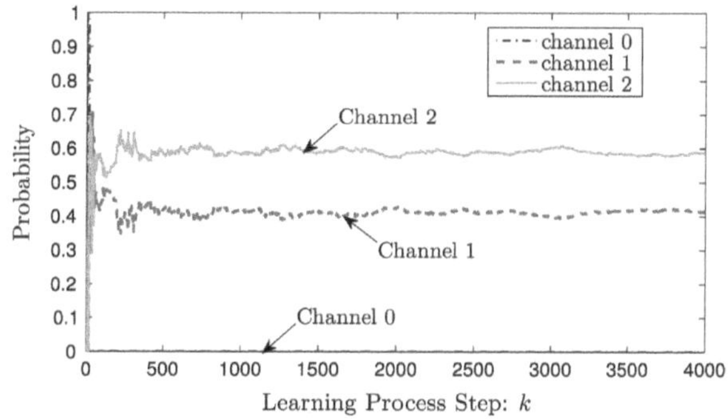

Figure 8.13 Strategy learning process for the sensor in state $\bar{P}$

Table 8.1 Optimal strategies in states $\bar{P}$ and $h(\bar{P})$.

State		**Channel**		
		0th	**1st**	**2nd**
$\bar{P}$	Sensor	0	0.42	0.58
	Attacker	0	0.71	0.29
$h(\bar{P})$	Sensor	0	0.44	0.56
	Attacker	0	0.63	0.37

stationary strategies for the sensor and the attacker under different states. We summarize the optimal strategies of each player under states $\bar{P}$ and $h(\bar{P})$ in Table 8.1.

- **Result 3.** Based on the NE strategies shown in Table 8.1, we compare $\text{Tr}(P_k)$ and its averaged form $\frac{1}{k}\sum_{l=0}^{k}\text{Tr}(P_l)$ under the defensive and the non-defensive case (in the latter the players choose channels following a uniform distribution). The corresponding simulation result is given in Fig. 8.14. It is easy to conclude that the defensive strategy provides a more accurate state estimate.

8.3.12.2 Multichannel Strategies and Power-Level Selection

We consider a system with the same parameters as the previous one, but as foreshadowed in Section 2.2.3, add new energy levels for the two channels, i.e., the total transmission power levels deployed by the sensor are $[0.3\ 0.5]'$ and $[0.3\ 0.6]'$, respectively, and the attacker power levels are $[0.15\ 0.3]'$ and $[0.1\ 0.2]'$. With extra energy levels, the pure strategy set for each player becomes the combination of the channel choice and power assignment. Applying the Nash Q-learning algorithm, we have similar conclusions as in

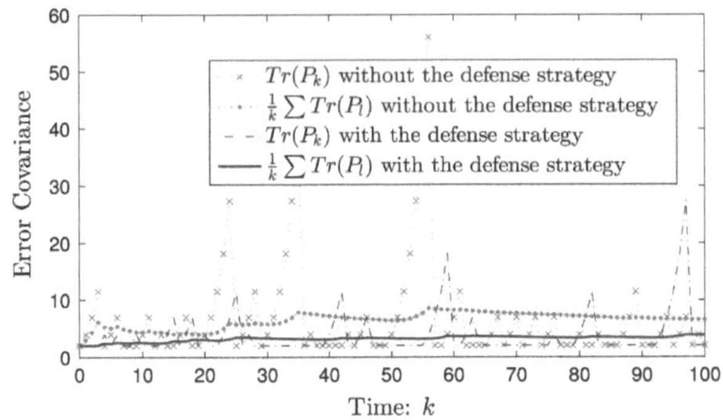

Figure 8.14 Comparison between the defense and no-defense case

Table 8.2 Settings and the optimal strategies.

Channel		0th	1st	2nd		
Sensor	Energy	0	0.2	0.5	0.2	0.6
	Probability	0	0.48	0.37	0.15	0
Sensor	Energy	0	0.1	0.3	0.05	0.3
	Probability	0.76	0.06	0.17	0	0.01

the previous example. The Q-values also converge and the optimal strategies for each player are illustrated in Table 8.2 (also taking state $\bar{P}$ as an example).

8.4 NOTES

In this chapter, we adopt a hybrid automaton approach to describe the dynamic transitions between the nominal CPS and the system under attack. The unified modeling framework of CPS under attack is described as a time-invariant system subject to unknown input. The monitor designed in this chapter is capable of detecting exogenous attacks and triggering the discrete event, *attack*, in the hybrid automaton. The defense mechanism is achieved via a passivation transformation M-matrix design. Passivity is guaranteed for the hybrid automaton under attack.

In the following section, the work continued to investigate security issues in CPSs with a multichannel network, where a malicious agent can launch DoS attacks by jamming communication channels between the sensor and remote estimator. Taking energy limitations into consideration, a two-player zero-sum stochastic game model was formulated and a Nash

Q-learning method was proposed to solve for the Nash equilibrium. As this work is limited to the case of a game between one sensor and one attacker, more research involving multiple sensors or multiple attackers can be done.

We have shown that the presented l_0-norm optimization procedure for state estimation can be formulated as a mixed integer linear program. Although there exist efficient MILP solvers, MILPs are effectively NP-hard.

Therefore, the natural next step would include transforming the l_0 state estimator into a convex program based on l_1/l_r optimization (e.g., $r = 2$) as done in [48]. In this case, providing a bound for the state-estimation error when the l_1/l_r convex relaxation is used will be an avenue for future work.

REFERENCES

[1] K.-D. Kim, P.R. Kumar, Cyber-physical systems: a perspective at the centennial, Proc. IEEE 100 (2012) 1287–1308.

[2] X. Cao, P. Cheng, J. Chen, Y. Sun, An online optimization approach for control and communication codesign in networked cyberphysical systems, IEEE Trans. Ind. Inform. 9 (1) (2013) 439–450.

[3] J. Lin, S. Sedigh, A. Miller, Towards integrated simulation of cyber-physical systems: a case study on intelligent water distribution, in: Eighth IEEE Int. Conference on Dependable, Autonomic and Secure Computing, DASC'09, 2009, pp. 690–695.

[4] A. Cardenas, S. Amin, B. Sinopoli, A. Giani, A. Perrig, S. Sastry, Challenges for securing cyber physical systems, in: Workshop on Future Directions in Cyberphysical Systems Security, DHS, July 2009 [Online]. Available: http://chess.eecs.berkeley.edu/pubs/601.html.

[5] A. Teixeira, D. Pérez, H. Sandberg, K.H. Johansson, Attack models and scenarios for networked control systems, in: Proceedings of the 1st International Conference on High Confidence Networked Systems, HiCoNS '12, ACM, New York, NY, USA, 2012, pp. 55–64 [Online]. Available: http://doi.acm.org/10.1145/2185505.2185515.

[6] C. Kiennert, Z. Ismail, H. Debar, J. Leneutre, A survey on game-theoretic approaches for intrusion detection and response optimization, ACM Comput. Surv. 51 (5) (Aug. 2018) 1–31.

[7] Y. Chen, S. Kar, J.M.F. Moura, Dynamic attack detection in cyber-physical systems with side initial state information, ArXiv e-prints, Mar. 2015.

[8] A.A. Cárdenas, S. Amin, Z.-S. Lin, Y.-L. Huang, C.-Y. Huang, S. Sastry, Attacks against process control systems: risk assessment, detection, and response, in: Proceedings of the 6th ACM Symposium on Information, Computer and Communications Security, ASIACCS '11, ACM, New York, NY, USA, 2011, pp. 355–366 [Online]. Available: http://doi.acm.org/10.1145/1966913.1966959.

[9] T.T. Tran, O.S. Shin, J.H. Lee, Detection of replay attacks in smart grid systems, in: Computing, Management and Telecommunications, 2013 International Conference on, ComManTel, Jan. 2013, pp. 298–302.

[10] F. Pasqualetti, F. Dörfler, F. Bullo, Cyber-physical attacks in power networks: models, fundamental limitations and monitor design, in: 2011 50th IEEE Conference on Decision and Control and European Control Conference, Dec. 2011, pp. 2195–2201.

[11] H. Fawzi, P. Tabuada, S. Diggavi, Secure estimation and control for cyber-physical systems under adversarial attacks, IEEE Trans. Autom. Control 59 (6) (June 2014) 1454–1467.
[12] W. Zhang, M. Branicky, S. Phillips, Stability of networked control systems, IEEE Control Syst. Mag. 21 (1) (2001) 84–97.
[13] Y. Li, L. Shi, P. Cheng, J. Chen, D.E. Quevedo, Game theory meets network security and privacy, ACM Comput. Surv. 44 (112) (Dec. 2011) 1–45.
[14] Y. Mo, B. Sinopoli, On the performance degradation of cyberphysical systems under stealthy integrity attacks, IEEE Trans. Autom. Control 61 (9) (Sept. 2016) 2618–2624.
[15] Y. Liu, P. Ning, M.K. Reiter, False data injection attacks against state estimation in electric power grids, in: Proceedings of the 16th ACM Conference on Computer and Communications Security, CCS '09, ACM, New York, NY, USA, 2009, pp. 21–32 [Online]. Available: http://doi.acm.org/10.1145/1653662.1653666.
[16] W.-H. Chen, Disturbance observer based control for nonlinear systems, IEEE/ASME Trans. Mechatron. 9 (4) (Dec. 2004) 706–710.
[17] Y. Mo, B. Sinopoli, False data injection attacks in control systems, preprints of the 1st Workshop on Secure Control Systems.
[18] C.Z. Bai, F. Pasqualetti, V. Gupta, Security in stochastic control systems: fundamental limitations and performance bounds, in: 2015 American Control Conference, ACC, July 2015, pp. 195–200.
[19] W.H. Chen, J. Yang, L. Guo, S. Li, Disturbance-observer-based control and related methods – an overview, IEEE Trans. Ind. Electron. 63 (2) (Feb. 2016) 1083–1095.
[20] J.C. Willems, Dissipative dynamical systems part ii: linear systems with quadratic supply rates, Arch. Ration. Mech. Anal. 45 (5) (1972) 352–393, https://doi.org/10.1007/BF00276494.
[21] Y. Yan, P. Antsaklis, Stabilizing nonlinear model predictive control scheme based on passivity and dissipativity, in: 2016 American Control Conference, ACC, July 2016, pp. 4476–4481.
[22] M. Xia, P.J. Antsaklis, V. Gupta, Passivity indices and passivation of systems with application to systems with input/output delay, in: 53rd IEEE Conference on Decision and Control, Dec. 2014, pp. 783–788.
[23] X. Chen, S. Komada, T. Fukuda, Design of a nonlinear disturbance observer, IEEE Trans. Ind. Electron. 47 (2) (Apr. 2000) 429–437.
[24] P. Colaneri, J.C. Geromel, A. Astolfi, Stabilization of continuous-time switched nonlinear systems, Syst. Control Lett. 57 (1) (2008) 95–103 [Online]. Available: http://www.sciencedirect.com/science/article/pii/S0167691107000953.
[25] T. Basar, P. Bernhard, A General Introduction to Minimax (H^∞ Optimal) Designs, Birkhäuser Boston, Boston, MA, 2008, pp. 1–32.
[26] J. Zhao, D.J. Hill, Dissipativity theory for switched systems, IEEE Trans. Autom. Control 53 (4) (May 2008) 941–953.
[27] L. Guo, Q. Liao, S. Wei, Y. Huang, A kind of bicycle robot dynamic modeling and nonlinear control, in: The 2010 IEEE International Conference on Information and Automation, June 2010, pp. 1613–1617.
[28] A. Agah, S.K. Das, K. Basu, A game theory based approach for security in wireless sensor networks, in: Proc. IEEE Conf. Performance, Computing, and Communications, 2005, pp. 259–263.
[29] B.D.O. Anderson, J.B. Moore, Optimal Filtering, vol. 1, Prentice-Hall, Englewood Cliffs, NJ, 1979.

[30] D.P. Bertsekas, Dynamic Programming and Optimal Control, vol. 1, Athena Scientific, 2005.
[31] J. Hespanha, P. Naghshtabrizi, Y. Xu, A survey of recent results in networked control systems, Proc. IEEE 95 (1) (2007) 138–162.
[32] P. Hovareshti, V. Gupta, J.S. Baras, Sensor scheduling using smart sensors, in: Proc. IEEE Conf. Decision and Control, 2007, pp. 494–499.
[33] J. Hu, M.P. Wellman, Nash q-learning for general-sum stochastic games, J. Mach. Learn. Res. 4 (2003) 1039–1069.
[34] O.D. Incel, A survey on multi-channel communication in wireless sensor networks, Comput. Netw. 55 (13) (2011) 3081–3099.
[35] S. Kar, B. Sinopoli, J.M. Moura, Kalman filtering with intermittent observations: weak convergence to a stationary distribution, IEEE Trans. Autom. Control 57 (2) (2012) 405–420.
[36] K. Leyton-Brown, Y. Shoham, Essentials of Game Theory: A Concise Multidisciplinary Introduction, Morgan & Claypool Publishers, 2008.
[37] H. Li, L. Lai, R. Qiu, A denial-of-service jamming game for remote state monitoring in smart grid, in: Proc. IEEE Conf. Information Sciences and Systems, 2011, pp. 1–6.
[38] Y. Li, D.E. Quevedo, S. Dey, L. Shi, A game-theoretic approach to fake-acknowledgment attack on cyber-physical systems, IEEE Trans. Signal Inform. Process. Netw. 3 (1) (2017) 1–11.
[39] Y. Li, D.E. Quevedo, S. Dey, L. Shi, SINR-based DoS attack on remote state estimation: a game-theoretic approach, IEEE Trans. Control Netw. Syst. 4 (3) (2017) 632–642.
[40] Y. Li, L. Shi, P. Cheng, J. Chen, D. Quevedo, Jamming attacks on remote state estimation in cyber-physical systems: a game-theoretic approach, IEEE Trans. Autom. Control 60 (10) (2015) 2831–2836.
[41] S. Liu, P.X. Liu, A. El Saddik, A stochastic game approach to the security issue of networked control systems under jamming attacks, J. Franklin Inst. 351 (9) (2014) 4570–4583.
[42] J. Proakis, Digital Communications, McGraw-Hill, 2001.
[43] D.E. Quevedo, A. Ahlén, A.S. Leong, S. Dey, On Kalman filtering over fading wireless channels with controlled transmission powers, Automatica 48 (7) (2012) 1306–1316.
[44] L. Shi, P. Cheng, J. Chen, Sensor data scheduling for optimal state estimation with communication energy constraint, Automatica 47 (8) (2011) 1693–1698.
[45] X. Song, P. Willett, S. Zhou, P. Luh, The MIMO radar and jammer games, IEEE Trans. Signal Process. 60 (2) (2012) 687–699.
[46] J. Wu, Y. Yuan, H. Zhang, L. Shi, How can online schedules improve communication and estimation tradeoff?, IEEE Trans. Signal Process. 61 (7) (2013) 1625–1631.
[47] Q. Zhu, T. Basar, Game-theoretic methods for robustness, security, and resilience of cyberphysical control systems: games-in-games principle for optimal cross-layer resilient control systems, IEEE Trans. Control Syst. 35 (1) (2015) 46–65.
[48] H. Fawzi, P. Tabuada, S. Diggavi, Secure state-estimation for dynamical systems under active adversaries, in: Proceedings of the 2011 49th Annual Allerton Conference on Communication, Control, and Computing, Allerton, 2011, pp. 337–344.
[49] Y. Mo, B. Sinopoli, Secure control against replay attacks, in: Proc. 47th Annual Allerton Conference on Communication, Control, and Computing, Allerton, Monticello, IL, 2009, pp. 911–918.

[50] Y. Mo, R. Chabukswar, B. Sinopoli, Detecting integrity attacks on SCADA systems, IEEE Trans. Control Syst. Technol. 22 (4) (July 2014) 1396–1407.

[51] Y. Mo, S. Weerakkody, B. Sinopoli, Physical authentication of control systems: designing watermarked control inputs to detect counterfeit sensor outputs, IEEE Control Syst. 35 (1) (Feb. 2015) 93–109.

[52] S. Weerakkody, Y. Mo, B. Sinopoli, Detecting integrity attacks on control systems using robust physical watermarking, in: Proc. 53rd IEEE Conference on Decision and Control, Los Angeles, CA, 2014, pp. 3757–3764.

[53] B. Satchidanandan, P.R. Kumar, Dynamic watermarking: active defense of networked cyber-physical systems, Proc. IEEE 105 (2) (Feb. 2017) 219–240.

CHAPTER 9

Cyberphysical Security Methods

9.1 A GENERALIZED GAME-THEORETIC APPROACH

Critical infrastructures, such as power grids and transportation systems, are increasingly using open networks for operation. The use of open networks poses many challenges for control systems. The classical design of control systems takes into account modeling uncertainties as well as physical disturbances, providing a multitude of control design methods such as robust control, adaptive control, and stochastic control. With the growing level of integration of control systems with new information technologies, modern control systems face uncertainties not only from the physical world but also from the cybercomponents of the system. The vulnerabilities of the software deployed in the new control system infrastructure will expose the control system to many potential risks and threats from attackers. Exploitation of these vulnerabilities can lead to severe damage, see the outlets in [1, 2]. The recent report in [3] and [4] announced that a computer worm, Stuxnet, was spread to target Siemens supervisory control and data acquisition (SCADA) systems that are configured to control and monitor specific industrial processes.

Uncertainties from the cybersystem are often unanticipated and more catastrophic for control systems in terms of their high impact and low effort as compared to those from the physical world. It is imperative to consider the cyberuncertainties in addition to the physical ones in the controller design. Those uncertainties can be caused by intentional malicious behaviors and/or by rare events, such as severe weather or natural disasters. Engineers are accustomed to designing systems to be reliable and robust, despite noise and disturbances. However, the cybersecurity aspect of control systems has posed new challenges for engineers and system designers.

The goal of this section is to introduce game-theoretic methods for resilient control design and develop a framework that studies the trade-off between robustness, security, and resilience. A hybrid dynamic game-theoretic approach is introduced that integrates the discrete-time Markov model for modeling the evolution of cyberstates with continuous-time dynamics for describing the underlying controlled physical process. The hybrid dynamic game model provides a holistic and crosslayer viewpoint in

Networked Control Systems
https://doi.org/10.1016/B978-0-12-816119-7.00017-4

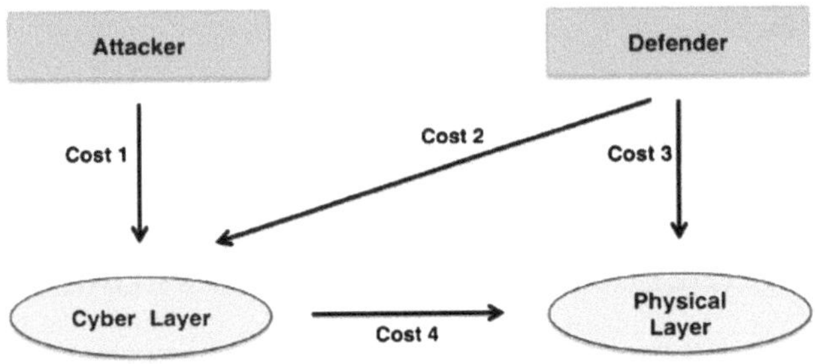

Figure 9.1 Cyberphysical game theory model

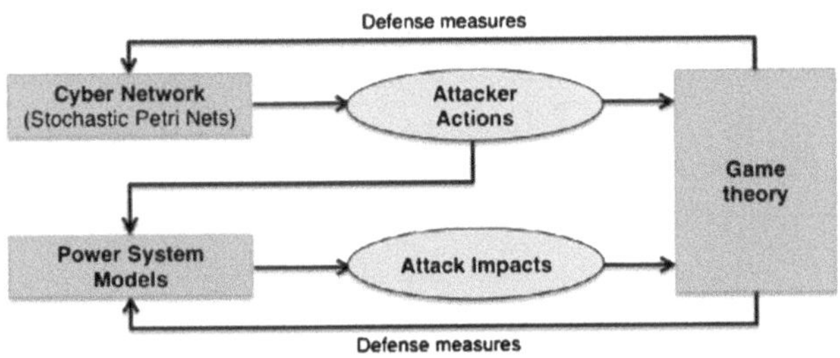

Figure 9.2 A game-theoretic framework for cybersecurity

the decision-making and design for cyberphysical systems. The continuous-time dynamics models the physical layer, that is, the plant, subject to disturbances and control efforts. The discrete-time dynamics models the cyberlayer of the system, which involves system configurations and dynamic human–machine interactions (HMIs). A zero-sum differential game is used for robust control design at the physical layer, while a stochastic zero-sum game between an administrator and an attacker is used for the design of defense mechanisms. The controlled transition between pre-event states to post-event states in the hybrid system framework leads to the design of the resilient hybrid dynamical system. The controller design at the physical layer and the security policy design at the cyberlayer of the system are intertwined. A policy made at the cyberlayer can influence the optimal control design for the physical system, and the optimal control design at the lower level needs to be taken into account when security policies are determined. For a class of system models, the overall optimal design of the cyberphysical system can be characterized by a Hamilton–Jacobi–Isaacs (HJI) equation together with a Shapley optimality criterion.

For a brief review of game theory, the reader is referred to the Appendix A at the end of the book. See Figs. 9.1 and 9.2.

9.1.1 Physical Layer Control Problem

Resilient control requires a crosslayer control design. The control problem at the physical layer of the system is described below. Consider a general class of systems subject to two types of uncertainty:

1. a continuous deterministic uncertainty that models the known parametric uncertainties and disturbances and
2. a discrete stochastic uncertainty that models the unknown and unanticipated events that lead to a change in the system operation state at random times.

Let the system state evolve according to the piecewise deterministic dynamics

$$\dot{x}(t) = f(t, x, u, w; \theta(t, a, l)), \qquad x(t_0) = x_0, \tag{9.1}$$

where $x(t) \in \mathbb{R}^n$, x_0 is a fixed (known) initial state of the physical plant at starting time t_0, $u(t) \in \mathbb{R}^r$ is the control input, $w(t) \in \mathbb{R}^p$ is the disturbance, and all these quantities lie at the physical and control layers of the entire system.

The state of the cybersystem is described by θ. The evolution of θ depends on the cyberdefense action l and the attacker's action a which are also functions of time; $\theta(t)$ is a shorthand notation in place of $\theta(t, a, l)$ if the pair of actions (a, l) is fixed. For a given pair (a, l), $\theta(t)$, $t \in [0, t_f]$, is a Markov jump process with right-continuous sample paths, with initial distribution π_0 and with rate matrix $\lambda = \{\lambda_{ij}\}_{i,j \in \mathcal{S}}$ where $\mathcal{S} := \{1, 2, \ldots, s\}$ is the state space; $\lambda_{ij} \in \mathbb{R}_+$ are the transition rates such that for $i \neq j$, $\lambda_{ij} \geq 0$, and $\lambda_{ij} = 1 - \sum_{j \neq i} \lambda_{ij}$ for $i \in \mathcal{S}$.

Transitions between the structural states are controlled by the attacker and the system administrator. An attacker can exploit the vulnerabilities in the control system software and launch an attack to bring down the operation. An example is Stuxnet, a Windows-based worm that was recently discovered, to target industrial software and equipment [3]. An administrator can enforce security by dynamically updating the security policy of the control systems [9,10]. Once an attack occurs, the administrator can restore the system back to normal operation. Different from conventional computer networks, control systems are reported to experience lower rates of attacks [11], and the software updates are less frequent than those in computer networks. Hence, the transition between structural states evolves at a different time scale from that of physical states. The systems are assumed to have reached their physical steady states when the structural transition happens. This assumption is validated by the fact that the attack rate on control systems is often lower than that on information systems [12,13] and the fact that the time scale of the failure rate of devices and components in control systems is higher than that of the system dynamics and operations [14].

9.1.2 Cyberstrategy

Let $\bar{k} = t/\varepsilon$, $\varepsilon > 0$ be the time scale on which cyberevents happen, which is often on the order of days, in contrast to the one of the physical systems which evolve on the time scale of seconds. Denote by $a \in \mathcal{A}$ a cyberattack chosen by the attacker from his attack space $\mathcal{A} := \{a_1, a_2, \ldots, a_M\}$ composed of all M possible actions. We let $l \in \mathcal{L}$ be the cyberdefense mechanism that can be employed by the network administrator, where $\mathcal{L} := \{l_1, l_2, \ldots, l_N\}$ is the set of all the possible defense actions. Without loss of generality, $\mathcal{A}$ and $\mathcal{L}$ do not change with time even though, in practice, they can change due to technological updates and advances. The mixed strategies $\{(k) = [f_i(k)]_{i=1}^{N} \in \mathcal{F}_k$, $g(k) = [g_j(k)]_{j=1}^{M} \in \mathcal{G}_k$ of the defender and the attacker, respectively, are considered here, where $f_i(k)$ and $g_j(k)$ are the probabilities of choosing $l_i \in \mathcal{L}$ and $a_j \in \mathcal{A}$, respectively, where $\mathcal{F}_k$ and $\mathcal{G}_k$ are sets of admissible strategies, defined by

$$\mathcal{F}_k := \left\{ \mathrm{f}(k) \in [0,1]^N : \sum_{i=1}^{N} f_i(k) = 1 \right\}, \tag{9.2}$$

$$\mathcal{G}_k := \left\{ \mathrm{g}(k) \in [0,1]^N : \sum_{i=1}^{N} g_j(k) = 1 \right\}. \tag{9.3}$$

The transition law of the cybersystem state $\theta(k)$ at time k depends on the actions of the attacker as well as the defense mechanism employed by the administrator. More precisely, the rate matrix has

$$\mathrm{Prob}\{\theta(k+\Delta) = j | \theta(k) = i\} = \begin{cases} \lambda_{ij}(\mathrm{f}(k), \mathrm{g}(k)), & j \neq i, \\ \lambda_{ij}(\mathrm{f}(k), \mathrm{g}(k)), & j = i, \end{cases} \tag{9.4}$$

where $\Delta > 0$, which is on the same time scale as k (for example, days), and $\lambda_{ij}(\mathrm{f}(k), \mathrm{g}(k))$ are the average transition rates in terms of the transition rates $\tilde{\lambda}_{ij}(a(k), l(k))$, $i, j \in \mathcal{S}$, defined by

$$\lambda_{ij}(\mathrm{f}(k), \mathrm{g}(k)) = \sum_{i=1}^{N} \sum_{j=1}^{M} f_i(k) g_j(k) \tilde{\lambda}_{ij}(a_i, l_j(k)). \tag{9.5}$$

Note that (9.1) and (9.4) describe hybrid systems [15–17] with both continuous and discrete states. Let $\mathcal{F}_t$ be the sigma-field generated by $\theta_{[t_0,t]} := \{\theta(s), s \leq t\}$. The admissible control and disturbance processes, $u(\cdot)$ and $w(\cdot)$, are taken to be $\mathcal{F}_t$ measurable and piecewise continuous, with the corresponding spaces denoted by $\mathcal{U}$ and $\mathcal{V}$, respectively; f is taken to be piecewise continuous in t and Lipschitz continuous in (x, u, w) for each fixed sample path of θ with probability one. The process θ models the unanticipated or rare uncertainties that arise from cyberattacks or component failure. These events result in random structural changes in the dynamics of the system. For each $u \in \mathcal{U}$, $w \in \mathcal{W}$, the state process $x(\cdot)$ is continuous with probability one, and if (u, w) is chosen to be memoryless, then the pair (x, θ) is a Markov process.

9.1.3 Perfect-State Feedback Control

A closed-loop, perfect-state information structure is considered for control design. The controller has access to $x[t_0, t]$ at time t which can be written as

$$u(t) = \mu(t, \chi_{x[t_0,t]}; \theta_{x[t_0,t]}), \qquad t \in x[t_0, t_f], \tag{9.6}$$

where μ is an admissible closed-loop control strategy, piecewise continuous in its first argument, and Lipschitz continuous in its second argument. The class of all such control strategies is denoted by $\mathcal{M}_{\mathrm{CL}} \subseteq \mathcal{U}$. Analogously, let $\mathcal{N}_{\mathrm{CL}} \subseteq \mathcal{U}$ denote the class of all closed-loop disturbance strategies

$$w(t) = \nu(t, \chi_{x[t_0,t]}; \theta_{x[t_0,t]}), \qquad t \in x[t_0, t_f]. \tag{9.7}$$

The performance index for the hybrid control system is given by the expected cost over the statistics of θ,

$$J(u, w) := \mathbb{E}_\theta\{L(x, u, w; \theta)\}, \tag{9.8}$$

with the cost function L given as

$$L(x, u, w; \theta) = q_f(x(t_f); \theta x(t_f)) + \int_{t_0}^{t_f} c(t, x(t), u(t), w(t); \theta(t))dt$$
$$+ c_0(x_0; \theta(t_0)), \tag{9.9}$$

where q_f is continuous in x, and g is jointly continuous in (t, x, u, w). In the infinite-horizon case, q_f is dropped out, and $t_f \to \infty$. For each $\mu_{\mathcal{CL}} \in \mathcal{M}_{\mathcal{CL}}$

and $\nu_{\mathcal{CL}} \in \mathcal{N}_{\mathcal{CL}}$, the stochastic differential equation will admit a well-defined solution (as a piecewise deterministic process), which will induce corresponding unique elements in $\mathcal{U}$ and $\mathcal{W}$ which provide the "open-loop representations" of μ and ν, respectively.

The H^∞-optimal control problem in the time domain is in fact a minimax optimization problem, and hence a zero-sum game, where the controller can be viewed as the minimizing player and the disturbance as the maximizing player [16,18]. Here, the objective is to find a minimax closed-loop controller $\mu^*_{\mathcal{CL}} \in \mathcal{M}_{\mathcal{CL}}$ that minimizes the supremum of J over all closed-loop disturbance policies

$$\sup_{\nu \in \mathcal{N}_{\mathrm{CL}}} J(\mu^*\mathrm{CL}, \nu) = \inf_{\mu \in \mathcal{M}_{\mathrm{CL}}} \sup_{\nu \in \mathcal{N}_{\mathrm{CL}}} J(\mu, \nu). \tag{9.10}$$

A cost structure of interest is a separable one, namely

$$c(t, x, u, w, ; \theta) = c_0(t, x, u; \theta) - \gamma^2 r(w; \theta). \tag{9.11}$$

The solution of (9.10) parameterized in γ is denoted by μ^*_γ, and $\gamma_{\mathcal{CL}}$ denotes the smallest value of $\gamma > 0$ such that for $\gamma > \gamma_{\mathcal{CL}}$ the right-hand side of (9.10) is bounded. Then μ^*_γ for $\gamma > \gamma_{\mathcal{CL}}$ is an H^∞ controller for the hybrid system, with respect to the performance index

$$\sup_{w \in \mathcal{W}} \left\{ \frac{\mathbb{E}_\theta\{q_f(x_f; \theta(t_f)) + \int_{t_0}^{t_f} c_0(t, x(t), u(t); \theta(t))dt\}}{\mathbb{E}_\theta\{\|w\|^2 + q_0(x_0; \theta(t_0))\}} \right\}, \tag{9.12}$$

where $\|\cdot\|$ denotes the $\mathcal{L}_2$-norm of w for each sample path of θ. The minimum value of (9.12) is γ^2_{CL}. It defines a measure of disturbance attenuation in the nonlinear hybrid system. Note that in (9.10), x_0 is considered as part of the disturbance.

What has been formulated above, and described by (9.10), is a differential game. Let $V(\cdot) : \mathbb{R} \times \mathbb{R}^n \times \mathcal{S}$ denote the cost-to-go function associated with this differential game, that is, $V(t, x, i)$ is the upper value of a similar game defined on the shorter interval $[t, t_f]$ with initial state x and initial structure $\theta(t) = i$. The following assumptions are quite standard.

Assumption 9.1. The differential game defined by (9.10) has an upper value VV for every initial time t state $x(t)$ and structure $\theta(t)$, which is jointly continuously differentiable in (t, x).

Under Assumption 9.1, the infinitesimal generator of the upper-value function is

$$\mathcal{L}V(t,x;\theta)|_{\theta=i} := \lim_{h\to 0}\frac{1}{h}\mathbb{E}\left\{V(t+h,x(t+h);\theta(t+h)) - V(t,x(t),\theta(t))|x(t)=x,\theta(t)=x,\theta(T)=i\right\}$$
$$= V_t(t,x;i) + V_x(t,x;i)f(t,x,u,w;i) + \sum_{j=1}^{s}\lambda_{ij}V_{t,x;j}, \quad (9.13)$$

with $u \in \mathcal{U}$ and $w \in \mathcal{W}$ chosen to be memoryless, which, in fact, is not a restriction as further elaborated below. From (9.13), the associated HJI equation is

$$-V_t^i(t,x) = \inf_{u\in\mathbb{R}^r}\sup_{w\in\mathbb{R}^p}\left\{V_x^i(t,x)f(t,x,u,w,i) + c(t,x,u,w,i) + \sum_{j\in\mathcal{S}}\lambda_{ij}V^j(t,x)\right\}, \quad (9.14)$$

$$V^i(t_f,x) = q_f(x(t_f);i), \qquad i\in\mathcal{S}, \quad (9.15)$$

where the simpler notation $V^i(t,x)$ is used in place of $V(t,x;\theta(t)=i)$. Denoting any such control by $\mu^{\mathrm{F}} \in \mathcal{M}_{\mathrm{CL}}$, (9.14) and (9.15) can be rewritten as

$$-V_t^i(t,x) = \inf_{u\in\mathbb{R}^r}\sup_{w\in\mathbb{R}^p}\left\{V_x^i(t,x)f(t,x,\mu^F(t,x,i),w,i) + c(t,x,\mu^F(t,x,i),w,i) + \sum_{j\in\mathcal{S}}\lambda_{ij}V^j(t,x)\right\}.$$

Furthermore, if the Isaacs condition (on interchangeability of infimum and supremum in (9.14)) holds and if there exists a disturbance policy $\nu^{\mathrm{F}} \in \mathcal{N}_{\mathrm{CL}}$ that achieves the maximum in (9.14), then ν^{F} is also a Markov policy, and $(\mu^{\mathrm{F}},\nu^{\mathrm{F}})$ are in saddle-point equilibrium. In this case, the upper value is also the value function, satisfying the partial differential equation (PDE)

$$-V_t^i(t,x) = V_x^i(t,x)f(t,x,\mu^F(t,x,i),\nu^F(t,x,i),i) + c(t,x,\mu^F(t,x,i),\nu^F(t,x,i),i) + \sum_{j\in\mathcal{S}}\lambda_{ij}V^j(t,x). \quad (9.16)$$

The preceding discussion and the ensuing result are now summarized in the theorem below.

Theorem 9.1. *Let the cyberstrategy pair* $(f(k), g(k))$ *be fixed, Assumption 9.1 hold, and* $\mu^F \in \mathcal{M}_{CL}$ *be defined as above. Then,* μ^F *is a closed-loop minimax controller. If, furthermore, the Isaacs condition holds and* $\nu^F \in \mathcal{N}_{CL}$ *is defined above, the pair of Markov policies* (μ^F, ν^F) *provides a saddle-point solution on the product space* $\mathcal{M}_{CL} \times \mathcal{N}_{CL}$*. The corresponding saddle-point value function solves* (9.16) *subject to* (9.15).

The optimal cost $V^i(t_0, x_0)$ yields the physical layer control performance under the minimax controller. Note that this cost depends on the cyberstrategy pair (f, g) since the transition rate λ_{ij} is a function of mixed strategies. The cyberstrategies are determined by analyzing a security game at the cyberlayer.

9.1.4 Cyberlayer Defense System

At the cyberlayer, a zero-sum game framework can be used to capture the strategic interactions between an attacker and a defender. The game takes different forms depending on the information available to the defender, the targets of the attacker, and the security mechanism. For example, a zero-sum stochastic game has been used for dynamic configurations of a network of IDSs [19,20], in which the state of the system evolves according to transition rules determined by the actions taken by the players, and dynamic system configuration policy has been developed for IDSs to optimally defend against intrusions. In [21] and [22], a multistage Stackelberg game has been studied for developing deceptive routing strategies for nodes in a multihop wireless communication network. The framework is convenient to model the scenario where the defender first deploys a proactive defense, and the attacker follows the protocol. A stochastic repeated game and an iterative learning mechanism have been adopted for moving target defense [23,24]. Due to the lack of complete information of the attacker and the system itself, the players update their strategies in a feedback manner driven by the data they have observed from the system.

The following theorem provides a convergence result on the iterative algorithm in (A.7) of Appendix A.

Theorem 9.2. *Let* $\{(F_n^*, G_n^*)\}$ *be the sequence of strategies in* $F_s \times G_s$ *produced by the value iteration scheme described in* (A.7)*. Then, any limit point* (F_n, G_n) *of the sequence is a pair of saddle-point equilibrium strategies. Moreover, the limit point yields the unique game value* $v_\beta^*(i)$, $i \in \mathcal{S}$.

Note that the equilibrium solution $(\mathrm{F}^*, \mathrm{G}^*)$ depends on the physical layer minimax control $(\mu^*_{\mathrm{CL}}, \nu^*_{\mathrm{CL}})$, while finding the control policy $(\mu^*_{\mathrm{CL}}, \nu^*_{\mathrm{CL}})$ using (9.14) and (9.15) relies on the security policy $(\mathrm{F}^*, \mathrm{G}^*)$ taken at the cyberlayer. The optimality criterion (A.6) in Theorem 8.2, together with the HJI equation in (9.14), defines a set of coupled optimality conditions that are used to solve for obtaining the cyberpolicy F^* and the robust controller μ^*_{CL} and its associated performance index γ^*.

The coupling between the cybersystem game (CSG) and the physical system game (PSG) captures the essential trade-offs between robustness, resilience, and security. To ensure that the system operates in a normal condition, either a perfect secure system is designed so that no attack can succeed or the system is capable of recovering back to its normal condition quickly once it fails. However, given limited resources, perfect security is not possible, and the solution to the cybergame (for good states) provides a fundamental limit for the best-effort security strategies. Hence, it is essential to allocate resources to the cybersystem to recover from the failure states. This security and resilience trade-off is captured by the stochastic CSG introduced in this section, which yields strategies balancing the level of security that prevents the cybersystem from failure, and the level of resilience that brings the system quickly back to its normal state of operation.

On the other hand, to achieve a higher level of robustness, the control effort has to be distributed across different cyberstates. Robustness of the control system is high if the system is perfectly secure, that is, the system does not fail and does not move to a compromised state, because control effort only needs to be expended for the good states. However, perfect security does not exist, and the control effort has to be expended on the bad states as well in case the system fails to ensure robustness at the failure states. Hence, there is a trade-off between security and robustness. The formulation of the PSG captures this trade-off.

In addition, the coupling between the CSG and PSG yields a design relationship between the level of robustness against w in the physical system and the cost for security defense against $\mathbf{G}$ in the cybersystem. A higher demand of robustness at the physical level will lead to a higher control cost and a higher impact cost for the cybersystem once the system is compromised, which in turn requires a stronger level of security and resilience to prevent or recover from the failure. Given limited resources for defense, physical-level robustness will dictate the trade-off relationship between security and resilience. Hence, as a result of this framework, a balance of security, resilience, and robustness is achieved for the cyberphysical control system.

9.1.5 Linear Quadratic Problem

The set of optimality equations can be simplified by considering the special case of the linear quadratic problem defined as

$$f(t, x, u, w; i) = A^i x + B^i u + D^i w, \tag{9.17}$$

$$q_f(t_f; i) = |x(t_f)|^2_{Q^i_{t_f}}, \tag{9.18}$$

$$q_0(x_0, i) = |x_0|^2_{Q^i_0}, \tag{9.19}$$

$$c_0(t, x, u, i) = |x|^2_{Q^i_{t_f}} + |u|_{R^i}, \tag{9.20}$$

$$r(w; \theta) = |w|^2, \tag{9.21}$$

where $i \in \mathcal{S}$, $|\cdot|$ denotes the Euclidean norm with appropriate weighting, and A^i, B^i, D^i, Q^i, R^i, are matrices of appropriate dimensions, whose entries are continuous functions of time t. Further, $Q^i(\cdot) \geq 0$, $R^i(\cdot) > 0$, and $Q^i_0 > 0$ and $Q^i_f \geq 0$.

By the minimax theorem [26], J admits a saddle point, which means that the matrix game A has a saddle point in mixed strategies, that is, there exists a pair (p_1^*, p_2^*) such that for all other probability vectors p_1 and p_2 of dimensions m and n, respectively, the following pair of saddle-point inequalities hold:

$$p_1^{*\prime} A p_2 \leq p_1^{*\prime} A p_2 \leq p_1^* A p_2.$$

The quantity $p_1^{*\prime} A p_2^*$ is the value of the game in mixed strategies. This result is now captured in the following theorem.

Theorem 9.3 (Minimax Theorem). *Every finite two-person, zero-sum game has a saddle point in mixed strategies.*

The extension of this result to N-player finite games was first obtained in [32], as captured in the following theorem.

Theorem 9.4. *Every finite N-player nonzero-sum game has a Nash equilibrium in mixed strategies.*

A standard proof for this result uses Brouwer's fixed point theorem; see [18].

Consider the infinite horizon case with the cost function defined by

$$L(x, u, w; \theta) = \mathbb{E} \int_{t_0}^{\infty} (|x(t)|^2_{Q^i} + |u(t)|R^{i^2} - \gamma^2 |w(t)|^2) dt. \tag{9.22}$$

Before stating Theorem 9.5, the following assumptions are made:

Assumption 9.2. The Markov chain θ is irreducible for any admissible strategies ([30], p. 78).

Assumption 9.3. The pair (A^i, B^i) is stochastically stabilizable ([31], see its definition on p. 59).

Assumption 9.4. The pair (A^i, Q^i) is observable for each $i \in \mathcal{S}$.

Theorem 9.5 ([16]). *Consider the soft-constrained, zero-sum differential game with perfect measurements in the infinite-horizon case defined by* (9.10), (9.17)–(9.21), (9.22), (9.6), *and* (9.7), *with* λ_{ij}*s fixed. Let Assumptions 9.2–9.4 hold. Then,* $\gamma_{CL,\infty^*} < +\infty$, *and for any* $\gamma_{CL} > \gamma^*_{CL,\infty}$, *there exists a set of minimal positive definite solutions* Z_i, $i \in \mathcal{S}$, *to generalized algebraic Riccati equations (GAREs),*

$$\begin{aligned} A^{iT} Z_i + Z_i A^i \quad Z_i \left(B^i (R^i)^{-1} B^{iT} - \frac{1}{\gamma^2} D^i D^{iT} \right) Z_i \\ + Q^i + \sum_{j=1}^{s} \lambda_{ij}(F, G) Z_j = 0; \quad i \in \mathcal{S}, \end{aligned}$$

which further satisfy the condition

$$\gamma^2_{CL} Q^i_0 - Z_i \geq 0, \quad i \in \mathcal{S}, \tag{9.23}$$

and a strategy $\gamma > \gamma^*_{\infty}$ *for player* P_i *that guarantees the zero upper value is*

$$u^*_{\gamma\infty}(t) = u^*_{\gamma\infty}(t, x(t), \theta(t)) = -(R^i)^{-1} B^{iT} Z_i x(t). \tag{9.24}$$

For almost all $\gamma < \gamma^*_{\infty}$, the jump linear system driven by both the optimal control and the optimal disturbance,

$$\dot{x}(t) = (A^i - (B^i (R^i)^{-1} B^{iT} - \frac{1}{\gamma^2} D^{i^{iT}}) Z_i) x(t), \tag{9.25}$$

is also mean-square stable, that is, $\lim_{t \to \infty} \mathbb{E}\{|x(t)|^2\} = 0$.

For $\gamma < \gamma^*_{\mathrm{CL},\infty}$, on the other hand, either condition (9.23) is not satisfied or the set of GAREs does not admit nonnegative definite solutions, and in both cases, the upper value of the game is $+\infty$.

On a longer time scale, the continuous-time, zero-sum game between the attacker and administrator has the stationary saddle-point equilibrium characterized by Theorem 8.2. Let $\tilde{g}^i = V^i$ be the cost function which describes the physical layer system performance. Then, the fixed-point equation (A.6) can be written as

$$\beta v^*_\beta(i) = x_0' Z_i(\mathrm{F}^*, \mathrm{G}^*) x_0 + \sum_{j \in \mathcal{S}} \lambda_{ij}(\mathrm{F}^*, \mathrm{G}^*) v^*_\beta(j). \qquad (9.26)$$

The optimal control u^* and the optimal defense strategy F^* need to be found by solving the coupled equations (9.26) and GAREs in Theorem 9.5.

9.1.6 Cascading Failures

Cascading failure is kind of failure in a system comprising interconnected parts, in which the failure of a part can trigger the failure of successive parts. Such a failure is common in computer networks and power systems. In the case of cascading failures, state $\theta = 1$ is the normal operating state and state $\theta = N$ is the terminal failure state. The states i, $2 \leq i \leq N-1$ are intermediate compromised states, in which one system component failure leads to another. The failure and compromised states are taken to be irreversible, that is, the system cannot be fixed or brought back to its normal state immediately after faults occur. This is usually due to the fact that the time scale for critical cascading failures is much shorter than the time scale for system maintenance. In our modeling framework, the transition between the failure states follows a Markov jump process with rate matrix $\lambda = \{\lambda_{ij}\}_{i,j \in \mathcal{S}}$ such that for $i \neq j$, $\lambda_{ij} \geq 0$, $\lambda_{ii} = 1 - \sum_{j \neq i} \lambda_{ij}$, and $i > j$ and $j > i+1$, $\lambda_{ij} = 0$. For simplicity, the notation $\lambda_{i,i+1} = p_i$, $1 \leq i \leq N-1$, denotes the transition rates between adjacent states, and hence $\lambda_{ii} = 1 - p_i$, $1 \leq i \leq N-1$, $p_N = \lambda_{NN}$. Here, p_i, $i = 1, \ldots, N-1$, are dependent on the cyberstrategy pair (F, G) which has been introduced earlier. An effective cyberdefense action will lead to lower transition rates, and a power cyberattack will increase them. The structure of state transition of cascading failures is depicted in Fig. 9.3.

Following (9.26), the optimality criteria for the cybersystem under cascading cyberstates can be further simplified to

$$\beta v^*_\beta(N) = V^N, \qquad (9.27)$$

$$\beta v^*_\beta(i) = \mathrm{val}\{c^i + p_i v^*_\beta(i+1) - p_i v^*_\beta(i)\}, \qquad (9.28)$$

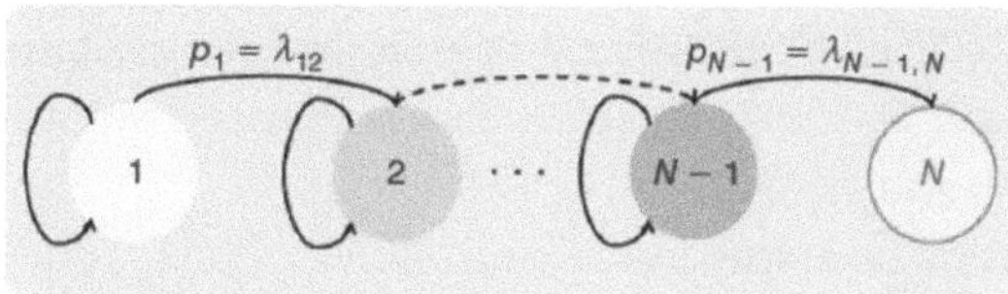

Figure 9.3 A system progresses from a normal operating state $\theta = 1$ to the failure state N

$$\begin{aligned}(\mathrm{f}_i, \mathrm{g}_i) &\in \arg\mathrm{val}\{c^i + p_i v^*_\beta(i+1) - p_i v^*_\beta(i)\},\\ i &= 1, \ldots, N-1,\\ c^i &= x_0^\top Z_i x_0.\end{aligned} \tag{9.29}$$

Here, $p_i = \lambda_{i,i+1}$, $1 \leq i \leq N-1$, and V^i is dependent on p_i through Z_i in Theorem 9.5. Note that (9.28) and (9.29) find the game value v^i_{beta} and stationary saddle-point equilibrium strategies (F, G), respectively. Since both players have a finite number of choices for each k the existence of a saddle-point solution is guaranteed for the zero-sum stochastic game [18,32].

In addition, the optimality criteria for the H^∞ optimal control in the linear quadratic case can be reduced to

$$\begin{aligned}&A^{N\top} Z_N + Z_N A^N\\ &\quad - Z_N \left(B^N (R^N)^{-1} B^{N\top} - \frac{1}{\gamma}^2 D^N D^{N\top} \right) Z_N + Q^N = 0,\end{aligned} \tag{9.30}$$

$$\begin{aligned}&A^{i\top} Z_N + Z_i A^i - Z_i \left(B^i (R^i)^{-1} B^{i\top} - \frac{1}{\gamma}^2 D^i D^{i\top} \right) Z_i\\ &\quad + Q^N = 0, \qquad i = 1, \ldots, N-1.\end{aligned} \tag{9.31}$$

Here, γ is a chosen level of attenuation. Under the regularity conditions in [29], there exists a finite scalar $\gamma^\infty > 0$ such that for all (9.30) and (9.31) admit unique minimal nonnegative definite solutions.

In (9.28), p_is are dependent on F and G. At the same time, as a result of solving (9.31), the value V^i is dependent on p_is and B_is, which are in turn functions of and V^N. The above set of coupled equations can be solved by starting with (9.30) for obtaining the value of the terminal state V^N. From (9.27), the value v^*_N is calculated and then in the next step use (9.28) and (9.31) to find the stationary saddle-point equilibrium strategies f^*_{N-1}, g^*_{N-1} at state $\theta = N-1$ their corresponding transition rate $p^*_{N-1} = \lambda_{N-1,N}(\mathrm{f}^*_{N-1}, \mathrm{g}^*_{N-1})$ and the Riccati solution Z_{N-1}. The process is iterated again by using Z_{N-1} in (9.28) for $i = N - 2$, and the obtained strategy pair (f^*_{N-2}, g^*_{N-2}) is used in (9.31) to solve Z_{N-2}. Hence backward induction is used to obtain Z_1 and (f^*_1, g^*_1).

Note that the coupling between (9.30)–(9.31) and (9.27)–(9.28) demonstrates the interdependence between security at the cyberlevel and the robustness at the physical level. The holistic viewpoint toward these system properties is essential in addressing the resilience of cyberphysical control systems. The coupling between cyber and physical levels of the system is not onesided but rather reciprocal. The upward resilience from the physical level to the cyberlevel results from the function c^i while the downward resilience from the cyberlevel to the physical level follows from the dependence of λ_{ij} on the cyberpolicies.

9.1.7 Games-in-Games Structure

The cross-layer, game-theoretic model captures the coupling between the cyber and the physical layers of the system dynamics. In the framework, robustness of the cyberphysical control system is studied under an H^∞ optimal control model, while its security is studied using a two-person zero-sum cybersecurity game. The control and defense strategy designs are extended to incorporate post-event system states, where resilient control and cyber-strategies are developed to deal with uncertainties and events that are not taken into account in pre-event robustness and security designs. Under the assumptions made in Theorems 9.1 and A.1, a secure, resilient, and robust control and cyberstrategy pair (μ_{CL}, F) has to satisfy the general optimality criteria (9.14), (9.15), and (A.6). In the linear quadratic problem with cascading states, they are reduced to (9.30), (9.31), (9.27), and (9.28). They are derived from the optimality criteria of two dynamic games. One is the zero-sum differential game for the H^∞ robust control design, and the other is the zero-sum stochastic game for equilibrium defense policy. Due to the layering architecture and the time-scale separation, the CSG can be seen as the one on top of the PSG. The two games are coupled and exhibit a games-in-games structure as illustrated in Fig. 9.4. The outcome of the PSG affects the cost structure of the CDG. In addition, the solution to the PSG depends on the equilibrium solution (F^*, G^*) from the CSG. Solutions to this game structure define the trade-off between robust and resilient control of cyberphysical control systems.

One interesting aspect of the games-in-games structure is that its solution is featured by *zooming-in* and *zooming-out* operations. The zooming-out operation refers to the fact that the solution of the PSG provides an input to

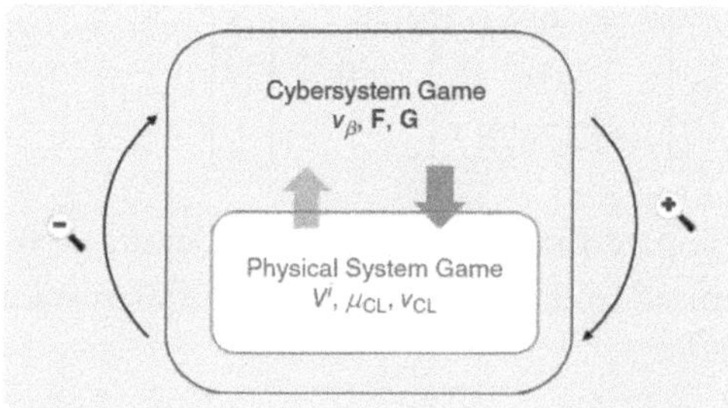

Figure 9.4 A games-structure for crosslayer resilient control design

the CSG, which leads to a solution of the CSG. The zooming-in operation refers to the reverse fact, that is, the PSG also affects the CSG. As depicted in Fig. 9.4, solutions to the coupled optimality criteria precisely involve these two procedures. Zooming in is defined as the operation of passing the parameters from higher level CSG to lower level PSG, and zooming out as the operation of passing the parameters from the lower-level PSG to the higher level CSG. A sequence of structured zooming-in and zooming-out operations is observed in the linear-quadratic problem with cascading states. The procedure for finding the solution starts with finding Z_{N-1} using (9.31) and then zooming out to the CSG to find $(v_\beta^*(N-), f_i, g_i)$. This is followed by zooming in to the PSG again and finding Z_{N-2}. The zooming-in and zooming-out operations alternate until reaching the initial state $\theta = 1$.

9.1.8 Simulation Example 9.1

Consider the following two-state linear system that arises from a single-machine infinite bus power system linearized around its operation point [34]. Let $x \in \mathbb{R}^3$ be the state vector that includes the power angle, the relative speed, and the active power delivered by the generator. Let $u \in \mathbb{R}^3$ be the control variable that determines the amplifier of the generator. At the normal operating state $\theta = 1$, its dynamics are described by

$$\dot{x} = A^1 x + B^1 u + D^1 w, \tag{9.32}$$

where

$$A^1 = \begin{bmatrix} 0 & 1 & 0 \\ 0 & -0.625 & -39.2699 \\ -0.156627 & 1.65884 & -0.738602 \end{bmatrix},$$

$$B^1 = \begin{bmatrix} 0 \\ 0 \\ -0.271287 \end{bmatrix}, \; D^1 = \begin{bmatrix} 0 \\ 0 \\ 1 \end{bmatrix}.$$

With an unanticipated fault caused by a cyberattack at the rate λ_{12}, the system is compromised and its dynamics at the failed state $\theta = 2$ are given by

$$\dot{x} = A^2 x + B^2 u + D^2 w,$$

where

$$A^2 = \begin{bmatrix} 0 & 1 & 0 \\ 0 & -0.625 & -39.2699 \\ -0.0691878 & 0.960155 & -0.407174 \end{bmatrix},$$

$$B^2 = \begin{bmatrix} 0 \\ 0 \\ -0.119837 \end{bmatrix}, \; D^2 = \begin{bmatrix} 0 \\ 0 \\ 1 \end{bmatrix}.$$

The design strategy based on the linear quadratic criterion described above could be used here, by choosing the weighting matrices

$$Q^1 = Q^2 = \begin{bmatrix} 1000 & 0 & 0 \\ 0 & 1 & 0 \\ 0 & 0 & 10 \end{bmatrix}, R^1 = 10, R^2 = 1,$$

where the weights of 1000 are used in Q_1 and Q_2 to emphasize the willingness to use more control in a post-attack state.

The $\tilde{\lambda}_{ij}, i, j = 1, 2$, take the following parameterized form: $\tilde{\lambda}_{12} = p$, $\tilde{\lambda}_{11} = -p$, $\tilde{\lambda}_{21} = \tilde{\lambda}_{22} = 0$, where it has been assumed that the operation after the attacker cannot be immediately recovered. At the cyberlayer, the administrator can take two actions, that is, to defend ($l_1 = \mathrm{D}$) and not to defend ($l_2 = \mathrm{ND}$) The attacker can also take two actions, that is, to attack ($a_1 = \mathrm{A}$) or not to ($a_2 = \mathrm{NA}$). Parameter p determines the probability transition law with respect to pure strategies and its values are tabulated as

	A	NA
D	0.1	0.05
ND	0.95	0.05

In the above table, a higher transition rate has been attached to a failure state if the attacker launches an attack while the cybersystem does not have proper measures to defend itself. On the other hand, the probability is lower if the cybersystem can defend itself from attacks. In the above table, a base transition rate of 0.05 has been assumed to capture the inherent reliability of the physical system without exogenous attacks. The optimality criterion (9.26) and GAREs in Theorem 9.5 are used to obtain the discounted value functions $v_{\beta}^{*}(i)$, $i = 1, 2$, with the discount factor chosen to be $\beta = 1$, and yield $V^2 = 7.2075 \times 10^4$ independent of the parameter p. Hence, $v_{\beta}^{*}(2) = V^2$, and v_{β}^{*} satisfies the following fixed-point equation:

$$v_{\beta}^{*}(1) = \text{val}\{\text{H} - v_{\beta}^{*}(1)\text{G}\}, \tag{9.33}$$

where

$$\text{H} = \begin{bmatrix} 1.4396 \times 10^4 & 0.9994 \times 10^4 \\ 8.4867 \times 10^4 & 0.9994 \times 10^4 \end{bmatrix}$$

and

$$\text{H} = \begin{bmatrix} 0.1 & 0.05 \\ 0.95 & 0.05 \end{bmatrix},$$

with val being the value operator for a matrix game [27,28]. Using value iteration, it is possible to compute $v_{\beta}^{*}(1) = 1.3087 \times 10^4$, and the corresponding stationary saddle-point strategy $\text{f}^* = [1, 0]'$, $\text{g}^* = [1, 0]'$ which is a pure strategy leading to an optimal value of $p = 0.05$. The stationary saddle-point equilibrium strategy informs that the defender should always be defending and the attacker should not be attacking. At $p = 0.05$, the physical layer robust feedback control at each state i is obtained by

$$u^F(t, x, 1) = -(R^1)^{-1} B^{1\top} Z^1,$$
$$u^F(t, x, 2) = -(R^2)^{-1} B^{2\top} Z^2,$$

where

$$\text{Z}^1 = \begin{bmatrix} 399.3266 & 31.8581 & -162.2334 \\ 31.8581 & 5.7083 & -15.2963 \\ -1622.2334 & -15.2963 & 149.7459 \end{bmatrix}$$

and

$$Z^2 = \begin{bmatrix} 2.8512 & 0.1066 & -2.8575 \\ 0.1066 & 0.0345 & -0.1041 \\ -2.8575 & 0.1041 & 4.1506 \end{bmatrix},$$

and the optimum performance index is $\gamma_\infty^* = 8.5$.

9.1.9 Defense Against Denial-of-Service Attack

The games-in-games principle for the special case of the linear–quadratic problem with cascading failures has been applied to study the resilience of the power energy system in [34]. The principle can be further extended to discrete-time systems where the physical layer game is a discrete-time minimax design problem with perfect state measurements and the cyberlayer game is a discrete-time stochastic Markov game. In parallel to the results developed for continuous-time systems, a similar set of coupled equations for discrete-time systems can be developed. Interested readers can refer to [29] and [37] for results on the discrete-time minimax design problem with perfect and imperfect state measurements. To further illustrate this, a case study of denial-of-service (DoS) attack is discussed below, which can cause delays and congestion in the communication channel of the cyberphysical systems.

9.1.10 Control System Model

A networked control system is vulnerable to different types of cyberattacks, including false data injection, DoS, and sensor node capture and cloning attack, as depicted in Fig. 9.5. Here, the games-in-games principle is used to study a class of DoS attacks on control systems. The controlled plant under DoS attacks is described by a discrete-time model for computational convenience

$$\begin{cases} x_{k+1} = Ax_k + B_2 u_{c,k} + B_1 \omega_k, \\ z_k = Dx_k, \end{cases} \tag{9.34}$$

where $x_k \in \mathbb{R}^n$ and $u_k \in \mathbb{R}^m$ are respectively the state variable and the control signal received by the actuator, ω_k is the disturbance belonging to $l_2[0, \infty)$; A, B_1, B_2, and D are matrices with appropriate dimensions. The measurement with randomly varying communication delays is described by

$$\begin{cases} y_k = Cx_k, \\ y_{c,k} = (1 - \delta^\theta y_k + \delta^\theta y_{k-1}), \end{cases} \tag{9.35}$$

where $\delta^\theta = 1$ is the measured output and $y_{c,k} = y_{k-1}$ is the actual output; $\delta^\theta = 0$ is the state space of the cybersystem. The stochastic variable δ^θ is distributed according to a Bernoulli distribution:

$$\bar{\delta}^\theta := \Pr\{\delta^\theta = 1\} = \mathbb{E}\{\delta^\theta\},$$
$$\Pr\{\delta^\theta = 0\} = 1 - \mathbb{E}\{\delta^\theta\} = 1 - \bar{\delta}^\theta. \tag{9.36}$$

When $\delta^\theta = 1$, the measured output is $y_{c,k} = y_{k-1}$, that is, the measured output has a one-step time delay. When $\delta^\theta = 0$, the measured output is $y_{c,k} = y_k$, that is, there is no delay between the measured output and the actual system output. An observer-based control strategy takes the form of

$$\begin{cases} \hat{x}_{k+1} = A\hat{x}_k + B_2 u_{c,k} + L^\theta (y_{c,k} - \bar{y}_{c,k}), \\ \bar{y}_{c,k} = (1 - \delta^\theta) C\hat{x}_k + \delta^\theta C\hat{x}_{k-1}, \end{cases} \tag{9.37}$$

$$\begin{cases} u_k = K^\theta \hat{x}_k, \\ u_{c,k} = (1 - \beta^\theta) u_k + \beta^\theta u_{k-1}, \end{cases} \tag{9.38}$$

where $u_k = \mathbb{R}^m$ is the control signal generated by the controller and $u_{c,k}$ is the signal received by the actuator; $K^\theta \in \mathbb{R}^{m \times n}$ and $L^\theta \in \mathbb{R}^{m \times n}$ denote the controller gains and observer gains to be designed. The stochastic variable β^θ, mutually independent of δ^θ, is also a Bernoulli distributed white sequence with expected value $\bar{\beta}^\theta$. Note that the sensor-to-controller (S–C) delay is described by the situation that $\delta^\theta = 1$, and the controller-to-actuator (C–A) delay is described by $\beta^\theta = 1$.

9.1.11 Intrusion Detection Systems

IDSs are deployed in communication networks for detecting unauthorized system access. They are passive devices that receive and evaluate information sent over a network against a set of signatures. IDS signatures have been developed for most published vulnerabilities and for potentially dangerous activity in common IT protocols. The C–A and S–C delays depend on the configurations of the IDSs. A configuration providing high information assurance can result in significant delays for control system applications since

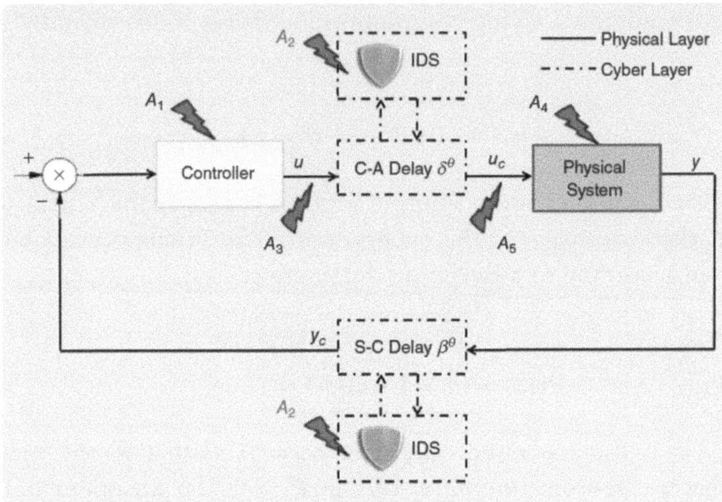

Figure 9.5 Detailed networked control system under cyberattacks

a large number of signatures have to be checked for each incoming packet. The configuration of IDSs is not a trivial task. The current version of the Snort IDS, for example, has approximately 10,000 signature rules located in 50 categories. Each IDS also comes with a default configuration to use when no additional information or expertise is available. It is not trivial to determine the optimal configuration of an IDS because of the need to understand the quantitative relationships between a wide range of analyzers and tuning parameters.

Fig. 9.5 demonstrates that many components of a networked control system are vulnerable to cyberattacks, including the controller, the physical plant, and the communication networks [34]. In the diagram, dotted blocks constitute the cyberlayer of the system while blocks with solid lines are components at the physical layer. Further, A_1 and A_4 represent direct attacks against the actuators or the plant; A_2 is the denial of service attack, where the controller is prevented from receiving sensor measurements and the actuator from receiving control signals; A_3 and A_5 represent deception attacks, where the false information $\tilde{y} \neq y$ and $\tilde{u} \neq u$ is sent from sensors and controllers. Intrusion detection systems (IDSs) are detection devices used to defend the control system from intruders but may cause C–A delay between controller (C) and actuator (A) and/or S–C delay between sensor (S) and controller (C). The probabilities of incurring a one time-step delay are denoted by parameters δ^θ and β^θ which depend on the cyberstate θ.

An IDS is configured optimally as a trade-off between physical layer control system performance and cyberlevel security enhancement.

For industrial control systems, a set of SCADA IDS signatures that parallel Snort rules for enterprise IT systems have been designed by Digital Bond's Quick-draw, which leverages the existing IDS equipment by developing signatures for control system protocols, devices, and vulnerabilities [33]. In a typical SCADA, the IDS rule is used to detect a buffer overflow attack. The rule is specifically designed for Siemens Tecnomatix FactoryLink software, which is used for monitoring, supervising, and controlling industrial processes. FactoryLink is commonly used to build applications such as HMI and SCADA systems. The logging function of FactoryLink is vulnerable to a buffer overflow caused by the usage of vsprintf with a stack buffer of 1024 B. The vulnerability can be exploited remotely in various ways like the passing of a big path or filter string in the file related operations [33].

The goal of the network administrator is to configure an optimal set of detection rules to protect the cybersystem from attackers. To model the interaction between an attacker and a defender, a dynamic game approach is used. Let $\mathcal{L}^*$ be a finite set of possible system configurations in the network and $\mathcal{A}$ be the finite action set of the attacker. The mixed strategies f(k) and g(k) are defined on the action spaces $\mathcal{L}^*$ and $\mathcal{A}$, respectively.

The distributions of random variables δ^θ and β^θ are dependent on the states and attack and defense mechanism in the cyberlayer. Let $\mathrm{H} \in \mathbb{R}^{N\times M}$ and $\mathrm{W} \in \mathbb{R}^{N\times M}$ be two state-dependent matrices whose entries H_{ij} and W_{ij} reflect the S–C and C–A delays for different attack and defense action pairs $(F_i, a_j.)$ The parameters of the Bernoulli random variables are determined by mixed strategies f_θ, g_θ as

$$\bar{\delta}^\theta = \mathrm{f}_\theta^T \mathrm{H}(\theta)\mathrm{g}_\theta, \qquad \bar{\beta}^\theta = \mathrm{f}_\theta^T \mathrm{W}(\theta)\mathrm{g}_\theta.$$

The cybersystem transitions between different states and its transition probabilities,

$$\mathbb{P}(\theta'(n+1)|\theta(n), a_j, F_i), \theta'(n+1), \quad \theta(n) \in \Theta,$$

are dependent on the defense and attack action pair (F_j, a_j) at time n, and

$$\sum_{\theta' \in \Theta} \mathbb{P}(\theta'|\theta(n), F_i, a_j) = 1.$$

9.1.12 Crosslayer Control Design

The H^∞ index is the expectation over f_θ and g_θ for a given θ. Without loss of generality, let $x_0 = 0$; then

$$\mathbb{E}^{f,g(\theta)}\left\{\sum_{k=0}^{\infty}\{\|z_k\|^2\}\right\} < \gamma_\theta^2\{\|\omega_k\|^2\} \tag{9.39}$$

for all $\theta \in \Theta$. The goal of the physical layer control design is to find optimal K^θ, L^θ, $\theta \in S$. The theorem below indicates how to convert the conditions satisfying the H^∞ index into linear matrix inequalities (LMIs) that are easy to solve numerically using available tools.

Theorem 9.6 ([37]). *Given scalars $\gamma_\theta > 0$ and a strategy pair (f_θ, g_θ) for all $\theta \in \Theta$, the hybrid model described by (9.34)–(9.38) is exponentially mean-square stable and the H^∞-norm constraint (9.10) is achieved for all nonzero ω_k if there exist positive definite matrices $P_{11}^\theta \in \mathbb{R}^{m\times m}$, $P_{22}^\theta \in \mathbb{R}^{(n-m)\times(n-m)}$, $S_1^\theta \in \mathbb{R}^{n\times n}$ and $P_2^\theta \in \mathbb{R}^{n\times n}$ and $S_2^\theta \in \mathbb{R}^{n\times n}$, and real matrices $M^\theta \in \mathbb{R}^{m\times n}$, $N^\theta \in \mathbb{R}^{n\times p}$ such that*

$$P_1^\theta := U_1^T P_{11}^\theta U_1 + U_2^T P_{22}^\theta U_2,$$

where $U_1 \in \mathbb{R}^{n\times n}$ and $U_2 \in \mathbb{R}^{(n-m)\times n}$ satisfy

$$\begin{bmatrix} U_1 \\ U_2 \end{bmatrix} B_2 V = \begin{bmatrix} \Sigma \\ 0 \end{bmatrix}, \ \Sigma = \mathrm{diag}\{\sigma_1, \sigma_2, \ldots, \sigma_m\},$$

and σ_i, $i = 1, 2, \ldots, m$, are eigenvalues of B_2. In addition, the controller gain and observer gain satisfy the following LMIs:

$$K^\theta = V\Sigma^{-1}P_{11}^{\theta-1}\Sigma V^T M^\theta, L^\theta = S_1^{\theta-1}N^\theta, \tag{9.40}$$

$$\Pi_{11}^\theta = \begin{bmatrix} \Pi_{11}^\theta & * \\ \Pi_{21}^\theta & \Pi_{22}^\theta \end{bmatrix} < 0, \tag{9.41}$$

where

$$\Pi_{11}^\theta = \begin{bmatrix} P_2^\theta - P_1^\theta & * & * & * & * \\ 0 & S_2^\theta - S_1^\theta & * & * & * \\ 0 & 0 & -P_2^\theta & * & * \\ 0 & 0 & 0 & -S_2^\theta & * \\ 0 & 0 & 0 & 0 & -\gamma_\theta^2 I \end{bmatrix},$$

$$\Pi^\theta_{22} = \begin{bmatrix} -P^\theta_1 & * & * & * & * \\ 0 & -S^\theta_1 & * & * & * \\ 0 & 0 & -P^\theta_1 & * & * \\ 0 & 0 & 0 & -S^\theta_1 & * \\ 0 & 0 & 0 & 0 & I \end{bmatrix}, \tag{9.42}$$

$$\Pi^\theta_{21} = \begin{bmatrix} \Pi^\theta_{21}(1,1) & \Pi^\theta_{21}(1,2) \\ \Pi^\theta_{21}(1,1) & \Pi^\theta_{21}(1,2) \end{bmatrix},$$

$$\Pi^\theta_{21}(1,1) = \begin{bmatrix} P^\theta_1 A + (1-\bar{\beta}^\theta)B_2M^\theta & -(1-\bar{\beta}^\theta)B_2M^\theta \\ 0 & S^\theta_1 A - (1-\bar{\delta}^\theta)N^\theta C(\theta) \end{bmatrix},$$

$$\Pi^\theta_{21}(1,2) = \begin{bmatrix} \bar{\beta}^\theta | B_2M^\theta & -\bar{\beta}^\theta | B_2M^\theta & P^\theta_1 B_1 \\ 0 & -\bar{\delta}^\theta N^\theta C(\theta) & S^\theta_1 B_1 \end{bmatrix},$$

$$\Pi^\theta_{21}(2,1) = \begin{bmatrix} \alpha^\theta_1 B_2 M^\theta & \alpha^\theta_1 B_2 M^\theta \\ \alpha^\theta_2 N^\theta C(\theta) & 0 \\ D & 0 \end{bmatrix},$$

$$\Pi^\theta_{21}(2,2) = \begin{bmatrix} -\alpha^\theta_1 B_2 M^\theta & \alpha^\theta_1 B_2 M^\theta & 0 \\ -\alpha^\theta_2 N^\theta C(\theta) & 0 & 0 \\ D & 0 & 0 \end{bmatrix},$$

$$\alpha^\theta_1 = [(1-\bar{\beta}^\theta)\bar{\beta}^\theta)]^{1/2},$$
$$\alpha^\theta_2 = [(1-\bar{\delta}^\theta)\bar{\delta}^\theta)]^{1/2}.$$

Note that (9.40) and (9.41) in Theorem 9.6 lead to a convex optimization problem

$$\hat{\gamma}_\theta := \min_{\substack{P^\theta_{11}>0,\, P^\theta_{22},\, P^\theta_2>0 \\ S^\theta_1>0,\, S^\theta_2>0,\, M^\theta,\, N^\theta}} \gamma_\theta \tag{9.43}$$

subject to (9.39).

Since γ_θ is influenced by the cyberstate and strategy, it is actually dependent on the triple $(\theta, \mathrm{f}(\theta), \mathrm{g}(\theta))$. Let $\mathrm{C}(\theta) \in \mathbb{R}^{N\times M}$ be the performance matrix with its entry $C_{ij}(\theta)$ corresponding to the physical layer H^∞ performance index under the action pair (F_i, a_j); $\hat{\gamma}_\theta$ can be seen as the value of the mapping

$$\hat{\gamma}_\theta = \mathbf{f}^T_\theta \mathbf{C}(\theta)\mathbf{g}_\theta.$$

The coupled design here means that the cyberdefense mechanism takes into account the H^∞ index, and the H^∞ optimal controller is designed with $\bar{\delta}^{\theta} = f_{\theta}^{*}H(\theta)g_{\theta}^{*}$ and $\bar{\beta}^{\theta} = f_{\theta}^{*}H(\theta)g_{\theta}^{*}$. Algorithm 9.1 is proposed for the coupled design. The algorithm involves a value iteration for computing the stationary mixed saddle-point equilibrium for the stochastic game, in which a linear program for matrix games (LPMG) is solved at each step. For computation of the saddle-point equilibrium using linear programming, see "Linear Programming for Computing the Saddle-Point Equilibrium." Since the game here is zero-sum and finite, the value iteration method converges to stationary saddle-point equilibrium strategies. Readers interested in a proof of convergence of value iteration in zero-sum finite games can refer to [27] and [28]. The algorithm also invokes the computational tools for solving a set of LMIs for obtaining the H^{∞} robust controller in the form of (9.37) and (9.38) that achieves optimal control system performance.

9.1.13 Simulation Example 11.2

An uninterrupted power system (UPS) model is used to illustrate the design procedures. A UPS usually provides uninterrupted, high quality, and reliable power for vital loads, such as life support systems, data storage systems, or emergency equipment. Thus, the resilience and robustness of the UPS are essential. An integrated design of the optimal defense mechanism for IDSs and the optimal control strategy for a pulse-width modulation (PWM) inverter is performed such that the output ac voltage can maintain its desired setting under the influence of DoS attacks. Let the system parameters be

$$A = \begin{bmatrix} 0.9226 & -0.6330 & 0 \\ 1.0 & 0 & 0 \\ 0 & 1.0 & 0 \end{bmatrix},$$

$$B_1 = \begin{bmatrix} 0.5 \\ 0 \\ 0.2 \end{bmatrix}, B_2 = \begin{bmatrix} 1 \\ 0 \\ 0 \end{bmatrix},$$

$$D = \begin{bmatrix} 0.1 & 0 & 0 \end{bmatrix},$$

$$C = \begin{bmatrix} 23.738 & 20.287 & 0 \end{bmatrix}.$$

For the cyberlayer, two states are considered: a normal state θ_1 and a compromised state θ_2. When there are no attacks and the system is in normal state, the communication network is taken to be delay free, that

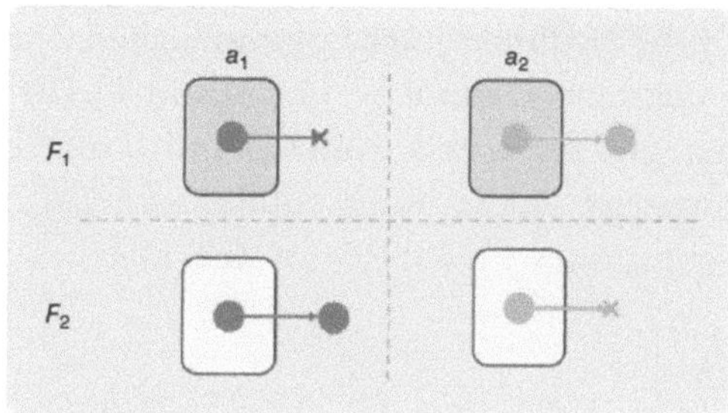

Figure 9.6 An example to illustrate the necessity of different intrusion detection system (IDS) configurations

is, $\bar{\delta}^\theta = \beta^\theta = 0$. The IDS system contains two libraries l_1, l_2 for defending against two attacks a_1, a_2. Library l_1 is used for detecting a_1 whereas library l_2 for a_2. Let $\mathcal{A} = \{a_1, a_2\}$, $L^* = \{F_1, F_2\}$, where F_1 is the configuration where l_1 is loaded and F_2 is the configuration where l_2 is loaded. Fig. 9.6 illustrates the performance of IDS configurations under different attack scenarios.

Remark 9.1. *Note in Fig. 9.6 that library l_1 is used to detect a_1 while library l_2 is used to detect a_2. Configurations $F_1 := \{l_1\}$ and $F_2 := \{l_2\}$ are used to detect a sequence of attacks composed of a_1 and a_2. Each configuration leads to a different physical layer probability of delay in S–C and C–A communication channels. The diagram shows IDS performance under four action pairs (a_1, F_1), (a_1, F_2), (a_2, F_1), and (a_2, F_2) in a two-by-two matrix style, where the row corresponds to configurations, while the column refers to different attack actions. A circle refers to an attack. A box refers to a configuration. A successful defense thwarts an attack. An X denotes a successful defense that prevents the attack from propagating further. The attacks a_1 and a_2 can be successfully detected in the case of (a_1, F_1), and (a_2, F_2), respectively. The attacks will penetrate the system for the scenarios corresponding to (a_1, F_2) and (a_2, F_1).*

9.1.14 Linear Programming for Computing Saddle-Point Equilibrium

Let A and B be two $(m \times n)$-dimensional matrices related to each other by the relation

$$A = B + c1_m 1'_n, \tag{9.44}$$

where 1_m stands for the m-dimensional column vector whose entries are all ones and c is some constant. Denote by $V_m(A)$ and $V_m(B)$ the saddle-point values in mixed strategies for matrix games A and B respectively. Then:

1. Every mixed strategy saddle-point equilibrium (MSSPE) for matrix game A also constitutes a MSSPE for the matrix game B and vice versa.
2. $V_m(A) = V_m(B) + c$.

Matrix games that satisfy (9.44) are strategically equivalent matrix games. For a given matrix game A, a strategically equivalent matrix game can be found with all entries positive by adding a constant c.

Based on this fact, the complete equivalence between a matrix game and a linear program (LP) is used to compute its MSSPE. The following proposition captures this result, a proof of which can be found in [18].

Proposition 9.1. *Given a zero-sum matrix game described by the $m \times n$ matrix A, let B be another matrix game (strategically equivalent to A), obtained from A by adding an appropriate positive constant to make all its entries positive. Introduce the two LPs:*

(Primal LP) $\max y'1_m$ *such that* $B'y \leq 1_n,\ y \geq 0;$

(Dual LP) $\min z'1_n$ *such that* $Bz \geq 1_m,\ z \geq 0,$

with their optimal values (if they exist) denoted by V_p and V_d, respectively. Then:

1. *Both LPs admit solutions, and $V_p = V_d = 1/V_m(B)$.*
2. *If (y^*, z^*) solves matrix game B, $y^*/V_m(B)$ solves the primal LP, and $z^*/V_m(B)$ solves the dual LP.*
3. *If $\tilde{y}^*$ solves the primal LP, and $\tilde{z}^*$ solves the dual LP, the pair $(\tilde{y}^*/V_p, \tilde{z}^*/V_d)$ constitutes an MSSPE for the matrix game B and hence for A and $V_m(B) = 1/V_p$.*

9.2 GAME-THEORETIC APPROACH

9.2.1 Introduction

Recent years have witnessed wide application of network technology to control systems. Integrated with a communication core, networked control systems (NCS) have increased mobility and interoperability, and low maintenance and installation cost [35]. The exposure to public networks renders the control system the target of potential cyberattacks. Being the connection of the information world and reality, control systems targeted by cyberattacks can have serious incidents, which have been verified during the past decade [36,37].

By targeting different components of the control system, the attacker can launch various types of attacks. Most of these control system-oriented attacks can be categorized as deception-like attacks [38,39], which compromise data integrity, and denial-of-service (DoS) attacks which compromise data availability. Compared with deception attacks, DoS attacks, which lead to congestions in the communication network, require little prior knowledge about the control system [40], but they seriously threaten the control systems operating in real time. For example, the control system using deadline corrective control may be driven to instability under DoS attacks [41]. Hence, such attacks have been listed as the most financially expensive security incidents [42] and raise major concerns. In the literature, some arguably representative theoretic tools have been employed in securing the NCS under DoS attacks, which are algebraic graph methods [43], game-theoretic methods [44], and control methods such as the linear–quadratic–Gaussian (LQG) optimal control method [45], model predictive control method [46], and event trigger control method [47].

There have been two major lines of research on securing the NCS under a DoS attack, which can be categorized as the attack-tolerant control methods and attack-compensation control methods. For the first category, the control strategies have been exploited, which can tolerate unwanted network-induced phenomenon caused by DoS attacks, such as extra packet dropout or time delays [46,48]. Note that overly long time delays or excessive loss of packets may exceed the ability of the attack-tolerant control method [45,46], and hence, the control system may still be driven to instability. Thus, some attack-compensation control methods have been employed to compensate for the control performance degradation caused by DoS attacks via contradicting undesirable attack-induced phenomenon. In [41,44,49], the intrusion detection system (IDS) has been deployed in the cyberlayer which can defend against DoS attacks and improve the performance of the underlying control system. The authors of [50] have developed data-sending strategies to contradict the packet dropout induced by DoS attacks, while [51] has introduced the concept of security investment into the field of resilient control, where the degradation of control performance can be minimized by finding the appropriate investment strategies. Unfortunately, to the best of the authors' knowledge, all the aforementioned resilient control schemes are developed in the continuous or discrete domain, and therefore ignore the possible numerical illness for the sampled system with a high sampling rate [52]. The purpose of this section is therefore to shorten such a gap.

In this section, a novel resilient control method against DoS attacks is proposed for NCSs by recurring to the delta operator, which has been well recognized to overcome numerical illness for the discrete-time system with fast sampling rate [53]. Two control structures are considered for NCSs, which are the multiple-tasking optimal control (MTOC) and central-tasking optimal control (CTOC) structures. The algorithms to obtain the optimal compensation strategy (defense strategy) and optimal attack strategy are provided, respectively.

9.2.2 Model of NCS Subject to DoS Attack

Consider the dynamics of the delta-domain NCS under the DoS attack as follows:

$$\delta x_k = A_\delta x_k + \sum_{i=1}^{S} \alpha_k^i B_\delta^i u_k^i, \tag{9.45}$$

where $x_k = x_{KT_s}$, T_s is the sampling interval, A_δ and B_δ^i are matrices in the delta domain with appropriate dimensions. The variable $\alpha_k^i = \alpha_k^{FN}\alpha_k^{BN}$, $i \in \mathbf{S} := \{1, 2, \ldots, S\}$ indicates the effect of DoS attacks on the control system. When DoS attacks are launched, there is a chance that the kth sensor packet is dropped and the "zero-control" input strategy is applied. We assume that α_k^i is a random variable which is distributed according to the Bernoulli distribution. Suppose that α_k^i are i.i.d. and, for $i \neq j$, $i, j, \in \mathbf{S}$, α_k^i is independent of α_k^j. Let us denote

$$\mathbb{P}\{\alpha_k^i = 0\} = \alpha^i, \quad \mathbb{P}\{\alpha_k^i = 1\} = 1 - \alpha^i = \bar{\alpha}^i,$$
$$\forall i \in \mathbf{S}, k \in \mathbf{K} := \{1, 2, \ldots, K\}.$$

Note that α^i here is viewed as intensity of attack (IoA). More intense DoS attack will lead to larger IoA α^i. Let us define the strategy set of the DoS attacker to be $\mathcal{G} := \{\alpha^i\}_{i=1}^S$.

Remark 9.2. 1. *System (9.45) is actually a delta-domain extension of the widely used discrete-domain model describing the control system under DoS attacks; see, e.g., [41,45,51].*

2. *Resilient control under DoS attacks should be able to address the problem of packet dropout which is also common in the traditional NCS [41,45,48,51]. However, it is worth mentioning that the packet dropout rate caused by inherent communication failure is much smaller comparing with the one induced by the* *malicious DoS attack [35]. This puts forward a higher requirement for resilient control, which should be able to tolerate serious congestions in the communication channel, not just occasionally occurring information loss.*

9.2.3 Optimal Tasking Design

In this part, we will consider the NCS with the structures of MTOC and CTOC, respectively. The CTOC structure requires that all the controllers be coordinated to reach a common objective, while those of the MTOC do not cooperate with each other and only minimize the individual cost function. Thus, MTOC can be used to model some complex and distributed systems, where each decision maker is selfish and only pursues its own interest. The CTOC structure, on the other hand, aims to address problems where central coordination and cooperation among decision makers are possible. Note that the controller design of MTOC falls within the framework of noncooperative dynamic game in the delta domain with each controller (player) minimizing the cost-to-go function

$$J^i = \mathbb{E}\left\{x_K^T Q_\delta^K x_K + T_s \sum_{k=0}^{K-1} \Psi_0\right\}, \quad i \in \mathbf{S}, \tag{9.46}$$

where $\Psi_0 = x_k^T Q_\delta^i x_k + \alpha_k^i u_k^{i\,T} R_\delta^i u_k^i$. We assume that $Q_\delta^K \geq 0$, $Q_\delta^i \geq 0$, and $R_\delta^i > 0$ for all $i \in \mathbb{S}$. In contrast, for CTOC, there is a central coordination such that all players cooperate to minimize a common cost-to-go function defined as $\tilde{J} = \sum_{i=1}^{S} \eta^i J^i$, where $\eta^i > 0$ is the weighting factor on player i's cost-to-go function and satisfies the normalization condition $\sum_{i=1}^{S} \eta^i = 1$. For both MTOC and CTOC, transmission control protocol (TCP) is applied where each packet is acknowledged, and the information set is defined as $\mathcal{I}_0 = \{x_0\}$, $\mathcal{I}_k = \{x_1, x_2, \ldots, x_k, \alpha_0, \alpha_1, \ldots, \alpha_{k-1}\}$. The admissible strategies μ_k^i for MTOC and $\tilde{\mu}_k^i$ for CTOC are seen as functional to map information set $\mathcal{I}_k$ to u_k^i, i.e., $u_k^i = \mu_k^i(\mathcal{I}_k)$ or $u_k^i = \tilde{\mu}_k^i(\mathcal{I}_k)$. Let us further define

$$\mu^i := \{\mu_0^i, \mu_1^i, \ldots, \mu_{K-1}^i\}, \quad \tilde{\mu}^i := \{\tilde{\mu}_0^i, \tilde{\mu}_1^i, \ldots, \tilde{\mu}_{K-1}^i\}.$$

Toward this end, two optimization problems should be addressed. For Problem 9.1, all the players are selfish and noncooperative. Let μ^{-1} denote the collection of strategies of all players except player i, i.e., $\mu^{-1} = \{\mu^1, \ldots, \mu^{i-1}, \mu^{i+1}, \mu^S\}$. Player i is faced with minimizing its own associated cost function by solving the following dynamic optimization problem.

Problem 9.1. Find control strategies μ^{i*}, $i \in \mathbf{S}$ such that the following optimization problem is solved for all $i \in \mathbf{S}$:

$$
\begin{aligned}
(\mathrm{OC}(i)) \ & \min_{\mu^i} \ J^i(\mu^i, \mu^{-i*}) \\
& := \mathbb{E}\left\{ x_K^T Q_\delta^K x_K + T_s \sum_{k=0}^{K-1} \Psi_0 \right\} \\
\text{s.t.} \quad & \delta x_k = A_\delta x_k + \sum_{i=1}^{S} \alpha_k^i B_\delta^i u_k^i. \quad \square
\end{aligned}
\tag{9.47}
$$

If the optimization is carried out for each i, Nash equilibrium (NE) will be obtained, which satisfies $J^{i*}(\mu^{i*}, \mu^{-i*}) \leq J^i(\mu^i, \mu^{-i})$. The corresponding total cost achieved is given by $J^* = \sum_{i=1}^{S} \eta^i J^{i*}$. In the same spirit, the associate optimization problem of CTOC is shown as follows.

Problem 9.2. Find control strategies $\tilde{\mu}^* := \{\tilde{\mu}^{i*}, \tilde{\mu}^{-i*}\}$ such that the following optimization problem is solved:

$$
\begin{aligned}
(\mathrm{COC}) \ & \min_{\tilde{\mu}} \ \tilde{J}(\tilde{\mu}^i, \tilde{\mu}^{-i}) \\
& := \sum_{i=1}^{S} \eta^i \mathbb{E}\left\{ x_K^T Q_\delta^K x_K + T_s \sum_{k=0}^{K-1} \Psi_0 \right\} \\
\text{s.t.} \quad & \delta x_k = A_\delta x_k + \sum_{i=1}^{S} \alpha_k^i B_\delta^i u_k^i. \quad \square
\end{aligned}
\tag{9.48}
$$

This minimization is essentially an optimal control problem. The optimal value $\tilde{J}^*(\tilde{\mu}^{i*}, \tilde{\mu}^{-i*}) \leq \tilde{J}(\tilde{\mu}^i, \tilde{\mu}^{-i})$ will be obtained if the optimization problem is solved.

9.2.4 Defense and Attack Strategy Design

In this section, we assume that the defender has the freedom to participate in choosing the weighting matrices Q_δ^i, $i \in \mathbf{S}$ to compensate for the performance degradation of NCS with the MTOC (CTOC, respectively) structure [55]. We define the set of defense strategies as $\mathcal{F} := \{Q_\delta^i\}_{i=1}^{S}$ and the optimal defense strategies are obtained by solving the following problem.

Problem 9.3. For the MTOC structure, find the optimal defense strategy $\mathcal{F}^*$ which is the solution of

$$
\min_{\mathcal{F}} \left| J_{\mathcal{G}_1}^* / \tilde{J}_{\mathcal{G}_0}^* - 1 \right|
$$

while for the CTOC structure, find the optimal defense strategy $\mathcal{F}^*$ which is the solution of

$$\min_{\mathcal{F}} \left| \tilde{J}^*_{\mathcal{G}_1} / \tilde{J}^*_{\mathcal{G}_0} - 1 \right|$$

where $\mathcal{G}_0 = \{\alpha^i = 0\}_{i=1}^S$ and $\mathcal{G}_1 = \{0 < \alpha^i < 1\}_{i=1}^S$. □

It is worth mentioning that $J^*_{\mathcal{G}_1} > \tilde{J}^*_{\mathcal{G}_1} > \tilde{J}^*_{\mathcal{G}_0} > 0$, since both the noncooperative behavior and packet dropout phenomenon can lead to performance degradation of the control system. On the other hand, for DoS attackers, a case is considered where the attacker is fully aware of the defender's compensation strategy $\mathcal{F}^*$, i.e., the attacker and defender possess asymmetric information. To save the attacking cost, the DoS attacker aims to drive the underlying NCS out of safety zone using attacking intensity $\mathcal{G} = \{\alpha^i\}_{i=1}^S$ as low as possible. The optimal attack strategies $\mathcal{G}^*$ are obtained by solving the following problem.

Problem 9.4. For given weighting factors $\{\rho^i\}_{i=1}^S$ and the MTOC structure, find the optimal attacking strategy $\mathcal{G}^*$ which is the solution of the following optimization problem:

$$\begin{aligned} &\min \alpha^1 \\ &\text{s.t.} \quad \left| J^*_{\mathcal{G}_1} / \tilde{J}^*_{\mathcal{G}_0} - 1 \right| > \mathcal{S}_o, \quad \alpha^1 = \rho^2\alpha^2 = \cdots = \rho^S\alpha^S \end{aligned} \tag{9.49}$$

while for the CTOC structure, find the optimal attacking strategy $\mathcal{G}^*$ which is the solution of the following optimization problem:

$$\begin{aligned} &\min \ \alpha^1 \\ &\text{s.t.} \quad \left| \tilde{J}^*_{\mathcal{G}_1} / \tilde{J}^*_{\mathcal{G}_0} - 1 \right| > \mathcal{S}_o, \quad \alpha^1 = \rho^2\alpha^2 = \cdots = \rho^S\alpha^S, \end{aligned} \tag{9.50}$$

where $\mathcal{S}_o$ is a scalar representing the safety zone. Without loss of generality, we assume that $\rho^i = 1$, $i = 2, \ldots, S$ subsequently. □

9.2.5 Tasking Control Strategies

In this section, the conditions and analytical form of optimal control strategies for the MTOC and CTOC structures are provided, respectively. Some preliminary notations are provided as follows:

$$\Theta_0 = (T_s A_\delta + I) - T_s \sum_{j \neq i}^{S} \bar{\alpha}^j B^j_\delta L^j_k,$$

$$\Theta_1 = A_\delta - \sum_{i=1}^{S} \bar{\alpha}^i B_\delta{}^i L_k^i,$$

$$\Theta_2 = (T_s A_\delta + I)x_k + T_s \sum_{i=1}^{S} \alpha_k^i B_\delta^i u_k^i,$$

$$\Theta_3 = (T_s A_\delta + I)x_k + T_s \sum_{j \neq i}^{S} \bar{\alpha}^j B_\delta^j u_k^j,$$

$$\Omega_0 = \sum_{i=1}^{S} \eta^i Q_\delta^i,$$

$$\Omega_1 = (T_s A_\delta + I),$$

$$\Omega_2 = \text{diag}\left(\left[\bar{\alpha}^1(1-\bar{\alpha}^1){B_\delta^1}^T \tilde{P}_{k+1} B_\delta^1, \bar{\alpha}^2(1-\bar{\alpha}^2) {B_\delta^2}^T \tilde{P}_{k+1} B_\delta^2, \ldots, \bar{\alpha}^S(1-\bar{\alpha}^S){B_\delta^1}^T \tilde{P}_{k+1} B_\delta^S\right]\right).$$

The following theorem is presented first to provide solution to Problem 9.1.

Theorem 9.7. *For NCS with the MTOC structure and given attack strategy $\mathcal{G}$, the following conclusions are true:*

1. *There exists a unique NE if*

$$R_\delta^i + T_s {B_\delta^i}^T P_{k+1}^i B_\delta^i > 0 \tag{9.51}$$

and the matrix Φ_k is invertible, where $\Phi_k(i,i) = R_\delta^i + T_s {B_\delta^i}^T P_{k+1}^i B_\delta^i$, $\Phi_k(i,j) = T_s \bar{\alpha}^j {B_\delta^i}^T P_{k+1}^i B_\delta^j$.

2. *Under Condition* **1.** *above, the optimal control strategies of MTOC are given by $u_k^i = \mu_k^{i*}(\mathcal{I}_k) = -L_k^i x_k$ for all $i \in \mathbf{S}$, where*

$$L_k^i = (R_\delta^i + T_s {B_\delta^i}^T P_{k+1}^i B_\delta^i)^{-1} B_\delta^{iT} P_{k+1}^i \Theta_0. \tag{9.52}$$

3. *The backward iterations are carried out with $P_K^i = Q_\delta^K$ and*

$$\begin{aligned} -\delta P_k^i = {} & Q_\delta^i + \bar{\alpha}^i {L_k^i}^T R_\delta^i L_k^i \\ & + T_s \Theta_1^T P_{k+1}^i \Theta_1 + \Theta_1^T P_{k+1}^i + P_{k+1}^i \Theta_1 \\ & + T_s \sum_{j=1}^{S} \left(\left(\bar{\alpha}^j - \bar{\alpha}^{j2}\right) {L_k^j}^T {B_\delta^j}^T P_{k+1}^i B_\delta^j L_k^j\right), \end{aligned} \tag{9.53}$$

$$P_k^i = P_{k+1}^i - T_s \delta P_k^i. \tag{9.54}$$

4. *Under Condition* **1.**, *the NE values under the MTOC are* $J^{i^*} = x_0^T P_0^i x_0$, $i \in \mathbf{S}$, *where* x_0 *is the initial value.*

Proof. An induction method is employed here. The claim is clearly true for $k = K$ with parameters $P_K^i = Q_\delta^K$. Let us suppose that the claim is now true for $k+1$, and the cost function at $k+1$ is constructed as

$$V^i(x_{k+1}) = \mathbb{E}\{x_{k+1}^T P_{k+1}^i x_{k+1}\} \tag{9.55}$$

with $P_{k+1}^i > 0$. According to [53, Lemma 1], the above equation can be rewritten in the delta domain as

$$\begin{aligned} V^i(x_{k+1}) &= T_s\delta(x_k^T P_k^i x_k) + x_k^T P_k^i x_k \\ &= T_s \delta x_k^T P_{k+1}^i x_k + T_s x_k^T P_{k+1}^i \delta x_k \\ &\quad + T_s^2 \delta x_k P_{k+1}^i \delta x_k + x_k^T P_{k+1}^i x_k. \end{aligned}$$

Using the dynamic programming method, the cost at time k is obtained by

$$\begin{aligned} V^i(x_k) = \min_{u_k^i} \mathbb{E}\Big\{ & T_s x_k^T Q_\delta^i x_k \\ & + T_s \alpha_k^i u_k^{i\,T} R_\delta^i u_k^i + V^i(x_{k+1})\Big\}. \end{aligned} \tag{9.56}$$

The cost-to-go function $V^i(x_k)$ is strictly convex in u_k^i, since the second derivative of (9.56) yields $R_\delta^i + T_s B_\delta^{i\,T} P_{k+1}^i B_\delta^i > 0$. The optimal control strategies can be obtained by solving $\partial V^i(x_k)/\partial u_k^i = 0$, which is set up for all players. Thus, there exists a unique NE if the invertible matrix Φ_k satisfies

$$\Phi_k \bar{L}_k = \Pi_k \tag{9.57}$$

with $\Phi_k(i,i) = R_\delta^i + T_s B_\delta^{i\,T} P_{k+1}^i B_\delta^i$, $\Phi_k(i,j) = T_s \bar{\alpha}^j B_\delta^{i\,T} P_{k+1}^i B_\delta^j$, $\Pi_k(i,i) = B_\delta^{i\,T} P_{k+1}^i (T_s A_\delta + I)$, and $\Pi_k(i,j) = 0$, $\bar{L}_k = \left[L_k^{1T}, L_k^{2T}, \ldots, L_k^{ST}\right]^T$. Substituting $u_k^{i^*} = -L_k^i x_k$ into (9.56), we have (9.53). This completes the proof. □

Next, the following theorem is given to provide solutions to Problem 9.2.

Theorem 9.8. *First, let us define*

$$\begin{aligned} \tilde{R}_\delta &:= \mathrm{diag}\left(\left[\bar{\alpha}^1 \eta^1 R_\delta^1, \bar{\alpha}^2 \eta^2 R_\delta^2, \ldots, \bar{\alpha}^S \eta^S R_\delta^S\right]\right), \\ \tilde{B}_\delta &:= \left[\bar{\alpha}^1 B_\delta^1, \bar{\alpha}^2 B_\delta^2, \ldots, \bar{\alpha}^S B_\delta^S\right], \end{aligned}$$

$$\Xi := \tilde{R}_\delta + \mathrm{diag}\left(\left[T_s\bar{\alpha}^1(1-\bar{\alpha}^1)B_\delta^{1\,T}\tilde{P}_{k+1}B_\delta^1,\right.\right.$$
$$T_s\bar{\alpha}^2(1-\bar{\alpha}^2)B_\delta^{2\,T}\tilde{P}_{k+1}B_\delta^2,\ldots,T_s\bar{\alpha}^S(1-\bar{\alpha}^S)$$
$$\left.\left.B_\delta^{S\,T}\tilde{P}_{k+1}B_\delta^S\right]\right) + T_s\tilde{B}_\delta^T\tilde{P}_{k+1}\tilde{B}_\delta.$$

For NCS with the CTOC structure and given attack strategy $\mathcal{G}$, the following conclusions are true:

1. *If we have invertible matrix $\Xi > 0$, there exists a unique optimal solution.*
2. *Let us denote $\tilde{u}_k = [u_k^{1T}, u_k^{2T}, \ldots, u_k^{ST}]^T$. Under Condition 1, the optimal control strategy of CTOC $\tilde{\mu}^*$ is given by $\tilde{u}_k = \tilde{\mu}^*(\mathcal{I}) = -\tilde{L}_k x_k$, where*

$$\tilde{L}_k = \Xi^{-1}\tilde{B}_\delta^T\tilde{P}_{k+1}\Omega_1. \tag{9.58}$$

3. *The backward recursions are carried out with $\tilde{P}_K = \sum_{i=1}^S \eta^i Q_\delta^K$ and*

$$-\delta\tilde{P}_k = \Omega_0 + T_s A_\delta^T\tilde{P}_{k+1}A_\delta + \tilde{P}_{k+1}A_\delta \tag{9.59}$$
$$+ A_\delta^T\tilde{P}_{k+1} - \Omega_1^T\tilde{P}_{k+1}\tilde{B}_\delta\Xi^{-1}\tilde{B}_\delta^T\tilde{P}_{k+1}\Omega_1,$$
$$\tilde{P}_k = \tilde{P}_{k+1} - T_s\delta\tilde{P}_k. \tag{9.60}$$

4. *Under Condition 1, the optimal value with the CTOC structure is $\tilde{J}^* = x_0^T\tilde{P}_0 x_0$, where x_0 is the initial value.*

Proof. The proof is similar to that of Theorem 9.7 and is omitted here. □

Remark 9.3. *Note that Theorems 9.7 and 9.8 differ from most of the existing literature on the delta operator system control due mainly to the following:*

1. *By referring to the dynamic programming method, the optimal solution can be obtained instead of suboptimal results obtained using the LMI method.*
2. *The results in Theorems 9.7 and 9.8 are readily extended to develop time-varying optimal control strategies for a time-varying system, while the LMI method normally obtains time-invariant control strategies [52,53].*

9.2.6 Defense and Attack Strategies

In this section, an intelligent but resource-limited attacker is considered. The attacker is intelligent in the sense that it can capture the defender's strategies and adopt countermeasures, i.e., the attacker and defender possess asymmetric information. The aforementioned interactions can be described by a Stackelberg game [41], where the attacker acts as the leader and the defender acts as the follower. The defender aims to minimize the performance degradation caused by the attacker. On the other hand, being fully aware of the defender's strategies, the attacker aims to drive the control system out of the safety zone while minimizing the attacking intensity. The details of the Stackelberg game and the derivation of the corresponding Stackelberg solutions are shown subsequently.

9.2.7 Development of Defense Strategies

According to the previous section, the NCS with the CTOC structure under no DoS attack is viewed as the nominal/ideal case. Thus, the desirable cost-to-go function yields $\tilde{J}^*_{\mathcal{G}_0}$. On the other hand, we denote the desired feedback gains as $\{\hat{L}^i_k\}_{i=1}^S$, which are "desirable" in the sense that they can adapt to the network communication environment better. The desirable feedback gains can be determined by using Theorems 9.7 and 9.8, or via a system identification method from the experimental data [54]. In what follows, we provide algorithms to achieve both desirable cost-to-go function and feedback gains.

Theorem 9.9. *Consider $\mathcal{G}_1 = \{\alpha'^i\}_{i=1}^S$ and desired strategies $\{\hat{L}^i_k\}_{i=1}^S$ to be given.*

1. *If positive matrices $\{Q^i_\delta\}_{i=1}^S$ and $\{P^i_0, P^i_1, \ldots, P^i_{K-1}\}_{i=1}^S$ exist such that (9.52) and (9.53) hold with $L^i_k = \hat{L}^i_k$, then $\{u^i_k = \hat{L}^i_k x_k\}$ is the optimal control strategies of MTOC.*
2. *If positive matrices $\{Q^i_\delta\}_{i=1}^S$ and $\{\tilde{P}_0, \tilde{P}_1, \ldots, \tilde{P}_{K-1}\}$ exist such that (9.58) and (9.59) hold with $\tilde{L}_k = \left[\hat{L}^{1T}_k, \hat{L}^{2T}_k, \ldots, \hat{L}^{ST}_k\right]^T$, then*

$$\tilde{u}_k = -\tilde{L}_k = \left[\hat{L}^{1T}_k \quad \hat{L}^{2T}_k \quad \ldots \quad \hat{L}^{ST}_k\right]^T x_k$$

is the optimal control strategy of CTOC.

Proof. The result is straightforward using Theorems 9.7 and 9.8.

According to Theorem 9.9, Problem 9.3 can be reformulated as the following convex optimization problems $\mathcal{P}_1$ and $\mathcal{P}_2$:

$$\begin{aligned} \mathcal{P}_1: \ \min \ & \left| J^*_{\mathcal{G}_1}/\tilde{J}^*_{\mathcal{G}_0} - 1 \right| \\ \text{s.t.} \ & \left(\{Q^i_\delta > 0\}_{i=1}^S, \ \{P^i_0 > 0, \ldots, P^i_{K-1} > 0\}_{i=1}^S \right) \in \mathcal{C}, \end{aligned} \tag{9.61}$$

$$\mathcal{C} := \Big\{ \Big(\big\{ Q_\delta^i \big\}_{i=1}^S, \ \big\{ P_0^i, \ldots, P_{K-1}^i \big\}_{i=1}^S \Big) : \\ L_k^i = \hat{L}_k^i, \ (9.52) \text{ and } (9.53) \text{ hold} \Big\},$$

or

$$\begin{aligned} \mathcal{P}_2: \ & \min \ \left| \tilde{J}_{\mathcal{G}_1}^* / \tilde{J}_{\mathcal{G}_0}^* - 1 \right| \\ & \text{s.t.} \ \Big(\big\{ Q_\delta^i > 0 \big\}_{i=1}^S, \ \big\{ \tilde{P}_0 > 0, \ldots, \tilde{P}_{K-1} > 0 \big\} \Big) \in \mathcal{C}, \\ & \mathcal{C} := \Big\{ \Big(\big\{ Q_\delta^i \big\}_{i=1}^S, \ \big\{ \tilde{P}_0, \ldots, \tilde{P}_{K-1} \big\} \Big) : \\ & \qquad \tilde{L}_k = \begin{bmatrix} \hat{L}_k^{1T} & \hat{L}_k^{2T} & \ldots & \hat{L}_k^{ST} \end{bmatrix}^T, \ (9.58) \text{ and } (9.59) \text{ hold} \Big\}. \end{aligned} \tag{9.62}$$

By solving $\mathcal{P}_1$ and $\mathcal{P}_2$, the players can achieve the desired strategies by just tuning the interface parameters Q_δ^{i*}. Furthermore, using the desired feedback gains from practical data to determine the optimality can be helpful to illustrate the goals of the players [54]. □

Remark 9.4. 1. *Note that $\mathcal{P}_1$ and $\mathcal{P}_2$ retain their convexity if weighting matrices R_δ^i and Q_δ^i are both viewed as decision variables. Thus, the weighting matrices R_δ^i and Q_δ^i can be determined simultaneously such that the players achieve desirable strategies.*

2. *The algorithm for the defense strategies can be seen as the pricing mechanism design [55] in game theory, i.e., by finding appropriate pricing parameters $(\{Q_\delta^{i*}\}_{i=1}^S)$, the final outcome of the game can be driven to a desired target. On the other hand, the proposed pricing mechanism in this section can be regarded as an extension to the inverse LQR method [54], where the desired control strategies are known in advance, and the weighting matrix Q or R is to be determined.*

9.2.8 Development of Attack Strategies

It should be noticed that the traditional work on NCS normally regards the packet dropout as a constraint [51] and seldom stands at the opposite side to exploit how to enhance the packet dropout rate to degrade the system performance. In this part, a strategic and resource-limited attacker is considered, and furthermore, the corresponding attacking strategy is provided. This actually provides a worst case for the defender and is helpful in the defense mechanism design [45].

In what follows, we provide Algorithm 9.1 to solve Problem 9.4, which is a standard dichotomy algorithm providing the optimal attacking strategy $\mathcal{G}^*$ such that the NCS with MTOC (CTOC, respectively) structure is driven out of the safety zone with the smallest IoA α^i, $i \in \mathbf{S}$.

Algorithm 9.1 The algorithm for optimal attacking strategy $\mathcal{G}^*$

Initialization: Set $a = 1$; $b = 0$; $S_o > 0$; $0 < \varepsilon \ll 1$.
1: Denote $\alpha = \bar{\alpha}^1 = \bar{\alpha}^2 = \cdots = \bar{\alpha}^S$, and $\alpha = (a + b)/2$.
2: **while** $|b - a| > \epsilon$ **do**
3: Calculate the optimal defense strategy $(\{Q_\delta^{i*}\}_{i=1}^S)$ by solving (9.61) (resp. (9.62)).
4: Calculate $\left|J^*_{\mathcal{G}_1}/J^*_{\mathcal{G}_0} - 1\right|$ (resp. $\left|\tilde{J}^*_{\mathcal{G}_1}/\tilde{J}^*_{\mathcal{G}_0} - 1\right|$) using Q_δ^{i*}, and compare it with $\mathcal{S}_o$.
5: **if** $\left|J^*_{\mathcal{G}_1}/\tilde{J}^*_{\mathcal{G}_0} - 1\right| < \mathcal{S}_o$ So (resp. $\left|\tilde{J}^*_{\mathcal{G}_1}/\tilde{J}^*_{\mathcal{G}_0} - 1\right| < \mathcal{S}_o$) **then**
6: $a = \alpha$, $b = b$
7: **else if** $\left|J^*_{\mathcal{G}_1}/\tilde{J}^*_{\mathcal{G}_0} - 1\right| > \mathcal{S}_o$ (resp. $\left|\tilde{J}^*_{\mathcal{G}_1}/\tilde{J}^*_{\mathcal{G}_0} - 1\right| > \mathcal{S}_o$) **then**
8: $a = a$, $b = \alpha$
9: **end if**
10: Set $\alpha = (a + b)/2$.
11: **end while**

Remark 9.5. 1. *The feasibility of Algorithm 9.1 is that the open-loop system $\delta x_k = A_\delta x_k$ should be unstable. The reasons are that the closed-loop system will be driven to instability again if all the control commands are lost, i.e., $\alpha^1 = \alpha^2 = \cdots = \alpha^S = 1$. Then, we will have $J^*_{\mathcal{G}_1}/\tilde{J}^*_{\mathcal{G}_0} \to \infty$ $(\tilde{J}^*_{\mathcal{G}_1}/\tilde{J}^*_{\mathcal{G}_0} \to \infty$, respectively) and the constraint (9.51) will be feasible.*

2. *It can be deduced that the time complexity of the proposed dichotomy algorithm is $O(\log_2 n)$, where $n = 1/\varepsilon$, while the time complexity of the exhaustive method is $O(n)$. Thus, the proposed attacking algorithm saves more time for the attackers to find out the attacking strategy online. For example, if the safety zone $\mathcal{S}_o$ for the control system changes, the attacker can quickly adapt to such a change and find out the corresponding attacking strategy.*

In what follows, we aim to demonstrate the validity and applicability of the proposed method. For this purpose, the proposed methodology is applied to the numerical simulations of the heating, ventilation, and air conditioning (HVAC) system, and the experimental verification on the ball and beam system is also provided.

9.2.9 Model Description

Consider the following dynamics of the HVAC system [55]:

$$\rho v_i C_p \frac{d\mathcal{T}_i}{dt} = \sum_{j \neq i}^{S} h_{i,j} a_{i,j} (\mathcal{T}_j - \mathcal{T}_i) + T_0$$

with $T_0 = h_{i,o} a_{i,o} (\mathcal{T}_\infty - \mathcal{T}_i) + \dot{m}_i C_p (\mathcal{T}_i^{sup} - \mathcal{T}_i)$, where $\mathcal{T}_i$ is the temperature in zone i, ρ is the density of air, C_p is the specific heat of air, and $\mathcal{T}_\infty$ is the outside air temperature. Parameter v_i is the volume of air in the ith zone, $a_{i,j}$ is the area of the wall between zone j and i, $a_{i,j}$ is the total area of the exterior walls and roof of zone i, $h_{i,j}$ and $h_{i,o}$ are the heat transfer coefficients of the wall between zone j and i and the heat coefficient of the exterior walls, respectively, $\dot{m}_i$ is the mass flow rate of air into zone i, and $\mathcal{T}_i^{sup}$ is the supply air temperature for zone i. The specific values of the parameters can be found in [55]. Note that $\mathcal{T}_{di}$ is the set temperature of zone i. Then, temperature error of zone i is denoted as $\mathcal{T}_{ei} = \mathcal{T}_i - \mathcal{T}_{di}$. Define $x = \begin{bmatrix} x^{1T} & x^{2T} & \dots & x^{ST} \end{bmatrix}^T \triangleq \begin{bmatrix} \mathcal{T}_{e1} & \mathcal{T}_{e2} & \dots & \mathcal{T}_{eS} \end{bmatrix}^T$ to be the vector of zone temperatures and $\tilde{\mathbf{u}} = \begin{bmatrix} \tilde{u}^{1T} & \tilde{u}^{2T} & \dots & \tilde{u}^{ST} \end{bmatrix}^T \triangleq \begin{bmatrix} \mathcal{T}_1^{sup} & \mathcal{T}_2^{sup} & \dots & \mathcal{T}_S^{sup} \end{bmatrix}^T$ to be the control input, i.e., vector of supply air temperatures. Then, the system dynamics are

$$\dot{x}_t = A_s x_t + \sum_{i=1}^{S} B_s^i \tilde{u}_t^i + d_s,$$

where

$$d_s := \begin{bmatrix} d_{s1}^T, & d_{s2}^T, & \dots, & d_{sS}^T \end{bmatrix}^T.$$

Matrices $A_s(i,j)$ and $B_s^i(j,1)$ are given as follows:

$$A_s(i,j) = \begin{cases} -\left(\sum_{j \in \mathcal{N}_i} \frac{h_{i,j} a_{i,j}}{\rho v_i C_p} + \frac{\dot{m}_i}{\rho v_i} + \frac{h_{i,o} a_{i,o}}{\rho v_i C_p} \right), & i = j, \\ \frac{h_{i,j} a_{i,j}}{\rho v_i C_p}, & j \in \mathcal{N}_i \text{ and } j \neq i, \\ 0, & \text{otherwise}, \end{cases}$$

$$B_s^i(j,1) = \begin{cases} \frac{\dot{m}_i}{\rho v_i}, & i = j, \\ 0, & \text{otherwise}, \end{cases}$$

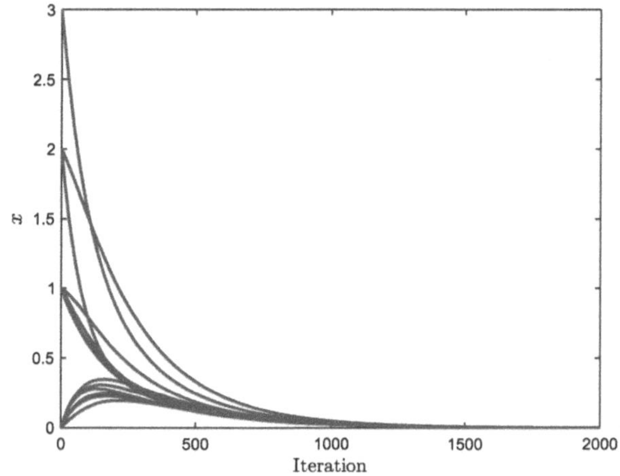

Figure 9.7 State trajectories of MTOC without compensation

where $\mathcal{N}_i$ denotes the neighborhood of player i. Denote $B_s = [B_s^1, B_s^2, \ldots, B_s^S] \in \mathbb{R}^{S\times S}$ and suppose that $\tilde{u}$ is composed of a feedback term u and a feedforward compensation term, which is $\tilde{u} = u - B_s^{-1}d_s$. Then, we have $\dot{x} = A_s x + B_s(u - B_s^{-1}d_s) + d_s = A_s x + B_s u$, where $u = [u^{1T}, u^{2T}, \ldots, u^{ST}]^T$. Considering the packet dropout and assign $T_S = 0.05$ s, one has the delta operator system

$$\delta x_k = A_\delta x_k + \sum_{i=1}^{S} \alpha_k^i B_\delta^i u_k^i.$$

9.2.10 Strategy Design

Let us denote IoA $\alpha^i = 0.2$ and $R_\delta^i = I$ for all $i \in \mathbf{S}$. We set the weighting matrices Q_δ^i, $i \in \mathbf{S}$ and Q_δ^K of the MTOC structure as identity matrix. The desired strategies are obtained using Theorem 9.8 with $\alpha^i = 0$, $Q_\delta^i = I$, $i \in \mathbf{S}$, and $Q_\delta^K = I$. Then, the state trajectories of MTOC with and without compensation are shown in Figs. 9.7 and 9.8, respectively. It is evident that the control performance improves since all the players are induced to cooperate.

Next, we stand at the side of the attacker and assign $\mathcal{S}_o = 0.38$ and $\varepsilon = 1 \times 10^{-6}$. By using Algorithm 9.1, we arrive at the results in Table 9.1, where we conclude that the optimal IoA for the attacker yields $\alpha^i = 0.4639$ if $\alpha^1 = \alpha^2 = \cdots = \alpha^S$.

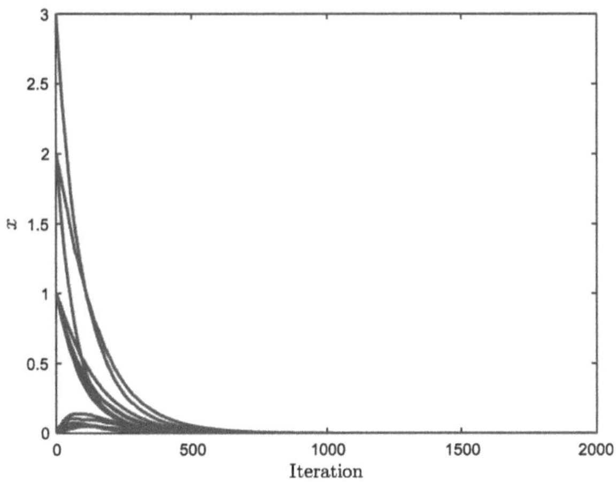

Figure 9.8 State trajectories of MTOC with compensation

Table 9.1 Optimal values for network security.

Parameter	Value
Cost of CTOC $\tilde{J}^*$	74.9575
Cost with defense strategy of MTOC J^*	103.4415
Overall performance degradation $J^*_{\mathcal{G}_1}/\tilde{J}^*_{\mathcal{G}_0}$	1.3800
Optimal attack intense α^{i*}	0.4639

Table 9.2 Comparisons for MTOC system with different estimation errors.

IoA	0	10%	20%	40%
J^*	117.1279	118.9997	120.9949	125.4236
J'^*	102.0808	102.0810	102.0832	102.0910

9.2.11 Robustness Study

In this part, the robustness of the proposed defense algorithm is verified. Note that the proposed defense algorithm and control schemes all require the estimation of IoA which is not an easy task in practice. Therefore, it is necessary to test the robustness of the defense algorithm by checking whether it still works when the estimation of IoA is not accurate. In Table 9.2, where J^* denotes the original cost of MTOC system and J'^* denotes the cost of MTOC system with defense strategies, we assume that the actual value of IoA is $\alpha^i = 0.2$, and the estimation error is described by a certain percentage of the actual value. From Table 9.2, we can see that

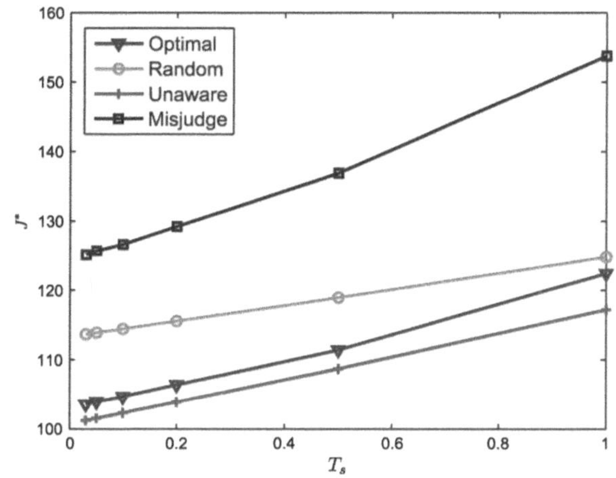

Figure 9.9 Performance comparisons among different scenarios

the proposed defense mechanism still works well even with the existence of the estimation error.

9.2.12 Comparative Study

In the sequel, the following four scenarios are considered for comparison.

1. *Optimal.* The optimal attack and defense strategies are used as proposed previously.
2. *Random.* Both the attacker and defender set their strategies randomly, regardless of the existence of the other.
3. *Unaware.* The attacker does not know about the existence of a defender and chooses IoA randomly, but the defender chooses an optimal strategy Q^{i*}, $i \in \mathbf{S}$ according to the IoA α^i the attacker chooses.
4. *Misjudge.* The attacker believes that a defender exists, but the defense strategy is just selected randomly.

The comparison results are shown in Fig. 9.9, where we can see that the cases of "optimal" and "unaware" are better than "random" and "misjudge." Thus, the proposed defense strategy can improve the system performance no matter what strategies the attacker adopts. On the other hand, it should be noticed that the proposed attacking strategy can degrade the system performance more, especially when there are no defense mechanisms.

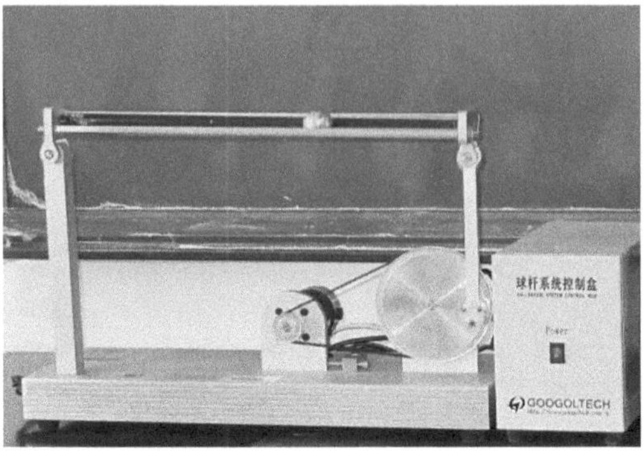

Figure 9.10 Ball and beam platform

9.2.13 Verification

To further illustrate the validity and applicability, the proposed methodology is applied to the ball and beam system in Fig. 9.10. Let us denote $x = [\gamma\ \dot{\gamma}\ \theta\ \dot{\theta}]^T$, where γ and θ are the ball position and the beam angle, respectively. Then, the state-space representation of the ball and beam system yields [52]

$$\dot{x}_t = A_s x_t + B_s u_t, \tag{9.63}$$

where

$$A_s = \begin{bmatrix} 0 & 1 & 0 & 0 \\ 0 & 0 & -7.007 & 0 \\ 0 & 0 & 0 & 1 \\ 0 & 0 & 0 & 0 \end{bmatrix}, \quad B_s = \begin{bmatrix} 0 \\ 0 \\ 0 \\ 1 \end{bmatrix}.$$

We set the initial value of the state as $x_0 = [0.2\ 0\ 0\ 0]^T$ and the sampling period as $T_s = 0.02$ s. We aim to reselect the weight matrices of the cost function such that the controller obtained using the traditional LQR method [56] can adapt to the network environment. The original controller using the LQR method is $u_k^* = [-9.2713\quad -8.47462\ \ 7.1400\ \ 7.2887]x_k$ without considering the packet dropout. The desired control strategy yields $u_k^* = [-9.3113\quad -8.52352\ \ 7.3359\ \ 7.3519]x_k$ which can be obtained from Theorem 9.8 with $\alpha = 0.01$, $Q_1 = \text{diag}([1000\ \ 0\ \ 10\ \ 0])$, and $R_1 = [10]$. Then, the following optimized weighting matrices can be obtained by solving $\mathcal{P}_2$ with $\bar{\alpha} = 1$:

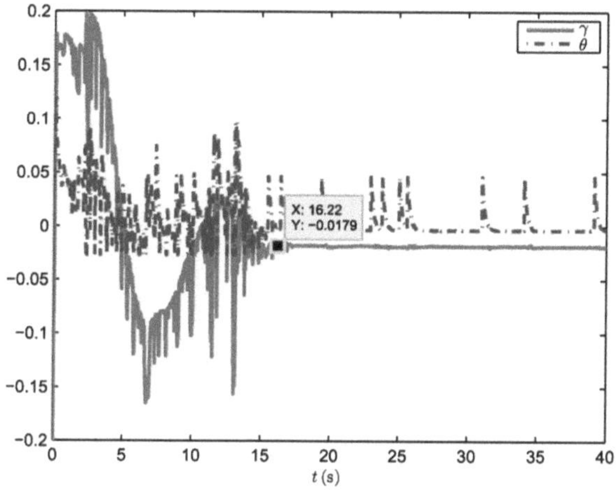

Figure 9.11 Experiment result using the LQR method with Q_1 and R_1

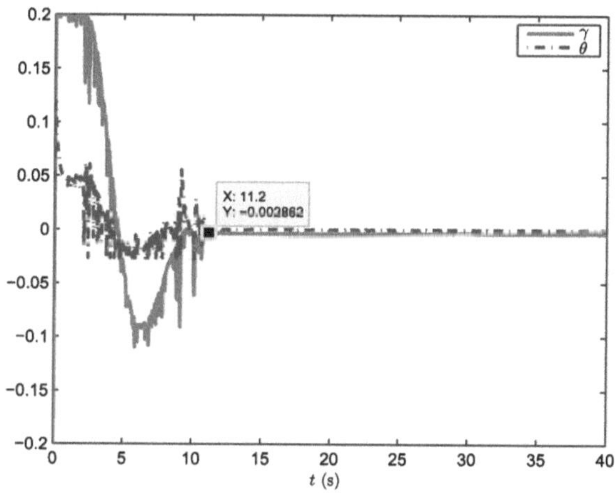

Figure 9.12 Experiment result using the LQR method with Q_2 and R_2

$$Q_2 = \begin{bmatrix} 181.0158 & 5.9030 & -48.0873 & -1.4436 \\ 5.9030 & 13.7252 & 2.6831 & -2.1865 \\ -48.0873 & 2.6831 & 29.8829 & -0.3409 \\ -1.4436 & -2.1865 & -0.3409 & 1.0521 \end{bmatrix},$$

$$R_2 = \begin{bmatrix} 1.7921 \end{bmatrix}.$$

The control results using LQR controller [56] with Q_1, R_1 and Q_2, R_2 are shown in Figs. 9.11 and 9.12, respectively. It is easy to see that the

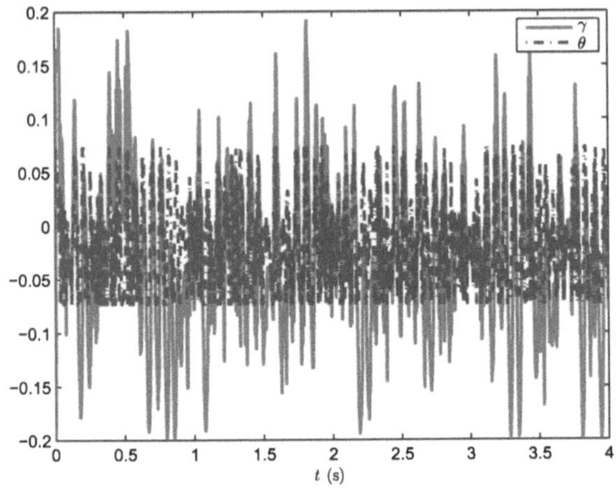

Figure 9.13 Experiment results using [58, Lemma 2]

LQR controller with optimized weighting matrices achieves better control performance.

Next, the control performance of the finite precision representation [57] of the system model (9.63) is exploited. Let us assign $T_s = 0.002$ s, $\alpha = 0.01$, $Q = \text{diag}([1000\ \ 0\ \ 10\ \ 0])$, and $R = [10]$. The finite precision representation of the system model (9.63) is derived with four significant digits in the traditional discrete and delta domains, respectively. Then, we apply the control strategies obtained from [58, Lemma 2] and Theorem 9.8 to such finite precision models, and the results are shown in Eqs. (9.51) and (9.52), respectively. It is not difficult to see that the proposed delta domain control scheme performs better under the finite word-length constraint. See Figs. 9.13 and 9.14.

9.3 CONVEX OPTIMIZATION PROBLEMS

Guaranteeing the security of cybermissions is a complex, multidimensional challenge that demands a multifaceted, strategic solution. The terminology cybermission refers to a set of computer transactions aimed at accomplishing a specific purpose or task, such as placing an online shopping order, submitting a paper to a conference through an online submission system, or printing a bank statement at an ATM machine. Cybermissions typically require a large number of computer services, including encryption services, authentication servers, database engines, web servers. We are es-

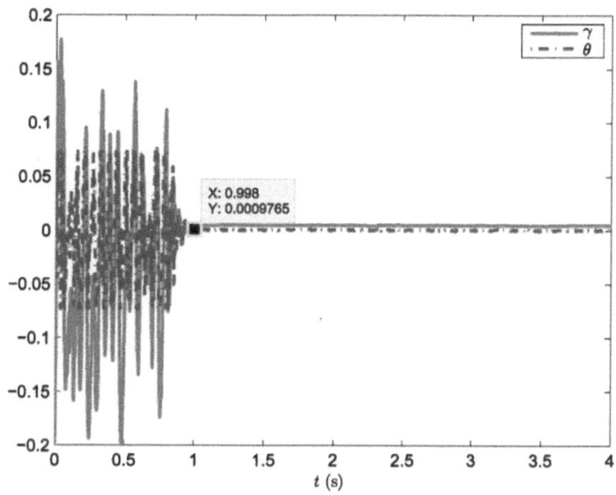

Figure 9.14 Experimental results using Theorem 9.8

pecially interested in cybermissions that go through several states, each of which may require one or more computer services. Cybermissions are especially vulnerable to attacks because it may be possible to prevent the mission's completion by compromising just one of the multiple services required by the mission, provided that the right service is compromised at the right time.

9.3.1 Introduction

Cybermissions are pervasive and can be found in trading, banking, power systems management, road traffic managements, health care, online shopping, business-to-business transactions, etc. The disruption to cybermissions can thus result in cyber or physical consequences that threaten national and economic security, critical infrastructure, public health, and welfare. Moreover, stealthy cyberattackers can lay a hidden foundation for future exploitation or attack, which they can later execute at a time of greatest advantage. Securing cyberspace requires a layered security approach across the public and private sectors.

In the cybermission security domain, a security analyst is interested in making decisions based on the potential damage that attacks can inflict to the mission and also on the probability that the potential damage is realized. To focus their attention and coordinate defensive actions, security professionals must be able to determine which attacks presents the biggest threat and prioritize which services to defend, a problem often referred to as cybersituation awareness. Situation awareness [59] is a common feature of many cybersecurity solutions but most of them are fragmented. In this paper, we present a model that can be used to predict how an attacker may try to compromise a cybermission with a limited amount of resources, based on a model that takes into account potential damage to the mission and probabilistic uncertainty.

This approach followed here is motivated by the need to avoid flooding the security analyst with raw data about complex missions and detailed logs from intrusion detection systems (IDSs). Instead, an automated or semi-automated system should process this data and present the analyst with high-level information about the computer services that are currently most crucial for mission completion and thus most likely to be the target of attacks, based on the current state of the mission and its future expected evolution. To achieve this, we propose a relatively general model to describe the damage to a cybermission caused by potential attacks. This model can be utilized in optimization schemes to discover optimal policies to distribute attack resources over time and over the different computer services relevant to the mission so as to maximize damage to the cybermission. The proposed models need mission parameters that typically vary in time according to complex dynamics, which are difficult to determine in an analytic fashion. To avoid this difficulty, we learn such parameters using system identification of low-order state-space models that are used to make predictions of the parameter evolution for a reasonable future time horizon.

Security competitions are exceptional venues for researchers to discover and validate novel security solutions. The international Capture The Flag (iCTF) [60] contest is a distributed wide-area security exercise whose goal is to test the security skills of the participants. The iCTF contest is organized by the Security Lab of the Department of Computer Science at UCSB and is held once a year. The Capture the Flag contest is a multisite, multiteam hacking contest, in which a number of teams compete independently against each other. The 2011 edition of iCTF was aimed at cybersituation awareness and, to the best of our knowledge, produced the first experimental dataset that includes mission descriptions as well as attack logs and the statuses of computer services required by missions [61, 59,62,60]. We have used this data to validate the algorithms presented in this paper and show their efficacy in predicting attacks to cybermissions by the human participants in the exercise.

The results presented in this paper were also used in the design of a high-level visualization tool to help security analysts to protect the computer systems under attack in the 2011 iCTF competition [62]. We are in the process of developing human subject experiments to demonstrate the benefits of using the predictions generated by the methodology proposed in this section, instead of searching through mission traces and security logs.

Our goal is to capture complex behaviors with a relatively simple model and incorporate them in a cybersecurity advisory system to show its effectiveness. This section presents a general framework to model mission-critical cybersecurity scenarios.

9.3.2 Cybermission Damage Model

Suppose that the (potential) damage that an attacker can inflict to a cyber-mission is quantified by a scalar $x_{PD} \geq 0$, which is a function of the level of attack resources $u_{AR} \geq 0$ devoted to the attack. The mapping from attack resources to potential damage is expressed by the so-called potential damage equation that we approximate by a linear map:

$$x_{PD} = f(u_{AR}) := a + bu_{AR}, \tag{9.64}$$

where $a \in \mathbb{R}^+$ can be viewed as the zero-resource damage level (damage achieved without an intended attack), and $b \in \mathbb{R}^+$ the marginal damage per unit of attack resources.

Whether or not the potential damage to the mission x_{pD} is realized is assumed to be a stochastic event that occurs with a given probability $\rho \in [0, 1]$, which also depends on the attack resources $u_{AR} \in \mathbb{R}^+$, according to the so-called uncertainty equation that we approximate by a linear map projected to the interval [0, 1]:

$$\rho = g(u_{AR}) := \Pi_{[0,1]}(c - du_{AR}), \tag{9.65}$$

where $\Pi_{[0,1]} : \mathbb{R} \to \mathbb{R}$ denotes the projection function

$$\Pi_{[0,1]}(x) = \begin{cases} 0 & x < 0, \\ x & x \in [0, 1], \\ 1 & x > 1, \end{cases}$$

the scalar $c \geq 0$ corresponds to a zero-resource probability of damage, and the scalar $d \geq 0$ to the marginal decrease in the probability of damage per unit of attack resources. We note that an increase in attack resources u_{AR} leads to an increase in the potential damage x_{PD} (expressed by the + sign before the b term in (9.64)), but may actually decrease the probability that the potential damage will actually be realized (expressed by the $-$ sign before the d term in (9.65)), which is motivated by the fact that a large-scale attack is more likely to trigger defense mechanisms that can prevent the potential damage from being realized.

The total expected damage y_{TD} to the mission can be found by multiplying equations (9.64) and (9.65), leading to the expected damage equation

$$YTD = f(u_{AR})g(u_{AR}). \tag{9.66}$$

In the context of cybermissions that evolve over time and require multiple computer services, the potential damage equation (9.64) and the uncertainty equation (9.65) need to be augmented with an index $t \in \{1, 2, \ldots, T\}$ that parametrizes mission time and an index $s \in \{1, 2, \ldots, s\}$ that parameterizes the required computer services, as in

$$x_{PDAR_t^s} = f_t^s(u_{ARAR_t^s}) = a_t^s + b_t^s u_{AR_t^s}, \tag{9.67}$$

$$\rho_t^s = g_t^s(u_{AR_t}^s) = \Pi_{[0,1]}(c_t^s - d_t^s u_{AR_t^s}, \tag{9.68}$$

where $u_{AR_t^s}$ denotes the attack resources committed to attack service s at time t, $x_{pD_t^s}$ the potential damage at time t due to an attack to the service s, and P_t^s the probability of realizing this damage. The corresponding expected damage equation then becomes

$$y_{TD} = \sum_{t=1}^{T} \sum_{s=1}^{S} f_t^s(u_{AR_t^s}) g_t^s(u_{AR_t^s}). \tag{9.69}$$

An intelligent attacker would seek to optimally allocate his/her available resources to maximize the total expected missing damage. We shall consider here several options for this optimization that differ on the information that is available to the attacker.

9.3.3 Known Mission Damage Data

When all the data $a_t^s, b_t^s, c_t^s, d_t^s, \forall s, t$ that define the potential damage and uncertainty equations are known a priori, optimal attack resource allocation can be determined by solving the following optimization problem:

$$\begin{aligned}
&\text{maximize} && \sum_{t=1}^{T}\sum_{s=1}^{S} f_t^s(u_{AR_t^s})g_t^s(u_{AR_t^s}) \\
&\text{subject to} && \sum_{t=1}^{T}\sum_{s=1}^{S} u_{AR_t^s} \le U_{TR} \\
&\text{w.r.t.} && u_{AR_t^s} \in [0,\infty),\ \ \forall t, \forall s,
\end{aligned} \tag{9.70}$$

where U_{TR} denotes the total budget of attack resources available to the attacker. As stated in the following proposition, this optimization can be converted into the following concave maximization problem.

Proposition 9.2. *When the functions f_t^s, g_t^s are of the form (9.67)–(9.68) with $a_t^s, b_t^s, c_t^s, d_t^s \ge 0$, $\forall t, s$, the value and optimum of (9.70) can be obtained through the following concave maximization problem:*

$$\begin{aligned}
&\textit{maximize} && \sum_{t=1}^{T}\sum_{s=1}^{S} (a_t^s + b_t^s u_{AR_t^s})(c_t^s - d_t^s u_{AR_t^s} - \sigma_t^s) \\
&\textit{subject to} && \sum_{t=1}^{T}\sum_{s=1}^{S} u_{AR_t^s} \le U_{TR},\ c_t^s - d_t^s u_{AR_t^s} - \sigma_t^s \le 1,\ \ \forall t, \forall s, \\
&\textit{w.r.t.} && u_{AR_t^s} \in \left[0, \frac{c_t^s}{d_t^s}\right] \ge 0,\ \ \forall t, \forall s.
\end{aligned} \tag{9.71}$$

When $c_t^s \in [0,1]$, one can set the corresponding optimization variable $\sigma_t^s = 0$ in (9.71). Moreover, when $c_t^s \in [0,1]$, $\forall t, s$, and all the constraints on the $u_{AR_t^s}$ are inactive, the solution to this optimization can be found in closed form and is equal to

$$u_{AR_t^s} = \bar{u}_t^s - \bar{\mu}_t^s \max\left\{0, \sum_{\bar{t}}\sum_{s=1}^{S} \bar{u}_{\bar{t}}^s - U_{TR}\right\},$$

$$\bar{u}_t^s := \frac{b_t^s c_t^s - a_t^s d_t^s}{2 b_t^s d_t^s}, \quad \bar{\mu}_t^s := \frac{\frac{1}{2 b_t^s d_t^s}}{\sum_{\bar{t}}\sum_{s=1}^{S} \frac{1}{2 b_{\bar{t}}^s d_{\bar{t}}^s}}.$$

Note that, if any of the constraints on the attack resources are active, a closed-form solution may not be easy and one would have to solve the optimization problem (9.71) instead.

Proof. To prove that (9.70) and (9.71) are equivalent, we start by noting that

$$g_t^s(u_{AR_t^s}) = \begin{cases} 0, & c_t^s - d_t^s u_{AR_t^s} < 0 \implies u_{AR_t^s} > \frac{c_t^s}{d_t^s}, \\ 1, & c_t^s - d_t^s u_{AR_t^s} > 1 \implies u_{AR_t^s} < \frac{c_t^s - 1}{d_t^s}, \\ c_t^s - d_t^s u_{AR_t^s}, & u_{AR_t^s} \in \left[\frac{c_t^s-1}{d_t^s}, \frac{c_t^s}{d_t^s}\right]. \end{cases}$$

Suppose, by contradiction, that (9.71) could lead to a larger maximum than (9.70). The condition $u_{AR_t^s} \in \left[0, \frac{c_t^s}{d_t^s}\right]$ guarantees that the same set of $u_{AR_t^s}$ satisfies the constraints of (9.70) and that

$$g_t^s(u_{AR_t^s}) = \begin{cases} 1, & c_t^s - u_{AR_t^s} > 1 \implies u_{AR_t^s} < \frac{c_t^s-1}{d_t^s}, \\ c_t^s - d_t^s u_{AR_t^s}, & c_t^s - d_t^s u_{AR_t^s} \leq 1 \implies u_{AR_t^s} \geq \frac{c_t^s-1}{d_t^s}, \end{cases}$$

and the condition $c_t^s - d_t^s u_{AR_t^s} - \sigma_t^s \leq 1$ guarantees that

$$\begin{cases} c_t^s - d_t^s u_{AR_t^s} - \sigma_t^s \leq 1 = g_t^s(u_{AR_t^s}), & u_{AR_t^s} < \frac{c_t^s-1}{d_t^s}, \\ c_t^s - d_t^s u_{AR_t^s} - \sigma_t^s = g_t^s(u_{AR_t^s}) - \sigma_t^s \leq g_t^s(u_{AR_t^s}), & u_{AR_t^s} \geq \frac{c_t^s-1}{d_t^s}, \end{cases}$$

which shows that $c_t^s - d_t^s u_{AR_t^s} - \sigma_t^s \leq g_t^s(u_{AR_t^s})$ and therefore (9.71) cannot lead to a larger maximum than (9.70). Suppose now, also by contradiction, that (9.70) could lead to a larger maximum than (9.71). First note that if a few of the $u_{AR_t^s}$ that maximize (9.70) were larger than $\frac{c_t^s}{d_t^s}$, for those $u_{AR_t^s}$ we would have $g_t^s(u_{AR_t^s}) = 0$ and the same exact cost could be obtained for (9.70) by replacing each of these $u_{AR_t^s}$ with $\frac{c_t^s}{d_t^s}$. So we may assume, without loss of generality, that all the $u_{AR_t^s}$ are smaller than or equal to $\frac{c_t^s}{d_t^s}$.

In this case, we could use the same $u_{AR_t^s}$ in (9.71) and set

$$\sigma_t^s = \begin{cases} 0, & c_t^s - d_t^s u_{AR_t^s} \leq 1, \\ c_t^s - d_t^s u_{AR_t^s} - 1, & c_t^s - d_t^s u_{AR_t^s} > 1. \end{cases}$$

This selection of at would satisfy the constraints of (9.71) and guarantee that

$$g_t^s(u_{AR_t^s}) = c_t^s - d_t^s u_{AR_t^s} - \sigma_t^s,$$

and therefore (9.70) and (9.71) would lead to the same maximum. This completes the proof that (9.70) and (9.71) are equivalent.

The optimization scheme just defined is a concave maximization problem (convex minimization) with linear constraints. The dual problem is given by,

$$
\begin{aligned}
J^{\perp} := & \max_{\lambda_1 \geq 0, \eta_t^s \geq 0, \xi_t^s \geq 0} \max_{u_{AR_t^s} \in \mathbb{R}} \sum_{t=1}^{T} \sum_{s=1}^{S} (a_t^s + b_t^s u_{AR_t^s})(c_t^s - d_t^s u_{AR_t^s}) \\
& - \lambda_1 \left(\sum_{t=1}^{T} \sum_{s=1}^{S} u_{AR_t^s} - U_{TR} \right) - \sum_{t=1}^{T} \sum_{s=1}^{S} \eta_t^s \left(u_{AR_t^s} - \frac{c_t^s}{d_t^s} \right) + \sum_{t=1}^{T} \sum_{s=1}^{S} \xi_t^s u_{AR_t^s} \\
= & \max_{\lambda_1 \geq 0, \eta_t^s \geq 0, \xi_t^s \geq 0} \max_{u_{AR_t^s} \in \mathbb{R}} \sum_{t=1}^{T} \sum_{s=1}^{S} \Big(a_t^s c_t^s - a_t^s d_t^s u_{AR_t^s} + b_t^s c_t^s u_{AR_t^s} - b_t^s d_t^s u_{AR_t^s}^2 \\
& - \lambda_1 u_{AR_t^s} - \eta_t^s u_{AR_t^s} + \xi_t^s u_{AR_t^s} \Big) + \lambda_1 U_{TR} + \sum_{t=1}^{T} \sum_{s=1}^{S} \eta_t^s \frac{c_t^s}{d_t^s} \\
= & \max_{\lambda_1 \geq 0, \eta_t^s \geq 0, \xi_t^s \geq 0} \max_{u_{AR_t^s} \in \mathbb{R}} \sum_{t=1}^{T} \sum_{s=1}^{S} \Big(a_t^s c_t^s - b_t^s d_t^s u_{AR_t^s}^2 \\
& + (b_t^s c_t^s - a_t^s d_t^s + \xi_t^s - \eta_t^s - \lambda_1) u_{AR_t^s} \Big) + \lambda_1 U_{TR} + \sum_{t=1}^{T} \sum_{s=1}^{S} \eta_t^s \frac{c_t^s}{d_t^s}.
\end{aligned}
$$

The inner maximization problem can be solved using standard calculus and is achieved for

$$
u_{AR_t^s} = \frac{b_t^s c_t^s - a_t^s d_t^s + \xi_t^s - \eta_t^s - \lambda_1}{2 b_t^s d_t^s}
$$

yielding

$$
J^{\perp} := \max_{\lambda_1 \geq 0, \eta_t^s \geq 0, \xi_t^s \geq 0} \sum_{t=1}^{T} \sum_{s=1}^{S} \left(\frac{(b_t^s c_t^s - a_t^s d_t^s + \xi_t^s - \eta_t^s - \lambda_1)^2}{4 b_t^s d_t^s} + \eta_t^s \frac{c_t^s}{d_t^s} \right) + \lambda_1 U_{TR}.
$$

For this problem the Karush–Kuhn–Tucker (KKT) conditions [63] lead to

$$
\begin{aligned}
\frac{\partial J^{\perp}}{\partial \lambda_1} = 0 & \implies \lambda_1 = \frac{\sum_{t=1}^{T} \sum_{s=1}^{S} \frac{b_t^s c_t^s - a_t^s d_t^s + \xi_t^s - \eta_t^s}{2 b_t^s d_t^s} + \eta_t^s \frac{c_t^s}{d_t^s}}{\sum_{t=1}^{T} \sum_{s=1}^{S} \frac{1}{2 b_t^s d_t^s}} & \text{or} \quad \lambda_1 = 0, \\
\frac{\partial J^{\perp}}{\partial \eta_t^s} = 0 & \implies \eta_t^s = \xi_t^s - a_t^s d_t^s - b_t^s c_t^s - \lambda_1 & \text{or} \quad \eta_t^s = 0, \\
\frac{\partial J^{\perp}}{\partial \xi_t^s} = 0 & \implies \xi_t^s = -b_t^s c_t^s + a_t^s d_t^s + \eta_t^s + \lambda_1 & \text{or} \quad \xi_t^s = 0.
\end{aligned}
$$

Let us assume that $u_{AR_t^s}$ is inside the interval $\left[0, \frac{c_t^s}{d_t^s}\right]$, which would lead to all the η_t^s and ξ_t^s equal to zero (inactive constraints), and therefore we would need

$$\sum_{t=1}^{T}\sum_{s=1}^{S}\frac{b_t^s c_t^s - a_t^s d_t^s - \lambda_1}{2b_t^s d_t^s} = U_{TR} \iff \\ \lambda_1 = \frac{\sum_{t=1}^{T}\sum_{s=1}^{S}\frac{b_t^s c_t^s - a_t^s d_t^s}{2b_t^s d_t^s} - U_{TR}}{\sum_{t=1}^{T}\sum_{s=1}^{S}\frac{1}{2b_t^s d_t^s}} \geq 0, \quad \text{or} \quad \lambda_1 = 0, \tag{9.72}$$

and

$$u_{AR_t^s} = \bar{u}_t^s - \bar{\mu}_t^s \max\left\{0, \sum_{\bar{t}}\sum_{s=1}^{S}\bar{u}_{\bar{t}}^s - U_{TR}\right\},$$

$$\bar{u}_t^s := \frac{b_t^s c_t^s - a_t^s d_t^s}{2b_t^s d_t^s}, \quad \bar{\mu}_t^s := \frac{\frac{1}{2b_t^s d_t^s}}{\sum_{\bar{t}}\sum_{s=1}^{S}\frac{1}{2b_{\bar{t}}^s d_{\bar{t}}^s}}.$$

We can view the term being subtracted from $u_{AR_t^s}$ as a normalizing term that makes sure the $u_{AR_t^s}$ add up to the constraint U_{TR}.

Note that if the closed-form formula shown above for $u_{AR_t^s}$ ever becomes negative, then the corresponding ξ_t^s will become active, and we must have

$$\frac{\partial J^{\perp}}{\partial \xi_t^s} = 0 \implies \xi_t^s = \lambda_1 + a_t^s d_t^s - b_t^s c_t^s \implies u_{AR_t^s} = 0.$$

Similarly, if the formula for $u_{AR_t^s}$ ever becomes larger than $\frac{c_t^s}{d_t^s}$, then the corresponding η_t^s will become active, and we must have

$$\frac{\partial J^{\perp}}{\partial \eta_t^s} = 0 \implies \eta_t^s = -b_t^s c_t^s - \lambda_1 - a_t^s d_t^s \implies u_{AR_t^s} = \frac{c_t^s}{d_t^s}. \quad \square$$

Remark 9.6. *Note that if any of the constraints on the attack resources are active, a closed-form solution is not possible, and one has to solve the optimization problem instead.*

9.3.4 Unknown Mission Damage Data

Often the mission-specific parameters $\{a_t^s, b_t^s, c_t^s, d_t^s : \forall s, t\}$ that define the potential damage and uncertainty equations are not known a priori and, instead, need to be estimated online. The estimation problem will be posed using a grey-box model that looks like a k-step ahead predictor. The problem is solved at each time instant to incorporate the new information that is obtained in the form of a new measurement of the process variables.

One approach that can be used to address this scenario is to assume that these parameters are generated by linear dynamics of the form

$$x^s_{at+1} = A^s_a x^s_{at} + B^s_a w^s_t, \qquad a^s_t = C^s_a x^s_{at}, \tag{9.73}$$

$$x^s_{bt+1} = A^s_b x^s_{bt} + B^s_b w^s_t, \qquad b^s_t = C^s_b x^s_{bt}, \tag{9.74}$$

$$x^s_{ct+1} = A^s_c x^s_{ct} + B^s_c w^s_t, \qquad c^s_t = C^s_c x^s_{ct}, \tag{9.75}$$

$$x^s_{dt+1} = A^s_d x^s_{dt} + B^s_d w^s_t, \qquad d^s_t = C^s_d x^s_{dt}, \tag{9.76}$$

where the $\{w^s_t, \forall s, t\}$ are sequences of zero-mean random processes with variances σ^s_t. One can then use historical data to estimate these dynamics using black-box identification techniques. Once estimates for the dynamics are available, one can use online data to predict future values for the mission-specific parameters $\{a^s_t, b^s_t, c^s_t, d^s_t : \forall s, t\}$, based on past observations.

It is crucial that the data collected from the system conveys continual information on the parameters to be estimated and the identification algorithm can rely on fresh information in forming reliable current estimates (persistence of excitation).

Suppose that at some time $k < T$ the attacker has observed the values of the past mission-specific parameters $\{a^s_t, b^s_t, c^s_t, d^s_t : \forall s, t \leq k\}$ and needs to make decisions on the future attack resources $u_{AR^s_t}$, $t \geq k$. One can use (9.73)–(9.76) to construct estimates $\{\hat{a}^s_t, \hat{b}^s_t, \hat{c}^s_t, \hat{d}^s_t : \forall s, t > k\}$ for the future mission-specific parameters and obtain the future $u_{AR^s_t}$, $t \geq k$ using the following optimization:

$$\text{maximize} \quad \sum_{t=1}^{T}\sum_{s=1}^{S} f^s_t(u_{AR^s_t}) g^s_t(u_{AR^s_t}) + \sum_{t=1}^{T}\sum_{s=1}^{S} \hat{f}^s_t(u_{AR^s_t}) \hat{g}^s_t(u_{AR^s_t}) \tag{9.77}$$

$$\text{subject to} \quad \sum_{t=1}^{T}\sum_{s=1}^{S} u_{AR^s_t} \leq U_{TR} \tag{9.78}$$

$$\text{w.r.t.} \quad u_{AR^s_t} \in [0, \infty), \forall t \in \{k, \ldots, T\}, \forall s, \tag{9.79}$$

where f^s_t and g^s_t denote the functions defined in (9.67) and (9.68), respectively, whereas $\hat{f}^s_t$ and $\hat{g}^s_t$ are estimates of these functions computed using the estimated mission-specific parameters $\{\hat{a}^s_t, \hat{b}^s_t, \hat{c}^s_t, \hat{d}^s_t : \forall s, t > k\}$. The optimization problem (9.77) can be solved at each time step $k \in \{1, 2, \ldots, T-1\}$, allowing the attacker to improve her allocation of attack resources as new information about the missing parameters becomes available. Note that one could remove from the (double) sums in (9.77) any terms that do not depend on the optimization variables.

9.3.5 iCTF Competition

The international Capture The Flag (iCTF) is a distributed wide-area security exercise to test the security skills of the participants. This contest is organized by the Security Lab of the Department of Computer Science at UCSB and it has been held yearly since 2003. In traditional editions of the iCTF (2003–2007), the goal of each team was to maintain a set of services such that they remain available and uncompromized throughout the contest. Each team also had to attempt to compromise the other teams' services. Since all the teams received an identical copy of the virtual host containing the vulnerable services, each team had to find the vulnerabilities in their copy of the hosts and possibly fix the vulnerabilities without disrupting the services. At the same time, the teams had to leverage their knowledge about the vulnerabilities they found to compromise the servers run by other teams. Compromising a service allowed a team to bypass the service's security mechanisms and to "capture the flag" associated with the service. During the 2008–2010 iCTFs, new competition designs have been introduced. More precisely, in 2008 a separate virtual network was created for each team. The goal was to attack a terrorist network and defuse a bomb after compromising a number of hosts. In 2009, the participants had to compromise the browsers of a large group of simulated users, steal their money, and create a botnet. In 2010, the participants had to attack the rogue nation Litya, ruled by the evil Lisvoy Bironulesk. The teams' goal was to attack the services supporting Litya's infrastructure only at specific times, when certain activities were in progress. In addition, an intrusion detection system would temporarily firewall out the teams whose attacks were detected. The 2011 iCTF competition is briefly summarized below from the perspective of one team playing against the rest of the world. The 2010 [61] and 2011 [60] iCTF competitions were designed closely match practical cybersecurity mission scenarios.

9.3.6 2011 iCTF

The 2011 iCTF was centered around the theme of illegal money laundering. This activity is modeled after cybercriminal money laundering operations and provided a perfect setting for risk-reward analysis, as the trade-offs are very intuitively understood.

The general idea behind the competition was the conversion ("laundering") of money into points. The money was obtained by the teams by solving security-related challenges (e.g., decrypting an encrypted message, find hidden information in a document, etc.) The conversion of money into points was performed by utilizing data captured from an exploited service. Therefore, first a team had to obtain money by solving challenges, and then the money had to be translated into points by exploiting the vulnerability in a service of another team. Successful conversion of money to points depended on a number of factors, calculated together as the "risk function", which is described in detail below. Note that at the end of the game the money had no contribution to the final stand of a team, only points mattered.

One challenge with the formulation "one-against-world" is that, in the 2011 iCTF game, winning was not just about maximizing points. Winning was about getting more points than each of the opponents (individually).

The game was played in 255 rounds (each takes about 2 min), but we only have data for 248 rounds since the logging server was temporarily down. Each team hosted a server that ran 10 services, each with its own (unknown) vulnerabilities. Each service $s \in \{1, 2, \ldots, 10\}$ of each hosting team was characterized by three time varying quantities $\forall t \in \{1, 2, \ldots, 248\}$:

- The cut C_t^s, which is the percentage of money that goes to the team when money is laundered through service s (same values for every team),
- The payoff P_t^s, which is the percentage of money that will be transformed into points for the team that launders the money (same value for every team);

$$P_t^s = 0.9e^{-\frac{TicksActive}{10}},$$

- The risk R_t^s, which is the probability of losing all the money (instead of getting a conversion to points).

The generation of the time series for the cuts, payoffs, and risks for the different services was based on an underlying set of cybermissions that were running while the game was played. Essentially, when the states of the cybermissions required a particular service, the cut, payoff, and risk would make that service attractive for attackers from the perspective of converting money to points. However, the players were not informed about the state of the cybermissions and, instead, at the beginning of each round t, a team was informed of the values of C_t^s, P_t^s, R_t^s, for every s, and t.

9.3.7 Actions Available to Every Team

A team (we) has the following key actions in the actual competition:

1. *Defensive actions.* Activate/deactivate one of its own services.
   In the iCTF competition a team could also correct any vulnerability that it discovered in its services. We assumed here that all known vulnerabilities had been corrected.
2. *Money laundering.* Select
   - **a.** a team to attack (mute decision within the "one-against-world" formulation);
   - **b.** a service s to compromise, which implicitly determines the payoff P_t^s, the risk R_t^s, and the cut C_t^s;
   - **c.** amount of money to launder, $u_{AR_t^s}$, at time t through the service s. This action results in a number of points given by

$$X_t^s = \begin{cases} P_t^s(1 - C_t^s)D_t u_{AR_t^s}, & \text{with probability } 1 - \min\{\rho_t^s, 1\}, \\ 0, & \text{with probability } \min\{\rho_t^s, 1\}, \end{cases} \tag{9.80}$$

where D_t is the team's defense level and ρ_t^s is the probability that the conversion of money to points will succeed, as given by the formula

$$\rho_t^s := \frac{R_t^s u_{AR_t^s}}{30} + \frac{1}{6}\left(\frac{N_t^j - 700}{300 + |N_t^j - 700|} + 1\right) + \frac{1}{6}\left(\frac{Q_t^s - 1500}{300 + |Q_t^s - 1500|} + 1\right),$$

where N_t^j is the overall amount of money that has been laundered by team j through a particular team being exploited and Q_t^s is the overall amount of money that has been laundered by the team through a particular service being exploited. Because we do not model each team individually, we will consider the "worst" case scenario for the following quantities: $N = 492$, $Q = 2257$ (according to data from the competition), and defense level of the team as $D = 1$.

To map this game into the general framework described before, we associate the money to launder $u_{AR_t^s}$ at time t through service s with the resources $u_{AR_t^s}$ devoted to attack service s at time t, and associate the points X_t^s in (9.80) with damage to the mission.

The total attack resources U_{TR} available to each team in the general framework described before now corresponds to the money available to each team. While we could model more accurately the process by which teams get money, for simplicity we assumed that each team had a fixed amount of money ($5060) available that could be spent throughout the duration of the game, which is given by the average money of all the teams during the competition. The results, of which services were attacked and when, proved to be relatively insensitive to this parameter.

9.3.8 Optimization Schemes and iCTF

In this section we apply the optimization schemes defined in Sections 9.3.3 and 9.3.4 to the iCTF game. We are seeking to optimally allocate our available resources in the competition such that the total number of points is maximized while meeting the specified constraints. The maximization of the expected reward by a team can be formulated as follows:

$$\begin{aligned} &\text{maximize} && \sum_{t=1}^{248}\sum_{s=1}^{10} \rho_t^s P_t^s (1 - C_t^s) D_t^s u_{AR_t^s} \\ &\text{subject to} && \sum_{t=1}^{248}\sum_{s=1}^{10} u_{AR_t^s} \leq U_{TR}, := 5060 \\ &\text{w.r.t.} && u_{AR_t^s} \in [0, \infty), \forall s \in \{1, 2, \ldots, 10\}, \;\; t \in \{1, 2, \ldots, 248\}, \end{aligned}$$

where

$$\rho_t^s = \min\{(\beta_t^s + \frac{R_t^s}{30} u_{AR_t^s}), 1\},$$

$$\beta_t^s := \frac{1}{6}\left(\frac{N_t - 700}{300 + |N_t - 700|} + 1\right) + \frac{1}{6}\left(\frac{Q_t - 1500}{300 + |Q_t - 1500|} + 1\right) = 0.4,$$

and the parameters ρ_t^s, C_t^s, D_t^s, β_t^s can either be considered known or unknown. By using Proposition 9.1, and setting the constraint $\sigma_t^s = 0$ in (9.80) (since $(1 - \beta_t^s) \in [0, 1]$), we can write the equivalent optimization problem as

$$\begin{aligned} &\text{maximize} && \sum_{t=1}^{248}\sum_{s=1}^{10} \left(1 - \beta_t^s - \frac{R_t^s}{30} u_{AR_t^s}\right) P_t^s (1 - C_t^s) \\ &\text{subject to} && \sum_{t=1}^{248}\sum_{s=1}^{10} u_{AR_t^s} \leq U_{TR} \end{aligned}$$

$$\text{w.r.t.} \qquad u_{AR_t^s} \in \left[0, \frac{1-\beta_t^s}{\frac{R_t^s}{30}}\right], \forall s \in \{1, 2, \ldots, 10\}, t \in \{1, 2, \ldots, 248\},$$

which is a concave maximization problem with linear constraints that is easy to solve numerically as described in Section 9.3.3.

The above optimization depends on the following assignments: $a_t^s := 0$, $b_t^s = P_t^s(1 - C_t^s)$, $c_t^s := 1 - \beta_t^s$, $d_t^s := \frac{R_t^s}{30}$. When these are not known, one can estimate $b_t^s := P_t^s(1 - C_t^s)$, $c_t^s := 1 - \beta_t^s$, $d_t^s := \frac{R_t^s}{30}$ using a low-order state-space models given by (9.74)–(9.76). By then applying the optimization scheme described in Section 9.3.4, with a horizon of $N = 5$, one can still make accurate predictions of when and how to distribute the available attack resources. The optimization model just described, results in an optimization to obtain the future $u_{AR_t^s}$, $\forall t \geq k$ and performed under a moving horizon of 5 ticks,

$$\begin{aligned}
\text{maximize} \quad & \sum_{t=1}^{k}\sum_{s=1}^{10} b_t^s \left(c_t^s - d_t^s u_{AR_t^s}\right) u_{AR_t^s} + \sum_{t=1}^{248}\sum_{s=1}^{10} \hat{b}_t^s \left(\hat{c}_t^s - \hat{d}_t^s u_{AR_t^s}\right) u_{AR_t^s} \\
\text{subject to} \quad & \sum_{t=1}^{248}\sum_{s=1}^{10} u_{AR_t^s} \leq U_{TR}, \\
& x_{bt+1}^s = A_b^s x_{at}^s + B_b^s w_t^s, \\
& \hat{b}_t^s = C_b^s x_{bt}^s, \\
& x_{ct+1}^s = A_c^s x_{ct}^s + B_c^s w_t^s, \\
& \hat{c}_t^s = C_c^s x_{ct}^s, \\
& x_{dt+1}^s = A_d^s x_{dt}^s + B_d^s w_t^s, \\
& \hat{d}_t^s = C_d^s x_{dt}^s, \\
\text{w.r.t.} \quad & u_{AR_t^s} \in \left[0, \frac{\hat{c}_t^s}{\hat{d}_t^s}\right], \forall t \in \{k, 2, \ldots, 248\}, \forall s \in \{1, 2, \ldots, 10\}.
\end{aligned}$$

9.3.9 iCTF Results

This section presents numerical results obtained from the optimizations described above to data from the attack logs of the 2011 iCTF competition. All the optimizations have been implemented through a MATLAB-based convex optimization solver such as CVX [63]. The optimization scheme described in Section 9.3.4 yielded very close results to the scheme described in Section 9.3.3 for a predicting horizon of $N = 5$.

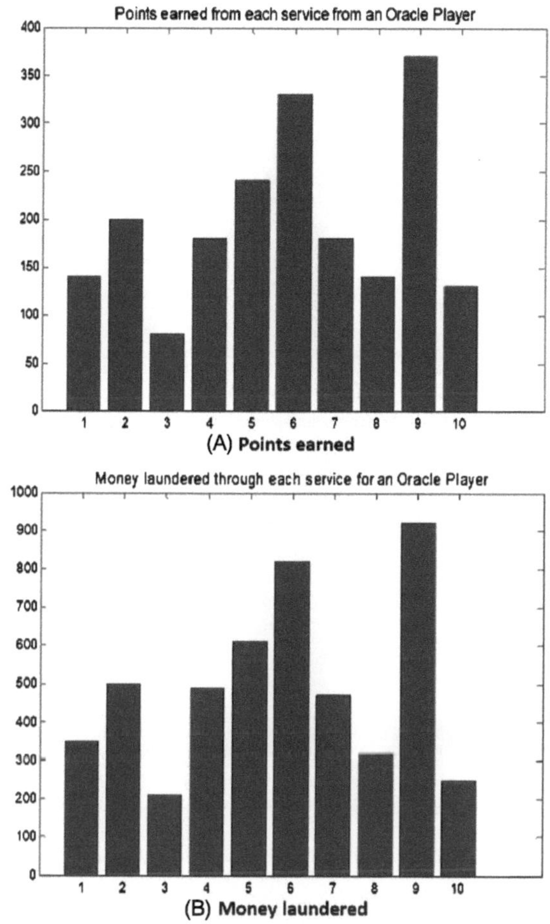

Figure 9.15 Behavior of an optimal "sophisticated" attacker able to attack all 10 services

Initially we will assume that a "sophisticated" attacker would be able to compromise any one of the 10 services. Fig. 9.15 shows the points and the money collected by such an optimal attacker, whereas Fig. 9.16 shows the same (aggregate) data for the teams that participated in the competition.

One can also consider attackers with different levels of sophistication, e.g., attackers that are only able to find vulnerabilities in a subset of the 10 services that the "sophisticated" attacker was able to attack. By observing the data of the top 20 teams in the competition, we were able to partition the sophistication in two levels. For comparison, we show the behavior of an attacker A that was only able to attack the services 1, 2, 4, 5, 6, 9

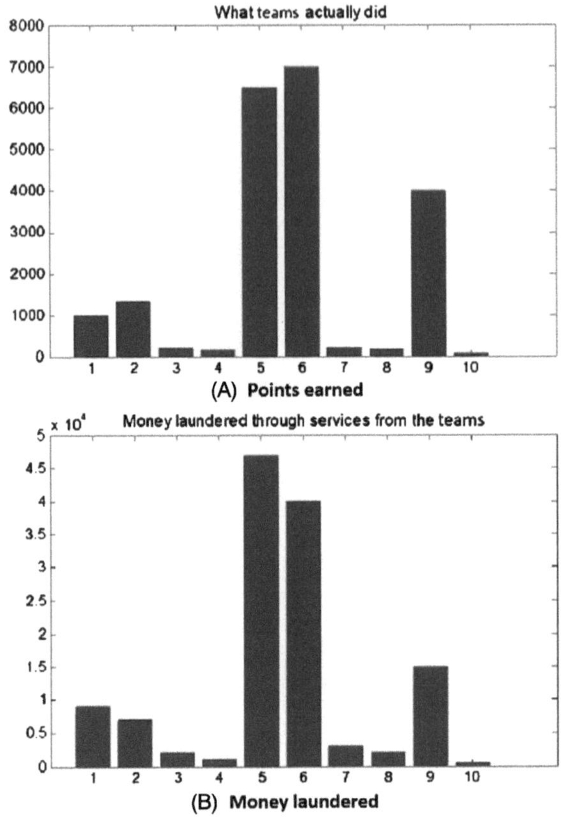

Figure 9.16 Aggregate behavior of all teams that participated in the competition

(similar to the first 10 teams in the competition); and another attacker *B* that was only able to attack services 1, 2, 5, 6, 7, 8 (similar to the teams from place 11 to 20 in the competition). The "sophisticated" attacker was able to gather 1987 points, whereas the two other attackers were able to get 1821 and 1721 points, respectively.

The results in Fig. 9.15A show that the most profitable services to attack were 5, 6 and 9. The top 10 teams in the competition attacked mostly 5 and 6 because 9 was a hard service to get into. Only the top 3 teams discovered how to attack service 9, and only at the end of the game, so they had relatively little time to explore that vulnerability. Aside from this, the prediction based on the optimization framework developed here qualitatively reflect the actions of the good teams. In fact, the top two teams

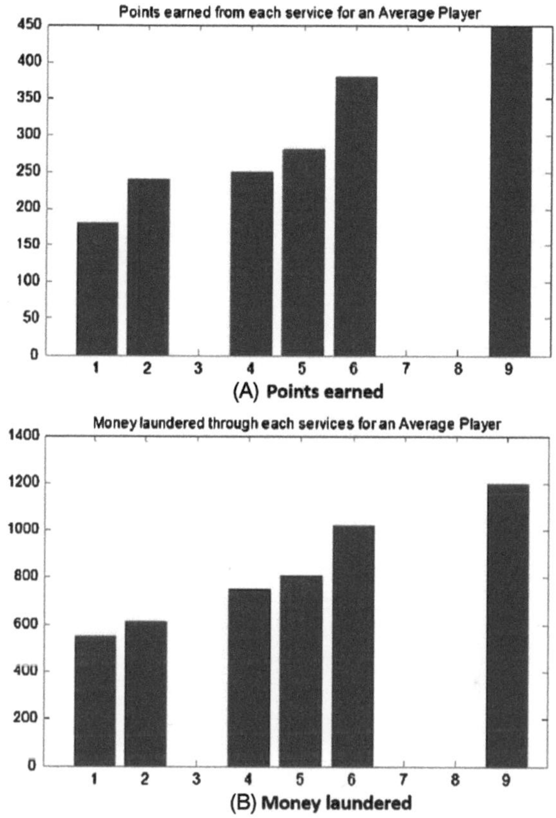

Figure 9.17 Behavior of an optimal attacker A able to attack services 1, 2, 4, 5, 6, 9

in the competition followed attack strategies qualitatively close to that of attacker A in Fig. 9.17 as seen in Fig. 9.19. See Fig. 9.18.

9.4 NOTES

With the increasing integration of information technologies into industrial systems and networks, such as the power grid, robust and resilient control system design is essential for assuring the robust performance of cyber-physical control systems in the face of adversarial attacks. This chapter has presented a hybrid game-theoretic framework whereby the occurrence of unanticipated events is modeled by stochastic switching, and deterministic uncertainties are represented by the known range of disturbances. The design of a robust controller at the physical layer takes into account risks of

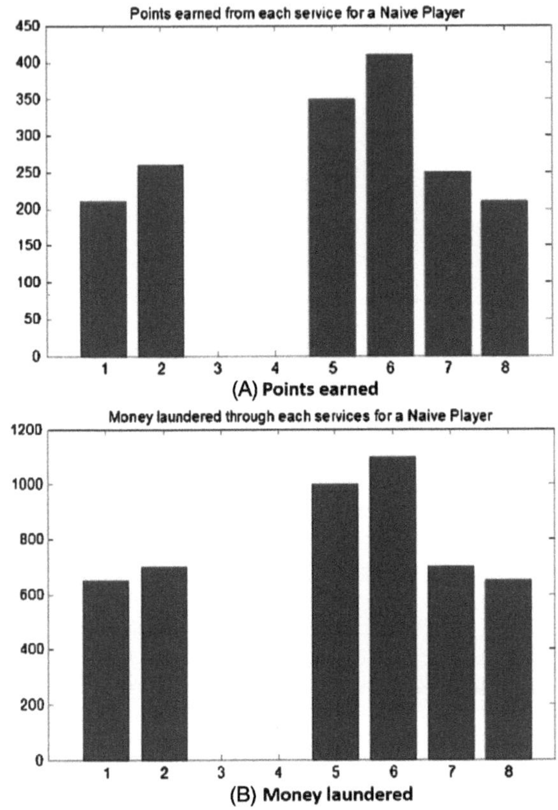

Figure 9.18 Behavior of an optimal attacker B able to attack services 1, 2, 5, 6, 7, 8

failures due to the cybersystem, while the design of the security policies is based on its potential impact on the control system. The crosslayer coupled design introduced in this chapter results in solving a zero-sum differential game for robust control coupled with a zero-sum stochastic game for the security policy. The two games are intertwined and coupled together through cyber and physical system variables. The solution to the two coupled games requires a zooming-in process, which uses variables from the cyberlevel to solve the physical layer game, and a zooming-out process, which uses physical system variables to solve the cyberlayer game. The joint design results in a robust and resilient controller switching between different modes for guaranteeing performance in the face of unexpected events.

The first section has presented a general class of system models, where the physical system is described by nonlinear ordinary differential equa-

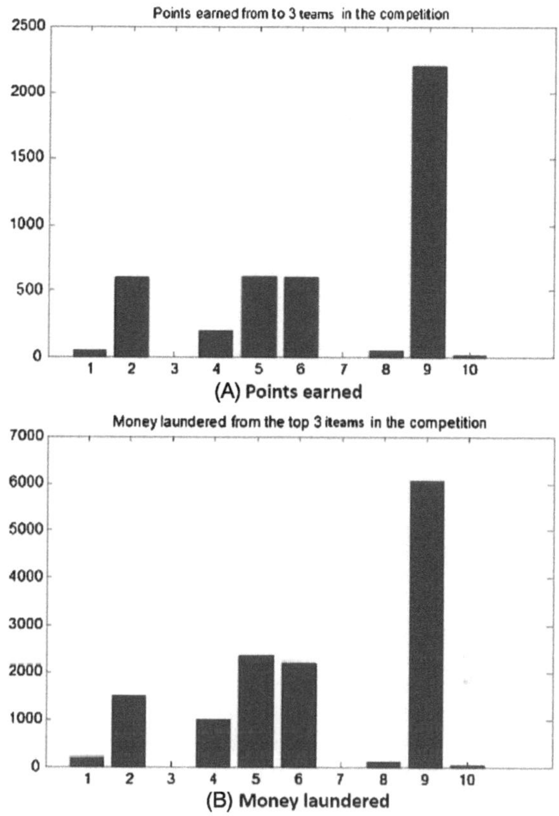

Figure 9.19 Behavior of the top 3 teams during the competition

tions, and the cybersystem is captured by Markov models. The framework can be further extended to other classes of systems including sampled-data systems, systems with delayed measurements, and model predictive control systems. The optimal design of new classes of systems can follow the same games-in-games principle discussed here and can be characterized by a new set of optimality conditions. Interesting future research includes the study of problems with stronger coupling, in which control and defense strategies depend on both cyberstates and physical states, and the development of advanced computational tools to compute the control and defense strategies. The article has also discussed offline computational methods to compute the equilibria. In addition, learning algorithms and adaptive mechanisms can be developed within this framework to provide online adaptation to changes, which will enhance the resilience of the system.

Study of a distributed network of cyberphysical systems is another possible research direction. The networking effects in the cybersystem can lead to performance interdependencies of distributed physical layer control systems

In the second section, the security issue and strategy design of NCS under DoS attacks have been analyzed. Two novel optimal control strategies have been developed in the delta domain using the game theoretic approach. The attacker and defender possessing asymmetric information pattern have been considered, and the algorithms to develop the optimal defense and attack strategies have been provided, respectively. The validity and advantage of the proposed method have been verified by both numerical simulations and practical experiments.

There are several ways that future work could be done based on the results of this chapter. Some specific problems can be enumerated as follows:

1. One can investigate the NCS with the CTOC and MTOC structures over the infinite time horizon. For example, one only needs to replace L_k^i by $\bar{L}^i$ and P_k^i and P_{k+1}^i by $\bar{P}^i$ in (9.52) and (9.53), and then the corresponding infinite-time horizon results for the MTOC structure can be obtained. However, this extension requires a stability analysis of the addressed system.
2. Another extension of this work is to exploit the pricing mechanisms in the cyber layer. The interactions of the DoS attacker and the cyberdefender (e.g., IDS [49]) can be captured by a cyberlayer game model [41], whose output determines the packet dropout rate in the communication channel of the control system and further determines the control performance. Thus, the pricing mechanisms can be exploited such that, by tuning the pricing parameters in the cyberlayer game model, the control system in the physical layer can achieve desirable performance.
3. The assumption made on the distribution of the stochastic variable α_k^i can be altered. The case of the cascading failure [44] can be considered where the link failure process $\{\alpha_k^i\}$ is correlated. The Markov chain can be employed to model this correlation.

As a future work in this area will be focused on developing analysis tools to explore what-if scenarios based on past data and the structure of the cybermission. To this end, we are developing optimization schemes for the defender's possible actions, such as taking a service offline when the service is not needed or extending the duration of a state that would be unable to progress if a certain service is compromised. We are also developing human–computer interfaces to demonstrate the useful of this type of analysis for security analysts. Moreover, this work can be extended to provide a method to analyze cybersecurity aspects of power system state estimators where the attacker has limited resources and an index is introduced to enable the operator to see what resources are the most important to protect.

REFERENCES

[1] S. Gorman, Electricity grid in U.S. penetrated by spies, Wall Street J. [Online]. Available: http://online.wsj.com/article/SB123914805204099085.html, Apr. 8, 2009.

[2] B. Krebs, Cyber incident blamed for nuclear power plant shutdown, Washington Post. [Online]. Available: http://www.washingtonpost.com/wp-dyn/content/article/2008/06/05/AR2008060501958.html, June 5,2008.

[3] S. Greengard, The new face of war, Commun. ACM 53 (12) (Dec. 2010) 20–22.

[4] R. McMillan, Siemens: Stuxnet worm hit industrial systems [Online]. Available: http://www.computerworld.com/s/article/print/9185419, Sept. 16, 2010.

[5] M. Ilic, From hierarchical to open access electric power systems, Proc. IEEE 95 (5) (2007) 1060–1084.

[6] Q. Zhu, T. Basar, A hierarchical security architecture for the smart grid, in: E. Hossain, Z. Han, H.V. Poor (Eds.), Smart Grid Communications and Networking, Cambridge Univ. Press, Cambridge, UK, 2012, ch. 18.

[7] H. Zimmermann, OSI reference model – the ISO model of architecture for open systems interconnection, IEEE Trans. Commun. 28 (4) (1980) 425–432.

[8] J.F. Kurose, K.W. Ross, Computer Networking: A Top-Down Approach, sixth edition, Pearson Education, Upper Saddle River, NJ, 2012.

[9] Q. Zhu, H. Tembine, T. Basar, Distributed strategic learning with application to network security, in: Proc. American Control Conf., San Francisco, CA, June 29–July 1, 2011, pp. 4057–4062.

[10] Q. Zhu, H. Tembine, T. Basar, Heterogeneous learning in zero-sum stochastic games with incomplete information, in: Proc. 49th IEEE Conf. Decision Control, 2010, pp. 219–224.

[11] M. Fabro, T. Nelson, Control Systems Cyber Security: Defense in Depth Strategies, INL Tech. Rep. INL/CON-07-12804, ISA Expo, Houston, TX, 2007.

[12] Q. Zhu, M. McQueen, C. Rieger, T. Basar, Management of control system information security: control system patch management, in: Proc. Workshop Foundations Dependable Secure Cyber-Physical Systems, CPSWeek, Apr. 2011, pp. 51–54.

[13] J. Eisenhauer, P. Donnelly, M. Ellis, M. O'Brien, Roadmap to Secure Control Systems in the Energy Sector, Energ. Incorp., U.S. Dept. Energy and the U. S. Dept. Homeland Secur., 2006.

[14] A. Dominguez-Garcia, J. Kassakian, J. Schindall, A generalized fault coverage model for linear time-invariant systems, IEEE Trans. Reliab. 58 (3) (2009) 553–567.

[15] A. Haidar, E.K. Boukas, Robust stability criteria for Markovian jump singular systems with time-varying delays, in: Proc. 47th IEEE Conf. Decision Control, 2008, pp. 4657–4662.

[16] Z. Pan, T. Basar, H^{∞} control of large scale jump linear systems via averaging and aggregation, Int. J. Control 72 (10) (1999) 866–881.

[17] T. Basar, Minimax control of switching systems under sampling, Int. J. Control 25 (5) (Aug. 1995) 315–325.

[18] T. Basar, G.J. Olsder, Dynamic Noncooperative Game Theory, second edition, Class. Appl. Math., Soc. Ind. Appl. Math., Philadelphia, PA, 1999.

[19] Q. Zhu, T. Basar, Dynamic policy-based IDS configuration, in: Proc. 48th IEEE Conf. Decision Control, Held Jointly with the 28th Chinese Control Conf., 2009, pp. 8600–8605.

[20] Q. Zhu, H. Tembine, T. Basar, Network security configurations: a nonzero-sum stochastic game approach, in: Proc. American Control Conf., 2010, pp. 1059–1064.

[21] Q. Zhu, A. Clark, R. Poovendran, T. Basar, Deceptive routing games, in: Proc. IEEE 51st Ann. Conf. Decision Control, 2012, pp. 2704–2711.

[22] A. Clark, Q. Zhu, R. Poovendran, T. Basar, Deceptive routing in relay networks, in: J. Grossklags, J.C. Walrand (Eds.), Proc. Conf. Decision Game Theory Security, in: Lect. Notes Comput. Sci., Springer-Verlag, Berlin Heidelberg, Germany, 2012, pp. 171–185.

[23] Q. Zhu, T. Basar, Feedback-driven multi-stage moving target defense, in: Proc. Conf. Decision Game Theory Security, in: Notes Comput. Sci., Springer-Verlag, Berlin Heidelberg, Germany, 2013.

[24] Q. Zhu, H. Tembine, T. Basar, Hybrid learning in stochastic games and its application in network security, in: F.L. Lewis, D. Liu (Eds.), Reinforcement Learning and Approximate Dynamic Programming for Feedback Control, in: Comput. Intell. Ser., IEEE Press, Piscataway, NJ, 2012, pp. 305–329, ch. 14.

[25] B. Randell, P. Lee, P.C. Treleaven, Reliability issues in computing system design, ACM Comput. Surv. 10 (2) (June 1978) 123–165 [Online]. Available: http://doi.acm.org/10.1145/356725.356729.

[26] L.S. Shapley, Stochastic games, Proc. Natl. Acad. Sci. 39 (10) (1953) 1095–1100.

[27] J. Filar, K. Vrieze, Competitive Markov Decision Processes, first edition, Springer-Verlag, Berlin Heidelberg, Germany, 1996.

[28] T.E.S. Raghavan, J.A. Filar, Algorithms for stochastic games – a survey, Methods Models Oper. Res. 35 (6) (2003) 437–472.

[29] O. Hernandez-Lerma, J. Lasserre, Zero-sum stochastic games in Borel spaces: average payoff criteria, SIAM J. Control Optim. 39 (5) (2000) 1520–1539.

[30] S. Meyn, R.L. Tweedie, Markov Chains and Stochastic Stability, second edition, Cambridge Univ. Press, Cambridge, UK, Apr. 2009.

[31] O. do Valle Costa, M. Fragoso, M. Todorov, Continuous-Time Markov Jump Linear Systems, Probab. Appl., Springer-Verlag, Berlin Heidelberg, Germany, 2012.

[32] J. Nash, Equilibrium points in N-person games, Proc. Natl. Acad. Sci. 36 (1) (1950) 48–49.

[33] D. Bond, Quickdraw SCADA IDS [Online]. Available: http://www.digitalbond.com/tools/quickdraw/, Feb. 20, 2012.

[34] R.A. Kisner, Cybersecurity Through Real-Time Distributed Control Systems, Tech. Rep. ORNL/TM-2010/30, Oak Ridge Natl. Lab., 2010, pp. 4–5.

[35] H. Li, M.Y. Chow, Z. Sun, Optimal stabilizing gain selection for networked control systems with time delays and packet losses, IEEE Trans. Control Syst. Technol. 17 (5) (Sept. 2009) 1154–1162.

[36] K.J. Park, J. Kim, H. Lim, Y. Eun, Robust path diversity for network quality of service in cyber-physical systems, IEEE Trans. Ind. Inform. 10 (4) (Nov. 2014) 2204–2215.

[37] F. Pasqualetti, F. Dorfler, Control-theoretic methods for cyberphysical security: geometric principles for optimal cross-layer resilient control systems, IEEE Control Syst. Mag. 35 (1) (Feb. 2015) 110–127.

[38] R.S. Smith, Covert misappropriation of networked control systems: presenting a feedback structure, IEEE Control Syst. Mag. 35 (1) (Feb. 2015) 82–92.
[39] Y.W. Law, T. Alpcan, M. Palaniswami, Security games for risk minimization in automatic generation control, IEEE Trans. Power Syst. 30 (1) (Jan. 2015) 223–232.
[40] A. Teixeira, I. Shames, H. Sandberg, K.H. Johansson, A secure control framework for resource-limited adversaries, Automatica 51 (2015) 135–148.
[41] Y. Yuan, F. Sun, H. Liu, Resilient control of cyber-physical systems against intelligent attacker: a hierarchical Stackelberg game approach, Int. J. Syst. Sci. 47 (9) (2015) 2067–2077.
[42] M. Long, C.H.J. Wu, J.Y. Hung, Denial-of-service attacks on network-based control systems: impact and mitigation, IEEE Trans. Ind. Inform. 1 (2) (May 2005) 85–96.
[43] Y. Wu, X. He, S. Liu, L. Xie, Consensus of discrete-time multiagent with adversaries and time delays, Int. J. Gen. Syst. 43 (3) (2014) 402–411.
[44] Q. Zhu, T. Basar, Game-theoretic methods for robustness, security, and resilience of cyberphysical control systems: games-in-games principle for optimal cross-layer resilient control systems, IEEE Control Syst. Mag. 35 (1) (Feb. 2015) 46–55.
[45] H. Zhang, P. Cheng, L. Shi, J.M. Chen, Optimal denial-of-service attack scheduling with energy constraint, IEEE Trans. Autom. Control 60 (11) (Nov. 2015) 3023–3028.
[46] M. Zhu, S. Martinez, On the performance analysis of resilient networked control systems under replay attacks, IEEE Trans. Autom. Control 59 (3) (Mar. 2014) 804–808.
[47] L. Li, B. Hu, M.D. Lemmon, Resilient event triggered systems with limited communication, in: Proc. IEEE Conf. Decis. Control, 2012, pp. 6577–6582.
[48] S. Amin, A.A. Cárdenas, S.S. Sastry, Hybrid Systems: Computation and Control, Springer, Berlin, Germany, 2009.
[49] Y. Yuan, F. Sun, Data fusion-based resilient control system under DoS attacks: a game theoretic approach, Int. J. Control. Autom. Syst. 13 (3) (2015) 513–520.
[50] Y. Li, L. Shi, P. Cheng, J. Chen, D.E. Quevedo, Jamming attack on cyber-physical systems: a game-theoretic approach, in: Proc. Annu. Int. Conf. Cyber Technol. Autom. Control Intell. Syst., 2013, pp. 252–257.
[51] S. Amin, S.A. Galina, S.S. Galina, Security of interdependent and identical networked control systems, Automatica 49 (1) (2013) 186–192.
[52] H. Yang, Y. Xia, Low frequency positive real control for delta operator systems, Automatica 48 (8) (2012) 1791–1795.
[53] Y. Xia, M. Fu, H. Yang, G.P. Liu, Robust sliding-mode control for uncertain time-delay systems based on delta operator, IEEE Trans. Ind. Electron. 56 (9) (Sept. 2009) 3646–3655.
[54] M.C. Priess, R. Conway, J. Choi, J.M. Popovich, C. Radcliffe, Solutions to the inverse LQR problem with application to biological systems analysis, IEEE Trans. Control Syst. Technol. 23 (2) (Mar. 2015) 770–777.
[55] S. Coogan, L. Ratliff, D. Calderone, C. Tomlin, S. Sastry, Energy management via pricing in LQ dynamic games, in: Proc. Amer. Control Conf., 2013, pp. 443–448.
[56] D.P. Bertsekas, Dynamic Programming and Optimal Control, Athena Scientific, Belmont, MA, USA, 1995.
[57] P. Suchomski, A J-lossless coprime factorization approach to H^{∞} control in delta domain, Automatica 38 (10) (2002) 1807–1814.
[58] E. Garone, B. Sinopoli, A. Goldsmith, A. Casavola, LQG control for MIMO systems over multiple erasure channels with perfect acknowledgment, IEEE Trans. Autom. Control 57 (2) (Feb. 2012) 450–456.

[59] M.R. Endsley, D. Garland, Theoretical underpinnings of situation awareness: a critical review, in: Situation Awareness Analysis and Measurement, 2000, pp. 3–32.
[60] G. Vigna, The 2011 ucsb ictf: description of the game (2011), http://ictf.cs.ucsb.edu, 2011.
[61] A. Doupé, M. Egele, B. Caillat, G. Stringhini, G. Yakin, A. Zand, L. Cavedon, G. Vigna, Hit'em where it hurts: a live security exercise on cyber situational awareness, in: Proceedings of the 27th Annual Computer Security Applications Conference, ACM, 2011, pp. 51–61.
[62] N. Stockman, K.G. Vamvoudakis, L. Devendorf, T. Höllerer, R. Kemmerer, J.P. Hespanha, A Mission-Centric Visualization Tool for Cybersecurity Situation Awareness, Tech. Rep., California Univ. at Santa Barbara, 2012.
[63] S. Boyd, L. Vandenberghe, Convex Optimization, Cambridge University Press, 2004.

APPENDIX A

Appendix

A.1 PRELIMINARIES AND NOTATIONS

The sets of real numbers and integers are denoted as $\mathbb{R}$ and $\mathbb{N}$, respectively. Strictly positive real numbers and integers are represented by $\mathbb{R}_+$ and $\mathbb{N}_+$, while the sets of positive real numbers and integers including zero are denoted as $\mathbb{R}_{\{0,+\}}$ and $\mathbb{N}_{\{0,+\}}$, respectively. The expected values are represented as $\mathsf{E}[\cdot]$.

A continuous function $\alpha : [0, a) \to [0, \infty)$ is said to belong to class $\mathcal{K}$ if it is strictly increasing and $\alpha(0) = 0$. It belongs to class $\mathcal{K}_\infty$, if $a = \infty$ and $\alpha(r) \to \infty$ as $r \to \infty$. Similarly, β is of class $\mathcal{L}$ if it is continuous and decreasing to zero. A function $\zeta : [0, \infty) \to [0, \infty)$ is said to be of class $\mathcal{G}$ if it is continuous, non-decreasing and $\zeta(0) = 0$. A continuous function $\gamma : [0, a) \times [0, \infty) \to [0, \infty)$ is said to belong to class $\mathcal{KL}$ if, for each fixed s, the mapping $\gamma(r, s)$ belongs to class $\mathcal{K}$ with respect to r and, for each fixed r, the mapping $\gamma(r, s)$ is decreasing with respect to s, and $\gamma(r, s) \to 0$ as $s \to \infty$. Class $\mathcal{KK}$ functions are defined in the same fashion.

Local stability is defined when the initial state of the system lies close to the equilibrium point. When it can lie anywhere in the state space, the stability is defined as global. A system is said to be uniformly stable if its stability is independent of the initial time $t_0 \geq 0$. A system is said to be stable if for each $\epsilon > 0$, there exists a $\delta = \delta(\epsilon) > 0$ such that if $||x(t_0)|| < \delta$ then $||x(t)|| < \epsilon$, $\forall t \geq 0$. It is said to be asymptotically stable if it is stable and δ can be chosen such that if $||x(t_0)|| < \delta$ then $\lim_{t\to\infty} x(t) = 0$. A system is said to be exponentially sable if there exist $\sigma, \lambda \in \mathbb{R}^+$ such that $\forall t \geq 0$ $||x(t)|| \leq \sigma ||x(t_0)|| e^{-\lambda t}$. The state of a system is said to be ultimately bounded if there exist constants $\varepsilon, \varrho \in \mathbb{R}^+$ (ε defined as the bound) and for every $\eta \in (0, \varrho)$ there is a constant $T = T(\eta, \varepsilon) \in \mathbb{R}^+$ such that if $||x(t_0)|| < \eta$ then $||x(t)|| \leq \varepsilon$, $\forall t \geq t_0 + T$. A system is said to be Input-to-State Stable (ISS) if there exist a class $\mathcal{KL}$ function γ and a class $\mathcal{K}$ function α such that for any initial state $x(t_0)$ and any bounded input $u(t)$, the state of the system satisfies $\forall t \geq t_0 \geq 0$ the following inequality:

$$||x(t)|| \leq \gamma(||x(t_0)||, t - t_0) + \alpha\left(\sup_{t_0 \leq \tau \leq t} ||u(\tau)|| \right).$$

Consider a system with input–output relation given as $y = Hu$ for some mapping H. This mapping is said to be $\mathcal{L}_p$ stable if there exist a class $\mathcal{K}$ function α, defined on $[0, \infty)$, and a nonnegative constant μ such that

$$||(Hu)_\tau||_{\mathcal{L}_p} \leq \alpha(||u_\tau||_{\mathcal{L}_p}) + \mu, \quad \forall\tau \in [0, \infty).$$

It is finite-gain $\mathcal{L}_p$ stable if there exist nonnegative constants ζ and μ such that

$$||(Hu)_\tau||_{\mathcal{L}_p} \leq \zeta||u_\tau||_{\mathcal{L}_p} + \mu, \quad \forall\tau \in [0, \infty).$$

Here $\mathcal{L}_p$ denotes the p-norm where $1 \leq p \leq \infty$.

Expectation operator and conditional expectation are denoted as $\mathrm{E}[\cdot]$ and $\mathrm{E}[\cdot|\cdot]$, respectively.

For a set $\mathcal{S}$, $|\mathcal{S}|$ denotes the cardinality (i.e., size) of the set, while for two sets $\mathcal{S}$ and $\mathcal{R}$, we use $\mathcal{S} \setminus \mathcal{R}$ to denote the set of elements in $\mathcal{S}$ that are not in $\mathcal{R}$. In addition, for a set $\mathcal{K} \subset \mathcal{S}$, we specify the complement set of $\mathcal{K}$ with respect to $\mathcal{S}$ as $\mathcal{K}^{\mathrm{C}}$, i.e., $\mathcal{K}^{\mathrm{C}} = \mathcal{S} \setminus \mathcal{R}$. We use $\mathbf{1}'_N$ to denote the row vector of size N containing all ones.

We use $\mathbf{A}^T$ to indicate the transpose of matrix $\mathbf{A}$, while the ith element of a vector $\mathbf{x}_k$ is denoted by $\mathbf{x}_{k,j}$. For vector $\mathbf{x}$ and matrix $\mathbf{A}$, we use $|\mathbf{x}|$ and $|\mathbf{A}|$ to denote the vector and matrix whose elements are absolute values of the initial vector and matrix, respectively. Also, for matrices $\mathbf{P}$ and $\mathbf{Q}$, $\mathbf{P} \preceq \mathbf{Q}$ specifies that the matrix $\mathbf{P}$ is *element-wise* smaller than or equal to the matrix $\mathbf{Q}$.

For a vector $\mathbf{e} \in \mathbb{R}^p$, the support of the vector is the set

$$\mathrm{supp}(\mathbf{e}) = \{i|\mathbf{e}_i \neq 0\} \subseteq \{1, 2, \ldots, p\},$$

while the l_0-norm of vector $\mathbf{e}$ is the size of $\mathrm{supp}(\mathbf{e})$, i.e., $||\mathbf{e}||l_0 = |\mathrm{supp}(\mathbf{e})|$. For a matrix $\mathbf{E} \in \mathbb{R}^{p\times N}$, we use $\mathbf{e}_1, \mathbf{e}_2, \ldots, \mathbf{e}_N$ to denote its columns and $\mathbf{E}'_1, \mathbf{E}'_2, \ldots, \mathbf{E}'_p$ to denote its rows. We define the row support of matrix $\mathbf{E}$ as the set

$$\mathrm{rowsupp}(\mathbf{E}) = \{i|\mathbf{E}'_i \neq 0\} \subseteq \{1, 2, \ldots, p\}.$$

As for vectors, the l_0-norm for a matrix $\mathbf{E}$ is defined as $\|\mathbf{E}\|_{l_0} = |\mathrm{rowsupp}(\mathbf{E})|$.

Given $\alpha \in \mathbb{R}$, we let $\mathbb{R}_{>\alpha}(\mathbb{R}_{\geq\alpha})$ denote the set of reals greater than (greater than or equal to) α. We define $\mathbb{N}_0 := \mathbb{N} \cup \{0\}$. Given a vector $v \in \mathbb{R}^n$, $\|v\|$ is its Euclidean norm. Given a matrix M, $M^\top$ is its transpose and $\|M\|$ is its spectral norm. Given a set A and a function $f : A \to \mathbb{R}_{\geq 0}$, we use the convention $\sup_{x \in A} f(x) = 0$ when A is empty. Given a measurable time function $f : \mathbb{R}_{\geq 0} \to \mathbb{R}^n$ and a time interval $[0, t)$ we denote the $\mathcal{L}_\infty$ norm of $f(\cdot)$ on $[0, t)$ by $\|f_t\|_\infty := \text{ess}\sup_{s \in [0,t)} \|f(s)\|$. Finally, we denote by $\mathcal{L}_\infty(\mathbb{R}_{\geq 0})$ the set of measurable and essentially bounded time functions on $\mathbb{R}_{\geq 0}$.

A.2 A BRIEF OF GAME THEORY

A.2.1 A Short Review

Game theory deals with strategic interactions among multiple decision makers, called players. Each player's preference ordering among multiple alternatives is captured in an objective function for that player. Players try to maximize (for utility or benefit functions) or minimize (for cost or loss functions) their respective objective functions. For a nontrivial game, the objective function of a player depends on the choices (actions or, equivalently, decision variables) of at least one other player, and generally of all the players, and hence players cannot simply optimize their own objective function independently of the choices of the other players. This introduces a coupling between the actions of the players and binds them together in decision making even in a noncooperative environment.

A noncooperative game is nonzero sum if the sum of the players' objective functions cannot be made zero even after appropriate positive scaling and/or translation that do not depend on the players' decision variables. A two-player game is zero sum if the sum of the objective functions of the two players is zero or can be made zero by appropriate positive scaling and translation that do not depend on the decision variables of the players. A game is a finite game if each player has only a finite number of alternatives, that is, the players pick their actions out of finite sets (action sets); otherwise the game is an infinite game; finite games are also known as matrix games. An infinite game is said to be a continuous-kernel game if the action sets of the players are subsets of finite-dimensional vector spaces, and the players' objective functions are continuous with respect to action variables of all players. A game is said to be deterministic if the players' actions uniquely determine the outcome, as captured in the objective functions, whereas if the objective function of at least one player depends on an additional variable (state of nature) with a known probability distribution, then the game is a stochastic game. A game is a complete information game if the description of the game (that is, the players, objective functions, and underlying probability distributions (if stochastic)) is common information to all players; otherwise it is an incomplete information game. Finally, a game is static if each player acts only once, and none of the players has access to information on the actions of any of the other players; otherwise it is a dynamic game. A dynamic game is said to be a differential game if the evolution of the decision process (controlled by the players over time) takes place in continuous time, and generally involves a differential equation.

A.2.2 General Game Model and Equilibrium Concept

Consider an N-player game, with $\mathcal{N} := \{1, \ldots, N\}$ denoting the players' set. The decision or action variable of player i is denoted by $x_i \in X_i$ where X_i is the action set of player i. Let X denote the N-tuple of action variables of all players, $x := (X_1, \ldots, X_N)$. Allowing for possibly coupled constraints, let $\Omega \subset X$ be the constraint set for the game, where X is the N-product of $X_1, \ldots, X_N$; hence for an N-tuple of action variables to be feasible, $x \in \Omega$. The players are minimizers, with the objective function (loss function or cost function) of player i denoted by $L_i(x_i, x_{-i})$, where x_{-i} stands for the action variables of all players except the ith one.

Now, an N-tuple of action variables $x^* \in \Omega$ is a *Nash equilibrium (or noncooperative equilibrium)* if, for all $i \in \mathcal{N}$, $x_i \in X_i$,

$$L_i(x_i^*, x_{-i}^*) \leq L(x_i, x_{-i}^*) \text{ such that } (x_i^*, x_{-i}^*) \in \Omega.$$

If $N = 2$ and $L_1 \equiv -L_2 =: L$, then the game is a two-player zero-sum game, with player 1 minimizing L and player 2 maximizing the same quantity. In this case, the Nash equilibrium becomes the saddle-point equilibrium, which is formally defined as follows, where the coupling constraint Ω is left out (or simply assumed to be equal to the product set $X := X_1 \times X_2$): A pair of actions $(x_1^*, x_2^*) \in X$ is in saddle-point equilibrium for a game with cost function L, if for all $(x_1, x_2) \in X$,

$$L(x_1^*, x_2) \leq L(x_1^*, x_2^*) \leq L(x_1, x_2^*).$$

This also implies that the order in which minimization and maximization are carried out is inconsequential, that is,

$$\min_{x_1 \in X_1} \max_{x_2 \in X_2} L(x_1, x_2) = \max_{x_2 \in X_2} \min_{x_1 \in X_1} L(x_1, x_2) \tag{A.1}$$

$$= L(x_1^*, x_2^*) =: L^*, \tag{A.2}$$

where the first expression in (A.1) is known as the upper value of the game, the second expression in (A.1) is the lower value of the game, and L^* is known as the value of the game. Upper and lower values are, in fact, defined in more general terms using infimum (inf) and supremum (sup) replacing minimum and maximum, respectively, to account for the facts that minima and maxima may not exist. When the action sets are finite, however, they always exist. Note that the value of a game, whenever it exists (which certainly does if there exists a saddle point), is unique. Hence, if there exists another saddle-point solution, say $(\hat{x}_1, \hat{x}_2)$, then $L(\hat{x}_1, \hat{x}_2) = L^*$. Moreover, these multiple saddle points are orderly interchangeable, that is, the pairs $(X_1^*, \hat{x}_2)$ and $(\hat{x}_1, x_2^*)$ are also in saddle-point equilibrium. This property of saddle-point equilibria does not extend to multiple Nash equilibria (for nonzero-sum games). Multiple Nash equilibria are generally not interchangeable, and furthermore, they do not lead to the same values for the players' cost functions, the implication being that when players switch from one equilibrium to another, some players may benefit from that (in terms of reduction in cost) while others may see an increase in their costs. Further, if the players pick randomly (for their actions) from the multiple Nash equilibria of the game, then the resulting N-tuple of actions may not be in Nash equilibrium.

A.2.3 A Stochastic Game Formulation

In this section, a general stochastic game formulation is introduced in which the state space coincides with $\mathcal{S}$. This class of models captures the uncertainties in cybersystem dynamics and the time evolution of system states and player strategies. Moreover, in the absence of decision making of the players (that is, when the costs and transitions are independent of player strategies), the framework would be reduced to a Markov-chain model which has been used for reliability analysis [2].

Let $\bar{k} = t/\varepsilon$, $\varepsilon > 0$ be the time scale on which cyberevents happen, which is often on the order of days, in contrast to the one of the physical systems which evolve on the time scale of seconds. Denote by $a \in \mathcal{A}$ a cyberattack chosen by the attacker from his attack space $\mathcal{A} := \{a_1, a_2, \ldots, a_M\}$ composed of all M possible actions. Let $\ell \in \mathcal{L}$ be the cyberdefense mechanism that can be employed by the network administrator, where $\mathcal{L} := \{l_1, l_2, \ldots, l_N\}$ is the set of all possible defense actions.

Without loss of generality, $\mathcal{A}$ and $\mathcal{L}$ do not change with time, even though, in practice, they can change due to technological updates and advances. The mixed strategies $\{(k) = [f_i(k)]_{i=1}^{N} \in \mathcal{F}_k$, $g(k) = [g_j(k)]_{j=1}^{M} \in \mathcal{G}_k$ of the defender and the attacker, respectively, are considered here, where $f_i(k)$ and $g_j(k)$ are the probabilities of choosing $\ell_i \in \mathcal{L}$ and $a_j \in \mathcal{A}$, respectively, where $\mathcal{F}_k$ and $\mathcal{G}_k$ are sets of admissible strategies, defined by $(\mathrm{f}(k), \mathrm{g}(k))$ as

$$\mathcal{F}_k := \left\{ \mathrm{f}(k) \in [0, 1]^N : \sum_{i=1}^{N} f_i(k) = 1 \right\}, \tag{A.3}$$

$$\mathcal{G}_k := \left\{ \mathrm{g}(k) \in [0, 1]^N : \sum_{i=1}^{N} g_j(k) = 1 \right\}. \tag{A.4}$$

The joint actions affect the transition rates λ_{ij} in (9.4) and also incur a cost $c^i(a, l; \mu^{\mathrm{CL}}, \nu_{\mathrm{CL}})$, where c^i is a bounded cost function that incorporates the physical layer control system performance under the closed-loop strategies $\mu_{\mathrm{CL}}, \nu_{\mathrm{CL}}$. The cost c^i has two components: the cost inflicted on the cyberlayer and the resulting impact-aware, physical-layer performance index from the action pair (a, ℓ).

The defense against attacks involves HMIs, which occur at the human and cyberlayers of the system. Hence, defense often evolves on a longer time-scale than the physical layer processes. Using time-scale separation, the optimal defense mechanism can be designed by viewing the physical control system at its steady state at each cyberstate θ at a given time k. The interaction between an attacker and a defending administrator can be captured by a zero-sum stochastic game with the defender aiming to maximize the long-term system performance or payoff function whereas the attacker is aiming to minimize it [3]. A discounted payoff criterion $\nu_\beta(i, \mathrm{f}, \mathrm{g})$ is used and defined as

$$\nu_\beta(i, \mathrm{f}, \mathrm{g}) := \int_0^\infty e^{-\beta k} \mathbb{E}_i^{\mathrm{f}(k)} c^i(a, l; \mu_{\mathrm{CL}}, \nu_{\mathrm{CL}}) dk,$$

where β is the discount factor. The operator $\mathbb{E}_i^{\mathrm{f,g}}$ is the expectation operator. Here a class of mixed stationary strategies $\mathrm{f}^i \in \mathcal{F}^i$ and $\mathrm{g}^i \in \mathcal{G}^i$, $i \in \mathcal{S}$, is considered that are only dependent on the current cyberstate i. Let $\mathrm{F} = \{\mathrm{f}_i\}_{i \in \mathcal{S}} \in \mathcal{F}_\mathcal{S}$ and $\mathrm{G} = \{\mathrm{g}i\}_{i \in \mathcal{S}} \in \mathcal{G}_s$, where $\mathcal{F}_s := \Pi_{i \in \mathcal{S}} \mathcal{F}^i$ and $\mathcal{G}_s := \Pi^i_{i \in \mathcal{S}\mathcal{G}}$. The following theorem characterizes the stationary saddle-point equilibrium of the stochastic zero-sum game in a similar fashion as in [3,4].

Theorem A.1 ([1]). *Let the strategy pair (μ_{CL}, ν_{CL}) be fixed. Assume that $\lambda_{ij}(k)$ are continuous in f^i, and g^i and the cost functions c^i are bounded. Then, there exists a pair of stationary strategies $(F^*, G^*) \in \mathcal{F}_s \times \mathcal{G}_s$ such that, for all $i \in \mathcal{S}$, the* *following fixed point equation is satisfied:*

$$\begin{aligned}
\beta v_\beta^*(i) &= \tilde{c}^i(F^*, G^*) + \sum_{j\in\mathcal{S}} \lambda_{ij}(F^*, G^*) v_\beta^*(j) \\
&= \sup_{F\in\mathcal{F}_s} \{\tilde{c}^i(F, G^*) + \sum_{j\in\mathcal{S}} \lambda_{ij}(F, G^*) v_\beta^*(j)\} \\
&= \inf_{G\in\mathcal{G}_s} \{\tilde{c}^i(F^*, G) + \sum_{j\in\mathcal{S}} \lambda_{ij}(F^*, G) v_\beta^*(j)\} \\
&= \sup_{F\in\mathcal{F}_s} \inf_{G\in\mathcal{G}_s} \{\tilde{c}^i(F, G) + \sum_{j\in\mathcal{S}} \lambda_{ij}(F, G) v_\beta(j)\} \\
&=: L_\beta(i) \\
&= \inf_{G\in\mathcal{G}_s} \sup_{F\in\mathcal{F}_s} \{\tilde{c}^i(F, G) + \sum_{j\in\mathcal{S}} \lambda_{ij}(F, G) v_\beta(j)\} \qquad \text{(A.5)} \\
&=: U_\beta(i), \qquad \text{(A.6)}
\end{aligned}$$

where $\tilde{c}^i(F, G)$ *is a shorthand notation for* $\mathbb{E}_{F,G} c^i(a, l; \mu_{CL}, \nu_{CL})$, *and* $L_\beta(i)$, $U_\beta(i)$ *are respectively defined to be the lower and upper values of the game. In addition,* (F^*, G^*) *from (A.6) is a pair of saddle-point equilibrium strategies, and the value of game* $v_\beta^*(i)$ *is unique and has the property that* $v_\beta^*(i) = L_\beta(i) = U_\beta(i)$.

The above result is also known as the *Shapley Optimality Criterion* for stochastic games. For more details on the properties of saddle points of zero-sum games, see "Minimax Theorem." The saddle-point equilibrium strategies can be computed using a value iteration scheme [2,3]. Let $\{v_\beta^n(i)\}_{n=1}^\infty$ be a sequence of values of the game which obeys the following update law:

$$\begin{aligned}
v_\beta^{n+1}(i) &= \tilde{c}^i(F_n^*, G_n^*) + \sum_{j\in\mathcal{S}} \lambda_{ij}(F_n^*, G_n^*) v_\beta^n(j) \\
&= \sup_{F\in\mathcal{F}_s} \{\tilde{c}^i(F_n, G^*) + \sum_{j\in\mathcal{S}} \lambda_{ij}(F_n, G^*) v_\beta^n(j)\} \\
&= \inf_{G\in\mathcal{G}_s} \{\tilde{c}^i(F_n^*, G) + \sum_{j\in\mathcal{S}} \lambda_{ij}(F_n^*, G) v_\beta^n(j)\}. \qquad \text{(A.7)}
\end{aligned}$$

A.2.4 Minimax Theorem

Consider two-person, zero-sum finite games, or equivalently matrix games, where player 1 is the minimizer and player 2 the maximizer. Let X_1 and X_2 be player 1's and player 2's action sets, respectively. Let $\text{card}(X_1) = m$ and $\text{card}(x_2) = n$ be the cardinality of action sets, that is, the minimizer has m

choices and the maximizer has n choices. The objective function $L(x_1, x_2)$ is defined on $X_1 \times X_2$. Equivalently, an $m \times n$ matrix A can be associated with this game, whose entries are the values of $L(X_1, X_2)$, following the same ordering as that of the elements of the action sets, that is, an (i, j)th entry of A is the value of $L(X_1, X_2)$ when x_1 is the ith element of X_1 and x_2 is the jth element of X_2. Player 1's choices are then the rows of the matrix and player 2's choices are its columns.

In general, a saddle point may not exist in pure strategies for all zero-sum games. One example is the game known as Matching Pennies. Each player has a penny and can choose heads or tails. The players then reveal their choices simultaneously. If the two choices are identical (that is, if they match), then player 1 wins and is given the other player's penny. If the choices do not match, then player 2 wins and is given the other player's penny. This is an example of a zero-sum game, where one player's gain is exactly equal to the other player's loss. The game matrix associated with the game is

$$A = \begin{pmatrix} -1 & 1 \\ 1 & -1 \end{pmatrix}.$$

The entries of this matrix are losses to player 1 (and thus gains to player 2). The first row corresponds to the choice of heads for player 1, and the second row corresponds to the choice of tails for him. Symmetrically, the first column corresponds to heads for player 1, and the second column is tails for him. Here there is no row–column combination, for which the players would not have an incentive to unilaterally deviate and improve their returns.

This opens the door for looking for a mixed-strategy equilibrium. A mixed strategy for player i is a probability distribution over his action set X_i, which is denoted by p_i for player i. If X_i is finite, which is the case here, then p_i will be a probability vector, taking values in the probability simplex determined by X_i, which is denoted by $\mathcal{P}_i$. A pair (p_1^*, p_2^*) constitutes a saddle point in mixed strategies (or a mixed-strategy saddle-point equilibrium), if for all $(p_1, p_2) \in \mathcal{P}$,

$$J(p_1^*, p_2) \leq J(p_1^*, p_2^*) \leq J(p_1, p_2^*),$$

where $J(p_1, p_2) = E_{p_1, p_2}[L(x_1, x_2)]$ and $\mathcal{P} := \mathcal{P}_1 \times \mathcal{P}_2$. Here $J^* = J(P_1^*, P_2^*)$ is the value of the zero-sum game in mixed strategies.

In terms of the matrix A and the probability vectors p_1 and p_2 (both column vectors), which were introduced earlier (note that in this case p_1 is of dimension m and p_2 is of dimension n, and components of each are non-negative and add up to one), the expected cost function can be rewritten as

$$J(p_1, p_2) = p_1' A p_2.$$

A.3 GATEAUX DIFFERENTIAL

We provide a definition of the Gateaux differential which was used repeatedly in Chapter 5.

Definition A.1 (Gateaux Differential [4]). Let X be a vector space and T a transformation defined on a domain $D \subset X$. Let $x \in D$, $\gamma \in \mathbb{R}$, and let h be arbitrary in X. If the limit

$$\delta T(x; h) = \lim_{\gamma \to 0} \frac{1}{\gamma}[T(x + \gamma h) - T(x)]$$

exists, it is called the Gateaux differential of T at x with increment h. If the limit exists for each $h \in X$, then T is said to be Gateaux differentiable at x.

Let $U(x_T)$ be the control sequence which drives the terminal state to

$$x_T, B_T U(x_T) = x_T.$$

Such a control sequence exists because of the controllability assumption. It should be noted that $U(\cdot)$ is a function of the terminal state. Consider the following cost functional:

$$\mathbb{J}(U(\cdot)) = \int_{\mathbb{R}^n} U(x_T)' \tilde{Q} U(x_T) p(x_T) dx_T. \tag{A.8}$$

The functions $U(\cdot)$ are $\mathbb{L}^2$ integrable and belong to the space $\mathbb{L}^2(\mathbb{R}^n, \mathcal{B}(\mathbb{R}^n), \mu_p)$. Note that $\mathcal{B}(\mathbb{R}^n)$ is the Borel σ-algebra on $\mathbb{R}^n$ and μ_p is the probability measure corresponding to the distribution of the terminal state. We use the same technique which we employed in the previous sections to remove the equality constraint, $B_T U(x_T) = x_T$. Let $\tilde{U}(x_T)$ be a given control function such that $B_T \tilde{U}(x_T) = x_T$, $\forall x_T \in \mathbb{R}^n$. We assume that $\tilde{U}(x_T)$ is linear in x_T. This is possible due to the controllability assumption, and one possible choice for this control function is $\tilde{U}(x_T) = B_T^+ x_T$, where

B_T^+ is the Moore–Penrose pseudo-inverse of B_T. Let $\tilde{B}$ be a basis for the null space of B_T. Then we can write:

$$U(x_T) = \tilde{U}(x_T) + \tilde{B}\eta(x_T), \eta(x_T) \in \mathbb{R}^q, \quad q = \dim(\text{Null}(B_T)). \tag{A.9}$$

Using (A.9), $\eta(\cdot)$ becomes our new optimization variable and the cost functional can be written as

$$\mathbb{J}(\eta(\cdot)) = \int_{\mathbb{R}^n} \left(\tilde{U}(x_T)\tilde{B}\eta(x_T)\right)' \left(\tilde{U}(x_T) + \tilde{B}\eta(x_T)\right) p(x_T) d_{x_T}. \tag{A.10}$$

We assume that the observation noise has a general finite mean noise distribution. Under the hypothesis H_{x_T}, the measurements made by the adversary are given by

$$H_{x_T} : Y_{0,k} = \bar{G}U(x_T)V_{0,k}, \quad \bar{G} = \bar{C}G. \tag{A.11}$$

A.4 LINEAR MATRIX INEQUALITIES

It has been shown in [5] that a wide variety of problems arising in system and control theory can conveniently be reduced to a few standard convex or quasiconvex optimization problems involving linear matrix inequalities (LMIs). The resulting optimization problems can then be solved numerically very efficiently using commercially available interior-point methods.

A.4.1 Basics

One of the earliest LMIs arises in Lyapunov theory. It is well-known that the differential equation

$$\dot{x}(t) = Ax(t) \tag{A.12}$$

has all of its trajectories converge to zero (stable) if and only if there exists a matrix $P > 0$ such that

$$A^t P + AP < 0. \tag{A.13}$$

This leads to the LMI formulation of stability, that is, *a linear time-invariant system is asymptotically stable if and only if there exists a matrix* $0 < P = P^t$ *satisfying the LMIs*

$$A^t P + AP < 0, \quad P > 0.$$

Given a vector variable $x \in \mathbb{R}^n$ and a set of matrices $0 < G_j = G_j^t \in \mathbb{R}^{n\times n}$, $j = 0, \ldots, p$, a basic compact formulation of a linear matrix inequality is

$$G(x) \triangleq G_0 + \sum_{j=1}^{p} x_j G_j > 0. \tag{A.14}$$

Notice that (A.14) implies that $v^t G(x) v > 0$, $\forall 0 \neq v \in \mathbb{R}^n$. More importantly, the set $\{x \mid G(x) > 0\}$ is convex. Nonlinear (convex) inequalities are converted to LMI form using Schur complements in the sense that

$$\begin{bmatrix} Q(x) & S(x) \\ \bullet & R(x) \end{bmatrix} > 0, \tag{A.15}$$

where $Q(x) = Q^t(x)$, $R(x) = R^t(x)$, and $S(x)$ depends affinely on x, is equivalent to

$$R(x) > 0, \quad Q(x) - S(x)R^{-1}(x)S^t(x) > 0. \tag{A.16}$$

More generally, the constraint

$$\mathrm{Tr}[S^t(x)P^{-1}(x)S(x)] < 1, \quad P(x) > 0,$$

where $P(x) = P^t(x) \in \mathbb{R}^{n\times n}$, $S(x) \in \mathbb{R}^{n\times p}$ depend affinely on x, is handled by introducing a new (slack) matrix variable $Y(x) = Y^t(x) \in \mathbb{R}^{p\times p}$ and the LMI (in x and Y)

$$\mathrm{Tr}\, Y < 1, \quad \begin{bmatrix} Y & S(x) \\ \bullet & P(x) \end{bmatrix} > 0. \tag{A.17}$$

Most of the time, our LMI variables are matrices. It should be clear from the foregoing discussions that a quadratic matrix inequality (QMI) in the variable P can be readily expressed as a linear matrix inequality (LMI) in the same variable.

A.4.2 Some Standard Problems

Here we provide some common convex problems that we encountered throughout the monograph. Given an LMI $G(x) > 0$, the corresponding LMI problem (LMIP) is to

find a feasible $x \equiv x^f$ *such that* $G(x^f) > 0$,
or *determine that the LMI is infeasible.*

It is obvious that this is a convex feasibility problem.

The generalized eigenvalue problem (GEVP) is to minimize the maximum generalized eigenvalue of a pair of matrices that depend affinely on a variable, subject to an LMI constraint. GEVP has the general form

$$\begin{aligned} &\text{minimize } \lambda \\ &\text{subject to } \lambda B(x) - A(x) > 0, \ B(x) > 0, \ C(x) > 0, \end{aligned} \tag{A.18}$$

where A, B, C are symmetric matrices that are affine functions of x. Equivalently stated,

$$\begin{aligned} &\text{minimize } \lambda_M[A(x), B(x)] \\ &\text{subject to } \quad B(x) > 0, \ C(x) > 0, \end{aligned} \tag{A.19}$$

where $\lambda_M[X, Y]$ denotes the largest generalized eigenvalue of the pencil $\lambda Y - X$ with $Y > 0$. This is problem is a quasiconvex optimization problem since the constraint is convex and the objective $\lambda_M[A(x), B(x)]$ is quasiconvex.

The eigenvalue problem (EVP) is to minimize the maximum eigenvalue of a matrix that depends affinely on a variable, subject to an LMI constraint. EVP has the general form

$$\begin{aligned} &\text{minimize} \quad \lambda \\ &\text{subject to} \quad \lambda I - A(x) > 0, \ B(x) > 0, \end{aligned} \tag{A.20}$$

where A, B are symmetric matrices that are affine functions of the optimization variable x. This is a convex optimization problem.

EVPs can appear in an equivalent form of minimizing a linear function subject to an LMI, that is,

$$\begin{aligned} &\text{minimize} \quad c^t x \\ &\text{subject to} \quad G(x) > 0, \end{aligned} \tag{A.21}$$

where $G(x)$ is an affine function of x. Examples of $G(x)$ include

$$PA + A^t P + C^t C + \gamma^{-1} PBB^t P < 0, \ P > 0.$$

It should be stressed that the standard problems (LMIPs, GEVPs, EVPs) are tractable, from both theoretical and practical viewpoints:

- They can be solved in polynomial-time.
- They can be solved in practice very efficiently using commercial software [6].

A.4.3 The S-Procedure

In some design applications, we faced the constraint that some quadratic function can be negative whenever some other quadratic function is negative. In such cases, this constraint can be expressed as an LMI in the data variables defining the quadratic functions.

Let $G_0, \ldots, G_p$ be quadratic functions of the variable $\xi \in \mathbb{R}^n$:

$$G_j(\xi) \triangleq \xi^t R_j \xi + 2u_j^t \xi + v_j, \quad j = 0, \ldots, p, \quad R_j = R_j^t.$$

We consider the following conditions on $G_0, \ldots, G_p$:

$$G_0(\xi) \leq 0 \ \forall \xi \quad \text{such that } G_j(\xi) \geq 0, \ j = 0, \ldots, p. \tag{A.22}$$

It is readily evident that if there exist scalars $\omega_1 \geq 0, \ldots, \omega_p \geq 0$ such that

$$\forall \xi, \quad G_0(\xi) - \sum_{j=1}^{p} \omega_j \, G_j(\xi) \geq 0, \tag{A.23}$$

then inequality (A.22) holds. Observe that if the functions $G_0, \ldots, G_p$ are affine, then Farkas lemma [5] states that (A.22) and (A.23) are equivalent. Interestingly enough, inequality (A.23) can be written as

$$\begin{bmatrix} R_0 & u_0 \\ \bullet & v_0 \end{bmatrix} - \sum_{j=1}^{p} \omega_j \begin{bmatrix} R_j & u_j \\ \bullet & v_j \end{bmatrix} \geq 0. \tag{A.24}$$

The foregoing discussions were stated for nonstrict inequalities. In case of strict inequality, we let $R_0, \ldots, R_p \in \mathbb{R}^{n \times n}$ be symmetric matrices with the following qualifications:

$$\xi^t R_0 \xi > 0 \ \forall \xi \quad \text{such that } \xi^t G_j \xi \geq 0, \ j = 0, \ldots, p. \tag{A.25}$$

Once again, it is obvious that if there exist scalars $\omega_1 \geq 0, \ldots, \omega_p \geq 0$ such that

$$\forall \xi, \quad G_0(\xi) - \sum_{j=1}^{p} \omega_j G_j(\xi) > 0, \tag{A.26}$$

then inequality (A.25) holds. Observe that (A.26) is an LMI in the variables $R_0, \omega_1, \ldots, \omega_p$.

It should be remarked that the **S**-procedure deals with nonstrict inequalities and allows the inclusion of constant and linear terms. In the strict version, only quadratic functions can be used.

A.5 SOME LYAPUNOV–KRASOVSKII FUNCTIONALS

In this section, we provide some Lyapunov–Krasovskii functionals and their time-derivatives which are of common use in stability studies throughout the text:

$$V_1(x) = x^t P x + \int_{-\tau}^{0} x^t(t+\theta) Q x(t+\theta)\, d\theta, \tag{A.27}$$

$$V_2(x) = \int_{-\tau}^{0} \left[\int_{t+\theta}^{t} x^t(\alpha) R x(\alpha)\, d\alpha \right] d\theta, \tag{A.28}$$

$$V_3(x) = \int_{-\tau}^{0} \left[\int_{t+\theta}^{t} \dot{x}^t(\alpha) W \dot{x}(\alpha)\, d\alpha \right] d\theta, \tag{A.29}$$

where x is the state vector, τ is a constant delay factor and the matrices $0 < P^t = P$, $0 < Q^t = Q$, $0 < R^t = R$, $0 < W^t = W$ are appropriate weighting factors.

Standard matrix manipulations lead to

$$\dot{V}_1(x) = \dot{x}^t P x + x^t P \dot{x} + x^t(t) Q x(t) - x^t(t-\tau) Q x(t-\tau), \tag{A.30}$$

$$\begin{aligned} \dot{V}_2(x) &= \int_{-\tau}^{0} \left[x^t(t) R x(t) - x^t(t+\alpha) R x(t+\alpha) \right] d\theta \\ &= \tau\, x^t(t) R x(t) - \int_{-\tau}^{0} x^t(t+\theta) R x(t+\theta) \Big] \, d\theta, \end{aligned} \tag{A.31}$$

$$\dot{V}_3(x) = \tau\, \dot{x}^t(t) W x(t) - \int_{t-\tau}^{t} \dot{x}^t(\alpha) W \dot{x}(\alpha)\, d\alpha. \tag{A.32}$$

A.6 SOME FORMULAE FOR MATRIX INVERSES

This section concerns some useful formulae for inverting matrix expressions in terms of the inverses of their constituents.

A.6.1 Inverses of Block Matrices

Let A be a square matrix of appropriate dimension and partitioned in the form

$$A = \begin{bmatrix} A_1 & A_2 \\ A_3 & A_4 \end{bmatrix}, \tag{A.33}$$

where both A_1 and A_4 are square matrices. If A_1 is invertible, then

$$\Delta_1 = A_4 - A_3 A_1^{-1} A_2$$

is called the Schur complement of A_1. Alternatively, if A_4 is invertible, then

$$\Delta_4 = A_1 - A_2 A_4^{-1} A_3$$

is called the Schur complement of A_4.

It is well-known from Chapter 2 that matrix A is invertible if and only if either

$$A_1 \text{ and } \Delta_1 \text{ are invertible}$$

or

$$A_4 \text{ and } \Delta_4 \text{ are invertible.}$$

Specifically, we have the following equivalent expressions:

$$\begin{bmatrix} A_1 & A_2 \\ A_3 & A_4 \end{bmatrix}^{-1} = \begin{bmatrix} \Upsilon_1 & -A_1^{-1} A_2 \Delta_1^{-1} \\ -\Delta_1^{-1} A_3 A_1^{-1} & \Delta_1^{-1} \end{bmatrix} \tag{A.34}$$

or

$$\begin{bmatrix} A_1 & A_2 \\ A_3 & A_4 \end{bmatrix}^{-1} = \begin{bmatrix} \Delta_4^{-1} & -\Delta_4^{-1} A_2 A_4^{-1} \\ -A_4^{-1} A_3 \Delta_4^{-1} & \Upsilon_4 \end{bmatrix}, \tag{A.35}$$

where

$$\begin{aligned} \Upsilon_1 &= A_1^{-1} + A_1^{-1} A_2 \Delta_1^{-1} A_3 A_1^{-1}, \\ \Upsilon_4 &= A_4^{-1} + A_4^{-1} A_3 \Delta_4^{-1} A_2 A_4^{-1}. \end{aligned} \tag{A.36}$$

Important special cases are

$$\begin{bmatrix} A_1 & 0 \\ A_3 & A_4 \end{bmatrix}^{-1} = \begin{bmatrix} A_1^{-1} & 0 \\ -A_4^{-1} A_3 A_1^{-1} & A_4^{-1} \end{bmatrix} \tag{A.37}$$

and

$$\begin{bmatrix} A_1 & A_2 \\ 0 & A_4 \end{bmatrix}^{-1} = \begin{bmatrix} A_1^{-1} & -A_1^{-1}A_2A_4^{-1} \\ 0 & A_4^{-1} \end{bmatrix}. \tag{A.38}$$

A.6.2 Matrix Inversion Lemma

Let $A \in \mathbb{R}^{n\times n}$ and $C \in \mathbb{R}^{m\times m}$ be nonsingular matrices. By using the definition of the matrix inverse, it can be easily verified that

$$[A + BCD]^{-1} = A^{-1} - A^{-1}B[DA^{-1}B + C^{-1}]^{-1}DA^{-1}. \tag{A.39}$$

A.7 PARTIAL DIFFERENTIATION

Definition A.2. *Consider a scalar multivariable function $f(x, \Theta) \in \mathbb{R}$ with $\Theta \in \mathbb{R}^{m\times n} = [\theta_{ij}]$. The partial derivative of $f(x, \Theta)$ with respect to matrix Θ is defined by*

$$\frac{\partial f(x,\Theta)}{\partial \Theta} \triangleq \left[\frac{\partial f(x,\Theta)}{\partial \Theta}\right] = \begin{bmatrix} \frac{\partial f}{\partial \theta_{21}} & \frac{\partial f}{\partial \theta_{11}} & \cdots & \frac{\partial f}{\partial \theta_{m1}} \\ \frac{\partial f}{\partial \theta_{12}} & \ddots & & \vdots \\ \vdots & & & \vdots \\ \frac{\partial f}{\partial \theta_{1n}} & \cdots & & \frac{\partial f}{\partial \theta_{mn}} \end{bmatrix}. \tag{A.40}$$

Lemma A.1. *Given a scalar multivariable function $f(x, y, \Theta) = x^T\Theta y \in \mathbb{R}$ with $\Theta \in \mathbb{R}^{m\times n}$, $x \in \mathbb{R}^m$, $y \in \mathbb{R}^n$, the partial derivative of $f(x, y, \Theta)$ with respect to matrix Θ is computed by*

$$\frac{\partial f(x,y,\Theta)}{\partial \Theta} = yx^T. \tag{A.41}$$

Proof. Rewriting $f(\cdot,\cdot,\cdot)$ as a sum gives

$$f(x,y,\Theta) = \sum_j^n \sum_i^n \theta_{ij}x_iy_j. \tag{A.42}$$

Applying Lemma A.1 to (A.42) yields

$$\frac{\partial f(x,y,\Theta)}{\partial \Theta} \triangleq \begin{bmatrix} x_1y_1 & x_2y_1 & \cdots & x_my_1 \\ x_1y_2 & \ddots & & \vdots \\ \vdots & & & \vdots \\ x_1y_n & \cdots & & x_my_n \end{bmatrix} = yx^T, \tag{A.43}$$

which completes the proof. □

REFERENCES

[1] J. von Neumann, Zur Theorie der Gesellschaftsspiele, Math. Ann. 100 (1) (1928) 295–320.
[2] B. Randell, P. Lee, P.C. Treleaven, Reliability issues in computing system design, ACM Comput. Surv. 10 (2) (June 1978) 123–165.
[3] L.S. Shapley, Stochastic games, Proc. Natl. Acad. Sci. 39 (10) (1953) 1095–1100.
[4] O. Hernandez-Lerma, J. Lasserre, Zero-sum stochastic games in Borel spaces: average payoff criteria, SIAM J. Control Optim. 39 (5) (2000) 1520–1539.
[5] S. Boyd, L. El Ghaoui, E. Feron, V. Balakrishnan, Linear Matrix Inequalities in Systems and Control Theory, SIAM Stud. Appl. Math., Philadelphia, 1994.
[6] P. Gahinet, A. Nemirovski, A.J. Laub, M. Chilali, LMI Control Toolbox, The Math-Works, MA, 1995.

INDEX

A

C

D

Q

R

S

FSC
www.fsc.org
MIX
Papier | Fördert gute Waldnutzung
FSC® C083411

Kolibri 360 GmbH
Ferdinand-Jühlke-Straße 7
99095 Erfurt, Deutschland
produktsicherheit@kolibri360.de